ESTATES GAZETTE LAW REPORTS

Estates Gazette Law Reports

1997

Volume 1

Edited by
Barry Denyer-Green LLM PhD FRICS
of the Middle Temple, barrister

Audrey Boyle BA
Publisher's editor

Estates Gazette

A member of Reed Business Information

ESTATES GAZETTE
151 WARDOUR STREET, LONDON W1V 4BN

First published 1997
ISBN for complete set of 2 volumes: 0 7282 0289 1
ISBN for this volume: 0 7282 0290 5

ISSN 0951-9289

Typesetting by Page Bros, Norwich
Printed and bound by Redwood Books, Trowbridge, Wiltshire

CONTENTS

SUPPLEMENT

Cases not reported in "Estates Gazette"

TABLE OF CASES

Titles of cases show in bold type
Names of parties reversed shown in ordinary type

INDEX OF SUBJECT-MATTER

AGRICULTURE

A Barrett and others *v* Morgan

CHANCERY DIVISION: NEWCASTLE UPON TYNE DISTRICT REGISTRY

September 27 1996

MR PW SMITH QC, sitting as a deputy judge of the division

Estates Gazette March 22 1997

B [1997] 12 EG 155

Agricultural holdings — Headlease — Notice to quit — Tenants failing to serve counternotice — Whether subtenancy determined — Whether landlord and tenants entitled to act collusively to destroy subtenancy — Whether arrangements amounted to surrender

The plaintiffs held the severed freehold of land subject to a yearly agricultural tenancy granted by a headlease dated March 1 1970 to the plaintiffs E and his two sons. The plaintiffs were either related to the tenants or were family trustees holding land for the benefit of relatives. By an agreement which was in form a partnership arrangement, but
C which was conceded on behalf of the plaintiffs at trial to be a subtenancy, the defendant farmed the land. On April 1 1992 a notice to quit was served on the remaining tenants holding the headlease (E having died in 1976) which terminated the headlease; in accordance with a scheme advised by solicitors, the tenants did not serve any counternotices claiming the protection of the Agricultural Holdings Act 1986. It was intended to terminate the defendant's subtenancy by this means. The plaintiffs claimed possession of the land against the defendant.

***Held*: The claim was dismissed. The plaintiffs acted collusively with their tenants under a contrived scheme to obtain vacant possession of the land: landlord and tenant cannot act collusively
D for the purpose of destroying a subtenancy. Alternatively the arrangements amounted to a surrender of the headlease. The notice to quit did not determine the derivative subtenancy of the defendant. If a tenant chooses not to act as one would expect a tenant to act to further the landlord's desire to recover possession and destroy the subtenancy, the court will be astute to prevent that happening. It will not interfere if the tenant has genuine grounds for so acting. Had *mesne* profits been awarded, they would have been assessed on the basis of a reasonable rent for the land.**

The following cases are referred to in this report.

E
Brown v *Wilson* (1949) 208 LT 144; 93 SJ 640; 156 EG 45
Harrison v *Wing* (1988) 56 P&CR 358; [1988] 2 EGLR 4; [1988] 29 EG 101
Magdalen College, Oxford v *Heritage* [1974] 1 WLR 441; [1974] 1 All ER 1065; (1974) 27 P&CR 169; [1974] EGD 1; 230 EG 219, CA
Mellor v *Watkins* (1874) LR 9 QB 400
Pennell v *Payne* [1995] QB 192; [1995] 2 WLR 261; [1995] 2 All ER 592, CA
Sparkes v *Smart* [1990] 2 EGLR 245

This was an application by the plaintiffs, Robert Asheton Barrett, Philip Adrian Scrope and Mary Isobel Scott, for possession of land against the defendant Robert Cecil Morgan.

John McGhee (instructed by Burges Salmon, of Bristol) appeared F for the plaintiffs; Shane Dougall (instructed by Smith Roddam, of Bishop Auckland) appeared for the defendant.

Giving judgment, **MR PW SMITH QC** said: By this action the plaintiffs claim possession and *mesne* profits in respect of the agricultural land known as Mordon North Farm, County Durham, and Mordon West Farm, County Durham ("the farm"), from the defendant.

The first and second plaintiffs' claim relates to 146.17 acres of the Mordon North land and 184.84 acres of the Mordon West land (total 331.01 acres). The third plaintiff's claim relates to 23.58 acres at G Mordon. The defendant has farmed the land since 1980.

On the first day of the trial the defendant applied for leave to amend his defence and to serve a counterclaim. I granted leave to amend the defence and counterclaim in a form to be produced (as a result of counsel for the defendant acknowledging that the claim based upon clause 36 of a headlease ("the headlease") dated March 1 1970 and made between the then fourth Earl of Eldon, the said Earl of Eldon, Simon Peter Scott and John Joseph Nicholas Scott (now the Fifth Earl) was misconceived). The counterclaim was then comprised by the parties and I made an order staying the counterclaim in *Tomlin* form on the first day of the trial. I also granted the plaintiffs leave to adduce the supplemental witness statement of the second plaintiff (Mr Scrope) H dated September 13 1996.

There was a substantial dispute over the value of the *mesne* profits claim if it was established by the plaintiffs. In the event the experts were able to agree the *mesne* profits as being £131,449 if calculated on the basis of the net profit made by the defendant in farming the land and £92,316 if the same is assessed on the basis of a reasonable rent for the same. Both sums are subject to deduction of the sums Mr Morgan has paid on account (he having tendered the sums he was previously paying and the same having been accepted as on account *mesne* profits). In addition there will be a claim for interest, but if that arises the sum can be calculated and agreed. J

Background

By the headlease the then fourth Earl let 938.14 acres of land (including the farm), as I have said to himself and his two sons. Although the headlease is stated to be an agreement, it was actually executed as a deed. A potential limitation point on the counterclaim has been avoided by the compromise. The letting was for a tenancy from year to year from March 1 1970 at an actual rent of £3,350. Clause 3 provides that the headlease could be terminated by giving not less than one year's notice expiring on May 13 in any year.

All parties are agreed that the headlease was protected under the K (then) Agricultural Holdings Act 1948. The land comprised in the headlease was actually farmed by the tenants.

Devolution of freehold

The fourth Earl died on October 20 1976 and probate of his will was granted to his two sons out of the Principal Probate Registry on April 1 1977. John Joseph Nicholas became the fifth Earl.

On April 3 1984 3.2 acres of the Mordon West land was vested in the first and second plaintiffs as trustees for the SP Scott Children (being the children of Mr Scott, the other son of the fourth Earl). On

July 25 1984 387.47 acres of land at Mordon was vested in Mr Scott by an assent from the executors of the will of the fourth Earl. On July 30 1984 23.58 acres was transferred to the third plaintiff (the wife of Mr Scott) by Mr Scott (being the land claimed by her in these proceedings).

On July 23 1986 146.17 acres of the Mordon North land and 181.64 of the Mordon West land were transferred by Mr Scott to the first and second plaintiffs to hold the same upon trust for Mr Scott's children. The latter land, together with the earlier 3.2 acres total 184.84 acres, the total of the land secondly claimed by the first and second plaintiffs. That led to a discrepancy in the notices to quit which were served, to which I make reference below. However, such discrepancy did not, in my view, affect the validity of the notices.

Thus by the time of the notices to quit the freehold reversion of the farm was severed between different freeholders. It is nevertheless accepted that by virtue of section 140 of the Law of Property Act 1925 each severed freehold reversioner can serve a notice to quit in respect of the severed part. Alternatively it is accepted that the freeholders can combine together to serve one notice to quit.

Occupation of the farm

The tenants under the headlease occupied the farm as part of the land comprised in the headlease from 1970. By 1980 the farm had become run down. I was told by Mr Scrope (and I accept his evidence in that regard) that the farm had become rundown. It was also the intention of Mr Scott in the medium term to sell the farm to raise capital for his children as and when they needed it. Mr Scrope acting for the Eldon family was anxious to secure an income from the farm in the interim, but equally he wished to ensure that any such income was derived from a relationship that did not give rise to an agricultural holding with security of tenure under what became the Agricultural Holdings Act 1986.

Both Lord Eldon and Mr Scott gave evidence to the effect that the desire to recover the land for sale purposes for the children was communicated to Mr Morgan and he understood and accepted it. They both acknowledged in cross-examination that any such communication must have been via Mr Scrope; they not having discussed the matter with Mr Morgan directly. Mr Scrope in his evidence was of the view that he first communicated this view in the mid-1980s, that is to say some years after Mr Morgan started farming the farm. He said he informed Mr Morgan of the desire and he made no comment. I doubt whether there was a clear communication of such a desire before that. It must be borne in mind that the children did not become beneficially interested in the freehold until July 23 1986. If there was any communication before then it would only have been in the most general of terms and would express a need for *Mr Scott* to sell land for his children.

Mr Morgan's evidence was unchallenged in that his witness statement only was put in. It does not deal with the agreement (which was called "the gentleman's agreement") in the trial. On instruction Mr Shane Dougall for Mr Morgan informed me that he did not recall any original discussion, but conceded it might have occurred later. His understanding was that he would occupy until 1987.

I find that the communication was made to Mr Morgan in the 1980s after he went into occupation and that he accepted it in the sense that he understood that was the wish of the freeholders. I do not find anything else and in particular I attach no significance to the gentleman's agreement. Like all such arrangements, it was not binding on Mr Morgan and there was nothing to stop him asserting his statutory rights, if any. It has been said that a gentleman's agreement is one made between persons who are not gentlemen. Certainly in this case the Eldon family could not have expected anything other than the promise being non-binding and dependent on Mr Morgan agreeing to give up whatever rights he had.

In this context I accept Mr Scrope's evidence that Mr Morgan was identified as a potential farmer of the farm lease, he was known to the Eldon family (being a tenant of another farm for them), was considered to be likely to accede to a gentleman's agreement as envisaged, had other farms and would not therefore be dependent on the farm for his existence.

Documentation

Mr Scrope would of course be failing in his duty to the Eldon family if he had allowed Mr Morgan into occupation under a gentleman's agreement to recover possession. He was an experienced land agent and would be well aware of the pitfalls presented by the Agricultural Holdings regime. The Acts provided tenants with a substantial security of tenure and severely restricted the ability of freeholders to recover possession. In particular if Mr Morgan had a tenancy protected under the 1986 Act, the gentleman's agreement could not be used as a basis for recovering possession. That perceived imbalance has been redressed by the Farm Business Tenancy regime created by the Agricultural Tenancies Act 1995. It does not apply to the instant case, however. Mr Scrope, therefore, to avoid giving Mr Morgan a protected tenancy under the Agricultural Holdings regime proposed a partnership arrangement with Mr Simon Scott and Lord Eldon. He sent a proposal to Mr Morgan under cover of a letter dated October 16 1980. In the letter he suggests a figure of "rent" at £1,000 per month. He explained in his evidence that the inverted commas were used because it was not rent, but a payment under the limited partnership.

The arranged terms would if implemented have constituted a partnership of which Mr Morgan was a partner and he would have had no separate tenancy. There is nothing wrong in drawing up arrangements with the intention of avoiding the statutory protection regime provided they are genuine and bona fide arrangements and not mere shams to avoid such regimes. Mr Morgan accepted the terms on December 5 1980 (having gone into occupation in November). On June 17 1983 a limited partnership agreement was drawn up and duly registered under the Limited Partnerships Act 1907. The purpose of the limited nature of the partnership is of course to avoid the potential difficulty of Mr Morgan otherwise being able to incur liabilities in the partnership, same for which Lord Eldon and Mr Scott would be jointly and severally liable.

However, even by 1983 it was all a pretence. None of the partnership obligations (save the capital contributions) had been implemented. Demands for money were issued, variously described as "rent", "amount due", "payment due". Mr Morgan never provided accounts. Instead of seeking to enforce the partnership obligations, Lord Eldon and Mr Scott accepted that he became a subtenant of theirs. The illusion was further revived in 1993 when Mr Scrope purported to serve a notice of dissolution of the partnership. In fact the document had been prepared I find in 1992 and its service was, in my view, held back deliberately to be served as late as possible on Mr Morgan.

Mr John McGhee properly for the plaintiffs accepted that Mr Morgan was a subtenant of the farm. That too was Mr Dougall's contention. In fact, that eventuality the agreement raised by Mr Dougall that Lord Eldon and Mr Scott were in breach of duties owed by them to Mr Morgan as co-partners was no longer sustainable and he abandoned it.

Need for possession

By 1991 the position had arisen that Mr Scott's children were desirous of realising the capital of the farm if it all possible. They were living in London in a damp houseboat and wished to return to terra firma. No reason was given as to why they could not like a large number of people simply move into rented accommodation. By this time I find Mr Morgan had been aware of the need and accepted that might be why possession might be needed.

Negotiations took place for a possible sale to Mr Morgan. However, he was advised quite properly that he was a tenant and was only prepared to offer to buy on the basis of the freehold being valued on a non-vacant possession basis (probably according to Mr Scrope, leading to a reduction by 30%). In effect in 1991 he maintained his claim to a subtenancy which claim was accepted by the freeholders.

A *Subsequent events*

Faced with this Mr Scrope sought advice probably from Burges Salmon. The position was considered I find exclusively with Mr Scott alone. This at first sight is curious. Mr Scott is after all one of the *mesne* tenants (with Lord Eldon his brother). Mr Scrope I find took no instructions from the children, nor from Mrs Scott, the other freeholder. Mr Scott in his evidence tellingly said he was glad the children had taken *his* advice.

The result of the advice was that a decision was made to serve notices to quit terminating the headlease. It is necessary to explain the thinking behind this arrangement. A tenant under the Agricultural B Holdings regime is in a strong position. A notice to quit of at least 12 months has to be served. If the notice does not specify one of the statutory grounds for termination the tenant can serve a counternotice within one month claiming the benefit of section 26 of the 1986 Act. If he does then the Agricultural Land Tribunal only can consent to the notice to quit taking effect. Alternatively, if grounds are specified the tenant can serve a counternotice challenging the grounds if he does so within one month. If he fails to serve the counternotice, the notice to quit operates (unless fraudulent) even if there was no actual ground: see *Magdalen College, Oxford* v *Heritage* [1974] 1 WLR 441. Thus a tenant served with a notice to quit if he wishes to remain a tenant C would invariably serve a counternotice. Anyone advising a tenant would so advise (a point Mr Scrope and Lord Eldon conceded).

Subtenants are in a much weaker position. Unlike the regime under the Rent Act 1977 and Part II of the Landlord and Tenant Act 1954, they have no security of tenure. They hold limited rights only if the *mesne* landlord serves a notice to quit. The position is summarised in *Woodfall Landlord and Tenant* para 21.158. If a headlease is determined either by notice given by the freeholder to the superior tenant or vice versa or by forfeiture, the derivative sublease also determines: see *Pennell* v *Payne* [1995] QB 192, CA. Conversely, if a surrender occurs the subtenancy remains and becomes binding on the freeholder under sections 139 and 150 of the Law of Property Act D 1925. Thus if a notice to quit was served terminating the headlease and it became effective, Mr Morgan's subtenancy would also come to end and his right to possession would determine. This is the scheme that Mr Scrope and Mr Scott embarked upon in my view.

The notices

On April 1 1992 Mr Scrope sent a letter to Mr Scott and Lord Eldon at Mr Scott's home address. It refers to an enclosed *notice* to quit. I find as a fact two notices were included, that the letter and the notices were sent by recorded delivery and were received by Mr Scott at his house. I also find as a fact that service of the notices was brought to the E attention of Lord Eldon.

The notice to quit in relation to the freehold land held by the first and second plaintiffs refers to different figures (184.84 acres in the title and 183.30 in the body of the notice). Mr Scrope said this was a clerical error (he was not sure now whose error). In my view, and I so find neither tenant (Mr Scott and Lord Eldon) was misled about it. They knew (because it was the scheme that Mr Scott and Mr Scrope had created) that the notices were intended to operate on all the land owned by the plaintiffs, included in the headlease and occupied by Mr Morgan under his subtenancy. Further the acreage is prefaced by the word "approximately" to allow a deviation from the exact acreage. I F find therefore that the notice was validly served and received. I also find that it operated to terminate the headlease in relation to the farm.

The next part of the scheme was *not* to serve a counternotice. Otherwise vacant possession could not be obtained. No counternotice was served. The only reason given was that Mr Scott (not surprisingly) and Lord Eldon (hardly less surprisingly) were anxious to assist the freeholders in the desire to recover possession from Mr Morgan by destroying his subtenancy. They could not do so because as against them he would be able to assert his rights. It is quite clear that Lord Eldon was willing to fall into any scheme which would give vacant possession to his nephews and nieces. He was not really consulted, he was informed of the proposed notice (and according to Mr Scrope took "it on the chin"). Lord Eldon said he was surprised by that comment; he was concerned about his nephews and nieces' plight and was G anxious they could get somewhere decent to live.

Mr Scrope barely consulted Mrs Scott, his client. All discussions were between him and Mr Scott. Further there appears to be little consideration (whether at the breakfast table or otherwise) between Mr Scott and his wife over the fact that she had served a notice to quit on him. Mr Scott and Lord Eldon were quite frank about this. Their stance is that the decision not to challenge the notice was to ensure the landlords recovered possession. Mr Scrope was equally frank in conceding that if there was no connection between the landlord and the tenant he would (if advising the tenant only) advise the service of a counternotice. Similarly Lord Eldon and Mr Simon Scott both H acknowledged that the desire not to challenge the notice was motivated solely to achieve a benefit for the landlords and I so find that was the position.

Equally, it was important that Mr Morgan did not find out about the notice until the one month had expired. I find as a fact that the existence of the notices was deliberately concealed from Mr Morgan as long as possible. Indeed he only became aware of them in March 1993 when he was given copies by Mr Scrope when he served the notice "dissolving" the partnership. This was despite the fact that he had visited Mr Morgan three to four times. He chose not to mention the notices. For Mr Scrope to say to tell Mr Morgan about the notice J would be "gratuitous" is specious in my view. It is quite clear by that time both sides were insisting on their "strict legal rights".

Challenge to notices to quit

Mr Dougall who appears for Mr Morgan contends that there was collusion between the tenants and the landlords to give notices to quit and not challenge them for the purpose of destroying the subtenancy to enable the landlords to obtain vacant possession. Mr McGhee attempted with great elegance to suggest that the tenants were merely making a decision not to challenge the landlords' unilateral decision to give notice. The submission was elegantly put with great force but it is with respect to Mr McGhee unreal. K

The following facts are relevant:

(1) The landlords wanted vacant possession and Mr Morgan was the only obstacle.

(2) The tenants wanted to assist in that aim.

(3) As tenants there were no grounds for *not* giving a counternotice, but it was never contemplated one would be served.

(4) The tenants never considered their position *qua* tenants (Lord Eldon especially so).

(5) The notice to quit would not have been served (because it would have achieved no purpose otherwise) had it not been known by the L landlords that no counternotice would be served.

(6) Mr Scrope advised the landlords and the tenants. He was taking his instructions from Mr Scott, a tenant who was acting in a way no tenant would and in a way Mr Scrope would not advise a tenant so to act.

(7) The purpose of the notice was to destroy the subtenancy.

(8) Part of the purpose required the suppression of the service of the notice and that was done.

(9) Mr Scott had advised the children to serve the notice to quit.

In my view, it cannot against that factual background be realistic to suggest the tenants were falling in line with a unilateral decision of the M landlords.

Authorities

It is necessary to analyse the effect of various authorities: in *Mellor* v *Watkins* (1874) LR 9 QB 400 the Court of Appeal decided that a surrender of a headlease did not determine a subtenancy. The plaintiffs' counsel at pp403-404 conceded that "no voluntary act of surrendering or *otherwise* putting an end to his tenancy can affect the interest of his other tenants". Cockburn CJ decided the case on the effect of a surrender. Blackburn J decided it on the basis that no voluntary act of the tenant could put on to the sublease. Lush J merely concurred with the result. There has apparently been no authority for the concession to be read in the *Mellor* case that a head tenant could not determine a subtenancy by giving notice to terminate his own tenancy; see *Pennell* v *Payne* [1995] QB 192 at p199, *per* Simon Brown LJ.

In *Brown* v *Wilson* (1949) 208 LT 144; 93 SJ 640 and 156 EG 45, Hilbery J decided there was a broad underlying principle that "will not allow a man by an act done between him and another . . . to destroy the right he has granted to a Third Party".

In *Pennell* v *Payne* the Court of Appeal considered these (and other cases) in the context of the head tenant giving notice to the freeholder to determine his lease. The Court of Appeal concluded he could and in so doing overruled *Brown* v *Wilson*. Counsel for the tenant argued successfully that the rule was that if the headlease was determined other than by surrender or merger, the derivative subleases were destroyed. Counsel for the landlord argued unsuccessfully in favour of the wider principle derived from *Mellor* v *Watkins* and *Brown* v *Wilson*. The Court of Appeal rejected that and thus applied the general principle, ie there was nothing to stop a head tenant destroying a subtenancy by giving notice to his superior landlord to determine his lease. If the decision had been confined to that I would have unhesitatingly found in favour of the plaintiffs on the basis that the headlease and sublease had been determined by the unchallenged notice to quit.

However, in the course of his judgment Simon Brown LJ said (p201c):

> [Counsel for the landlord] envisages landlords letting to compliant tenants who would then underlet and then when the landlord desired it, serve an upwards notice to quit with a view to destroying the only true interest in the underlease. I find this wholly unpersuasive. That there is ample scope for collusion *whatever the outcome of this appeal* (emphasis supplied) is plain from existing authorities [such as *Sparkes* v *Smart* [1990] 2 EGLR 245] . . . [Further the fact that the tenant might incur liability on his covenant for quiet enjoyment to his subtenant] would make the possibility of the collusion difficult indeed — more difficult certainly than for example were he to seek the same collusive end by failing to serve an appropriate notice on his landlord.

Accordingly, in my view, the general principle of the effect of how the headlease is terminated may be affected if there is collusion between the head landlord and the head tenant. Such collusion can be either by the head tenant *serving* a notice or *failing* to serve a notice.

The case of *Sparkes* v *Smart* [1990] 2 EGLR 245, CA is both relevant and instructive. An aged agricultural tenant had permitted his youngest son, the defendant, to help with his farm. Over the years it developed into a subtenancy (in breach of the headlease, but acquiesced in by the freeholders). There was a prospect of the freehold being acquired. All the family except an uncle wanted to destroy the subtenancy. A son-in-law bought the freehold for £92,000 which reflected a vacant possession price. He served a notice to quit on the father-in-law specifying an irremediable breach of covenant, namely the subletting to the son.

The tenant deliberately chose not to serve a counternotice. The service of the notice was concealed from the son. The reason was to destroy the tenants' headlease and the son's derivative sublease. The judge at first instance found the actions of the tenant were collusive with the freeholder with a view to putting an end to the sublease to award the freeholder with vacant possession. Warnings about *Mellor* v *Watkins* from the solicitors and counsel were ignored. The effect was held to be the same as the voluntary surrender of a headlease. An appeal was dismissed by the Court of Appeal.

Purchas LJ (p250F) upheld the judge's decision based on the freeholder and the head tenant acting collusively in the service of a notice to quit and not challenging it to destroy the subtenancy rights. Ralph Gibson LJ did likewise (p251M). Stuart-Smith LJ did likewise (p253J) on the basis that the scheme was to provide the freeholder with vacant possession.

Application of law

In my view, the facts of the present case as set out above show that the landlords acted collusively with the tenants. The purpose was a contrived scheme to deliver up vacant possession for the children. It is stronger than the case of *Sparkes* v *Smart* which was expressly approved of in *Pennell* as an exception to the actual principles of the consequences of a notice to quit determining the derivative subleases.

I was much pressed by Mr McGhee that there must be a surrender finding only to deprive the notices of effect. I do not accept that is correct. It seems to me that the principle is that the landlord and tenant cannot act collusively *for the purpose* (emphasis supplied) of destroying the subtenancy. If they do so the effect will be that the sublease will survive, ie it will be *as if* it was a surrender of the headlease. That is, in my view, simply an analogous description of the result of the finding of collusion. However, the court acts in the case of collusion to protect the subtenant from such collusion.

Harrison v *Wing* [1988] 2 EGLR 4 was relied upon by Mr McGhee but I find it does not assist. First it is clear that the judgments are founded on the so called wider principle in *Mellor* v *Watkins* and *Brown* v *Wilson* which was rejected by the later decision in *Pennell* v *Payne*. Second, there was no argument of collusive conduct. The tenant had an interest in two capacities, that of co-executor in respect of the freehold and tenant. He was entitled to give paramountcy to his duty as executor by not challenging the notice. That there is no analysis of what is meant by collusion. In my view, the clearer decision of *Sparkes* is more pertinent and I can see no basis for departing from it.

If I am wrong in my understanding of the principle I would say that the arrangements are, in my view, the equivalent to a surrender (*Mellor*) or tantamount to one (*Harrison* v *Wing* at p5C) in reality so that Mr Morgan's subtenancy still subsists.

Conclusion

The plaintiffs' claim, in my opinion, fails and should be dismissed. I do not accept Mr McGhee's contention that this puts tenants in an impossible position. It means the court should look to the reality. If a tenant chooses not to act as one would expect a tenant to act (as happened here) to further the landlord's desire to recover possession and destroy the subtenancy, the court will be astute to prevent that happening. It will not interfere if the tenant has genuine independent grounds for so acting.

I accept the distinction can be fine. Thus if Mr Simon Scott had remained executor he could have served a notice and not challenged it, thus presenting a position identical to that of *Harrison* v *Wing*. That, however, is not the case. It may be that *Harrison* v *Wing* will have to be reconsidered in the light of *Pennell* v *Payne* or that it should be confined to its special fact. It does not, in my view, have any relevance for the reasons I have given.

If contrary to my view the plaintiffs are entitled to possession, I would have awarded *mesne* profits at the lower level. While the claim is restitutionary to take all the profits rather than a reasonable rent, would overcompensate them for the trespass.

The decision will I have no doubt cause the plaintiffs distress. The Eldon family, however, seem to have placed too much reliance on a gentleman's agreement and then compounded it by failing to operate an agreement which would have ensured possession, in my view, because Mr Morgan would have obtained no tenancy. However, if devices are to be resorted to they must be genuinely acted on. That did not happen here.

I will hear representations as to costs.

Judgment for the defendant.

A
Cork *v* Cork

CHANCERY DIVISION

August 22 1996

Knox J

Estates Gazette April 19 1997

[1997] 16 EG 130

B ***Agricultural tenancy — Joint tenancy — Notice to quit — Whether one joint tenant can compel other tenant to join in service of counternotice — Whether interlocutory mandatory injunction can be granted***

By an agreement dated March 25 1974 the plaintiff and defendant, who are brothers, were granted an annual agricultural tenancy of a holding. In 1977 the defendant ceased working on, and in 1983 he ceased to reside at, the holding. The plaintiff farmed the holding at first in partnership with his father, then in partnership with his mother, and latterly on his own account. In 1991 the family agreed not to deal with the arrangements of the defendant's beneficial interest in the tenancy.
C A notice to quit dated July 31 1996 was served by the landlord on August 7 on both brothers requiring possession of the holding on Lady Day 1998. The plaintiff issued proceedings and, in applying for an interlocutory mandatory injunction ordering the defendant to join in the service of a counternotice under section 26(1) of the Agricultural Holdings Act 1986, offered an undertaking that if at trial it was found that the defendant should not have been compelled to join in the service of the counternotice, the plaintiff would take steps to surrender or otherwise bring the tenancy to an end.

***Held*: The application was allowed. It was arguable that consequent to the family agreement of 1991 it was agreed that the**
D **plaintiff would remain as the occupying tenant. The plaintiff therefore arguably had an equitable interest to be protected. It was therefore effectively agreed that the defendant would remain locked into the tenancy and its obligations. The balance of convenience required the counternotice to be served to prevent the notice to quit taking effect save with the consent of the agricultural lands tribunal. The fact that the defendant had been a no-rent paying absentee tenant since 1977, and that the plaintiff had offered the undertaking, were factors in exercising the balance of convenience.**

The following cases are referred to in this report.

E
Bull v *Bull* [1955] 1 QB 234; [1955] 2 WLR 78; [1955] 1 All ER 253
Dennis v *McDonald* [1982] Fam 63; [1982] 2 WLR 275; [1982] 1 All ER 590, CA
Doe d Aslin v *Summersett* (1830) 1 B&Ad 135
Featherstone v *Staples* [1986] 1 WLR 861; [1986] 2 All ER 461; (1986) 52 P&CR 287; [1986] 1 EGLR 6; 278 EG 867, CA
Hammersmith and Fulham London Borough Council v *Monk* [1992] 1 AC 478; [1991] 3 WLR 1144; [1992] 1 EGLR 65; [1992] 09 EG 135, HL
Harris v *Black* (1983) 46 P&CR 366, CA
Henderson v *Eason* (1851) 17 Ad & Ell R QBD 701
Sykes v *Land* [1984] 2 EGLR 8; (1984) 271 EG 1264, CA
Warren, Re [1932] 1 Ch 42

F
This was an application by the plaintiff, Roger Stanley Cork, by notice of motion in proceedings brought by him against the defendant, his brother and co-joint tenant, Anthony George Cork.

Martin Rodger (instructed by Lanyon Bowdler, of Shrewsbury) appeared for the plaintiff; Piers Feltham (instructed by Keely Smith Primmer & Partners, of Lichfield) represented the defendant.

Giving judgment, **Knox J** said: This is an application by the plaintiff, Mr Roger Stanley Cork ("Mr Roger Cork") for an interlocutory mandatory injunction against his brother, Anthony George Cork ("Mr Anthony Cork"), requiring the latter forthwith to join with Mr Roger Cork in signing a counternotice under section 26(1) Agricultural Holdings Act 1986 in a form which is scheduled to G
the notice of motion and is in standard form in regard to a counternotice to a notice to quit. It is common ground between the parties the notice to quit was dated July 31 1996 and served by the landlord on August 7 following. The notice will in fact take effect, if it is valid and not successfully the subject matter of a counternotice, on Lady Day of the year after next, that is to say 1998.

This concerns a Lady Day annual tenancy ("the tenancy") in standard form for agricultural lettings of a farm called Harriets Hayes Farm, Cotshall Wood in Shropshire. It would appear that originally the farm was let to the parties' father, Mr Eric Alfred Cork ("the father") but that was superseded — and the original tenancy does not now H
matter — by a tenancy agreement which is in evidence and is the operative one dated March 25 1974 in favour of the father, Mr Roger Cork, and Mr Anthony Cork, who are together called "the tenant".

It was, as I have indicated, an annual tenancy from Lady Day 1974 until the tenancy should be determined at the end of any year of the tenancy by either party giving to the other not less than 12 months' previous notice in writing, at a rent. There was the usual form of division of responsibility for repairs between landlord and tenant, which I need not go into in any detail. There was the usual absolute covenant against assignment in agricultural tenancies. There was a requirement that: J

> . . . At all times during his tenancy [the tenant] shall personally reside on the farm

The indications are that Mr Anthony Cork did continue to reside at the farm until 1977 and worked there at the same time. Thereafter it appears that he did not leave the farm so far as his residence was concerned, but he had ceased to work on it. He went to work elsewhere on land that he farmed. But in 1983 he also ceased to reside at the farm. There is some obscurity in the evidence about the basis upon which Mr Roger Cork went on farming. He certainly did go on farming throughout, but whether there was a partnership from 1980 K
onwards with his father and mother is not very clearly set out in his evidence before me. The father died in 1983, and there certainly was a partnership which arose thereafter between the father's widow and Mr Roger Cork. That went on until 1991 when there ensued negotiations for dealing with a variety of questions concerning notably the father's estate but also, perhaps somewhat peripherally, what should be done with regard to Mr Anthony Cork's interests in the tenancy.

In the midst of correspondence between the solicitors about the various issues that were being discussed regarding the other aspects of the administration of the father's estate in particular, there was a correspondence about what, if anything, should be done about Mr L
Anthony Cork's interests in the tenancy. That correspondence is in fact in evidence, but the correspondence was inconclusive. It included originally a proposal on September 10 1991, made on behalf of Mr Roger Cork to the solicitors for Mr Anthony Cork and indeed their mother, in the following terms:

> Mr AG Cork [Mr Anthony Cork] To assign his share in the tenancy of the adjoining farm [the farm with which I am concerned] To Mr RS Cork.

The reply to that was:

> You do not say how much Mr RS Cork is offering for Mr AG Cork's share. Please clarify. M

The reply to that was:

> We do not see there is any basis for a payment.

Without going through the rest of the correspondence, the effect clearly was that the parties agreed to leave things as they stood. In particular, one of the proposals which was finally taken up and this is from Mr Anthony Cork's solicitors to Mr Roger Cork's solicitors:

> Mr Anthony Cork will not be required as part of the settlement to assign his share in the tenancy of the adjoining farm to Mr RS Cork.

That was effectively agreed upon. The actual answer was:

A It is agreed that we are not to be concerned with this at the present time.

So at that stage it seems to me that the parties through their solicitors agreed to leave the beneficial interest as it was. The evidence, apart from the inter-solicitors' correspondence, is primarily contained in an affidavit sworn by Mr Roger Cork in support of this application. Para 6 reads as follows:

> There had been no discussion in 1977 of any continued involvement by my brother in the tenancy of the holding.

I pause to observe that "the holding" means the farm with which I
B am concerned. The paragraph continues:

> Discussions concerned the dissolution of the partnership which began in 1991, encompassed other family property issues which had been left unresolved following the death of my father in 1983. Anthony sought to use his status as joint tenant of the holding (which it was agreed I would continue to farm in my own right) to improve his bargaining position in relation to the purchase by him of other land belonging to me but occupied by the partnership at Poplars Farm. An assignment of the tenancy was not agreed at that stage although the dissolution of the partnership was agreed in principle with effect from 31st May 1991.

He then goes on to state what is not in any dispute, namely that the
C partnership between Mr Roger Cork and his mother was finally dissolved by a deed of retirement which is also in evidence, and one of the provisions regarding that operation was that Mrs Cork senior, Mr Roger Cork's mother, was described as "the retiring partner"; Mr Roger Cork was described as "the continuing partner", and clause 3.4 of this deed of retirement reads:

> The retiring partner hereby declares that she holds all and any interest in the partnership's tenancy of Harriets Hayes Farm on trust for the continuing partner and will continue to comply with all her obligations under the tenancy.

The affidavit that has been sworn by Mr Anthony Cork challenges
D the claim that is made elsewhere in an affidavit of Mr Roger Cork that he, Mr Roger Cork, was beneficially entitled to all or more of the beneficial interest in the tenancy of the farm. What Mr Anthony Cork says on that score is:

> I do not accept that Roger owns all or most of the beneficial interest in the tenancy of the farm. On 25th March 1974 we entered into a joint tenancy with our father of the farm. Our father farmed the farm as a sole trader under his name EA Cork. He employed Roger and me in the farming business. I left in 1977 and the land was thereafter farmed, I believe, by Roger and our father in partnership, and by Roger and our mother in partnership, and then by Roger. However, I have never formally or informally assigned my interest in the
> E tenancy, and have continued to maintain that I own a share of it. I have not contributed to the rent since 1977, but I have not had the benefit of occupation of the farm, nor have I received any account of the profits that have been made by Roger on his own and in partnership with each of my parents. I maintain that I continue to be entitled to an equal share of the tenancy, and I am likely to benefit if anything from the taking of an account in respect of Roger's outgoings on the farm, occupation thereof and profits therefrom.

Mr Feltham prudently did not seek to support that last proposition that after, as he accepts, he left in 1977 and permitted his brother Roger to continue farming, at first in partnership and subsequently solo, now in 1996 or hereafter it would be open to Mr Anthony Cork to
F claim an account of Mr Roger Cork's farming enterprises. That, Mr Feltham recognised rightly would be clearly flying in the face of what was said in *Dennis* v *McDonald* [1982] Fam 63 by Sir John Arnold; and, indeed, if one wants ancient authority on the subject, what was said by Baron Clark in *Henderson* v *Eason* (1851) 17 Ad & Ell R QBD 701. I need not pursue that aspect of the matter.

The other thing that is to be observed with regard to Mr Anthony Cork's affidavit is that he does not challenge the passage in brackets which I read from Mr Roger Cork's affidavit that it was agreed that he, Mr Roger Cork, would continue to farm the holding in question in his own right. Indeed, as a matter of commonsense, it is fairly clear that the parties must in practice have come to such an agreement. There is no problem, so far as I am concerned, in accepting that proposition.

The resistance that is made to the application by Mr Roger Cork is G
primarily based, so far as a matter of law is concerned, on the decision of the House of Lords in *Hammersmith and Fulham Borough Council* v *Monk* [1992] 1 AC 478*, which conclusively decided that a single joint tenant could bring a periodic tenancy to an end without the concurrence of the other joint tenant by alone serving a notice to quit. I was referred in particular to the passage in the speech of Lord Bridge at p490 in a judgment with which three of their lordships, that is to say Lords Brandon, Ackner and Jauncey, agreed. The passage reads as follows:

> The second submission for the appellant is that, whatever the law may H
> have been before the enactment of the Law of Property Act 1925, the effect of that statute, whereby a legal estate in land vested in joint tenants is held on trust for sale for the parties beneficially entitled, coupled with the principle that trustees must act unanimously in dealing with trust property, is to reverse the decision in *Summersett* 's case and to prevent one of two joint tenants determining a periodic tenancy without the concurrence of the other. It is unnecessary to reconsider the position where the parties beneficially entitled are different from those who hold the legal interest. But where, as here, two joint tenants of a periodic tenancy hold both the legal and the beneficial interest, the existence of a trust for sale can make no difference to the principles applicable to the termination of the tenancy. At any given moment the extent of the interest to which the trust relates extends no further than the end of the period of the tenancy which will next expire on a date for which J
> it is still possible to give notice to quit. If before 1925 the implied consent of both joint tenants, signified by the omission to give notice to quit, was necessary to extend the tenancy from one period to the next, precisely the same applies since 1925 to the extension by the joint trustee beneficiaries of the periodic tenancy which is the subject of the trust.

Lord Browne-Wilkinson, who did give a judgment with reasons, unlike the other members of their lordships' House than Lord Bridge, said at the end of his speech at p493:

> The trust property in question was a periodic tenancy. As between the lessor and the lessees the nature of the contract of tenancy cannot have been altered
> by the fact that the lessees were trustees. The tenancy came to an end when K
> one of the lessees gave notice to quit. It may be that, as between the lessees, the giving of the notice to quit was a breach of trust, theoretically giving rise to a claim by Mr Monk against Mrs Powell for breach of trust. Even this seems to me very dubious, since the overreaching statutory trusts for sale imposed by the Law of Property Act 1925 do not normally alter the beneficial rights *inter se* of the concurrent owners: . . .

He cites *Re Warren* [1932] 1 Ch 42 and *Bull* v *Bull* [1955] 1 QB 234 and goes on:

> But even if, contrary to my view, the giving of the notice to quit by Mrs L
> Powell was a breach of trust by her, the notice to quit was not a nullity. It was effective as between the lessor and the lessees to terminate the tenancy. The fact that a trustee acts in breach of trust does not mean that he has no capacity to do the act he wrongly did.

He finds that the notice was good.

That of course is unchallengeable authority for the proposition that, as a matter of landlord and tenant law, a joint lessee can give a valid notice without the concurrence of his fellow lessee in a periodic tenancy.

What their lordships did not in terms deal with is a trilogy of cases M
in the Court of Appeal in which this was contemplated at least as a possibility — and in one case at least the actual event occurred — of the court directing that one of several trustees who were co-owners of a tenancy should concur in serving a counternotice of one sort or another in relation to the tenancy. Not all three cases were concerned with agricultural properties. One was a business tenancy case. That was *Harris* v *Black* (1983) 46 P&CR 366. The term there was a term of 14 years, so we are not dealing with a periodic tenancy, but the headnote states the conclusion of the Court of Appeal, which consisted of Slade and Robert Goff LJJ, in the following terms:

---

*Editor's note: Also reported at [1992] 1 EGLR 65

> Although the Court had jurisdiction in an action by one trustee to make an order compelling his co-trustee to join him both in signing a counter-notice and applying to the Court for a new lease, there was no principle requiring that order to be made in every case concerning business tenancies in that the matter was one for the discretion of the Court in the exercise of its equitable jurisdiction depending on the particular circumstances of the particular case.

In that case the parties who were at law the joint tenants had been in partnership but had dissolved the partnership and had retired, so to speak, into separate parts of the demised premises. In the exercise of the court's discretion the court decided, more especially as the one who did not want to go on was in fact paying rather more rent than the one who did want to go on as a tenant, that the jurisdiction to make an order compelling the co-trustee to join in serving the counternotice in that case, under the Landlord and Tenant Act 1954 (Part II) would not be exercised. But the significance of the case, of course, is limited for my purposes to the recognition of the existence of the jurisdiction in the first place.

The second case was *Sykes* v *Land* [1984] 2 EGLR 8. This was an Agricultural Holdings Act case. There is not strictly speaking a headnote but it is the sort of heading that one finds in *Estates Gazette Law Reports*, the operative part of which reads as follows:

> Plaintiff and defendant carried on a farming business in partnership in pursuance of an agreement which provided *inter alia* that on the dissolution of the partnership otherwise than by death the plaintiff was to be entitled to purchase the defendant's share in the capital and profits of the partnership — The farm which was to be the subject of the present litigation was demised to the plaintiff and defendant and became a partnership asset — Subsequently the defendant became entitled, on the death of her husband, to the freehold reversion expectant on the determination of this tenancy — In her capacity as landlord she gave the partnership notices to quit, the first of which was invalid as not purporting to terminate the tenancy on the correct date, but the second was correct in point of form — She also gave notice of intention to terminate the partnership — The plaintiff gave notice of his intention to exercise his rights under the partnership agreement to purchase the defendant's share of the partnership assets, including the tenancy if it continued — No counternotice was served in response to the second notice to quit but the parties agreed that no objection would be taken on the ground of the time-limit.

The issue then arose as to whether or not there should be an order on the lady, who had become entitled to the reversion, to serve a counternotice, then under the Agricultural Holdings (Notice to Quit) Act 1977, but it is the same legislation as now enacted in the Act with which I am concerned. In that case the judgment was given by Fox LJ. At p10J it includes the following:

> By serving a notice to quit, therefore, Mrs Land [the lady who had become entitled to the reversion and whom it was sought to compel to serve the notice] is not of her own motion bringing about an impossibility of performance. The crucial matter is not whether Mrs Land can serve notice to quit (I think she can) but whether she can refuse to join in the service of a counternotice under the Act of 1977; it is common ground that a counternotice served by one only of the joint tenants would be invalid. To that question I now turn.
>
> Upon Mr Sykes' completing the purchase of Mrs Land's share under the provision of clause 10 he will, on the view which I have taken as to the construction of clause 10, become the absolute beneficial owner of the entirety of the assets of the partnership including the lease. Section 26(3) of the Law of Property 1925 provides that trustees for sale under statutory trusts (which the trusts of the lease are)
>
> "shall so far as practicable consult with persons of full age for the time being beneficially interested in possession in the rents and profits of the land until sale and shall, so far as consistent with the general interest of the trust, give effect to the wishes of such persons . . ."
>
> In pursuance of that provision it seems to me that Mrs Land and Mr Sykes as trustees would, as from the completion of the purchase, be required to give effect to the wishes of Mr Sykes as to the service of a counternotice. Such service would certainly be in the general interests of the trust since it would preserve trust property.

So in that case the order was made. That also recognises of course the jurisdiction in the court to do this in a much closer set of circumstances to the present, in that we are dealing here with the Agricultural Holdings (Notices to Quit) Act 1977.

The third and last of the trilogy of cases in the Court of Appeal, which I need not go into in any detail, is *Featherstone* v *Staples* [1986] 1 WLR 861*. In that case it was also held that the counternotice under the Agricultural Holdings (Notices to Quit) Act 1977, as it then was, needed to be served by all the joint tenants, and a special decision (which does not impinge on this case) decided the case, namely that where the tenants included the landlord public policy required that tenants other than the landlord should have authority to serve an effective counternotice. That is not a point that arises in the present case but what it does do, in my judgment, is to review and recognise the earlier cases, not of course *Monk*'s case which had not then yet been decided, but it does recognise and treat as valid the case which *Monk*'s case confirmed as binding, namely *Doe d Aslin* v *Summersett* (1830) 1 B&Ad 135.

So the principles upon which the House of Lords decided the *Hammersmith and Fulham Borough Council* case were recognised in *Featherstone* v *Staples* and were not treated by Slade LJ as being inconsistent with the possibility of the court, in an appropriate case, directing one of two or more joint tenants at law, who were also trustees for sale, to serve a counternotice, if the equitable considerations under the trust for sale so dictated.

On this application, which is an interlocutory one, at a fairly late hour of the day, I do not propose to enter into a lengthy debate on the extent to which the *Hammersmith and Fulham Borough Council* v *Monk* case is consistent with those three authorities in the Court of Appeal, which recognised the jurisdiction that I am being asked to exercise in this case. In my view, it is at least arguable that they can perfectly well coexist. It seems to me that the speech of Lord Browne-Wilkinson recognises explicitly in the *Hammersmith and Fulham* case the possibility of there being trust considerations existing which might in an appropriate case (and if the court had an opportunity of intervening at an early enough stage) have an effect upon the way in which the trustees for sale should be directed to behave. *Monk*'s case was concerned with what the results of what one of the two trustees for sale had actually done *vis à vis* the landlord. That is a different situation, in my judgment, from what I am asked to look at, which is how the trustees for sale should be directed to behave *vis à vis* the landlord. So I see no incompatibility between those two lines of authority.

On the other hand, I think Mr Feltham is right in his submission that the plaintiff does not establish even a serious issue to be tried on the question of whether he has the sole beneficial interest in the tenancy of this farm. He certainly started out life just as a beneficial joint tenant in 1974 when the tenancy was first granted. Although there have been subsequent dealings, the express agreement not to deal with Mr Anthony Cork's beneficial interest in the tenancy when the family arrangement was arrived at in 1991 seems to me very strongly against any proposition that Mr Anthony Cork went so far as to divest himself of that beneficial interest.

That, however, does not conclude the question because, in my judgment, what the court has to look at is whether there are equitable interests which operate to vary the *prima facie* position which certainly obtained in 1974, namely that either of the two (or any of the three, as it was in those days) joint tenants could serve a tenant's notice to quit at an appropriate time under the tenancy at law, and, had a notice under the relevant legislation been served by the landlord, could have declined to serve a counternotice.

On that issue, are there any such equitable interests which arise, it seems to me that it is arguable that a result was achieved, consequent upon the family discussions in 1991 and thereafter, leading up to the deed of retirement, which positively had the effect of an agreement that Mr Roger Cork should remain as the occupying farming tenant of this farm. It is true that the evidence on this score is thin but this is an interlocutory application, and the evidence is there and is not actually in terms disputed, that there was an agreement that Mr Roger Cork would continue to farm this farm in his own right. Mr Feltham

*Editor's note: Also reported at [1986] 1 EGLR 6

submitted to me that that was not enough to override the *prima facie* situation which I have described, governed by what we now know from *Hammersmith and Fulham Borough Council* v *Monk*, and that the effect of such an event would have been that Mr Anthony Cork would be locked into the tenancy and the obligations that it imposes on the tenants, notably to pay the rent and do the repairs that are required, for as long as the plaintiff, Mr Roger Cork, thought fit. It seems to me that the answer to that argument may well be that that was what was effectively agreed, because the agreement as it is stated very succinctly in the affidavit is without any sort of qualification or provision for determination. Just as the agreement, as reported, does not say that it was agreed that Mr Roger Cork would remain the occupying tenant farming in his own right so long as he wanted, so equally it does not say that Mr Roger Cork would only continue to farm as the occupying tenant so long as Mr Anthony Cork saw fit to allow that state of affairs to continue. The matter simply was not gone into so far as one can tell from the evidence.

In those circumstances, the door seems to me wide open to considerations of balance of convenience. I think there is absolutely no doubt where the balance of convenience lies because it is perfectly obvious that, if the counternotice is not served, signed by both tenants, the notice to quit will take effect; whereas, if the counternotice is served, there is not in fact any evidence before me that the Agricultural Land Tribunal, which would be seized of the matter if the landlord wished to enforce the notice to quit, would have material upon which to allow that to happen. That is a matter which is not by any means certain, but what is clear is that, if the counternotice is not served, there will undoubtedly be a risk of determination.

I appreciate of course that all this hinges on the true construction of a somewhat vague family arrangement some few years ago, and that at this present interlocutory stage and without the benefit of seeing any of the witnesses it is by no means certain that what I have held is an arguable proposition will turn out at the end of the day to be actually the correct and precise result of what this family did agree. In those circumstances I should contemplate the possibility that it will turn out at the end of the day, when the trial of this matter comes (if the parties cannot agree), that Mr Anthony Cork is right and that he is in the unqualified position of a joint tenant, as was the tenant in question in the *Hammersmith and Fulham Borough Council* v *Monk* case.

That is covered, fortunately, by an undertaking which was offered on behalf of Mr Roger Cork, and which I am prepared to accept, that, if at the trial it is felt that Mr Anthony Cork was not compellable to sign the counternotice, the plaintiff, Mr Roger Cork, will either concur in effecting a surrender or take such other steps as may be necessary to bring the tenancy to an end, when it would, had no counternotice been served, have determined, that is to say Lady Day of the year after next. That seems to me to take the sting out of the possibility, which I cheerfully accept as being a lively one, that I may have reached the wrong tentative conclusion in this matter, and will foreseeably protect Mr Anthony Cork against the more serious consequences of my granting an injunction against him, should it at the trial turn out that the injunction should not have been made.

I am fortified in taking what may on the face of it be a rather robust attitude of granting a mandatory interlocutory injunction by the additional factor that Mr Anthony Cork has, so far as one knows, without suffering any hardship, in fact been an absentee non-rent-paying joint tenant in respect of this farm since 1977. That too I put in the scales on questions of balance of convenience.

For all those reasons, I propose to grant the injunction that is sought, but on the undertaking that I have mentioned that was offered me by the plaintiff.

ESTATE AGENTS

A

Espley *v* Williams

COURT OF APPEAL

October 18 1996

HENRY and ALDOUS LJJ

Estates Gazette February 22 1997

[1997] 08 EG 137

B ***Estate agents — Covenant restrictive of practice within two-year two-mile radius — Whether valid and enforceable***

The plaintiff and defendant entered into an agreement as partners to run an estate agency. By the partnership agreement the defendant covenanted that he would not within a period of two years from the date of termination of the partnership practice as an estate agent either on his own account or in conjunction with any person, firm or company within two miles of their premises. In May 1995 the defendant left the partnership and later in 1995, obtained employment with a rival estate agent within the two-mile radius. In the court below the recorder upheld the validity of the restrictive covenant; the defendant appealed.

C ***Held*: The appeal was dismissed. The goodwill of the partnership was a legitimate interest which the plaintiff was entitled to have protected. The restrictive covenant was not a covenant against competition. On the evidence before him, of the state of the market and estate agency in the locality, the recorder was entitled to find that without the enforcement of the covenant, there was a danger that the goodwill would be injured or damaged. It was a covenant which was no more than adequate to protect the goodwill of the interest.**

The following cases are referred to in this report.

D *Stenhouse Australia Ltd* v *Phillips* [1974] AC 391; [1974] WLR 134; [1974] All ER 117; [1974] 1 Lloyd's Rep 1, PC

This was an appeal by the defendant, Mark Stuart Williams, from a decision of Mr Recorder Axtell in Bournemouth County Court, in proceedings by the plaintiff, John Gordon Espley, for breach of a partnership agreement.

Ian Peacock (instructed by Ellis Jones, of Bournemouth) appeared for the appellant; Paul Cairnes (instructed by Richards & Morgan, of Bournemouth) represented the respondent.

E Giving judgment, **HENRY LJ** said: By a judgment delivered on April 10 1996, Mr Recorder Axtell upheld a restrictive covenant in a partnership agreement restraining an estate agent in certain ways. His judgment is now appealed to us.

The history of the matter is this. The plaintiff, Mr Espley, was at the time of the hearing 51 years old. He had been an estate agent between the years 1985 and 1987. He worked with the defendant, who, at the time of the judgment, was 31 years old, at an estate agency firm. That firm was taken over by the Prudential. When he left that firm in 1989 he was bound by a restrictive covenant imposed by the Prudential that restricted him for a period of two years from working within two miles and dealing in residential property. By the end of 1991 that covenant F was about to run out. He thought that he would set up his own firm with the defendant. The defendant would specialize on the residential side. He would not deal with the residential side until the two-year period of his covenant had expired and would then deal with both the commercial and the residential sides. They set up in Christchurch. Both of them knew the area well. It was the plaintiff's evidence that he had said from the word go that he intended the defendant to be his partner. However, no final partnership agreement was drawn up until February 15 1994. The nature of that agreement was this. Only the plaintiff had put money into the business and that is reflected by the terms of the agreement. First: G

Financial

3.1 The profits of the Partnership shall (unless the Partners otherwise agree in writing) belong to the Partners in the following shares

Mr. Espley 65%

Mr Williams 35%

and they shall bear all losses in the same proportions.

And then:

The goodwill of the Partnership business and the firm name shall belong to Mr Espley alone and Mr Williams shall not acquire any share or interest therein. H

Then there comes the three covenants relied on by the plaintiff. Chapter 7 in the agreement:

Restrictions following termination

7.1 Mr Williams covenants with Mr Espley: —

7.1.1 that he will not within a period of two years from the date of termination of the partnership either on his account or for or jointly or in conjunction with or on behalf of any other person firm or company whether directly or indirectly practice as an estate agent or be in any way involved in the business of an estate agency in either case within a radius of two miles of the premises.

That was in Christchurch. Effectively that would have covered all of J Christchurch, and within that radius there were another 23 estate agencies:

7.1.2 he will not within a period of two years from the date of termination of the partnership either on his own account or for or jointly or in conjunction with or on behalf of any other person firm or company whether directly or indirectly solicit business entice clients or interfere with the relationship between Mr Espley and any former customers of the Partnership and

7.1.3 he will not at any time after the date of termination of the partnership use the name Espley Brooks or [any other like name].

What happened then was this. Business had been difficult over all of K these years because of the state of the housing market, and in May 1995 the plaintiff had talks with another small estate agent with the prospect of a merger between them being under consideration. The defendant was not attracted by this idea and he decided to leave the partnership and did so in that May. He went to a firm which unfortunately went into liquidation or a receiver was appointed, and so from October 1995 he was out of work. His response to that was to get work as a negotiator with Slades, a rival estate agent in Christchurch within the two miles. As a result of his taking up employment there,

A injunctive proceedings were commenced against him, culminating in the judgment under appeal before us today.

However, from the beginning he gave an undertaking agreeing to be bound by the non-solicitation of customers clause in 7.1.2 and by the clause as to use of name in 7.1.3, and so the live dispute was the two-year/two-mile restrictive covenant restricting him from being employed or acting as an estate agent within the period of that space or time.

It is conceded that the judge correctly set out the law on the second page of the judgment. He defined the three questions that he had to decide. He said:

B

> The essence of those authorities I take to be as follows: The plaintiff must establish that: —
>
> (1) He has a legitimate interest capable of being protected — here that is the "good will" of "Espley Brooks" which continues to trade as a firm of estate agents.
>
> (2) The restrictive covenants are no more than "adequate" to protect the interest meaning not excessive as regards area, duration or prohibited activities, and
>
> (3) Without the enforcement of such covenants the interest or good will could be injured or damaged.

C Those are the three covenants to be looked at, and in relation to the first of those the judge approached the matter in this way. He said:

> It has to be remembered that in 1993/1994 as now the residential market was fragile. In 1991 the plaintiff and defendant were starting "cold" in a highly competitive environment. In Christchurch with a population of 40,000 or so and with 23 or so agents within several miles of the town centre and a limited supply of houses for sale at any one time, the number of instructions from would-be sellers would be highly contested. The plaintiff, I find, had put up to £20,000 of his own money into the business of Espley Brooks. By 1994 he had been trading in the most difficult circumstances for only 2 1/2 to 3 years. He was vulnerable then to the loss of a key figure who knew the Christchurch market D and regardless of the actual value of Espley Brooks "goodwill" he had, I find a legitimate interest to be protected. Goodwill of course is that intangible asset that is the good name and reputation of a firm that encourages the public to use the services. By 1994 the defendant had helped to establish that in a two-partner firm.

He said in answer to the first question, that the goodwill was a legitimate interest capable of being protected. This was the goodwill of a two-man firm of which the plaintiff was the commercial partner and the defendant was the residential partner. They were a natural combination from the beginning, as the judge found, particularly as the plaintiff was banned by the tag end of the restrictive covenant the Prudential had imposed on him from dealing in residential property at E the beginning of the partnership.

Aldous LJ has defined goodwill, on the basis of earlier authority, as being the attractive force which brings in custom. There were only two men bringing in custom into this firm, the plaintiff and the defendant. That goodwill would seem to me to be *prima facie* capable of protection, and the judge so found. However, his finding is attacked by Mr Ian Peacock, for the appellant, on the basis that this covenant 7.1.1 was, he says, in reality a covenant against competition. He submits that what the plaintiff was entitled to protect was the customer base of his firm and nothing else. He cites the case of *Stenhouse Australia Ltd* v *Phillips* [1974] AC 391, a Privy Council case. In the speech of Lord F Wilberforce at p400, one gets the passage that Mr Peacock relies on. Lord Wilberforce says:

> The accepted proposition that an employer is not entitled to protection from mere competition by a former employee means that the employee is entitled to use to the full any personal skill or experience even if this has been acquired in the service of his employer: it is this freedom to use to the full a man's improving ability and talents which lies at the root of the policy of the law regarding this type of restraint.

I break off the quotation at that point to comment that, so far from Mr Peacock's point of view, so good. However, Lord Wilberforce continues in this way:

> . . . the employer's claim for protection must be based upon the identification G of some advantage or asset inherent in the business which can properly be regarded as, in a general sense, his property, and which it would be unjust to allow the employee to appropriate for his own purposes, even though he, the employee, may have contributed to its creation.

It seems to me that the advantage or asset here is the goodwill of the partnership to which the defendant had contributed, obviously because, as I have said, he was numerically half of the attractive force of two people who were there. That being so, the judge was quite entitled to find that this was not a covenant against competition, but that this goodwill amounted to a legitimate interest that the plaintiff was entitled to protect to make sure that half of the attractive force did H not move down the road to open up shop in a competitor's close by. That was a finding of fact made by the judge, one that he was quite entitled to make. Though that alone would be sufficient to sustain that finding, in my judgment, he was also right to have made it.

The linked point to that, and perhaps a rather stronger point from Mr Peacock's point of view, is the argument that what was rightly capable of protection, so far as the plaintiff was concerned, was their customer base. He submits, therefore, that they were adequately protected by the covenant at 7.1.2 against soliciting business, and that any covenant wider than that expressed in terms of 7.1.1 was unnecessary and so too wide. His argument was not based on attacking J either the two years or the two miles but attacking fundamentally the existence of that clause at all.

The judge was in a much better position to assess the answers to all of these three questions than we are. The reason for that is that in the case before him three Christchurch estate agents gave evidence, the plaintiff, the defendant and the defendant's new employer, Mr Slade. We do not have the notes of the evidence that they gave. That would make us doubly reluctant to interfere with findings of fact arrived at by the judge.

Against that background, we look at the answers to the second and third questions. It is convenient to take the third question relating to K vulnerability first. Clearly, one would start from the position that it would be likely, if half the firm moved down the road, and there was no covenant preventing that, that the interest or goodwill could be injured or damaged, and so the judge found. He dealt with the state of the market. He asked himself the right question, and he found that such damage could arise. He said how in these words:

> I accept the argument that an intimate knowledge on the defendant's part of the residential market in Christchurch coupled with experience could give a competitive edge to a rival. Such knowledge helps "pitch" prices at the right level, that gives rise to sales and more "sold" boards and thereby the public perception so important in estate agency, that the firm to instruct is the one that L seems to be doing the most business is both created and enhanced. Recommendations to third parties to instruct the defendant based on past dealings with him could also damage the plaintiff and in that instance the defendant would not have solicited or canvassed those instructions. Christchurch is a small place and knowledge of where the defendant is working is likely to be known by some solicitors, building societies, employers or friends who could all choose to give recommendations.

And then this important sentence:

> I emphasize that Mr Slade employed the defendant because of his knowledge of the area and an undertaking not to solicit or entice does not, in my judgment, go far enough to protect the plaintiff's interest. M

Having made those findings and having heard these three experts, the judge's findings cannot, in my judgment, be criticised or upset by this court.

I turn last to the second question, namely whether the restrictive covenants are no more than adequate to protect the interests, meaning not excessive, as regards area, duration or prohibited activities. First, in relation to the two miles and two-years covenant, as I have said, these figures have not been specifically attacked. It has been the principle that has been attacked. It is unsurprising that it has not been specifically attacked because it seems that they were not unknown in Christchurch at the time. One has the hard evidence of Mr Espley's

A covenant with the Prudential and one has the finding by the judge that Mr Slade in his evidence admitted that he had a similar restrictive covenant with Mr Kay, his partner there, and the judge was to find that he accepted the plaintiff's evidence that such a covenant is customary for estate agencies in the area both as to time and distance. The fact that it is customary in the area does not tie this court's hands, nor does it tie the judge's hands. He could have found that it was more than required. Had he thought that the covenant against non-solicitation would work in practice, it may be that he would have viewed the former covenant in a different light from that which he did, but he found in terms, and realistically, that that covenant against non-
B solicitation did not go far enough to protect the plaintiff's interests. The reason for that seems to me to be clear, namely, that, in practice, C
such covenants are very difficult to enforce, if not impossible, as Mr Paul Cairnes made clear before us.

In all those circumstances, it seems to me that the judge was quite entitled to come to the conclusion that he did and answer this question as well, having taken into account the public interest in the way that he did, favourably to the plaintiff. It seems to me that there are no grounds on which we can upset his decision, and, for those reasons, I would dismiss this appeal.

Aldous LJ agreed and did not add anything.

Appeal dismissed. D

LANDLORD AND TENANT
GENERAL

Church Commissioners for England *v* Ibrahim and another

COURT OF APPEAL

February 9 1996

BUTLER-SLOSS, ROCH and HOBHOUSE LJJ

Estates Gazette January 18 1997

[1997] 03 EG 136

Landlord and tenant —Arrears of rent — Tenancy provides that tenant pays landlord's costs on indemnity basis in any proceedings — Whether indemnity costs payable in proceeding for arrears of rent

In 1993 the appellant landlords granted the respondents an assured shorthold tenancy for a term of two years at a rent of £9,000 pa. By clause 9 of the agreement the tenants agreed to pay and compensate the landlords fully for any cost, expense, loss or damage by reason of any breaches of the agreement, and to indemnify the landlords in respect of all actions, claims and liabilities. In proceedings for possession based on arrears of rent, which by the hearing were about £8,000, the landlords sought an order for costs on an indemnity basis; they relied on clause 9. The landlords appealed the decisions of the recorder, who declined to make such an order.

***Held*: The appeal was allowed. A successful litigant's contractual rights to recover costs of any proceedings to enforce his primary contractual rights is a highly relevant factor when it comes to making a costs order. Such a litigant is not to be deprived of his contractual rights to costs unless there is a good reason; this applies to the making of a costs order in his favour and to the extent that costs are to be paid to him. Where the parties have agreed the basis of taxation, it would be an improper exercise of the court's discretion to direct taxation on some other basis. *Gomba Holdings (UK) Ltd* v *Minories Finance Ltd (No 2)* [1993] Ch 171 followed. The bargaining strength of the landlord is not a good reason. On an undertaking by the landlords not to enforce any order as to costs, the recorder's order as to costs was amended to an order that the landlords were entitled to an order for costs on County Court Scale 2 on an indemnity basis.**

The following cases are referred to in this report.

Gomba Holdings (UK) Ltd v *Minories Finance Ltd (No 2)* [1993] Ch 171; [1992] 3 WLR 723; [1992] 4 All ER 588

This was an appeal by the plaintiff landlords, Church Commissioners for England, against a decision of Mr Recorder Serota QC in Central London County Court in proceedings against the respondent tenants, Kamal Ibrahim and Abd-Al-Alah-Dauod.

Edwin Johnson (instructed by Radcliffes Crossman Block) appeared for the plaintiffs; Paul Randolph (instructed by William Heath & Co) represented the first respondent; the second respondent did not appear and was not represented.

Giving the first judgment, at the invitation of Butler-Sloss LJ, **ROCH LJ** said: The appellants own the freehold of flat 20, The Quadrangle, Cambridge Square, London. They granted the respondent an assured shorthold tenancy for a term of two years from June 24 1993 to June 23 1995 of those premises at a rent of £9,000 pa. Clause 9 of the second schedule of the tenancy agreement is one of the tenant's covenants. It is headed: "*Indemnity*", although it has to be observed that clause 1.13 of the agreement provides the paragraph headings in the schedules to the agreement are for ease of reference only and have no effect on the construction of the agreement. Clause 9 reads:

> To pay and compensate the Landlords fully for any cost expense loss or damage incurred or suffered by the Landlords as a consequence of any breach of the agreements on the part of the Tenant in this Agreement and to indemnify the Landlords from and against all actions claims and liabilities in that respect.

The respondent fell into arrears with payment of rent. He and the appellants reached an agreement for the clearance of those arrears by instalments. The appellants had prior to that arrangement commenced proceedings for possession in Central London County Court.

In the particulars of claim it is alleged that the arrears at the time of the particulars of claim were a little over £6,000 and it appears from the judgment of the recorder that at the time of the arrangement between the appellants and the respondent the amount of the arrears was a little over £8,000. In the particulars of claim the appellants set out clause 9 of the second schedule to the agreement and sought an order that the respondent should pay their costs of making the application for possession on the basis of clause 9 of the second schedule.

That claim for possession came before Mr Recorder Serota QC on October 6 1994. The recorder was informed of the arrangement as to the payment of arrears and made a suspended order for possession with the consent of the respondent. The appellants then asked for an order that the respondent should pay their costs on an indemnity basis. The appellants submitted that the respondent was bound to indemnify them for the costs incurred by them in seeking to recover possession of the premises and the arrears of rent because these were costs and expense incurred by the landlords as a consequence of a breach of the agreement on the part of the tenant. The recorder proceeded on the basis that that was the true construction of clause 9 of the second schedule of the tenancy agreement. Nevertheless he declined to make the order that the appellants sought holding that the case was a perfectly straightforward claim for possession, that the respondent had done nothing to justify an award for costs on an indemnity basis; that costs were always in the discretion of the court and that parties to litigation could not fetter the court's discretion as to costs by agreement. The recorder made an order that the appellants should have their costs on scale 1 of the county court scale. The recorder in reaching this decision observed that, and I quote from his judgment:

> The courts have been careful to control costs in undefended possession claims.

A little later he observed:

> . . . it would be most unfortunate, as a matter of principle, to hold that a landlord of residential premises, who is generally in a position to dictate the

terms of a tenancy agreement to his tenant, should be able to inflate the amount of costs that he can recover by providing that, in the event of his having to take proceedings for breach, he should be entitled to recover costs on an indemnity basis.

The recorder held that the appellants' entitlement to costs was a matter for the court's discretion and that the clause in the tenancy agreement was merely a matter for the court to take into account when deciding how to exercise its discretion.

The practical effect of the recorder's decision was that in Central London County Court it is the almost invariable practice in cases of undefended possession actions to allow costs of £225 together with the court fee of £70 in the case of private landlords or rather less in the case of local authorities and housing associations.

The sum to which the appellants would be entitled if costs were to be awarded to them on an indemnity basis would be in this case of the order of £550. The court's power to award costs is founded on section 51 of the Supreme Court Act 1981. Section 51(1) provides:

> Subject to the provisions of this or any other enactment and to rules of court, the costs of and incidental to all proceedings, in —
> (a) the civil division of the Court of Appeal, and
> (b) in the High Court, . . . [and
> (c) any County Court] shall be in the discretion of the court.

Subsection (3) of the section provides:

> The court shall have full power to determine by whom and to what extent the costs are to be paid.

In subsection (6) it is provided that:

> In any proceedings mentioned in subsection (1), the court may disallow, or (as the case may be) order the legal or other representative concerned to meet, the whole of any wasted costs or such part of them as may be determined in accordance with rules of court.

Subsection 7 defines "wasted costs". Subsection 8 requires the person responsible for determining the amount of costs which is to be awarded to the person who has started the proceedings in the High Court, which should have been started, in the court's opinion, in the county court, to have regard to that fact when assessing the amount of costs to be awarded.

The statutory provisions, the rules of court and the decided cases, both in this country and in certain Commonwealth jurisdictions, were considered by this court, consisting of Sir Stephen Brown P, Stocker and Scott LJJ in *Gomba Holdings (UK) Ltd* v *Minories Finance Ltd (No 2)* [1993] Ch 171. In a reserved judgment delivered by Scott LJ the court said (and I refer to p194A of the report):

> In our opinion, the following principles emerge from the cases and dicta to which I have referred.
>
> (i) An order for the payment of costs of proceedings by one party to another party is always a discretionary order: section 51 of the Act of 1981.
>
> (ii) Where there is a contractual right to the costs, the discretion should ordinarily be exercised so as to reflect that contractual right.
>
> (iii) The power of court to disallow a mortgagee's costs sought to be added to the mortgage security is a power that does not derive from section 51 but from the power of courts of equity to fix the terms on which redemption will be allowed.
>
> (iv) A decision by a court to refuse costs, in whole or in part, to a mortgage litigant may be a decision in the exercise of the section 51 discretion or a decision in the exercise of the power to fix the terms on which redemption will be allowed or a decision as to the extent of a mortgagee's contractual right to add his costs to the security or a combination of two or more of these things. The pleadings in the case and the submissions made to the judge may indicate which of the decisions to which we have referred has been made.
>
> (v) A mortgagee is not, in our judgment, to be deprived of a contractual or equitable right to add costs to the security merely by reason of an order for payment of costs made without reference to the mortgagee's contractual or equitable rights and without any adjudication as to whether or not the mortgagee should be deprived of those costs.

The recorder is, in my view, correct that parties to litigation cannot tie the hands of the court on the question of costs by agreement whether that agreement is one made after the commencement of proceedings or in the contract, breach of the terms of which gives rise to the proceedings. The court's power to decide by whom costs should be paid could probably not be fettered by a prior contract between the parties to the effect that a successful litigant should have to pay costs to an unsuccessful litigant. Clearly it would be contrary to the public interest that the court should be deprived of the powers given under section 51(6) to disallow wasted costs. Further, section 51(8) requires the person responsible for determining the amount of costs to take account of the factor there mentioned if it exists and that duty placed on that person cannot, in my view, be abrogated by a term in the contract. Whether the court's discretion to decide by whom the costs of proceedings should be paid could be fettered by a contractual agreement made before the litigation is started is a more difficult question which does not arise in this appeal.

Having made these observations, in my judgment, the statements of principle in the *Gomba Holdings* case are not confined to mortgage cases and have a wider application. The successful litigant's contractual rights to recover the costs of any proceedings to enforce his primary contractual rights is a highly relevant factor when it comes to making a costs order. He is not, in my view, to be deprived of his contractual rights to costs where he has claimed them unless there is good reason to do so and that applies both to the making of a costs order in his favour and to the extent that costs are to be paid to him. Indeed I would adopt the citation in the *Gomba Holdings* case from the judgment of Vinelott J which appears at p193A, namely:

> If the parties have agreed the basis of taxation it would, I think, be an improper exercise of the court's discretion to direct the taxation on some other basis, unless satisfied that there had been some conduct on the part of the mortgagee disentitling him to costs or to costs on the agreed basis.

A good reason for depriving a successful litigant to part of the costs to which the contractual term would entitle him would be that that part of the costs came within the definition of wasted costs in section 51(7), that is to say they were costs incurred by him as a result of improper, unreasonable or negligent conduct on his part or that of his legal or other representatives. There may well be other sufficient reasons for interfering with the basis of taxation.

In my opinion, it is not a proper exercise of a judge's discretion to refuse to allow a successful litigant to recover his contractual entitlement to costs because the judge considers that a lessor has an unfairly strong bargaining position or it is desirable that the courts keep a careful control of costs in undefended possession claims. Of course a landlord cannot by contract provide that he should recover a greater sum by way of costs than the costs that he has actually and reasonably incurred.

In this appeal Mr Paul Randolph, for the respondent, has raised the question of the construction of clause 9 of schedule 2 of the tenancy agreement. His submission is that on a proper reading of that clause the appellants are only entitled to costs on the standard basis and that the clause does not entitle them to costs on an indemnity basis. This interpretation, in my judgment, does not attribute meaning to the words "fully for any costs" or to the later words "to indemnify" which appear in this clause. In my judgment, the view accepted by the recorder of the proper construction of clause 9 is the correct interpretation of that provision.

The conclusion I have reached is that the recorder misdirected himself. The proper exercise of his discretion was to award the appellants their costs on the scale applicable to a claim for over £6,000, namely scale 2, and on an indemnity basis unless there was a good reason for not making that order or there was a good reason for depriving them of part of their costs, such reason to be found in the conduct of the appellants. The fact that this was a straightforward possession action was not such a reason. The appellants were clearly justified in bringing the proceedings and had been reasonable in coming to an accommodation with the tenant which entitled him to stay in the premises.

In their skeleton argument the appellants' position is summarised in this way. If the appeal is successful the appellants' position on costs is as follows, in view of the nature of the issue raised on this appeal the appellants do not seek their costs of this appeal and would not propose to enforce an award of the indemnity costs against the first defendant. Mr Edwin Johnson, who appears for the appellants, having taken instruction has given this court an undertaking on behalf of the appellants that they would not enforce an order for costs below against the respondent. The position that they are not seeking their costs of the appeal remains. In those circumstances I would allow this appeal. I would amend the costs order made below to an order that the appellants should have their costs on scale 2 on an indemnity basis and I would make no order as to the costs of this appeal.

Agreeing, **Hobhouse LJ** said: The order made by the recorder included a monetary judgment in favour of the plaintiffs for £8,277 plus costs. He also made, in favour of the plaintiffs, a possession order which was suspended upon terms that the defendants do pay by instalments the amount of the judgment plus the rent that was continuing to accrue meanwhile. Therefore the plaintiffs had succeeded in the action. They had obtained a substantial money judgment. They had also obtained a conditional possession order. However, the recorder only ordered that the plaintiffs' costs be taxed on scale 1. The point of principle raised upon this appeal is whether he was right and whether he had any grounds for doing what he did. He was asked to make an order for indemnity costs in favour of the plaintiffs on the basis of the clause in the lease by which the tenant covenanted:

> to pay and compensate the landlords fully for any cost, expense, loss or damage incurred or suffered by the landlords as a consequence of any breach of the agreements on the part of the tenant in this agreement and to indemnify the landlords from and against all actions, claims and liabilities in that respect.

The recorder simply said:

> However, I can see nothing special in the circumstances of this case to set it apart from the ordinary run of the mill undefended possession claim so as to justify the making of an order for payment of costs on the indemnity basis.

Implicitly he was also refusing to order costs other than on scale 1.

This case is governed by the decision of the Court of Appeal in *Gomba Holdings (UK) Ltd* v *Minories Finance Ltd (No 2)* [1993] Ch 171. The judgment of the court, delivered by Scott LJ, includes statements of general principle. The first two are (at p194):

> (i) An order for the payment of costs of proceedings by one party to another party is always a discretionary order: section 51 of the Act of 1981.
>
> (ii) Where there is a contractual right to the costs, the discretion should ordinarily be exercised so as to reflect that contractual right.

The recorder did not follow the guidance given by that case which was binding on him as it is on us. He did not give reasons for departing from what the Court of Appeal has said is the manner in which the discretion should ordinarily be exercised.

The position, translating what was said by the Court of Appeal to the County Court Rules, can usefully be outlined by reference to Ord 38, r 1(2). It provides:

> The costs of and incidental to all proceedings in a county court shall be in the discretion of the court.

That is an overriding and general provision. R 2:

> (1) Subject to this Order, where by or under these rules or any order or direction of the court costs are to be paid to any person, that person shall be entitled to his taxed costs.

In other words there is a right to have the costs taxed unless some different order is made and nothing in *Gomba Holdings* or in what we are saying on this occasion means that a plaintiff or other person who is the beneficiary of an order for costs may recover costs in the absence of some special order to that effect without their claim being subject to the assessment of a taxing master on a taxation.

R 3 provides for the scales of costs and r 4 provides that where there is a monetary claim and a sum of money is recovered by the judgment in the action the prescribed scales shall be observed. The recorder did not observe this provision. By para (3) of r 3:

> In relation to a sum of money only, the scales shall apply as follows: — . . .
> Exceeding £3,000 Scale 2

As appears from paras 3B and 3C scale 2 is equivalent to what would be allowed in the High Court on a taxation in the High Court proceedings.

The judge therefore clearly was at fault in limiting the costs that were recoverable to costs under scale 1. He gave no special reason for departing from the provisions of the rules or for depriving the plaintiffs of the award of costs to which they were entitled under the rules.

The other aspect of this appeal is the contractual point which was specifically the subject-matter of the *Gomba Holdings* case. Here again the judge was at fault in restricting the costs that were recoverable to scale 1, directly contrary to the contractual provision. There was no justification for departing from the contractual provision. He also erred in his approach to the basis of taxation.

The basis of taxation is to be decided after the scale to be applied has been determined. If there was any doubt about that, it is made clear in the notes to the County Court Practice.

The difference between the bases of taxation is not a difference between the criterion to be applied. The criterion still remains whether the costs were reasonably incurred and reasonable in amount. The difference between the two bases is that in one any doubt that a taxing master may feel is resolved in favour of the claiming party and in the other in favour of the paying party. The use of the phrase "indemnity basis" has a historical origin and is misleading. Under both bases the claimant at the end of the day may only recover reasonable costs. The difference relates to the grey area where there is room for doubt on the one side or the other.

The terms of a contract may be material to the basis which is to be ordered. That was the view of the Court of Appeal in the *Gomba* case. The judgment of Scott LJ includes an analysis of the contractual provisions which existed in that case and how they were to be construed. There were claims both under a mortgage and under a guarantee. As regards the language used in the mortgage, Scott LJ said at p186G of the report:

> The language used does, in our opinion, justify an approach that would hold the mortgagee prima facie entitled to recover or retain the full amount of its actual costs, charges and expenses; but the language leaves open, in our opinion, the right of the mortgagor to have excluded any costs, charges and expenses that were incurred in bad faith or were unreasonably incurred or were unreasonable in amount.

With regard to the guarantee, Scott LJ said at p187F:

> Reading the two clauses together, the quantification direction did, in our opinion, entitle the defendants to quantification on an indemnity basis but did not entitle the defendants to recover costs, charges and expenses that had been unreasonably incurred or were of an unreasonable amount.

In carrying those contractual provisions into effect the court ordered under heading (4) on p194:

> The Chancery master taking the account has power under Ord 62, r 24 to request a taxing master to tax any of the items of costs, including fees, charges, disbursements, expenses and remuneration, contained in the account. The taxation must be on the contractual basis, that is to say, the indemnity basis.

Therefore it is clear that in deciding what should be the basis of the taxation the court has a discretion but it should ordinarily exercise it in a way that carries into effect the contractual provision.

I agree with Roch LJ that in the present case the recorder failed to follow *Gomba* and that the appellants are entitled to the order for which they ask, in other words, an order for a taxation of their costs on scale 2 and the basis of taxation to be the indemnity basis.

On other contractual provisions, or in other circumstances, it may be appropriate for the court to make some different order. If some question has already been identified about the recovery of specific items of costs (this is unlikely to happen in cases such as this but it may happen) then the court may in its discretion choose to make a different order with regard to the basis of taxation in order to ensure that the claiming party places the appropriate material before the taxing master so that the relevant issue will be addressed in the most appropriate fashion. But in the present case there is no basis for exercising a discretion which departs from that indicated by the case of *Gomba*.

As regards the particular financial difficulties of the defendants in the present case, an undertaking has been given to this court as referred to already by Roch LJ and there is no need to say any more about it. I agree that this appeal should be allowed as he proposes.

Butler-Sloss LJ agreed and did not add anything

Appeal allowed.

Hunts Refuse Disposals Ltd *v* Norfolk Environmental Waste Services Ltd

COURT OF APPEAL

May 22 1996

Neill, Morritt and Hutchison LJJ

Estates Gazette January 18 1997

[1997] 03 EG 139

Landlord and tenant — Right to deposit refuse — Whether lease or licence — Whether payment to be made for capping materials as well as dry waste

By clause 1 of a deed dated May 15 1974 the appellants were granted by the respondents' predecessor in title an "exclusive licence and full liberty to use the [subject] Site for depositing waste . . . for a period of 21 years". The site was a large area of land on parts of which quarrying was taking place. In the court below the judge decided that the appellants held a licence and not a tenancy, and were therefore not entitled to the protection of the Landlord and Tenant Act 1954. He also decided that under the rent formula, which provided for a rate per cubic yard of dry waste, rent was payable in respect of the deposit of capping or covering materials. The appellants appealed contending that there were clauses in the deed which were only consistent with the grant of exclusive possession, and therefore of a tenancy.

***Held*: The appeal was dismissed. The deed plainly did not evidence an intention on the part of the grantor to grant to the grantee exclusive possession for 21 years. Clause 1 was the crucial clause; other clauses were at best equivocal. The only exclusive quality of the grant was that the grantor could not grant a competing right in the same site. One does not expect a person given the right to tip rubbish in the excavated parts of a large plot of land, on other parts of which quarrying was continuing, to have exclusive possession. The rent payments were based on a simple formula by which used airspace was to be calculated; this therefore included space filled with engineering materials needed for lining, capping and covering tipping waste.**

The following cases are referred to in this report.

Addiscombe Garden Estates Ltd v *Crabbe* [1958] 1 QB 513; [1957] 3 WLR 980; [1957] 3 All ER 563, CA

Ashburn Anstalt v *Arnold (No 2)* [1989] Ch 1; [1988] 2 WLR 706; [1988] 2 All ER 147; (1988) 55 P&CR 137; [1988] 1 EGLR 64; [1988] 23 EG 128, CA

Clays Lane Housing Co-operative Ltd v Patrick (1984) 49 P&CR 72; 17 HLR 188

Clore v *Theatrical Properties Ltd* [1936] 3 All ER 483

Shell-Mex & BP Ltd v *Manchester Garages Ltd* [1971] 1 WLR 612; [1971] 1 All ER 841; [1971] EGD 303; (1971) 218 EG 285, CA

Street v *Mountford* [1985] AC 809; [1985] 2 WLR 877; [1985] 2 All ER 289; [1985] 1 EGLR 128; (1985) 274 EG 821, HL; reversing (1984) 49 P&CR 324; [1984] 2 EGLR 119; 271 EG 1153, CA

This was an appeal by the defendants, Hunts Refuse Disposal Ltd, from a decision of Judge Marr-Johnson, who on May 5 1995, dismissed a summons issued by the defendants under Ords 14 and 14A of the RSC in their counterclaim in proceedings brought by the plaintiffs, Norfolk Environmental Waste Services Ltd.

David Elvin (instructed by Mills & Reeve) appeared for the appellants; Peter Crampin QC (instructed by McKenna & Co) represented the respondents.

Giving judgment, **Hutchison LJ** said: This is an appeal by the defendants, Norfolk Environmental Waste Services Ltd, from the judgment of Judge Marr-Johnson of May 5 1995. The appellant company was set up by Norfolk County Council pursuant to section 32 of the Environmental Protection Act 1990 and is, in effect, the waste disposal arm of the council. The respondents to the appeal are not the owners of the land with which it is concerned but are, *vis à vis* the appellants, in an equivalent position. Putting the matter in this way avoids the need to describe a number of antecedent dispositions which are not material to the issues to which the appeal gives rise. These involve the proper construction of deeds of May 15 1974 and August 11 1986, the terms of which are for practical purposes identical, and which expired on the same date, May 15 1995. The issues before the judge were: (1) whether these deeds conferred on the appellants a tenancy or merely a licence of land at Priory Farm, Blackborough End, Kings Lynn in Norfolk; and (2) what was the correct construction of the provision in the deeds as to payment for deposit of dry waste. On the resolution of the first issue depended the question whether the appellants were entitled to the protection of Part II of the Landlord and Tenant Act 1954. The second question involved whether the appellants were obliged to pay, at the stated rate per cubic yard of solid waste, only in respect of actual waste materials or in respect also of space occupied by materials used for covering and capping the waste, lining the edge of the infill and finishing with topsoil. The judge decided both questions in the respondents' favour — that is to say he construed the deeds as granting licences only and he held that payment had to be made in respect of all space filled, whether by actual waste or such additional materials.

The matter came before the judge on a summons issued by the appellants for summary judgment on their counterclaim, under Ord 14; or alternatively under Ord 14A. It was agreed that there was no substantial issue of fact to be tried and that the case turned on construction of the deeds. The judge's decision effectively terminated the proceedings in the respondents' favour.

All the references I am about to make are to the deed of May 15 1974, which was incorporated in the later deed. I begin with the second of the recitals.

Whereas:—

(2) The Licensee has applied to the Grantor for a Licence to deposit refuse and waste matter (collectively referred to as "Waste") at the Site upon the terms hereinafter contained

Now this deed witnesseth as follows:—

1. In consideration of the payments hereinafter mentioned or made payable and the covenants on the part of the Licensee hereinafter contained the Grantor hereby grants unto the Licensee

(1) exclusive licence and full liberty to use the Site for depositing waste without restriction as to amount for the period of 21 years from the date hereof ("the Licence Period")

(2) full liberty for the Licensee from time to time and at all times during the continuance of this Licence to enter upon the Site with or without vehicles for the purpose of exercising the Licence and liberty hereby granted.

A Clause 2 contains covenants by the licensee, which included the following:

> (1) to be responsible for all Site operations and restoration of the Site as required by the Local Planning Authority and the Town and Country Planning Act 1971 . . . and to keep the Grantor indemnified . . . in respect thereof . . .
>
> (4) to observe all planning requirements and statutory provisions regulations and orders relating to the deposit of waste at the Site during the Licence period
>
> (5) not to assign or otherwise dispose of the benefit of this Licence or any part or interest therein without the previous written consent of the Grantor (which consent shall not be unreasonably withheld) . . .
>
> B (6) to keep a proper account of all waste deposited at the Site and at all reasonable times to allow such accounts to be inspected and copied by the Grantor or its duly authorised agent and to permit the same to be audited from time to time as reasonably required by the Grantor at its expense . . .
>
> (8) to allow the Grantor reasonable access for the purpose of extracting minerals from the site subject to such extraction and access not impeding or interfering with the Licensees normal operations of waste disposal at the site provided that top-soil and overburden shall not be deemed to be minerals for the purposes of this Licence.

Clause 3 contains provisions as to payment, distinguishing between liquid waste which is to be paid for at stated rates for each 1,000 gallons deposited "as calculated from the records and accounts referred
C to in Clause 2(6)"; and dry waste, in respect of which the clause requires payment of:

> a rent at the rate of 10p per cubic yard of dry waste deposited at the site during each year of the Licence Period the volume to be calculated at the end of each year in accordance with the provisions of Clause 3 (2) hereof . . .

Clause 3(2), which is of crucial importance in relation to the second issue, reads:

> For the purpose of calculating the amount of dry waste so deposited a survey of the available air space shall be taken at the commencement of the Licence Period and at each anniversary thereof by a firm of Chartered Surveyors to be
> D appointed jointly by the Grantor and the Licensee the Licensee to be responsible for the Surveyors' charges. In the event of any dispute between the parties as to the amount of dry waste deposited in any year of the Licence Period the decision of such Chartered Surveyors based on the difference between the available air space at the beginning and end of each year shall be final and binding upon the parties. In calculating the difference in such air space the Surveyors shall have due regard to any additional air space created in connection with the extraction of minerals during the relevant year

Clause 4 is a provision entitling the grantor, if any part of "the rent hereby reserved" is unpaid for 30 days or there is a breach of licensee's covenants, to serve notice requiring payment or compliance, failing
E which the grantor is entitled to determine the licence. It does not stipulate for a right of re-entry.

The judge, in deciding the first issue in the respondents' favour, in effect accepted the arguments of Mr Peter Crampin QC on their behalf. Before the judge as before this court, the argument on this issue concentrated on the question whether the deed conferred exclusive possession on the appellants, and it is common ground that that is the vital question: if it did, then the appellants have a tenancy: if it did not they have only a licence. This is clear from a number of authorities, in particular *Street* v *Mountford* [1985] AC 809*. At p818 in the speech of Lord Templeman appears this passage:

> F There can be no tenancy unless the occupier enjoys exclusive possession; but an occupier who enjoys exclusive possession is not necessarily a tenant . . . To constitute a tenancy the occupier must be granted exclusive possession for a fixed or periodic term certain in consideration of a premium or periodical payments.

Later in his speech Lord Templeman said at p826:

> My Lords, the only intention which is relevant is the intention demonstrated by the agreement to grant exclusive possession for a term at a rent. Sometimes it may be difficult to discover whether, on the true construction of an agreement, exclusive possession is conferred. Sometimes it may appear from the surrounding circumstances that there was no intention to create legal relationships. Sometimes it may appear from the surrounding circumstances that the right to exclusive possession is referable to a legal relationship other than a tenancy. Legal relationships to which the grant of exclusive possession might be referable and which would or might negative the grant of an estate or interest in the land include occupancy under a contract for the sale of the land, occupancy pursuant to a contract of employment or occupancy referable to the holding of an office. But where as in the present case the only circumstances are that residential accommodation is offered and accepted with exclusive possession for a term at a rent, the result is a tenancy. G

It is not suggested in the present case that, if there was a grant of exclusive possession, there exists some special reason for none the less regarding the appellants as being only licensees. H

The judge, having rehearsed the arguments of counsel on each side, expressed his conclusion that the deed was what it purported to be, a licence to enter upon land. He said:

> The essential words of transfer in clause 1(1) are that the grantor grants the grantee "exclusive licence and full liberty to use the Site for depositing waste without restriction as to amount", and clause 1(2) permits the grantee at any time "to enter upon the Site with or without vehicles for the purpose of exercising the Licence". I do not accept that the word "Licence" in the expression "exclusive licence and full liberty" is a mere label which should be J ignored for construction purposes. It forms part of the operative words of transfer and in the context in which it appears it must mean "permission". Thus by subclause (1) the grantee is given permission to *use* the site for depositing waste and by subclause (2) he is given permission to *enter* the site for the stated purpose.
>
> By implication it seems to me that the grantee is not to use the site or to enter it for any other purpose. If the grantor were granting the grantee a tenancy, and therefore *ex hypothesi* exclusive possession of the land, there would be no necessity, in my judgment, to state that the grantee could deposit waste without restriction as to amount, nor would there be any necessity to state that he was entitled to enter the site with or without vehicles. The right to do those things and more would be a necessary incident of the tenancy, although of course one might find words of restriction limiting the grantee's K use of the land at a later point in the tenancy agreement.
>
> I conclude therefore that clause 1 does not by its terms operate so as to confer exclusive possession of the land on the grantee.

It is clear from the following pages of the judge's judgment that he had carefully considered arguments addressed to him by Mr David Elvin, counsel for the appellants, based on other provisions in the deed (which I too shall consider) and he gave his reasons for holding that they did not invalidate his conclusion.

Mr Elvin submitted that construing the deed as a whole and considering what it permitted the grantee to do led inescapably to the L conclusion that it conferred a tenancy, albeit one with a limited user provision. The judge gave too much weight to the "labels" used in clause 1. The significance the judge attached to that clause was misplaced: see *Addiscombe Garden Estates Ltd* v *Crabbe* [1958] 1 QB 513. There the equivalent clause was in these terms:

> The grantors hereby license and authorise the grantees to enter upon use and enjoy the following: (a) the club house [etc] . . . (b) the tennis courts . . . (c) the use of all articles of household use and ornament at present in and upon the said club house and also the use of all nets machines tools and equipment in and about the said premises . . .

This was, he submitted, a clause arguably more apt to confer a M licence than that in the present case, and yet the Court of Appeal held that the agreement constituted a tenancy.

Then Mr Elvin pointed to various clauses in the deed which he submitted supported the view that it conferred exclusive possession and were consistent only — or at least consistent — with a tenancy. I shall summarise these submissions:

Clause 1(1), granting *exclusive* licence and *full* liberty: it thus amounted to the grant of exclusive possession with a limited user restriction.

Clause 1(2), granting *full* liberty at *all* times: a right unfettered save for the user restriction.

---

*Editor's note: Also reported at [1988] 1 EGLR 64

The prohibition of assignment save with the grantor's consent: clause 2(5). This, he argued, was consistent with a tenancy — an interest in land — and quite inconsistent with a mere personal right, which cannot be assigned. In this connection he relied on *Clore* v *Theatrical Properties Ltd* [1936] 3 All ER 483 and *Ashburn Anstalt* v *Arnold (No 2)* [1989] Ch 1*.

Clause 2(8) allowing the grantor reasonable access for the purpose of extracting minerals — something the grantor did not need, it was argued, if he had not granted exclusive possession.

Clause 4, which Mr Elvin submitted was in substance a forfeiture provision entitling the grantor to re-enter. He relied on *Clays Lane Housing Co-operative Ltd* v *Patrick* (1984) 49 P&CR 72 where a question arose as to whether a provision entitling the landlord of premises let under a weekly tenancy to give four weeks' notice to terminate the tenancy for various stated reasons including arrears of rent, unauthorised alterations, etc, was a forfeiture provision. Holding, contrary to the judge's view, that it was not, Fox LJ, at p78 said:

> We accept, for present purposes, the submission . . . that a right to determine a lease by a landlord is a right of forfeiture if (a) when exercised, it operates to bring the lease to an end earlier than it would "naturally" terminate; and (b) it is exercisable in the event of some default by the tenant.

Clause 4, Mr Elvin submits, fulfils this test: and he relies on the words of Buckley LJ in *Shell-Mex & BP Ltd* v *Manchester Garages Ltd* [1971] 1 WLR 612, from which I cite a rather more extensive passage than that on which he relied:

> During the course of his argument Mr Dillon has taken us through the detailed provisions of the document with which we are here concerned and has pointed out that many of the clauses in it are clauses which could appropriately find their place in a tenancy agreement. That I think is perfectly true; but it is not to say that they do not equally appropriately find their place in a licence. One must look at the transaction as a whole and at any indications that one finds in the terms of the contract between the two parties to find whether in fact it is intended to create a relationship of landlord and tenant or that of licensor and licensee. There are in this case none of those features present which in *Addiscombe Garden Estates Ltd* v *Crabbe* [1958] 1 QB 513 led the court there to conclude that the intention of the parties was to create a tenancy. They are referred to succinctly by Parker LJ in his judgment, at p529. In that case there was a clause expressly permitting the grantors to enter on the premises to inspect the plant, which would have been unnecessary had they had a right to enter on the premises apart from the agreement. There was a covenant for quiet and uninterrupted enjoyment by the grantee — a covenant appropriate to a lease but inappropriate, I think, to a licence where the licensor had a right of possession in respect of the subject-matter; and, perhaps most cogent of all, there was a term for re-entry upon breach of covenant, which is a term of a kind quite inappropriate, in my judgment, to a licence.

The onerous obligations imposed by clause 2(1) and, in particular, clause 2(4) are said to show the considerable extent of the rights granted to the grantees and to be entirely consistent with the grant of exclusive possession. Indeed, it is argued, labels apart there is nothing in the deed which is inconsistent with the grant of an interest in land; and it lacks altogether the sort of provision one might expect to find where exclusive possession was not given, such as that in the *Shell Mex* case "Not to impede in any way the officers servants or agents of the company in the exercise by them of the company's rights of possession and control of the premises . . . ".

These arguments have not persuaded me that, on this issue, the judge reached the wrong conclusion, and I consider that Mr Crampin is right to submit that the judge was correct essentially for the reasons he gave. At a stage in his judgment when he was summarising the principles to be derived from *Street* v *Mountford* and the other authorities to which he had been referred, and had accepted (rightly) that "similar principles apply in the case of both residential and business premises" he went on to observe:

> It is right to say however that while general principle is immutable, its effect in practice may be very different when one is considering an agreement relating to the use of a 31-acre rubbish tip as opposed to the occupancy of a two-bedroomed residential flat.

The judge was here recognising, realistically and correctly, that the analysis to determine whether an agreement constitutes a licence or a tenancy, while of course it involves giving close attention to the terms of the agreement, is not to be undertaken in a vacuum but rather with a proper regard to the context in which the issue arises. Thus, while one would ordinarily expect that someone in occupation of a small house for a fixed term at a rent had exclusive possession, one would I suggest have no such preconceptions about a person given the right to tip rubbish in the excavated parts of a large plot of land, on other parts of which, it seems, quarrying was continuing.

Mr Crampin poses the question: Does the deed evidence an intention on the part of the grantor to grant to the grantee exclusive possession of the entirety of the site for 21 years? It is plain from clause 1, he submits, that the answer must be that it does not. It is not by accident that the words of the grant in clause 1 are coupled with reference to user for the purpose of depositing waste — an activity which involves the use of the site in stages. The appellants' only rights are to be on the site for the purpose of tipping or carrying out ancillary engineering or maintenance works. They have no right to go, for example, on to a part of the site not yet excavated or where quarrying is in progress. It is plain that the grantor was not (subject only to clause 2(8)) accepting exclusion from the entire site for the whole 21 years, whether or not tipping was taking place. The provisions of the clauses relied on by the appellants are incapable of bearing the weight which the appellants' submissions require them to bear. Of the majority of them, the most that can be said is that they are as consistent with a tenancy as a licence. Thus — to take clause 4 — there is nothing inconsistent with a licence about a stipulation that upon failure to pay the sums due or breach of the terms of the licence the grantor may, if the grantee fails to comply with a notice requiring the money to be paid or the breach remedied, determine the licence. Similarly, the obligations in clause 2(1) and (4) are as consistent with a licence to tip refuse as with a tenancy.

As to the clause relating to assignment, Mr Crampin submits that there is nothing inherent in a licence which prevents its being assigned, provided it contains within it something indicating that it is intended to be assignable. If and to the extent that clause 2(5) assumes wrongly that this licence was assignable, that affords some support for the argument that this was a tenancy not a licence — but is also explicable on the basis that the parties used inapt words to restrict assignment of something they assumed to be assignable. On any view, Mr Crampin submits, this provision is of little weight, when set against clause 1.

As to clause 2(8), which the judge accepted was the high point of the appellants' case, Mr Crampin makes what I regard as a telling point — namely that the clause does not contain a reservation of the right to extract minerals but, proceeding on the footing that the grantor has that right, is a covenant to allow him to exercise it. Nevertheless, he concedes that the judge was right to regard it as a high point of the appellants' argument. Maybe, he accepts, the grantor did not need to stipulate for such permission — but it is perfectly understandable that the parties should wish sensibly to regulate matters between the two competing users of the site. It certainly does not suffice to create a right to exclusive possession of the site. If anything, the assumption underlying this clause of there being a right in the grantor to extract minerals points the other way.

I accept the central thrust of Mr Crampin's argument, that the crucial clause in this agreement is clause 1, particularly subclause (1); and that there is nothing in the other clauses relied upon of sufficient weight to displace the clear indication in that clause that the grantee had only limited rights falling short of exclusive possession. I accept his submission that all those other clauses are equivocal, in that they are consistent with either a licence or a tenancy. I accept the submission that it is not surprising to find the draftsman of what was obviously intended to be a licence borrowing from the language of landlord and tenant.

*Editor's note: Also reported at [1988] 1 EGLR 64

I reject Mr Elvin's suggested answer to the clause 1 point which, it will be remembered, is that it cannot stand in the face of the decision in *Addiscombe Gardens Estates Ltd* v *Crabbe*. It seems to me that that case, and the wording of the comparable clause, were entirely different from the present. The clause itself pointed much less strongly towards a licence than does the clause in the present case — in particular it altogether lacked the ingredient of being permitted to enter only for a limited purpose. Further, the other provisions of the agreement were overwhelmingly inappropriate to a licence. As Parker LJ said (p529):

> . . . though the agreement . . . is labelled a licence, almost all, if not all the clauses in it are appropriate, and more appropriate, to a tenancy agreement. Not only that, but three, in my view, are completely inconsistent with the document being a licence . . .

If one stands back and looks at the deed in the present case, construing it in its context, everything, in my view, points to its being what it purports to be. This was the judge's view, and I agree with him. There is here no suggestion of a sham devised to circumvent statutory provisions, or anything of that sort. On the contrary, the grant of a licence to the grantee giving the exclusive right to tip refuse is a perfectly natural and logical thing to do. That it should be a licence rather than a tenancy causes one no surprise. In this respect also the present case is quite unlike *Addiscombe Gardens Estates* v *Crabbe*.

I would accordingly uphold the judge's conclusion on the first issue.

I now turn to the second issue which is one involving the construction of clause 3(2); though of course regard must be had to other provisions of the agreement in order to understand it in its proper context.

The importance of the point appears from the pleadings and evidence before us, to which I make brief reference. The defendants say that under current planning and waste disposal requirements they have to line the base and sides of any part of the quarry into which they are tipping with low permeability materials 1m thick and, when it is filled to the permitted level, cap it with a 30cm low-permeability layer, a 30cm drainage layer and a layer of topsoil of about 20cms. They say that the fill involved in these engineering works represents rather more than one-third of the total volume for the year in question; and they say that the cost of the engineering works, inclusive of materials, has been £1.6m — a vastly greater sum than the plaintiffs claim by way of rent. It is not difficult to understand why it is that the defendants feel aggrieved at having to pay, as though it were filled by refuse, for the space occupied by these engineering materials.

These considerations are, it must be accepted, immaterial if, on its true construction, the licence deed obliges the defendants to pay for all space that they use up in carrying out their dry waste tipping operations. The licence agreement has to be interpreted in the light of the surrounding circumstances as they were in 1974 when it was executed. Then, it is common ground, requirements as to what I have for convenience called engineering works were significantly less stringent. I am satisfied, however, that even then there existed statutory controls the implementation of which must have given rise to the same problem, though in a less acute form.

I begin my consideration of the true extent of the defendants' obligations in relation to payments for dry waste by gathering together what seem to me to be the parts of the licence deed which bear on the problem.

The second recital records that the licensee seeks a licence "to deposit refuse and waste matter (collectively referred to as 'Waste') at the Site . . .". It is suggested that this is a definition of waste, but I do not think it is properly to be regarded as such: certainly it is not permissible to treat those words, appearing where they do and directed to nomenclature rather than definition, as having any more than marginal relevance to the question to which this second issue gives rise. What can be said however is that, reading clause 1(1) of the agreement in accordance with the recital, the defendants' entitlement is to use the site for depositing refuse and waste matter.

Clause 2(1) imposes an obligation on the licensee to restore the site in accordance with local authority and Town and Country Planning Act requirements; and clause 2(4) a complementary obligation to observe all planning requirements and statutory obligations relative to the deposit of waste. Undoubtedly, therefore, it was appreciated by grantor and grantee at the date of the agreement that it was obligatory for the licensees to undertake these matters, and at their own expense.

Clause 3(1)(a) to (c) lays down what consideration is to be paid for the exercise of the right to deposit refuse and waste matter. In the case of liquid waste payment is to be made on the basis of a unit price per thousand gallons to be calculated from the accounts and records which under the agreement the licensees are obliged to keep. It is perhaps worth observing that the requirement of clause 2(6) is "to keep a proper account of all waste deposited at the site" — that is to say liquid and dry waste: whereas the provisions as to payment for dry waste do not depend on these accounts which they presumably could have done — for example by charging by weight or by volume for dry waste brought on to the site. Instead dry waste is charged on a quite different basis.

Clause 3(i)(c) — the clause which defines what has to be paid in respect of the deposit of dry waste — is perfectly plain. It requires rent to be paid at the stated rate "per cubic yard of dry waste deposited at the site during each year of the Licence Period the volume to be calculated at the end of each year in accordance with the provisions of Clause 3(2)". In the course of the hearing it emerged that (leaving aside the issue of engineering materials) the parties were at one in treating this provision as obviating the need to take account of any question such as the extent to which the dry waste had compressed or degraded or would in the future do so: the measurement on which payment is based falls to be calculated on the basis of a "snapshot" of the tip at the end of the year in question. I am satisfied that this common understanding, as well as having the merit of simplicity, correctly reflects the provisions of clause 3(1)(c).

So we come to the all-important question — what is the basis of the year-end calculation in accordance with which the volume of dry waste deposited is to be determined? Not (as with liquid waste) on the basis of volume coming into the site, but rather on the basis of space occupied at a particular date by dry waste deposited during the preceding year. However, clause 3(2) does not say that the amount of dry waste deposited in the preceding year shall be *deemed to be* the difference between the airspace available at the beginning of the year and at the end. There is in my judgment a real difference between the words:

> The amount of dry waste deposited shall be deemed to be the difference between

and the words:

> For the purpose of calculating the amount of dry waste deposited a calculation shall be made of the difference between . . .

The former, a deeming provision, is plainly laying down a formula which (if properly employed) results in a conclusive figure for dry waste, notwithstanding that (for example by reason of degradation) it may be very different from the volume of dry waste deposited. The latter is at least open to the construction that it is simply defining a first step to be taken towards arriving at a figure representing the amount of dry waste deposited. However, depending on the context in which the words have to be construed, they *could* fall to be interpreted as a deeming provision.

Turning to the other two sentences of clause 3(2), they, in my view, do have a significant impact on the meaning of the first. Logically, I consider, the third sentence ought to precede the second and for convenience I repeat the whole clause here, with the second and third sentences transposed:

> For the purpose of calculating the amount of dry waste so deposited a survey of the available airspace shall be taken at the commencement of the Licence Period and at each anniversary thereof by a firm of Chartered Surveyors to be appointed jointly by the Grantor and the Licensee . . .
>
> In calculating the difference in such airspace the surveyors shall have due regard to additional airspace created in connection with the extraction of

minerals during the relevant year. In the event of any dispute between the parties as to the amount of dry waste deposited in any year of the Licence Period the decision of such Chartered Surveyors based on the difference between the available airspace at the beginning and the end of each year shall be final and binding upon the parties.

All the first sentence requires the surveyors to do is by survey to ascertain the available airspace at the beginning and at the end of the relevant year. The second (previously third) sentence requires due regard to be had to the effect of extraction of minerals; which, in the context, must involve adding the volume of space produced by mineral workings in that year to the volume at the beginning of the year (or subtracting it from the year-end figure). The third (previously second) sentence envisages that when all that has been done there may be a dispute "*between the parties as to the amount of dry waste deposited*": not, be it noted, a dispute as to the difference between the space available at the beginning and the end of the year after making allowance for mineral extraction.

It might I suppose be said that the contest on this issue resolves itself into the question whether the rent for dry waste is payable in respect of the volume of airspace within the tip that is used up or the volume of dry waste that is deposited. There is, in my view, nothing to indicate that the former is the correct measure. It would I suggest, require clear wording to impose on the grantees an obligation to pay for something which is plainly not dry waste. I find no such words in clause 3. To my mind, the wording points the other way. Mr Crampin's argument, (which is that the words "based on" are sufficiently accounted for by the potential for dispute about the measurement of unoccupied airspace at the beginning and end of each year, the arithmetical calculation of the difference and the adjustment in respect of mineral abstraction), in my view, does not meet the point that the dispute referred to in clause 2(3) is "any dispute . . . as to the amount of dry waste deposited", not "any dispute as to the above measurements".

In the circumstances I cannot accept that the (adjusted) volume of airspace is to be equated with the volume of dry waste deposited. I do not attach to the words "based on" as much significance as in argument the appellants did — though those words are of course material. I am more influenced by the consideration that it is the very same surveyors who will have calculated the adjusted difference whose decision on any dispute is required to be treated as final. They will thus (if the plaintiffs' construction be correct) in effect already have determined the only thing about which there could be any dispute before that dispute is referred to them. It is, in my view, no answer to say that there could be room for argument as to the adjustment for mineral extraction: that is something for the surveyors, not the parties, to have due regard to — ie decide. There is in reality, on the matter of the adjusted airspace difference, no scope for dispute between the parties or between a party and the surveyors. The dispute which is envisaged is one *between the parties as to the amount of dry waste deposited*: a dispute on which the surveyors have to give a real decision ie not one which involves their merely saying: "We have already given you the answer to that, and that is our decision which binds you".

It is instructive to see how the judge reasoned in arriving at the opposite conclusion. He took the first sentence of clause 3(2) to be "essentially introductory". Then he said that the words "In the event of any dispute . . ." were mere surplusage because, whether or not there was a dispute between the parties, the calculation had to be made and was indeed the very mechanism whereby the rent was to be ascertained. I do not consider that either of these conclusions is correct; and the second, I think, begs the question to which this issue gives rise — whether the calculated difference is to be deemed to be the volume of dry refuse deposited, or is merely a step on the road to ascertaining the latter volume.

The judge's construction then obliged him to hold that:

The parties must be taken to have agreed therefore that any matter which entered the ground along with the dry waste would itself become waste for the purpose of the calculation of rent.

It seems to me that what I have called engineering materials — for example specially imported top soil — are plainly not (in the ordinary way) refuse or waste matter, and I would not attribute to the parties such an implied agreement unless compelled to do so.

It can of course be said that in favour of the respondent's and the judge's construction of clause 3(2) is simplicity and the elimination of much room for dispute — but that is at best a marginal consideration on an issue of construction.

I would accordingly hold that on its proper construction clause 3(2) involves that, once the difference between adjusted airspace has been ascertained, as it must be by the surveyors, it is for the parties to endeavour to agree the volume of dry waste, using the surveyors' figure as a base; and that if they cannot agree the surveyors must finally determine the matter.

I would therefore, while rejecting the appeal on the first question, allow it on the second.

Agreeing, **Morritt LJ** said: The issues which arise on this appeal and the circumstances in which they arise have been fully described by Hutchison LJ. I gratefully adopt his account of them.

On the first issue, namely whether the two deeds created leases or licences, I agree with him in concluding that they created licences. Counsel for the appellants placed considerable emphasis on the wording of clause 1 of the deeds and clause 1 of the relevant document in *Addiscombe Garden Estates Ltd* v *Crabbe* [1958] 1 QB 513, both of which have been quoted by Hutchison LJ. In my view, that comparison does not assist the appellants.

This case concerns a 31-acre site on which, at the time of the execution of the deeds, quarrying or mineral extraction operations were taking place and likely to continue. The deeds did not grant the exclusive licence and full liberty to use the site at all times and for any purpose for which it might be suitable. Rather it granted the exclusive licence and full liberty to use the site for depositing waste. Thus it was exclusive only for that purpose. It prevented the grantor from granting a competing right to any one else, but did not exclude the grantors from continuing with their quarrying and mineral extraction operations on the site or from using it or allowing others to use it for any purpose which did not inhibit the exercise by the appellants or their predecessor of the exclusive rights granted.

By contrast, in *Addiscombe Garden Estates Ltd* v *Crabbe* any limitation on the use to which the grantee might put the property was imposed by the nature of the property and not the form of grant. The grant itself was for the unrestricted use and enjoyment of specified property. Thus if the comparison of this case with that is of assistance, which I doubt, it does not, in my judgment, lead to the conclusion for which counsel for the appellants contends. I agree with Hutchison LJ, for the reasons which he gives, that the deeds were licences and not leases for they did not confer the right to exclusive possession on the grantee.

I acknowledge the force of the points on which Hutchison LJ relies for his conclusion on the second point, however I am unable to agree with him. The provisions of the licence relevant to this point have been quoted by him and I do not need to repeat them.

At the time the licences were executed it was known to the parties that the regulations dealing with the deposit of waste required some engineering works as described by Hutchison LJ. It was also then envisaged that the quarrying and mineral extraction operations on the site would continue, thereby creating further areas suitable for the deposit of waste. These were the circumstances in which the formula for the measurement of the waste deposited for rental purposes had to be agreed.

In my view, the provisions of clause 3(1) show what the parties chose. Thus in the case of liquid waste the volume deposited is to be calculated from the records and accounts referred to in clause 2(6) but in the case of dry waste it is "to be calculated at the end of each year in accordance with the provisions of clause 3(2)". Thus the provisions of clause 3(2) are to be resorted to for the purpose of the calculation of volume. The first and third sentences are in their terms to be applied for the purpose of that calculation.

The two surveys directed to be made for the purpose of such calculation equate the amount of dry waste so deposited with the amount of the airspace filled at the end of the year as compared with the amount filled at the beginning of the year. This will necessarily include the material used in the engineering works as described by Hutchison LJ. In my view, other provisions of the licences indicate that this was the intention.

First, it is hard to envisage how the accounts required to be kept by clause 2(6) could assist in resolving any dispute as to the volume of dry waste deposited if, as the parties rightly accept, the relevant measure is the volume on site as compacted or degraded rather than the volume delivered. Second, it is noteworthy that in connection with the creation of additional airspace in the course of the year by the extraction of minerals the proviso to clause 2(8) provides expressly that "the soil and overburden shall not be deemed to be minerals for the purposes of this licence". Had it been intended that the airspace occupied by the materials required for the engineering works should not count for rental purposes then it is surprising that the licence contains no comparable exclusion. Third, to require the airspace occupied by such materials to be ignored would sacrifice simplicity for a fertile source of disagreement. I do not think that there is any context requiring such an intention to be ascribed to the parties. Fourth, although the second sentence of clause 3(2) envisages a difference between the parties as to the amount of the dry waste deposited so as to require the intervention of the surveyors their decision has to be based on the difference between the available airspace at the end of the year compared with that available at the beginning of the year. It is true that the decision of the surveyors is to be "based on" such difference but there is no suggestion that they would be entitled to treat airspace which was unavailable because occupied by the material required for the engineering works as being available for the occupation of dry waste in the future. In my view, the use of the words "based on" is sufficiently accounted for by the recognition that the measurements might not be exact.

I agree with Hutchison LJ that no reliance can be placed by either side on the wording of the second recital and that the order of the sentences in clause 3(2) is illogical. It is also true that the second sentence is not entirely consistent with the first in that the decision of the surveyors is to be based on the survey which they have already carried out. Nevertheless those considerations are not, in my judgment, sufficient to displace the requirement to measure the volume of dry waste deposited by reference to the airspace occupied in consequence of that deposit, the failure to include any provision similar in this respect to the proviso to clause 2(8) so as to exclude the non-waste material used in the engineering operations and the good sense in providing for the simplest possible formula.

For these reasons I would dismiss the appeal on the second point as well as the first.

Also agreeing, **Neill LJ** said: I have had the advantage of reading in draft the judgments of Hutchison and Morritt LJJ.

On the first issue I agree that for the reasons given in these judgments the deeds created licences rather than leases. On this issue there is nothing that I can usefully add.

On the second issue, however, I am unable to agree with the conclusion reached by Hutchison LJ.

By clause 2(6) of the deed dated May 15 1974 the licensee was required to keep a proper account of all waste deposited at the site. But, in contrast to the method of calculation of the rent payable for the liquid waste which was expressed to be at a specified rate per 1,000 gallons deposited to be calculated from the records and accounts referred to in clause 2(6), a quite different method was laid down for the dry waste.

Clause 3(1)(c) of the deed provided for the payment of

> A rent at the rate of 10p per cubic yard of dry waste deposited at the site during each year of the licence period the volume to be calculated at the end of each year in accordance with the provisions of clause 3(2) hereof . . .

Clause 3(2) provided:

> For the purpose of calculating the amount of dry waste so deposited a survey of the available air space shall be taken at the commencement of the licence period and at each anniversary thereof by a firm of chartered surveyors to be appointed jointly by the grantor and the licensee the licensee to be responsible for the surveyors' charges. In the event of any dispute between the parties as to the amount of dry waste deposited in any year of the licence period the decision of such chartered surveyors based on the difference between the available air space at the beginning and end of each year shall be final and binding upon the parties. In calculating the difference in such air space the surveyors shall have due regard to any additional air space created in connection with the extraction of minerals during the relevant year.

I agree with the criticisms which have been made of the arrangement of the sentences in clause 3(2). Nevertheless, it seems to me that the provision in clause 3(2) for the calculation of the difference between the available airspace at the beginning and end of each year and the requirement that account is to be taken of any additional airspace created by the extraction of minerals are clear indications that the rent for the dry waste is to be calculated by reference to the space occupied by it. The space will be reduced by compaction but will be increased by any necessary protective layers of other materials. I am unable to find in the words "based on" in clause 3(2) any sufficient support for the argument that the materials used in any engineering works or for the protective layers are to be excluded.

In my view, clause 3(2) was intended to provide a simple formula. It is to be remembered that the formula was adopted at a time when less stringent rules were in force for the protection of the environment from seepage and other contamination. Moreover, in the absence of an express provision for finality one can imagine that the measurement of the amount of airspace in an irregularly shaped quarry could give rise to disputes.

For these reasons and for the reasons more fully explained by Morritt LJ. I agree that this issue too should be resolved in favour of the respondent company.

I would therefore dismiss the appeal.

Appeal dismissed with costs.

Kened Ltd and another *v* Connie Investments Ltd

COURT OF APPEAL

May 11 1995

Nourse, Millet and Otton LJJ

Estates Gazette January 25 1997

[1997] 04 EG 141

Landlord and tenant — Assignment — Landlord and Tenant Act 1988 — Whether landlord entitled to rely on reasons not advanced in refusing consent — Whether landlord entitled to request details of assignment including premiums — Whether consent unreasonably withheld

Following the default of the previous tenant on its mortgage, the receivers appointed by the second plaintiff bank assigned an underlease dated September 27 1988 of hotel premises to the first plaintiff, a wholly owned subsidiary of the bank formed for the purpose of holding the lease. The bank was a party to the licence to assign, and provided a form of guarantee until a further assignment subject to an acceptable replacement surety of sufficient financial status being provided. The defendant acquired the reversion to the lease in September 1991. In November 1992, following an application made in August 1992 by the first plaintiff for consent to assign the lease to R, the defendant refused consent stating that they were not being offered a guarantee of a surety with sufficient assets in the country. In proceedings brought by the plaintiffs under the Landlord and Tenant Act 1988, in which they contended that the defendant was unreasonably withholding consent to the assignment to R, the judge held that the defendant had withheld consent unreasonably and that R's parent company was an adequate replacement surety.The defendant appealed.

***Held*: The appeal was dismissed. A failure to state a reason in compliance with the obligation under the Landlord and Tenant Act 1988, unless adequately explained, must at least cast doubt on its importance and may put in doubt whether it was operating on the landlord's mind at all. The judge was well aware that it was not sufficient for the plaintiffs to demonstrate that a reasonable landlord might have granted consent; what had to be shown was that no reasonable landlord would have withheld it. The judge was not wrong in concluding that the defendant was not entitled to particulars of the assignment, and of the amount of any premium; a landlord who is asked for consent to an assignment is not concerned with the terms of the assignment but with the identity and character of the assignee. In terms of the licence to assign to the first plaintiff, "an acceptable replacement surety" means a replacement surety which is objectively suitable for acceptance by a reasonable landlord.**

The following cases are referred to in this report.

British Bakeries (Midlands) Ltd v *Michael Testler & Co Ltd* [1986] 1 EGLR 64; (1986) 277 EG 1245

Tollbench Ltd v *Plymouth City Council* [1988] 1 EGLR 79; [1988] 23 EG 132, CA

Venetian Glass Gallery Ltd v *Next Properties Ltd* [1989] 2 EGLR 42; [1989] 30 EG 92

This was an appeal by the defendant, Connie Investments Ltd, from a decision of John Mowbray QC (sitting as a deputy judge of the Chancery Division) in proceedings by the plaintiffs, Kened Ltd and Den Norske Bank plc, under the Landlord and Tenant Act 1988.

Edward Denehan (instructed by Hugh Cartwright & Amin) appeared for the appellant; Christopher Pymont (instructed by Stephenson Harwood) represented the respondents.

Giving judgment at the invitation of Nourse LJ, **MILLET LJ** said: This is an appeal by the defendant, Connie Investments Ltd, from a judgment of Mr John Mowbray QC, sitting as a deputy judge of the Chancery Division on November 17 1993, when he declared that the refusal or failure of the defendant to grant a licence to the first plaintiff, Kened Ltd ("Kened"), to assign an underlease to Realco plc ("Realco") was unreasonable; that Kened was entitled, upon the performance of certain specified conditions, to assign the underlease to Realco without the defendant's consent; and that Realco's parent company, Kuwaiti Algerian Investment Company SA ("KAIC") would be an acceptable replacement surety within the meaning of that expression in a previous licence to assign granted by the defendant's predecessor in title.

The case concerns an underlease of the Kensington Edwardian Hotel in Harrington Gardens, Kensington. The underlease was dated September 27 1988 and was granted by Scan-Alfa Ltd to Turpix Ltd. It was for a term of 32 years at an initial rent of £300,000 pa rising to £375,000 for the year from June 1991 to June 1992, with upwards-only rent reviews every four years. The rent was not increased at the rent review in June 1992, and from this and other evidence the judge inferred that the current market rent of the premises was below the rent payable under the underlease.

The underlease was mortgaged to the second plaintiff, Den Norske Bank plc ("the bank"), an English subsidiary of a Norwegian bank. There were three original sureties, all of them individuals. They joined in the underlease to guarantee performance of the tenant's covenants and undertook to accept a new lease themselves in similar terms if the underlease was disclaimed by a liquidator of the tenant or the tenant was wound up or otherwise ceased to exist.

Clause 5(23) of the underlease contained a covenant by the tenant not to assign the underlease without the previous written consent of the landlord, such consent not to be unreasonably withheld. Para (c) of that clause was as follows:

> *Assignees*
>
> If so required by the Landlord on any assignment of the Demised Premises to procure that the Assignee enters into a covenant with the Landlord to pay the rents reserved by and perform and observe the covenants on the part of the Tenant contained in this Deed and further if the Landlord shall reasonably so require to obtain acceptable Guarantors for any assignee or proposed assignee to whom this Lease has been or is about to be assigned.

By way of anticipation, it should be said that it was common ground before the deputy judge and before us that it was reasonable for the defendant to require a direct covenant to be given to it by Realco and that it was also reasonable for the defendant to insist upon a guarantor joining in the licence to assign.

Turpix Ltd defaulted on its mortgage. The bank appointed receivers, and in April 1991 the receivers assigned the underlease to Kened. Kened was a wholly owned subsidiary of the bank which was formed or obtained by the bank for the purpose of acquiring the underlease. The object of the transaction was to crystallise the bank's loss and avoid the need to run the business of the hotel in receivership. A formal licence to assign was obtained from the landlord on April 2 1991. The bank joined in the licence to guarantee performance of the tenant's covenants. Clause 9 of the licence to assign, however, qualified the bank's liability as a surety. It read as follows:

> IT IS HEREBY AGREED AND DECLARED by the parties hereto that on an assignment of the Lease by the Assignee in accordance with the terms of the Lease the liabilities of the Surety herein contained shall cease and determine absolutely (but without prejudice to any liability of the Surety in respect of any antecedent breaches of the covenants and conditions on the part of the Assignee or Surety herein contained) PROVIDED NEVERTHELESS THAT prior to any such assignment the Assignee or the Surety shall if required by the Lessor provide an acceptable replacement surety who shall be of sufficient financial status and standing as shall be able to comply with the covenants on the part of the Assignee herein contained and who shall enter into direct covenants with the Lessor in the form of clause 6 hereof.

The defendant acquired the reversion on the underlease on September 13 1991. It is the present landlord.

On August 19 1992 Kened applied to the defendant for its consent to assign the underlease to Realco. Realco offered direct covenants with the defendant and the covenants of its parent, KAIC, as replacement surety in substitution for the Bank. The defendant refused its consent on November 11 1992 in a letter written by the defendant's solicitors which contained the following passage:

> . . . so far as the prospective assignee is concerned the covenant is of little or no value. What effectively is being offered therefore is the guarantee . . . [of KAIC]. . . . [KAIC] is a company incorporated in Luxembourg, apparently mainly concerned with the promotion of projects in Algeria . . . our clients are concerned that they are not being offered the guarantee for the remainder of the term of the lease of a surety with adequate assets in this country visibly being of sufficient financial status and standing able to comply with the covenants required of an assignee in accordance with the leasehold documentation governing this property. For these reasons our clients do not feel able to accept . . . the prospective [assignee].

Kened renewed its application and attempted to meet the various stated objections of the defendant. The deputy judge found that there was a continuing request by Kened for the defendant's consent to the proposed assignment and a continuing withholding of consent by the defendant down to the date on which the originating summons was issued on March 15 1993. He also found that the factors which were relied upon by the defendant on November 11 1992 were still operating on its mind on March 15 1993. There is no appeal from that part of the judge's decision.

There were two issues before the judge:

1. whether the defendant's withholding of consent to the proposed assignment of the underlease to Realco was reasonable; and

2. whether KAIC was "an acceptable replacement surety" for the bank within the meaning of clause 9 of the 1991 licence to assign.

The judge found:

1. that the defendant's withholding of its consent to the proposed assignment to Realco was unreasonable; and

2. that KAIC was "an acceptable replacement surety" for the bank within the meaning of the licence to assign.

The defendant now appeals against both decisions.

The relevant principles of law are not in dispute, with one possible exception:

1. By virtue of clause 5.23(c) of the underlease it was reasonable for the defendant to require a direct covenant from the proposed assignee and the defendant reasonably required the provision of an acceptable guarantor for Realco.

2. Under clause 9 of the 1991 licence to assign the bank's liabilities would only cease and determine if the proposed replacement surety was both:

(a) acceptable to the defendant, and

(b) of sufficient financial status and standing to be able to comply with the covenants on the part of the proposed assignee.

3. The sufficiency of the proposed surety's financial status and standing has to be considered with reference to the residue of the term granted by the underlease.

4. By virtue of section 1(3) of the Landlord and Tenant Act 1988, when withholding its consent to the proposed assignment the defendant was under a statutory duty to state its reasons for withholding its consent.

5. By virtue of section 1(6) of the 1988 Act the burden lay on the defendant to show that it was reasonable for it to withhold its consent.

6. Whether a landlord's decision to withhold consent to a proposed assignment is reasonable can only be decided by reference to the circumstances existing and known to the landlord at the time the decision to withhold the consent is made.

7. A landlord can only rely upon reasons which actually influenced him at the time that he withheld his consent to the proposed assignment: see *Tollbench Ltd* v *Plymouth City Council* [1988] 1 EGLR 79*, at p81F.

8. It is not necessary for a landlord to prove that the conclusions which led him to refuse consent were justified, if they were conclusions which might be reached by a reasonable man in the circumstances.

9. It is only if no reasonable landlord could have withheld consent for the reasons stated by the landlord that a decision to withhold consent will be held to be unreasonable; the court is not entitled to substitute its own view for that of the landlord: see *British Bakeries (Midlands) Ltd* v *Michael Testler & Co Ltd* [1986] 1 EGLR 64† and *Venetian Glass Gallery* v *Next Properties Ltd* [1989] 2 EGLR 42‡.

Before the passing of the Landlord and Tenant Act 1988, the burden lay on the tenant to show, first, what reason the landlord actually had for withholding his consent, and, second, that that reason was unreasonable: see *Tollbench Ltd* v *Plymouth City Council* (*supra*). The Landlord and Tenant Act 1988 has made two material changes to the law in this respect: (1) it has required the landlord to state his reasons in writing if he decides to withhold his consent; and (2) it has reversed the burden of proof, so that it is now for the landlord to show that his reasons for withholding consent were reasonable. The 1988 Act gives the tenant a right to damages for breach of the landlord's obligations, including his obligation to state his reasons. I desire to leave open for future consideration the question whether, since the passing of the 1988 Act, a landlord can rely upon reasons which, although operating on his mind, he failed to state in writing in breach of his statutory duty. It is sufficient, for the purposes of the present case, to say that a failure to state a reason in compliance with the obligation under the Act, unless adequately explained, must at least cast doubt on its importance, and may put in doubt whether it was operating on the landlord's mind at all.

*Editor's note: Also reported at [1988] 23 EG 132
†Editor's note: Also reported at (1986) 277 EG 1245
‡Editor's note: Also reported at [1989] 30 EG 92

Before us, the defendant analysed its reasons for withholding consent as follows:

1. Realco's covenant was of little or no value, so that in reality what was being offered was the covenant of KAIC.

2. KAIC was a company registered in Luxembourg.

3. Its activities were for the most part carried on outside the United Kingdom (although not, in fact, in Algeria, as the letter of November 11 1992 stated).

4. It did not have adequate assets in the United Kingdom and was not of sufficient financial status and standing to satisfy a reasonable landlord that it would be able to comply with the tenant's obligations in the underlease.

5. Kened had refused to provide the defendant with details of the proposed assignment to Realco, and in particular of the amount of the premium or reverse premium, if any, which was payable.

That last reason is not mentioned at all in the letter of November 11 1992, even though by that date the defendant had asked to be supplied with details of the proposed assignment to Realco and had been told that that information was confidential and would not be supplied.

The defendant criticised the judge's conclusions on three grounds:

1. There was no evidence to support the judge's conclusion that no reasonable landlord could properly have withheld his consent to the proposed assignment; he must therefore have applied the wrong test or substituted his own view for that of the defendant.

2. The judge was wrong to hold that the landlord was not entitled to be informed of the details of the proposed assignment, and in particular of the amount of any premium or reverse premium.

3. The judge misconstrued clause 9 of the 1991 licence to assign and was in error in holding that KAIC was an acceptable replacement surety for the bank within the meaning of that clause.

1. *Value of Realco's covenant*

Realco was a company incorporated in England in April 1992. It was a newly formed company with no trading history. It was formed for the purpose of investing in hotels, including the hotel with which the present case is concerned and a nearby hotel called the Albany Hotel. The two hotels together have over 300 rooms; the plan was to try to attract a major international tour operator called Contiki (Europe) Ltd to enter into a contract to use the rooms for tourists coming to London. Realco proposed to spend £100,000 on improving the hotel, mainly the bathrooms, in order to make it more attractive to tourists.

By March 15 1993 Realco had acquired the Albany Hotel for £2.3m and had not mortgaged or charged it. The inference must be that its parent company had provided it with the purchase price, either by way of an unsecured loan or in exchange for shares. If the latter, of course, this would provide Realco with substantial net assets; but the defendant was given no information whether this was the case or not. Accordingly, the fact that Realco had not mortgaged or charged the Albany Hotel was not necessarily evidence that it had anything of substance in the way of net assets. In the circumstances I consider that the contention that Realco's covenant was of little or no value was made out.

2. *KAIC's covenant*

Realco was owned 85% by KAIC and 15% by a Swiss company of which no information was provided. KAIC itself is equally owned by the Kuwaiti Investment Authority and the Algerian Treasury. It has substantial assets. The judge had before him, as did the defendant, the audited accounts of KAIC for 1990 and 1991 and the unaudited accounts for 1992. Each of them showed very substantial net assets and a substantial amount of cash at bank. In 1990, for example, the amount of

A cash at bank was £5.5m. That had increased by up to £6.5m by 1991 and to nearly £7m according to the 1992 unaudited accounts. Moreover, those accounts showed an improving position.

KAIC offered to deposit the amount of a year's rent (that is to say, £375,000) with the defendant as security for Realco's obligations under the underlease. That offer was volunteered by KAIC as a means of overcoming the defendant's objections and the judge made it a condition of his order.

3. *KAIC's country of registration*

KAIC is registered in Luxembourg. The plaintiff attempted to meet B the concerns which the defendant may have felt by reason of that fact. First, KAIC had a limited life of 30 years under the terms of its incorporation. This apparently is normal in Luxembourg. There was expert evidence that this is the way in which companies are normally incorporated in Luxembourg but that it is also normal for the life of a company to be subsequently extended. Kened volunteered KAIC's undertaking to extend its life for a period which would cover the residue of the term granted by the underlease, and the judge imposed that as a condition of the order.

Second, Kened volunteered that KAIC would provide its address for service in England, and again the judge made that a condition of C the order. The judge took it that the address for service would be provided in the licence to assign and would therefore become irrevocable.

Third, Kened put expert evidence before the court concerning the enforcement of judgments obtained in Luxembourg to show that English judgments could easily be enforced in that country. The judge found that a judgment in England could be enforced in Luxembourg with little more difficulty than it could in England. Luxembourg is of course a signatory of the Brussels Convention. This finding of the judge was strongly criticised before us. It was submitted that even if the inconvenience was only a minor one, there remained an D inconvenience; it was not as easy to enforce a judgment in Luxembourg as in England. Accordingly, it was submitted, in dismissing this as a reasonable ground of objection the judge was necessarily substituting his own view for that of the defendant. I do not accept that criticism. The test is an objective one, and the fact that the defendant himself wants to avoid a minor inconvenience is irrelevant. The question is whether a reasonable landlord would regard this minor inconvenience as sufficient to justify withholding consent. The judge found that he would not, and I can see no error in his conclusion.

4. *KAIC's main activities*

KAIC's main activities were conducted abroad. That is another E aspect of enforcement and of no separate significance in itself.

5. *KAIC's assets in England*

The defendant objected that KAIC did not have adequate assets in this country and was not visibly of sufficient financial status and standing to be able to comply with covenants required of the assignee. I have already described the circumstances and the evidence of the financial status of KAIC. The judge's conclusion was that while KAIC was not a bank or a subsidiary of a bank, nevertheless it was a substantial company owned equally by the Kuwaiti and Algerian Governments, both heavily engaged in international trade and unlikely F to wish to allow their subsidiaries to default on their commercial undertakings. In my judgment, that was a conclusion which the judge was entitled to reach, and it cannot be criticised.

6. *Particulars of the assignment*

I will deal with this later as a separate matter.

In the course of his judgment the judge dealt with what might have been a separate point of objection altogether, and one which was not directly concerned with the financial standing of either Realco or KAIC. He pointed out that Realco was not an ideal tenant of an hotel. It was a newly formed company with no trading history; its directors had no experience in running hotels. The judge found that it was not proposed that Realco or KAIC should engage an hotel management G company to manage the hotel on Realco's behalf. The judge pointed out, correctly, that it would be a concern of the landlord of an hotel to ensure that the hotel was well run, as the success of the hotel would affect goodwill and would reflect on the value of the reversion. But he was wrong to make that finding. There was no evidence one way or the other whether Realco intended to engage a management company.

The judge said that there was evidence that Contiki, a company partly-owned by the owner of a substantial American hotel chain and a large tour company, was prepared to enter into a contract to take rooms in the two hotels for tourists producing a profit of approximately £800,000 for the two hotels per year. The judge discounted this H evidence because it was not clear whether the proposed contract would require Contiki to take the rooms or merely give Contiki an option to do so. The evidence as to the prospects of occupancy levels at either hotel was unclear; nothing like a business plan had been produced, and the landlord was not given any information as to how the profit figure of £800,000 was calculated. In my view, the judge was entitled to discount the evidence about Contiki. Perhaps more to the point, the evidence did not begin to meet the possible criticism that, even if KAIC and Realco were of sufficient financial standing to meet their obligations under the underlease, there was no evidence that the hotel would be properly managed. Contiki was not proposing to manage the J hotel; it would be a wholesale customer.

But this ground of objection was not mentioned in the letter of November 11 1992, and there was no evidence before the judge that it had operated on the defendant's mind at all. The whole thrust of the defendant's case throughout was that Realco and KAIC were not of sufficient financial standing to be able to comply with the obligations under the underlease. Even before us counsel for the defendant, in dealing with the evidence about Contiki, prayed it in aid as throwing doubt upon KAIC's ability to make a success of the hotel and therefore to be in a position to comply with the covenants under the underlease, rather than as evidence that Realco was insufficiently experienced to K preserve the goodwill of the hotel.

Accordingly, the point was not one on which the judge could properly have found in favour of the defendant. Had the matter been properly raised the defendant might well have been able to deal with it. They could have given evidence, if such evidence was available, as to their plans for running the hotel and whether or not they did intend to employ a management company, or how otherwise they could overcome the difficulty that their own directors were not experienced in the hotel trade. There was evidence before the judge from which he could infer that KAIC, which was an international investment vehicle for two overseas governments, would act sensibly when venturing into L the hotel business. Realco had already embarked on the hotel business by the acquisition of the Albany Hotel; the two hotels evidently had sufficient potential at least to be offered to a major tour operator; and Realco had been put in funds to modernise and improve the present hotel.

I turn to the three grounds upon which the defendant criticise the judge's conclusions. First, they complain that the judge misdirected himself and applied the wrong test or, alternatively, that he substituted his own view for that of the defendant. He stated the test correctly in his judgment. The defendant submits that, although he stated the test correctly, in fact he applied a slightly different test. The defendant M relies upon a passage in the judgment in which the judge is dealing with an argument by counsel for Kened that the comparison should be made between what was proposed and the original position under the lease when originally granted, when there was a £100 company as tenant which had only £2 paid up share capital and three individuals as guarantors all of whom proved worthless. The judge dealt with that argument as follows:

> I do not consider that to be correct, though it shows what a previous landlord was prepared to accept. I consider that the correct comparison is between the landlord's present position and the position that it will be in if the assignment goes through.

Counsel for the defendant correctly pointed out that this also is not the correct test. The proposal has to be considered on its own merits or demerits; whether it leaves a landlord better off or worse off than immediately before the assignment is neither here nor there. But, in my judgment, the judge was not applying a different test. He was dealing with a particular argument that he should compare the proposals with the original position, and he rightly rejected it. In my view, he was doing no more than saying that if a comparison had to be made, the correct comparison was between the present position and the proposal; but he dealt with the proposals on their own merits.

It is always an easy criticism to make of a judge that he has substituted his own view for that of the proper party, but, reading the judgment as a whole, I am unable to see any basis upon which that criticism can properly be levelled in the present case. The judge was well aware that the test was an objective one. The question was what a reasonable landlord would do. The judge was well aware, because he expressly stated, that it was not sufficient for Kened to demonstrate that a reasonable landlord might have granted consent; what had to be shown was that no reasonable landlord would have withheld it.

I can see no basis for saying that the judge was wrong in concluding that the defendant was not entitled to particulars of the assignment and of the amount of any premium or reverse premium paid for the assignment. Of course the landlord is entitled to be informed of anything which bears upon the giving of his consent. If there was anything in the assignment which would affect him, then the tenant was obliged to provide that information to him. But I can see no reason why the landlord would be interested in the details of the assignment. There have been cases in which the court has held that a landlord who is asked for consent to a subletting is entitled to know the terms of a proposed underlease. That is obviously correct. A landlord must be entitled to know what it is to which his consent is being sought. A landlord who is asked to consent to the grant of an underlease is concerned with the terms of the underlease. But a landlord who is asked for consent to an assignment is not concerned with the terms of the assignment but with the identity and character of the assignee. The relationship between the landlord and the assignee will be governed by the terms of the underlease and not by the terms of the assignment. In an ordinary case where the assignment is made at a premium the landlord is not normally told, and does not normally inquire, as to the amount of the premium, even though the payment of the premium by the assignee will to that extent diminish the assignee's financial resources and hence his ability to comply with the covenants in the lease.

In the present case the judge inferred that there was a reverse premium, and the case has been argued before us on the basis that there was a reverse premium, the existence and amount of which were concealed from the landlord. Counsel for the defendant submitted that if there was a reverse premium then that would assist the assignee to comply with the obligations of the underlease, and that the landlord was therefore interested and concerned to know the amount of that reverse premium so that he could calculate how long it would take before it was exhausted in complying with the covenants in the underlease. But since Kened did not inform the landlord of the existence or amount of any reverse premium, Kened cannot rely upon the existence of its receipt by Realco to satisfy the defendant that it has the means to comply with its covenants in the underlease. Having failed to provide the landlord with that information, Kened must seek to establish the ability of Realco and KAIC to comply with their covenants without reliance on any reverse premium. It has done so; and that is an end of the point.

The third criticism made of the judge is that he misconstrued clause 9 of the licence to assign and wrongly found that KAIC was an acceptable replacement surety for the bank. Counsel submitted that the word "acceptable" was subjective and meant "acceptable to the landlord". The judge held that it was objective and meant "reasonably suitable for acceptance by a landlord". I agree with the judge. It appears to me plain beyond argument that in the context in which the expression is found "an acceptable replacement surety" means a replacement surety which is objectively suitable for acceptance by a reasonable landlord. The landlord was not seeking the right to refuse an offer of a replacement surety arbitrarily. It must be borne in mind that the licence to assign was entered into between the landlord and the assignee and the bank. It was an agreement for novation in certain circumstances. If the landlord was intending to reserve the right arbitrarily to refuse an offer of a substitute surety, then there was no point in including the clause; it would be otiose. A landlord is always entitled, even in the absence of such a clause, to agree to the substitution of a new surety. The whole purpose of clause 9 of the licence to assign was to give the bank a right to offer a replacement surety and to commit the landlord to accept a substitute surety provided that certain objective criteria were met. The objective criteria were that the replacement surety should be suitable for acceptance by a reasonable landlord.

So far as the word "replacement" is concerned, there is nothing in the submission that this means that the substitute surety should be in every respect as good as, or better than, the bank. The word merely means that it should be reasonably acceptable as a substitute for the bank. If the landlord had intended to insist that the substitute surety should be in every respect as good as, or better than, the bank, there would have been nothing easier than to include such a requirement in the licence to assign.

Despite the careful arguments put before us by the defendant, I remain entirely unpersuaded that the deputy judge made any discernible error in the course of his judgment. He came to a decision on the facts. It was a decision to which he was entitled to come. I can find no error of law in reaching his decision. He took into account nothing which he ought not to have taken into account, and he omitted nothing which he ought to have taken into account. For my part, I would dismiss the appeal.

Nourse and **Otton LJJ** agreed and did not add anything.

Appeal dismissed with costs.

Boyle *v* Verrall

COURT OF APPEAL

July 26 1996

AULD and THORPE LJJ

Estates Gazette January 15 1997

[1997] 04 EG 145

Landlord and tenant — Assured tenancies — Failure to serve section 20 notice — Whether "just and equitable" to dispense with written notice in claim for possession under ground 1 of Schedule 2 to the Housing Act 1988

The appellant, who with her husband rented a country cottage, acquired the subject flat in 1987 intending to keep it for use as a London home for themselves. Prior to the grant of an assured tenancy of the flat in April 1993 to the respondent, there was some conversation between the parties about the appellant's intentions for occupation of the flat. The respondent knew that the appellant intended to serve a notice under section 20, to ensure the grant of an assured shorthold tenancy, but had failed to do so. The appellant appealed the decision of the judge who, in refusing her claim for possession under ground 1 of Schedule 2 to the Housing Act 1988, in that she required the flat for her husband who had found employment in London, decided that she had not given even oral notice prior to the grant of the tenancy as required by this ground.

A *Held*: **The appeal was allowed. It was "just and equitable" to dispense with the requirement of a written notice under ground 1 of Schedule 2 to the 1988 Act. If oral notice had been given prior to the grant of a tenancy, that may be an important factor in favouring dispensation of a written notice. But an oral notice is not a prerequisite of such dispensation. A proper comparison of the circumstances of the parties was such that the respondent's position was not such as to outweigh the error of the appellant in failing to serve a section 20 notice of which the respondent had taken advantage. The respondent's persistent late payment of rent was also a relevant circumstance to be taken into account.**

B
The following cases are referred to in this report.

Bradshaw v *Baldwin-Wiseman* (1985) 49 P&CR 382; [1985] 1 EGLR 123; 274 EG 285; 17 HLR 260, CA

Fernandes v *Parvardin* [1982] 2 EGLR 104; (1982) 264 EG 49; 5 HLR 33

Kennealy v *Dunne* [1977] QB 837; [1977] 2 WLR 421; [1977] 2 All ER 16; (1976) 34 P&CR 316; [1977] 1 EGLR 50; [1977] EGD 155; 242 EG 623, CA

This was an appeal by Rita Madeleine Boyle from a decision of Judge Maher in Wandsworth County Court in proceedings for possession against the defendant, Richard Hugh Verrall, under the
C Housing Act 1988.

The parties appeared in person.

Giving judgment, **Auld LJ** said: This is an appeal from an order of Judge Maher in Wandsworth County Court on August 3 1995 dismissing the plaintiff's, Mrs Rita Boyle's, claim for possession of a flat at 207b Castelnau, Barnes, London SW13 9EA.

Mrs Boyle let the flat to Mr Verrall on March 30 1993 at a rent of £650 per month, later increased to £700 per month.

There were two main alternative issues before the judge.

The first was whether the tenancy was an assured shorthold tenancy under section 20 of the Housing Act 1988, terminable by Mrs Boyle on
D notice at or after the end of the agreed period, or an assured tenancy under section 1 of the Act, terminable by Mrs Boyle only on one of a number of grounds specified in Schedule 2 to the Act. The judge held that it was an assured tenancy, and there is no appeal against that ruling.

The second issue was whether Mrs Boyle was entitled to terminate the tenancy on ground 1 in Schedule 2, namely that she required the property as a principal home for her husband. A precondition of such entitlement under that ground was that she had, not later than the beginning of the tenancy, notified Mr Verrall in writing that she might require possession for that reason, or, failing such notice, that she
E could satisfy the judge that it was just and equitable to dispense with it.

The circumstances in a little more detail are as follows. In 1987 Mrs Boyle bought the freehold of the flat. She and her husband, both of whom had worked as television or radio journalists in London, were unemployed. They had a tenancy of a small country cottage and intended to keep the flat for use as their London home if and when one or both of them secured work again in the capital. Mrs Boyle decided to let it furnished until they needed to use it again. In early 1993 she advertised it as "a long let". Mrs Verrall responded to the advertisement and, on March 29 1993, the two ladies met at the flat. Mrs Verrall liked it and Mrs Boyle handed her a signed and witnessed
F form of agreement for an assured shorthold tenancy and a blank notice under section 20 of the 1988 Act stating that it was to be such a tenancy. The form of agreement was the Oyez standard form for letting a furnished dwelling-house on an assured shorthold tenancy and was completed then or later so as to run for a term certain of one year from April 4 1993 to April 4 1994 at a rent of £650 per month. Clause 2 provided:

> This Agreement is intended to create an assured shorthold tenancy as defined in section 20 of the Housing Act 1988 and the provisions for the recovery of possession by the Landlord in section 21 thereof apply accordingly.

Mrs Verrall took away both documents, and two or three days later Mr Verrall signed the agreement and returned it to Mrs Boyle. On about April 6 Mrs Boyle and he met. They agreed a deposit of £300 G
which he paid, and Mrs Boyle gave him the keys. According to Mrs Boyle, she served on him on that occasion a completed section 20 notice. Mr Verrall has denied that, and the judge, while accepting the genuineness of Mrs Boyle's evidence on the point, was "not convinced" that she had done so. However, it is plain from Mrs Boyle's evidence and the tenancy form and blank notice used for the transaction that she intended to create an assured shorthold tenancy; and it is plain from Mr Verrall's evidence that he knew that that was her intention and that she had made a mistake in not completing the notice. He said that he knew if he signed the tenancy agreement before receiving a completed notice he could claim an assured tenancy. When H
asked why he did not draw Mrs Boyle's attention to her mistake, he said: "Why should I? You have to look after yourself in this world".

According to Mrs Boyle, she made plain that she and her husband were expecting to need the flat at some time in the future and Mr Verrall indicated that he and his wife would be using it as a base to look for a property in the area to buy. The judge dealt with this matter in the following way:

> I am unable to find that the plaintiff effectively brought to the attention of Mr and Mrs Verrall orally that the flat would be required for herself and her husband. Deciding what was said was very difficult. Different people J
> remember in different ways . . .
>
> When cross-examined as to why she did not give written notice Mrs Boyle said: "I thought I had set up an Assured Shorthold Tenancy. It was not intended to be a Ground 1 lease" . . .
>
> She left me with the impression that what was said was by way of conversation only. If I stand back and ask the question, it falls short of giving oral notice. One must be careful before finding that oral remarks would have brought home to a tenant that he was embarking on an exercise where he would get limited security of tenure. In general terms I see no reason not to accept the impression given by Mr Verrall and fortified by the evidence of Mrs Boyle.

K

Mr Verrall had been employed as a housing officer for a local authority and was familiar with the different security of tenure given by assured shorthold and assured tenancies. His wife was a solicitor. However, at the time of entering into the agreement he was unemployed, and he and his wife were apparently dependent on housing benefit for payment of the rent.

There were disputes throughout the first 12 months of the tenancy about Mr Verrall's late payments of, and deductions without Mrs Boyle's agreement from, the agreed rent. On March 4 1994, Mrs Boyle, believing that the agreement between them was for an assured shorthold tenancy, served notice purporting to terminate it and offering L
a new one at a rent of £700 per month. At about the same time Mr Verrall deducted £250 from the rent for the purchase of a new bed, and wrote to her pointing out for the first time that he had an assured tenancy because she had not completed the form of section 20 notice. They exchanged much correspondence on that issue, but in the meantime Mr Verrall agreed to pay an extra £50 per month. Mrs Boyle, while clearly unhappy about the legal relationship between them, took no steps to enforce her notice of termination.

Mr Verrall continued to pay the rent late, and Mrs Boyle wrote many letters of complaint to him about it. One such letter was on M
September 26 1994. In it she made plain that she and her husband would soon need the flat back for their own use as her husband's work was about to change. This is how she put it:

> Providing you pay your rent in future in full, and on time, I am still prepared to let you stay in the flat until the end of your second year of occupancy — April 1995. However, as I explained when you moved in, we will need the flat thereafter for our own use because my husband has now completed negotiations to join the staff of Reuters radio service as their political correspondent and will in future be based at Reuters' premises in Hammersmith. We will therefore not be making available to you a third year's Lease. I am giving you lengthy notice of this fact in order that you and your wife will have plenty of time to find somewhere else to live.

By letter of October 3 1994 Mr Verrall replied saying that in his present circumstances, namely of unemployment and dependence on housing benefit to pay the rent:

> . . . I cannot say whether or not I will wish to remain after April 1995

Mr Verrall continued to pay his rent late, and Mrs Boyle, as before, wrote to him complaining about it. On March 20 1995 she wrote to him asserting that his lease would end on April 4 1995 and asking him if he intended to leave the flat voluntarily. On March 27 1995 Mr Verrall replied by letter stating that he could not say when he and his wife would move and that her husband could satisfy his need for accommodation in London by renting a *pied-à-terre* or staying somewhere overnight as necessary.

Mrs Boyle then instituted these proceedings for possession. At the hearing, the judge, having found that there was no assured shorthold tenancy because of the lack of a section 20 notice and that, in consequence, there was an assured tenancy, went on to consider Mrs Boyle's entitlement to recover possession under ground 1(b) in Part 1 of Schedule 2 to the 1988 Act. As I have indicated, that provision requires the court to grant possession if the landlord "requires the dwelling house as [her] or [her] spouse's . . . principal home", but only if not later than the beginning of the tenancy the landlord has notified the tenant in writing

> that possession might be recovered on this ground or the court is of the opinion that it is just and equitable to dispense with the requirement of notice . . .

Both Mr and Mrs Boyle gave evidence of Mr Boyle's urgent need for a London home to enable him effectively to work the long and unsocial hours of a political editor for Reuters Radio. Mr and Mrs Verrall gave evidence about their need to remain in the property. Mr Verrall acknowledged that he did not have to live in that area and that he could probably rent property elsewhere, but said that he and his wife had to live on only £48 per week in the form of income support, and that he could not afford the expense of moving home. Mrs Verrall, however, said that she earned some money as a translator and as a solicitor working part-time. The judge accepted that Mr Boyle's new job requiring him to live in London in the week was a genuine reason for seeking possession and that in that respect circumstances had changed since the grant of the tenancy. He said:

> I acquit Mrs Boyle of any bogus desire to recover possession. I accept that Mrs Boyle's husband has undertaken a responsible job which requires him to work long hours in London. I need not rehearse this contention. He is clearly in a very unhappy position. He leaves early, gets back late and his social and domestic life is dealt a severe blow. If he were able to live at the flat this would be improved. I accept this.
>
> The conclusion I have come to is that the wish to assert the need for possession has arisen because of a new job that Mr Boyle has undertaken. This could be described as a supervening factor not present when the tenancy was entered into.

The judge went on to hold that it would have to be an "exceptional case" to justify dispensing with the written notice when, as he found, no oral notice had been given. He expressly contrasted the positions of the Boyles and the Verralls. As to the Boyles, he observed that they had another home, the country cottage which, by then, they had purchased and did not intend to sell. As to the Verralls, he noted that the flat was their only home, that they relied on income support and that they had no savings. He said:

> It would be wholly exceptional to grant possession in these circumstances

Before determining whether this was an exceptional case justifying dispensing with notice, the judge considered and rejected two other, discretionary, grounds for possession under Part II of Schedule 2 to the 1988 Act, namely unpaid rent by reason of Mr Verrall's deduction of £250 for the new bed and his persistent delay in paying rent. As to the former, he found that the deduction was properly made. As to the latter, he found that Mr Verrall had persistently and unjustifiably paid the rent late, but never by more than a month, and that his conduct was not such as to justify an order for possession.

The judge then returned briefly to record his conclusion on the issue of dispensing with notice under ground 1. This is how he put it:

> I at one stage thought carefully about dispensing with service of notice on the basis that it would prevent the tenant taking advantage of an error made by the landlord. But having given it further thought I decided it would be wrong to visit the consequences of homelessness . . . I have considered all the circumstances. It would be an improper exercise of discretion.

Mrs Boyle, who argued her own appeal, has submitted that the judge failed, when considering the exercise of his discretion, to consider or consider sufficiently all the material circumstances of the case. She referred to the following matters in support of her contention that he should have exercised his discretion to dispense with the requirement of service of a written notice:

1. the fact that she had intended to create an assured shorthold tenancy, and that Mr Verrall had known that when entering into the tenancy;
2. that that intention was her reason for not having treated it as an assured tenancy and for not having served a notice under ground 1;
3. that Mr Verrall had realised her mistake at the time and had decided to take advantage of it;
4. that Mr Boyle, as the judge found, had a genuine need to occupy the flat;
5. that, as Mr Verrall acknowledged in evidence, he and his wife were not tied by work commitments or otherwise to the Castelnau area of London and that he could rent property elsewhere, and that his protestations about being too poor to move were contradicted by his wife's evidence that she was practising part-time as a solicitor and earned money as a translator;
6. that Mr Verrall had behaved badly as a tenant in making unauthorised deductions from the rent and in persistently paying it late; and
7. that she had given the Verralls early notice of her husband's need for the flat.

Mr Verrall, who also argued his own case, submitted in reply that the judge did consider all the circumstances and that this court should not interfere with his exercise of discretion. He pointed out that the judge in his judgment expressly referred to the question whether he should be allowed to take advantage of Mrs Boyle's mistake about the section 20 notice, but did not describe his conduct as sharp practice. He said that the judge weighed the evidence on both sides on the question of need and found in his favour. As to his persistent late payment of rent, he submitted that it was irrelevant to the question whether the requirement of notice should be dispensed with. He submitted that where a landlord has given no oral notice when granting a tenancy of a possible future requirement of the property a judge is entitled to require it to be an exceptional case before dispensing with notice. He maintained that the judge was entitled to find in the circumstances of this case that it was not exceptional and, therefore, not deserving of such dispensation.

Ground 1 of Schedule 2 to the 1988 Act, if established by a landlord, entitles him to possession. The judge has no discretion. The landlord does not even have to show that his requirement of the property for use as his or his spouse's principal home is reasonable. All that he has to establish is that he bona fide wants, and genuinely has an immediate intention, of using the property for that purpose. *Kennealy* v *Dunne* [1977] QB 837*, CA. Discretion only comes into play where a landlord did not serve the requisite written notice at the time of entering into the tenancy and the court has to consider whether it is "just and equitable" to dispense with the requirement.

Bradshaw v *Baldwin-Wiseman* 49 (1985) P&CR 382†, CA is authority for the proposition that in determining what is just and equitable for this purpose the court should look at all the circumstances of the case.

---

*Editor's note: Also reported at [1977] 1 EGLR 50
†Editor's note: Also reported at [1985] 1 EGLR 123

A Griffiths LJ, with whom Browne-Wilkinson and Walker LJJ agreed, said, at p388:

> . . . The words "just and equitable" are of very wide import, and I can see nothing in the context that justifies giving them any restrictive meaning.
>
> I would regard the use of those words as directing the court to look at all the circumstances of the case. Those would embrace the circumstances affecting the landlord, or his successors in title, the circumstances of the tenant and, of course, the circumstances in which the failure to give written notice arose.

Clearly, if oral notice was given when a tenancy was granted, it B may, with or without other circumstances, be an important factor favouring dispensation: see, for example, *Fernandes* v *Parvardin* (1982) 5 HLR 33*, CA. However, it does not follow that oral notice is a prerequisite of such a decision. Nor, in my view, is the absence of oral notice a reason for restricting dispensation to circumstances where an "exceptional case" for it can be shown — the judge's stated approach here.

I acknowledge that this court should only interfere with a judge's decision on such an issue where he has failed to consider all the relevant circumstances or, if he has considered them, where his decision on those circumstances is plainly wrong. In my view, the C judge has had regard to most, but not all, the relevant circumstances; has plainly given the wrong weight to some of them and has, in any event, wrongly applied a test of exceptionality to them.

This dispute has arisen only because Mrs Boyle did not serve a ground 1 written notice because she mistakenly thought she was granting an assured shorthold tenancy, and because Mr Verrall knew at the time of her mistake and deliberately took advantage of it. While the judge found that she did not give any oral notice either, he clearly accepted, in the passage from his judgment that I have set out, that there was some conversation about her and her husband's intentions for occupation of the flat, albeit falling short of formal notice. There was also the early written notice that Mrs Boyle gave to Mr Verrall in D the September 26 1994 letter, and in other letters at about that time, that her husband would need the flat by April 1995.

As to the respective needs of the parties, the judge appears to have regarded the Verralls as in greater need and to have treated that factor as decisive when weighed against Mr Verrall's conduct in taking advantage of Mrs Boyle's error. However, it does not seem to me that he correctly analysed the position by comparing the availability to the Boyles of another home with the proposition that the Verralls were without money and that the flat was their only home. The Boyles' country cottage did not meet Mr Boyle's genuine — as the judge found — requirement for a London home in the week for his work. E The Verralls were not as poor as Mr Verrall had maintained, as the evidence of his solicitor wife indicated. They did not need to live in that particular area of London, and the housing benefit from which they were paying their rent would no doubt have been available to them at other rented accommodation. In short, they would not be homeless, as the judge suggested, if Mrs Boyle were granted possession of her property. In my view, on a proper comparison of the circumstances of the two parties, the Verralls' position was not such as to outweigh the error of Mrs Boyle of which Mr Verrall had taken advantage.

Finally, as to Mr Verrall's persistent late payment of rent, I do not F see why, in the light of Griffiths LJ's words in *Bradshaw* v *Baldwin-Wiseman*, it should not be a relevant circumstance on this issue. Having read the voluminous correspondence on this and other difficulties between the parties, all of which was before the judge, it is plain that Mr Verrall behaved in an unco-operative and obstructive manner throughout. The judge clearly did not fully accept his attempted explanation for the persistent late payments. He described his behaviour as selfish and as "peevish resistance" to paying promptly.

If the judge had applied the correct test, unglossed by any G consideration of exceptionality because of the lack of formal oral notice, I do not see how he could possibly have come to any other conclusion than that it was just and equitable to dispense with the requirement of written notice. Even applying his test of exceptionality, I would agree with Mrs Boyle that this case, in the circumstances in which the assured tenancy came into being, would plainly satisfy it.

I would, therefore, allow the appeal, and substitute for the order made on the claim an order for possession.

Thorpe LJ agreed and did not add anything.

Appeal allowed. H

Blockbuster Entertainment Ltd *v* Leakcliff Properties Ltd

CHANCERY DIVISION J

October 7 1996

Neuberger J

Estates Gazette February 22 1997

[1997] 08 EG 139

Landlord and tenant — Covenant against subletting without landlord's consent — Whether landlord entitled to withhold consent where reasonable belief held that subletting at rent less than open market value and subletting would detrimentally harm reversionary interest K

By an underlease dated September 16 1994 the plaintiff tenant holds a 15-year term of commercial premises at a rent of £115,000 pa from the defendant landlord, subject to covenants against alienation. These provide that the tenant may not underlet without the landlord's consent "such consent not to be unreasonably withheld or delayed", and that the rent of every underlease shall not be less than the open market rental value to be approved by the landlord. On the tenant's application to underlet at a rent of £75,000 pa, the landlord refused consent on the grounds that: (1) the proposed underletting would detrimentally affect the reversionary value; and (2) the rent of £75,000 L is, or is reasonably believed by the landlord to be, less than the open market rental value. The tenant applied for a declaration that the landlord had unreasonably withheld consent.

***Held*: The tenant was entitled to a declaration that the landlord had unreasonably withheld consent. (1) The proposed subletting would not detrimentally affect the reversionary interest of the landlord. It was not reasonable for the landlord to require the tenant to postpone subletting until rents improved so as to increase the value of the landlord's reversion. (2) On the evidence, the rent of £75,000 was not below the open market rental value. But on the proper construction of the alienation covenants, the landlord was M entitled to withhold consent if it believed at the time it withheld consent that the rent was less than the open market rental value; its "approval" of the rent was required. However the landlord could not rely on this aspect of the second ground because it was not a ground in its mind at the time of its refusal, it had not stated the ground in its letter of refusal, and, in any event, if the landlord had had such view in mind at the time, it was not a reasonable view.**

The following cases are referred to in this report.

Air India v *Balabel* [1993] 2 EGLR 66; [1993] 30 EG 90, CA

*Editor's note: Also reported at [1982] 2 EGLR 104

A *Bromley Park Garden Estates Ltd* v *Moss* [1982] 1 WLR 1019; [1982] 2 All ER 890; (1982) 44 P&CR 266; [1983] 1 EGLR 65; [1983] EGD 492; 266 EG 1189, CA

International Drilling Fluids Ltd v *Louisville Investments (Uxbridge) Ltd* [1986] Ch 513; [1986] 2 WLR 581; [1986] 1 All ER 321; (1985) 51 P&CR 187; [1986] 1 EGLR 39; 277 EG 62, CA

Ponderosa International Development Inc v *Pengap Securities (Bristol) Ltd* [1986] 1 EGLR 66; (1985) 277 EG 1252

This was an application by the tenant, Blockbuster Entertainment Ltd, for declaratory relief and damages in proceedings against the landlord, Leakcliff Properties Ltd.

B Alexander Hill-Smith (instructed by Brookstreet des Roches, of Witney) appeared for the tenant; Michael Driscoll QC (instructed by Pulver & Co, of Watford) represented the landlord.

Giving judgment, **Neuberger J** said: The issue in this case is whether the defendant, Leakcliff Properties Ltd ("the landlord") is entitled to withhold its consent to the proposed subletting of premises known as The Pyramid, High Street, Watford ("the premises"), by the plaintiff, Blockbuster Entertainment Ltd ("the tenant"), to Byte Computer Stores Ltd ("Byte").

The landlord, who holds the premises under a headlease, granted an underlease ("the lease") of the premises to the tenant on September 16
C 1994 for a term of 15 years at a rent of £115,000 pa subject to upward-only reviews every five years, with a one month rent-free period. Clause 5.8 of the lease is a covenant by the tenant regulating alienation and two of its subclauses are directly relevant to the point at issue. Clause 5.8.3. is a covenant by the tenant

> not to assign, underlet or charge the whole of the premises without the prior written consent of the landlord, such consent not to be unreasonably withheld or delayed.

Clause 5.8.6 is a covenant by the tenant:

> that each and every permitted underlease should be granted without any fine or
> D premium at a rent not less than the then open market rental value of the premises to be approved by the landlord prior to any such underlease, such approval not to be unreasonably withheld or delayed (save that in the case of the renewal of any permitted underlease under the provisions of the 1954 Act . . . the prior approval of the landlord to the rent payable under the renewed underlease shall not be required where the rent is determined by the court . . .) . . .

On July 1 1996 the tenant agreed to grant an underlease of the premises to Byte, subject to obtaining landlord's consent within a specified period, at an initial rent of £75,000 pa, but otherwise on terms which effectively reflect the terms of the lease.

E The landlord has refused consent to this proposed underletting and puts its case effectively on two grounds. The first ground is that the proposed underletting will, or the landlord reasonably believes that it will, detrimentally affect the value of the landlord's reversionary interest in the premises. The second ground is that the rent of £75,000 pa to be reserved under the proposed underlease is, or is reasonably believed by the landlord not to be, less than the "open market rental value of the premises".

It is common ground between the parties that the proposed underletting would be in breach of covenant if it is at a rent lower than the open market rental value of the premises. That appears to me to be
F right in light of the words "at a rent not less than the open market rental value of the premises" in clause 5.8.6.

It is convenient, in my judgment, to consider initially the second ground advanced by the landlord for two reasons. First, the second ground involves issues of fact and expert evidence, but does not raise any question of construction. Second, if I am in the landlord's favour on this second ground, it is really an end of the matter; conversely, if I am in favour of the tenant on this second ground, my conclusion will, I think, have some part to play when I turn to consider the first ground.

Open market rental value of the premises

The case for the tenant is in summary form that the premises were extensively marketed by the tenant and by experienced and competent surveyors, Johnson Fellows & Co, in early April 1995 and that this G
marketing campaign resulted in a number of bids. The highest of these bids was the £75,000 pa received from Byte at the beginning of June 1996. (It is right to say that there was another rental bid at that level, but it was for a use not permitted under the lease). The tenant also relies on expert valuation evidence of the surveyor, namely Mr Timothy Frost [FRICS] of Brasier Harris, who valued the premises at £60,000 pa.

The landlord's contention that the open market rental value of the premises is significantly more than £75,000 pa rests on the following factors. First, the rent of £115,000 pa payable under the lease was negotiated in the open market and is therefore good evidence of the H
rental value of the premises when the parties bindingly agreed to enter into the lease, which was in December 1993. Basing its argument on rental evidence in central Watford, the landlord contends that it is inconceivable that the rental value of the premises can have fallen by one-third from December 1993 to 1995 or 1996.

Second, the landlord, partly through cross-examination of the tenant's witnesses and partly through the expert evidence of Mr Colin Buckle [FRICS FSVA ACIArb] of Colin Buckle & Co, contends that the marketing campaign conducted by and on behalf of the tenant was effected in a manner which was not appropriate for the particular premises, and which cannot, at its lowest, be relied on to have produced the best rent J
which could have been obtained from the market. The landlord's case is that the marketing campaign has produced rental bids, all of which, including Byte's offer, are well below the best rent which could have been obtained for the premises.

Third, the landlord relies on the expert opinion of Mr Buckle, which is to the effect that the open market rental value of the premises is £95,000 pa.

I have reached the clear conclusion that the tenant's best evidence of what a particular property is worth on the open market is the rent payable under a binding agreement for the letting of the property after it has been properly exposed to the market. Of course, if it can be shown that the K
property has been marketed incompetently or that, for some special reason of the intending lessor, the property has not been properly exposed to the market, one may well conclude that the highest firm bid received as a result of the marketing exercise does not in fact represent the market value of the property concerned.

In this case I am not impressed by the criticisms which have been levelled against the marketing exercise conducted by and on behalf of the tenant in 1995 and 1996. The way in which a property is marketed is very much a matter of judgment and, unless the result of a particular marketing exercise is unusually good, it is no doubt always possible, with the wisdom of hindsight, to criticise aspects of the exercise. In the present case that L
observation is, I believe, reinforced when one takes into account the fact that, as all the surveyors who gave evidence before me agreed, the premises are very exceptional. This is partly due to their location, an island site at the southern end of the High Street and abutting the ring road, and partly due to the nature of the premises, a striking steel-framed glass pyramid on a brick base with a ground floor of just over 5,000 sq ft and a mezzanine floor of just over 1,350 sq ft. The premises would not appeal to most normal retailers and, particularly in light of the fact that the premises only include nine car parking spaces, they are not well suited for retail warehouse use.

It is said that it was inappropriate for the tenant to have employed M
Johnson Fellows & Co, in part because they are based in Birmingham and in part because they are not a firm which specialise in marketing first class or flagship properties such as the premises. I do not accept that. I heard evidence from Mr Andrew Heaney [ARICS], the partner at Johnson Fellows & Co who handled the marketing. He appeared to me to be competent and to have given the marketing of these premises a high priority, not least because he valued the tenant as an important client. His marketing campaign appeared to me to have method and it produced a number of rental bids. He contacted a very large number of retailers as well as agents and restaurant and brewery business operators. It is also said that the tenant should not have marketed the premises on the basis of a use which required planning permission and licensed change of the use under the

lease, namely the restaurant or brewery use. It seemed to be common ground that, if a restaurant or brewery had been prepared to bid for the premises, the chances are that it would have bid more than a retail user. I cannot therefore see why it was unwise to proceed in that way. In any event, I am not persuaded that the initial marketing of the premises for a restaurant or brewery purposes resulted in retail users not being prepared to bid for the premises thereafter, or lowering their bid for the premises thereafter.

The tenant was also criticised for advertising the premises from January 1996 at too low an asking rent, namely £75,000 pa. I am not impressed with that point either. The figure of £75,000 pa was arrived at by Mr Heaney after it was clear to him from various bids which have been put forward that the market was thinking in terms of around £50,000 pa as the appropriate rent for the premises. Further, as is clear from Mr Buckle's own evidence relating to another property in Watford (153 Harlequin Centre) if there is a demand for a property, it will often be let at a rent substantially higher than the asking rent. The evidence on that transaction involved two interested parties effectively bidding the property up against each other. It is interesting to note that it was on a somewhat similar basis that Byte was persuaded in the present case to go as high as £75,000 pa.

Various other criticisms were made of the marketing campaign such as the fact that the wrong retailers were targeted and that the marketing campaign was too impersonal. I am not impressed with these criticisms. There is no evidence to suggest that anyone would have been prepared to pay more than £75,000 pa. While I quite see the point made by Mr Buckle to the effect that it is not for the landlord to market the premises for the tenant, I consider that the criticisms made of the marketing exercise fall well short of causing me to doubt its effectiveness. It does not seem to me that Mr Buckle identified any significant number of organisations which should have been contacted, let alone any organisation which might have been interested in bidding for the premises, and certainly none which might have been prepared to contemplate paying more than £75,000 pa.

As to the expert opinions as to the open market rental value of the premises, I regret to say that I did not find the evidence of Mr Buckle or Mr Frost particularly helpful. They both accepted that the combination of location, construction and shortage of car parking spaces rendered the premises not so much unusual as unique in valuation terms. Accordingly, the valuations were all effected very much on the basis of general experience and, having decided that the marketing campaign was an effective one that produced the open market rent, it seems to me that the weight to be given to the expert evidence is comparatively slight on this issue.

In making these observations about the evidence of Mr Buckle or Mr Frost, I am not in any way adversely criticising them. They were both doing their best in difficult circumstances. However, when one has premises for which there are no comparables because of location, size, construction and lack of car parking spaces, the evidence of the rent thrown up by what I conclude was competent marketing, provides a more useful guide to market value than the opinion evidence of surveyors, however experienced they might be.

The reliance placed by the landlord, and indeed by Mr Buckle in his expert evidence, on the rent of £115,000 pa agreed for the subject lease in December 1993 appears to me to be ill founded. The rental value of any particular property must, as a matter of economic logic, ultimately be based on the profit which a retailer assesses he can make from that property. When the rent of £115,000 pa was bindingly agreed in December 1993, the premises were in the process of construction and the site had little if any, relevant history. It is fair that it had retail purposes in the past, but the buildings on the site prior to the premises being constructed were antiquated.

By March 1995, on the other hand, when the tenant decided to seek to sublet the premises, the tenant had been trading from the premises for several months. It appears to me that there is a considerable difference between assessing the profit-making potential of a property which is not yet built and assessing its potential once it has been built and traded from. That is all the more true in a case of premises where the site is as unusual as the present one. It is clear from the evidence before me that the tenant has not been able to trade profitably from the premises and, while the precise figures showing the tenant's turnover and losses may well not be known in the market, it would have been obvious, it seems to me, to anybody considering taking an underlease of the premises after March 1995 that the previous occupier had not found the premises profitable. Quite apart from that, Mr Buckle's contention that rents have not fallen in Watford over the period 1993 to 1996 appears to me to be an unhelpful generalisation. Rents may well have risen quite significantly over that period in parts of the Harlequin Centre, Watford, but it does not seem to me, from the evidence I have seen, that rents in the High Street, Watford, and indeed in some other parts of the Harlequin Centre, have risen. On the contrary, although it is impossible to identify with accuracy the degree and times, it does appear that the market has fallen between the date the lease was bindingly agreed and the date on which the landlord refused its consent. Furthermore, I accept the evidence of Mr Frost, who was the only surveyor witness who specialised in letting and acquiring properties specifically in the Watford area, that the fall in rents probably occurred after the opening of one phase of the Harlequin Centre. The opening took place in 1992 and he said that the market rents fell after that period in parts of the High Street, and in particular in the part near the subject premises. Quite apart from that, Mr Buckle himself appears to think that the premises have fallen in value by nearly 20% in the relevant period, given that he values the premises in terms of current rental value at £95,000 pa. Once one accepts that there has been a fall of that magnitude, it is difficult to see how one can place much weight at all on the £115,000 pa as a guide to the current level of rental value.

In these circumstances, it follows that I reject the landlord's case so far as it rests on what I have called the second ground, because I am of the view that the rent under the proposed underlease, namely £75,000 pa, is the open market rental value of the premises.

Of course, the concept of an open market rental value of the property does not necessarily involve a specific figure. Normally, there is a band of value. It is right to say that, on the evidence I have heard, I am not merely of the view that the rent of £75,000 pa is within the band of what properly could be described as the open market rental value of the premises; I think that it is towards the higher end of that band. I am of that view for three reasons. First, £75,000 pa is substantially higher than any bid which was received prior to June 1996 for any retail user. Second, the bid is as high as the highest bid put forward on behalf of a restaurant or brewery use which, as is agreed by all parties, would be likely to command a higher rent than retail use. Third, the £75,000 pa was a bid made by Byte in circumstances where there were two interested competitors in the premises, that being a circumstance which the experts who gave evidence were inclined to think would be conducive to obtaining a high rent.

Landlord's view of the open market rental value

This is not necessarily the end of the second ground, because the landlord contends that, even if £75,000 pa is the open market rental value of the premises, it is none the less entitled to withhold its consent to the proposed subletting on the basis that, at the time it refused its consent, the landlord reasonably believed, and indeed the landlord still reasonably believes, that £75,000 pa is not the open market rental value of the premises. This aspect of the second ground involves the resolution of an issue between the parties relating to the construction of the two provisions of the lease which I have quoted. I therefore turn to that issue of construction.

The tenant contends that, on the true construction of clause 5.8.6, once it has been established that the rent to be paid under the proposed underlease is the open market rental value of the premises, it is no longer open to the landlord to object to the proposed subletting. The landlord contends that clause 5.8.6 effectively lays down two conditions which have to be satisfied by a subletting. The first is that the subletting must be at a rent at least equal to the open market rental value of the premises; the second is that the rent payable under the proposed underlease must first be approved by the lessor, such approval not to be unreasonably withheld. If that is right, then, even

A though I have decided that £75,000 pa is the open market rental value of the premises, the landlord is still entitled to object to the proposed subletting if he can show that he objects to that level of rent and that his objection is reasonable.

The drafting of clause 5.8.6 is not conspicuously elegant. However, it does seem clear that what has to be "approved by the landlord" is the "rent" to be reserved under a proposed underlease. Further, it is difficult to see what word other than "rent" is governed by "to be approved by the landlord". The latter part of the clause emphasises the point. I think that the natural way in which to read clause 5.8.6 is, as Mr Michael Driscoll QC contends, as if the words "not less than the B open market rental value of the premises" were, as it were, in parentheses. Accordingly, I accept the landlord's contention on this issue of construction.

It is fair to say that this reading of clause 5.8.6 appears to make it somewhat onerous. The lessee would appear to be rather disadvantaged by the lessor having the right to maintain an objection to the proposed rent, even if it were at market levels. None the less, the parties may have envisaged that circumstances could arise where the lessor would have reasonable grounds to objecting to the level of rent under a proposed underlease, notwithstanding that level of rent reflected the market. Indeed, the landlord contends that the present C facts provide an example of why they should be so. Whether or not the landlord is right in the present case, I do not think the result of the natural meaning of the word used in clause 5.8.6 is so surprising or unfair that I should depart from the natural meaning. After all, whatever the ambit of the lessor's right to object, he is constrained by reasonableness.

Having reached the conclusion that, even if the rent payable under the underlease is the market rental value, the landlord can object to the subletting on the basis that he reasonably believes that it is not, I consider that on the facts of this case the landlord's reliance on that contention faces insuperable obstacles.

D The first problem for the landlord is that at the time of the refusal he did not consider whether or not £75,000 was the market rental value of the premises in any sort of way other than a most perfunctory one. Shortly before his consent was sought to the underletting to Byte, the landlord was asked to consent to another underletting, but at £60,000 pa. The surveyor then advising him, a Mr Peter Brown [ARICS] of Aitchisons Raffety Buckland, thought that £60,000 pa was probably below the market rental value of the premises but stated that further investigation was required.

In relation to that proposed subletting and in relation to the E proposed subletting to Byte at £75,000 pa, the landlord sought the advice of Mr Rodney Symondson [FSVA] of Smith Melzack. He gave evidence before me and stated that he was an investment surveyor and, although he had a view as to the rental value of the premises, he very fairly said that it would be wrong to place much reliance on it as he does not pretend to be an expert on rental values in Watford. His advice to the landlord in connection with the proposed subletting to Byte was based on the effect of the subletting at £75,000 pa on the value of the landlord's interest and not as to the rental value of the premises.

In these circumstances, I do not consider that it is now open to the F landlord to rely on this ground as a reason for refusing consent, as it was not a ground in his mind at the time of his refusal: see *Bromley Park Garden Estates Ltd* v *Moss* [1982] 1 WLR 1019*.

A second possible problem for the landlord is that, in giving its reasons for refusing consent to the underletting, the landlord stated in terms in a letter dated July 4 1996 that consent was being withheld under clause 5.8.3 and not clause 5.8.6. In those circumstances, it seems to me that there are difficulties in the landlord seeking to justify refusing consent under clause 5.8.6, although I am attracted to the point made on behalf of the landlord in argument, namely that it would be unfortunate if this case had to be decided on that sort of technical G point.

The third problem, in my view, for the landlord is that I do not think that the landlord's view (if he had one) to the effect that £75,000 pa was less than the open market rental value of the premises was or would have been a reasonable view. At the time its consent was refused, namely in June 1996, the landlord had not obtained full and proper advice as to the rental value of the premises from a surveyor who claimed expertise in that field. He had received from the tenant copies of three independent rental valuations commissioned from Chestertons, Richard Ellis and Hillier Parker, all of whom had arrived at a rental value for the premises of between £50,000 and £63,000 pa. H Further, the landlord was no doubt aware of the fact that the premises had been extensively marketed and that the tenant obviously wished to get as high a rent as he possibly could on a subletting.

Based on these factors, I conclude that, if the landlord had refused consent in June 1996 to the proposed underletting on the grounds that he believed that the rent of £75,000 pa was less than the market rental value of the premises, that ground would have been unreasonable. It may be contended that the landlord's view cannot have been unreasonable because it was supported by Mr Buckle's evidence as to the open market rental value. I do not accept that contention. Clearly, a landlord cannot say that his view on valuation matters is reasonable J merely because it is supported by the evidence or advice of a valuation surveyor. One has to consider whether the surveyor's view is reasonable. Although I have no doubt that Mr Buckle is an honest and experienced surveyor, I consider that on this occasion his view as to the rental value of the premises is an unreasonable one. I have already given my reasons for that conclusion. Accordingly, although I accept the landlord's case on the construction of clause 5.8.6, I reject the landlord's case on this aspect of the facts.

Diminution in the value of the landlord's interest

The landlord wishes to sell its head leasehold interest and is concerned that the proposed underletting to Byte of £75,000 pa will K diminish the value of its interest and hence the amount which the landlord will realise on such a sale. Accordingly, the landlord argues that it is entitled to withhold consent under clause 5.8.3 and/or under clause 5.8.6.

The landlord relies on the following three principles. First, it is normally reasonable for a lessor to object to a proposed subletting in circumstances where he reasonably believes that the proposed transaction will reduce the market value of his interest: see *Ponderosa International Development Inc* v *Pengap Securities (Bristol) Ltd* [1986] 1 EGLR 66. The landlord accepts that the position may be different where the lessor does not in fact wish to sell his interest: see L *International Drilling Fluids Ltd* v *Louisville Investments (Uxbridge) Ltd* [1986] Ch 513*. But, says the landlord, that is not the case here on the facts.

Relying on *International Drilling* at p520D as followed and applied in *Air India* v *Balabel* [1993] 2 EGLR 66 at p69A-F, the landlord's second contention is that it is enough for his view to be one which a reasonable lessor could have, a lessee does not succeed in showing the lessor to be unreasonable merely because some, or even many, people in the lessor's position may have held a different view from that of the lessor.

The third contention of the landlord is that the fact that a lessor's concern about the depreciatory effect of the proposed transaction is M illogical or is founded on a misunderstanding of the law does not render the lessor's concern unreasonable; provided the lessor can show that, because of the market's attitude, there actually will, or there is a real chance that there will, be a depreciatory effect on its interest as a result of the transaction that is sufficient. For this contention the landlord relies again on *Ponderosa*.

In my judgment, each of these contentions is right in principle. However, although the landlord puts its case on this aspect in a number of ways, I do not consider that they justify the refusal of consent of the

---

*Editor's note: Also reported at [1983] 1 EGLR 65

*Editor's note: Also reported at [1986] 1 EGLR 39

proposed subletting in the instant case. The landlord's first argument is that, even if £75,000 pa is the open market rental value of the premises, a subletting at £75,000 pa will make it clear to any potential purchaser of the landlord's interest that only £75,000 of the current rent of £115,000 pa is "hard core", and the balance of £40,000 is "froth". When assessing the value of the landlord's interest the market will apply a lower yield or a higher year's purchase to the "hard core" than to the "froth" rent. If no underletting takes place, then it is possible that someone will be found who assumes that more than £75,000 of the current rent of £115,000 is hard core and therefore the landlord may well realise a higher price for his reversionary interest than if the froth is, to use Mr Michael Driscoll QC's words, crystallised by an actual subletting.

Quite apart from its unattractive nature, that argument is not supported by the facts. (I do not need, therefore, to decide whether it is a good argument in principle.) After the landlord's interest was initially marketed, it was withdrawn because of poor market sentiment. Once sentiment improved it was not remarketed because it was learned that the tenant wished to sublet. Quoting from the landlord's own surveyor's evidence, Mr Symondson said:

> When it was learned that the tenant was looking to underlet the premises, it was still not practicable to market the premises until the details of the subletting were known. The reason that this is so is because of the uncertainty created not only as to who the occupying tenant would be and the strength of its covenant, but also the fact that the tenant was seeking to move out of the premises so soon after taking up occupation. Uncertainty affects the ability to dispose of the property at the optimum price. Accordingly, the landlord has not been able to date to seriously embark on the marketing of its reversion. This is unfortunate because the landlord is anxious to sell its reversion as soon as I am able to advise that the market conditions are such that that the optimum price for the landlord's reversion might be obtained.

This is consistent with the evidence of Mr Pulver, the managing director of the landlord, who in cross-examination somewhat hyperbolically described the landlord's interest as "unsaleable" in light of the uncertainties described in the passage of Mr Symondson's evidence which I have quoted.

The effect of the evidence, in my view, is that an underletting of the premises at £75,000 pa would actually increase the value of the landlord's interest or, to use Mr Pulver's language, would transform the landlord's interest from its present unsaleable state to a state where it could be sold. It is also consistent with the evidence of Mr Simon Taylor [FRICS], an investment surveyor and a partner in the firm of Richard Ellis, who gave evidence to the same effect and which I accept. Even Mr Buckle, at the end of his cross-examination, appeared to accept that a subletting on the premises at the market rental value would not detrimentally affect the value of the landlord's interest.

The landlord's second argument is that, on the assumption that the tenant cannot get more than £75,000 pa in the current market, it is entitled to require the tenant to postpone any subletting until the market improves because this would enable the tenant to get a higher rent on a subletting which in turn would increase the value of the landlord's reversion.

I reject that argument because I do not consider that this is a reasonable ground in principle upon which a lessor can withhold consent to a subletting. It would mean that in any case where a lessor wished to sell its interest in the future and anticipated that the market was going to rise, he could effectively prevent the lessee from subletting. The lessee could not even sublet at a rent higher than the market rent with a reverse premium, because any prospective purchaser would appreciate that such an arrangement would demonstrate the open market rental value of the premises just as surely as a straightforward underletting with no reverse premium and at a lower rent. It seems to me that the landlord's argument on this point runs into the same sort of problem as that identified by Mr Edward Nugee QC, sitting as a deputy judge at first instance, and approved by the Court of Appeal in *International Drilling* at pp521H to 522B. It effectively deprives the lessee of the right to sublet for the foreseeable future.

The third contention of the landlord is really based on the proposition that the investment value of its interest would be damaged because the rent payable under the proposed underlease will be less than the market rent. I have already concluded that the proposed rent of £75,000 pa is the market rental value of the premises and, accordingly, this argument fails.

Conclusion

The tenant is entitled to the declaratory relief it seeks. There is also a claim for damages under the Landlord and Tenant Act 1988 which the parties have very sensibly agreed should await the outcome of this hearing. I will hear counsel on the form of order and any directions which should be made.

Burrows *v* Brent London Borough Council

HOUSE OF LORDS

October 31 1996

LORDS BROWNE-WILKINSON, KEITH OF KINKEL, GRIFFITH, JAUNCEY OF TULLICHETTLE and STEYN

ESTATES GAZETTE MARCH 15 1997

[1997] 11 EG 150

Landlord and tenant — Secure tenancy — Order for possession — Whether agreement not to execute order while conditions complied with by tenant creates new secure tenancy

In August 1984 the appellant council granted the respondent and her husband a secure tenancy of a flat. They fell into arrears with the rent. In January 1992 the council obtained a final order requiring the respondent and her husband to give possession of the flat by February 12 1992. On February 5 1992 the council and the respondent agreed that the arrears of rent were to be paid off at a specified weekly amount and that if the payments ceased or became irregular the council would seek to evict the respondent. The respondent failed to make the payments and in June 1994 the council issued a warrant for possession which was executed. The county court judge ordered that the respondent be reinstated in the flat. The council appealed the decision of the Court of Appeal which had dismissed their appeal from the judge.

***Held*: The appeal was allowed. By the agreement of February 5 1992 the parties did not intend to create a tenancy or licence but only to defer the execution of the order for possession so long as the respondent complied with the agreement. The agreement should take effect as the parties intended; the former tenant could be characterised as a trespasser who the council had agreed not to evict pending the revival of the old tenancy or the breach of the agreed conditions. The agreement therefore did not create a secure tenancy or licence under Part IV of the Housing Act 1985. There was no reason why the agreement should not be given the same effect as an order of the court in similar terms suspending execution of the original order for possession.**

The following cases are referred to in this report.

Greenwich London Borough Council v *Regan* (1996) 28 HLR 469
Thompson v *Elmbridge Borough Council* [1987] 1 WLR 1425; (1987) 86 LGR 245; 19 HLR 526, CA

This was an appeal by Brent London Borough Council from a decision of the Court of Appeal which had dismissed the council's appeal from a decision of Judge Finestein QC in Willesden County

Court, who had allowed the application of the respondent, Miss AP Burrows, to be reinstated in premises following the execution of a writ of possession.

Ashley Underwood (instructed by the solicitor to Brent London Borough Council) appeared for the appellants; David Neuberger QC and William Geldart (instructed by Daniel & Harris) represented the respondent.

Giving the first speech, **Lord Browne-Wilkinson** said: This appeal raises a question as to the rights of a tenant of a dwelling-house let by a local authority. Under Part IV of the Housing Act 1985 such tenancy is a "secure tenancy". A landlord cannot terminate a secure tenancy otherwise than by obtaining from the court an order for possession. Local authority landlords often obtain an order for possession against a secure tenant but then agree with the tenant that, while certain conditions are observed, the order will not be executed. The question is whether such an agreement operates so as to create a new secure tenancy which can only be terminated if the landlord obtains a further order from the court.

On August 6 1984 Brent London Borough Council ("Brent") granted a tenancy of the basement and ground-floor flat at 28 Oxford Road, London NW6, jointly to the respondent, Miss Burrows, and her husband Richard Allen. In January 1986, Mr Allen left the premises leaving Miss Burrows with her children. She fell behind with her rent.

On January 29 1992 Brent obtained a final order for possession against Miss Burrows and Mr Allen in Willesden County Court. The order directed that the plaintiff should recover the sum of £2,427.63 being arrears of rent and costs. It further ordered: "that the defendant do give the plaintiff possession of the said land on 12 February 1992". Thus, the order was not suspended: it was an immediate order for possession on February 12.

On February 5 1992 Brent came to an arrangement with Miss Burrows which was recorded in writing. It was made on a form which was designed for use by tenants prior to a possession hearing. The memorandum of the arrangements stated:

> I acknowledge that there are arrears of £2,313.41 on my rent account as at 27.1.92 . . . I understand that in those circumstances the London Borough of Brent will seek a possession order and judgment for the full amount of the arrears from the County Court . . . I agree to pay the rent charge of £2.67 due every week and, in addition, to reduce the arrears by regular instalments of £3.00 per week.

There was then a section of the document which was only applicable when the document was to be signed before an order was made which was crossed out.

> If payments cease or are irregular the Council will seek to evict.

Miss Burrows failed to make the payments in accordance with the agreement of February 5 1992. Brent issued a warrant for possession informing her that she would be evicted on June 8 1994. Miss Burrows moved out on June 7. The order was executed. She returned on June 8 1994 to find herself locked out and the premises boarded up. She thereupon applied to Willesden County Court seeking to have the bailiff's warrant and the possession order set aside, but those proceedings were dismissed.

On July 4 1995 Miss Burrows commenced this action seeking a declaration that she remained a tenant and damages for unlawful eviction. She also sought a mandatory injunction to be allowed back into the premises. At the heart of her case was the contention that the effect of the arrangements made on February 5 1992 was to create a new tenancy or licence by operation of law. By his judgment of August 19 1994 Judge Finestein QC agreed with Miss Burrows' submissions and ordered her reinstatement. He made the declaration and injunction asked and referred the matter to the district judge for an assessment of damages.

In the course of his judgment, the judge found that:

(1) Miss Burrows understood, and this was part of the express agreement, that eviction would be a penalty if she was in default;

(2) it was not contemplated by her that when she entered into the arrangement on February 5 she was entering into a tenancy;

(3) when she left the premises on June 7 1994 it was not her state of mind that a new tenancy had come into being as a result of the arrangement of February 5 1992.

Brent appealed to the Court of Appeal (Butler Sloss, Otton and Auld LJJ) (1995) 27 HLR 748 who dismissed the appeal. Brent appeal to your Lordships' House.

Sections 79, 80 and 81 of the Act of 1985 define the conditions which have to be satisfied in order to constitute a "secure tenancy". Those conditions were satisfied in the present case. It is important to note that under section 79(3) the provisions of Part IV of the Act apply to a "licence to occupy a dwelling-house . . . as they apply in relation to a tenancy". Therefore nothing in this case turns on the distinction between a licence and a tenancy; if, by making the agreement not to enforce the possession order, the local authority are to be taken to have granted a licence for the tenant to continue in occupation the position will be just the same as if they had granted a tenancy.

Section 82 provides as follows:

> (1) A secure tenancy which is either —
> (a) a weekly or other periodic tenancy, or
> (b) a tenancy for a term certain but subject to termination by the landlord,
> cannot be brought to an end by the landlord except by obtaining an order of the court for the possession of the dwelling-house or an order under subsection (3).
> (2) Where the landlord obtains an order for the possession of the dwelling-house, the tenancy ends on the date on which the tenant is to give up possession in pursuance of the order.

It is important to note that the secure tenancy ends, not on the date on which possession is in fact given up, but on the date on which the order requires the defendant to give up possession.

Section 84 provides that the court shall not make an order for possession save on one of the grounds mentioned in Schedule 2, which include non-payment of rent. In addition, in the case of non-payment of rent the court must also be satisfied that it is reasonable to make the order. An order for possession cannot be made unless the ground on which an order is to be sought has been specified in a prior notice to be served on the tenant: sections 83 and 84(3).

Section 85 is central to the argument in this case. It provides:

> (1) Where proceedings are brought for possession of a dwelling-house let under a secure tenancy [for the non-payment of rent] . . . the court may adjourn the proceedings for such period or periods as it thinks fit.
> (2) On the making of an order for possession of such a dwelling-house on any of those grounds, or at any time before the execution of the order, the court may —
> (a) stay or suspend the execution of the order, or
> (b) postpone the date of possession, for such period or periods as the court thinks fit.
> (3) On such an adjournment, stay, suspension or postponement the court —
> (a) shall impose conditions with respect to the payment by the tenant of arrears of rent (if any) and rent or payments in respect of occupation after the termination of the tenancy (*mesne* profits), unless it considers that to do so would cause exceptional hardship to the tenant or would otherwise be unreasonable, and
> (b) may impose such other conditions as it thinks fit.
> (4) If the conditions are complied with, the court may, if it thinks fit, discharge or rescind the order for possession.

The argument for Miss Burrows, which the Court of Appeal accepted, is as follows. The order of January 29 1992 directed that possession should be given on February 12 1992. Therefore, by virtue of section 82(2), Miss Burrows' original secure tenancy terminated on that day. Yet, under the agreement of February 5 1992, Miss Burrows remained in occupation of the house paying a "rent charge" of £2.67 per week for such occupation. That agreement could not be effective to alter or vary the order itself, but could only take effect as an agreement to permit Miss Burrows to stay on after her existing tenancy had terminated on February 12 1992. This right of continued occupation can only be explained on the ground that the agreement conferred on Miss Burrows a new right of occupation, either by way of a new tenancy or as a licensee, it mattered not which. If it was a new tenancy, it was a new secured tenancy; if it was only a licence, by virtue of section 79(3) Miss Burrows enjoyed the same protection as if it were a tenancy. In either event, the new secure tenancy or new licence could only be terminated by Brent applying to the court for a further court order terminating the new right: section 82(1).

The argument for Brent before the Court of Appeal was that the agreement of February 5 was simply an agreement by Brent not to execute the possession order, provided that Miss Burrows complied with the agreed conditions. As the judge's findings demonstrated, Miss Burrows never intended that the agreement should create a tenancy and it was absurd to imagine that Brent, by granting Miss Burrows an indulgence in relation to a possession order which they had only just obtained, should have intended to create a new right of occupation necessitating a further application to the court in order to obtain possession. In the period during which Brent agreed to forbear from enforcing the order, Miss Burrows was a mere "tolerated trespasser".

The Court of Appeal, while accepting that the crucial factor in determining Miss Burrows' rights was the intention of the parties, rejected Brent's contention on the ground that it gave rise to manifest absurdities. At the rate for payment of arrears stipulated by the agreement of February 5 1992, it would have taken Miss Burrows 14 years to pay them off during which time, if the argument of Brent were correct, she would be a mere trespasser. As a trespasser she would enjoy none of the rights of a tenant. Thus she could not require Brent to repair the house. She would have no rights under the Defective Premises Act, 1972. She would qualify as a homeless person under section 58 of the Act of 1985. Brent would have no right to evict her for any breach of the covenants in her tenancy, but only for breach of the conditions contained in the agreement of February 5 1992.

On the case as presented to the Court of Appeal their conclusion was inevitable. But the significance of section 85 was not drawn to their attention. As a result of the recent decision of the Court of Appeal in *Greenwich London Borough Council* v *Regan* (1996) 28 HLR 469 Brent advanced before your lordships a far more compelling argument, *viz*:

(1) Although under section 82(2) the original tenancy came to an end on February 12 1992 (being the date fixed by the order for giving possession) that was not necessarily the final position;

(2) Under section 85(2) the court has power to postpone the date of possession;

(3) This power to postpone the date for possession is exercisable by the court "at any time before the execution of the order". This shows that the power can be exercised even after the date for possession specified in the order has passed and the tenancy has thereby been terminated by virtue of section 82(2);

(4) This conclusion is re-enforced by section 85(3)(a) which postulates that the court can make a suspended order conditional on making "payments in respect of occupation after termination of the tenancy (*mesne* profits)". This demonstrates that there can come a time when, although the old tenancy has terminated and the former tenant has remained in possession for which he is liable to pay *mesne* profits as a trespasser, the court can until the original order is executed make an order varying the date for the giving of possession thereby reviving the previously defunct tenancy.

(5) Therefore, so far as the tenant is concerned, the crucial event is the execution of the order for possession. Down to that date the tenant can apply to the court for a variation of the original order substituting a new date on which possession is to be given thereby reviving the old secured tenancy. This revived tenancy will not be terminated under section 82 until the new date for giving possession occurs.

(6) That such revival of the old tenancy is possible is demonstrated by section 85(4) which plainly assumes that on discharge or recision of the original order for possession, the old secure tenancy will revive.

I accept this analysis of the effect of section 85 which is largely derived from the judgment of Millett LJ in *Greenwich London Borough Council* v *Regan*. In that case, an order for possession was made against a secure tenant, the order (as construed by the Court of Appeal) providing for the giving of possession to be postponed so long as arrears of rent were paid by instalments and the current rent was paid. The tenant having failed to comply with the conditions in the order, the landlord and the tenant agreed variations in the amount of the payments which were to be made. The tenant having breached the terms of the order and the agreed terms, the landlord applied for a warrant of execution. The tenant sought a stay of execution. He submitted that the old tenancy had come to an end when he failed to comply with the conditions imposed by the order: see *Thompson* v *Elmbridge Borough Council* [1987] 1 WLR 1425. He then submitted, in reliance on the decision of the Court of Appeal in the instant case, that by agreeing to allow the former tenant to remain in possession the landlord had created a new tenancy or licence and therefore could not obtain possession under the old order.

The Court of Appeal in *Regan*, after analysing section 85 in much the same terms as I have summarised above, posed the question whether the parties could, by agreement, revive the expired tenancy without an order of the court. The Court of Appeal held (in my view rightly) that the parties could not by agreement vary the terms of the court order. But they held that by agreeing the new conditions, the landlord waived the right to complain that failure to comply with the conditions specified in the order constituted a breach of those latter conditions. Consequently, there being no breach of the conditions imposed by the order upon which the landlord could rely, the order remained in force, the date for giving possession had not passed and therefore the old tenancy had not been terminated. They distinguished the decision in the present case on the ground that they were dealing with a suspended conditional order of the court whereas in the instant case there is an immediate, unconditional order for possession.

One factor which weighed heavily with the Court of Appeal in *Regan* (to which I also attach importance) is the practical effect of the decision under appeal, ie any consensual variation of an order for possession produces a new secure tenancy or licence. Local authorities and other public housing authorities try to conduct their housing functions as humane and reasonable landlords. In so doing they frequently need to grant indulgences to their tenants to reflect changes in the tenants' circumstances. When applying for possession orders for non-payment of rent local authorities agree to the order being suspended upon the payment of arrears, the rate of payment being adjusted to meet the means of the tenant at the date of the order. If the tenant subsequently loses his job, the landlords will often be willing to reduce the rate of payment of arrears. Why should this not be done by agreement? Yet the effect of the local authority agreeing to such a reduction will be that the tenant, while keeping up his payments at the agreed reduced rate, will be in breach of the conditions specified by the order at the higher rate. If so his old tenancy will be terminated. On the view of the law adopted by the Court of Appeal in the present case, a new secure tenancy requiring a new order will come into existence. Similarly where, as with Miss Burrows, the court makes an immediate order for possession but the landlords grant an indulgence by agreeing not to execute the order immediately: if the Court of Appeal decision is correct, the effect of granting the indulgence is to create a new tenancy or licence and the local authority will have to obtain a new possession order. The practical result therefore will be either that the local authority will be reluctant to make reasonable and humane concessions by agreement or in every case will have to make an application to the court to vary the existing order so as to ensure that the old tenancy is not brought to an end. I find it impossible to believe that parliament intended to produce such an unreasonable regime, penalising sensible agreements out of court and requiring repeated applications to an already overstretched court system.

What, then, is the correct legal analysis? I start from the proposition that where a former tenant is by agreement allowed to remain in possession of the demised property after the termination of the tenancy, the question in each case is *quo animo* the parties have so acted: depending upon the circumstances, their conduct may give rise

A to a new tenancy, a licence or some other arrangement. In the present case, on February 5 1992 the parties plainly did not intend to create a new tenancy or licence but only to defer the execution of the order so long as Miss Burrows complied with the agreed conditions. It cannot be right to impute to the parties an intention to create a legal relationship such as a secure tenancy or licence unless the legal structures within which they made their agreement force that conclusion.

A secure tenancy protected by Part IV of the Act of 1985 is not like an ordinary tenancy. It can only be terminated by an order of the court ordering possession to be given on a particular date or in a particular B event. But even determination by order of the court is not final. Until the possession order is executed, the court can by variation of its order change the date on which possession is to be given and thereby revive a secure tenancy which has already been terminated. During the period between the date specified by the order for the giving of possession and the date on which the order is executed there is a period of limbo: the old tenancy has gone but may yet be revived by a further order of the court varying the date for possession. If the parties reach an agreement as to the continued occupation of the premises by the tenant during that limbo period, what intention is to be imputed to them?

In my judgment, little guidance is to be obtained from the cases C where a tenant holds over after the termination of an ordinary tenancy where there is no possibility that the expired tenancy can revive. The position in relation to secure tenancies is *sui generis*. In my judgment, the agreement can and should take effect in the way the parties intend, ie it is an agreement by the landlords that, upon the tenant complying with the agreed conditions, the landlords will forbear from executing the order, ie from taking the step which would finally put an end to the tenant's right to apply to the court for an order reviving the tenancy. There is no need to impute to the parties an intention to create a new tenancy or licence: the retention of possession and the payment of rent relate to occupation under the old tenancy which is in limbo, but which D may be revived. In these circumstances I think it is fair to characterise the former tenant as a trespasser whom the landlords have agreed not to evict, a — "tolerated trespasser" — pending either the revival of the old tenancy or the breach of the agreed conditions.

Once the effect of section 85 is appreciated, the absurdities which led the Court of Appeal not to accept that Miss Burrows could be a tolerated trespasser disappear. Technically the old secure tenancy is, during the limbo period, no longer in existence and therefore neither the repairing covenants in the tenancy nor the Defective Premises Act 1972 apply. But the tenant can at any time apply to the court for an order varying the date on which possession is to be given and thereby E retrospectively revive the old secure tenancy, together with its covenants. If the tenant has complied with the agreed conditions, there can be little doubt that the court would make the required order. Moreover, the tenant will not be a homeless person within section 58(2) of the Act of 1985 because the tenant will be occupying the residence by virtue of "any rule of law giving him the right to remain in occupation": see section 58(2)(c). If the tenant were in breach of any of the covenants in the old secure tenancy, Brent could apply to vary the order so as retrospectively to revive the old tenancy together with its covenants.

Finally, there is a method (albeit a clumsy one) whereby the order F for possession even if an immediate unconditional order, can be discharged or rescinded if so desired under section 85(4). The power in that subsection to discharge or rescind only arises "if the conditions are complied with", a requirement which cannot be satisfied in the case of an unconditional order. But there is no reason why the order cannot be discharged by consent or, if such consent is not forthcoming, by the court varying the original order so as to impose the agreed conditions and then discharging the varied order.

It was submitted that the fact that the tenancy was granted to Miss Burrows jointly with Mr Allen whereas the agreement of February 5 1992 was made with Miss Burrows alone, indicated that the agreement must have given rise to a new tenancy with Miss Burrows alone. Therefore there must be a new tenancy. However, since, in my view, G on its proper analysis the arrangement contained in the agreement of February 5 1992 gave rise to no new tenancy with anyone, that factor is irrelevant.

I therefore reach the conclusion that, in the absence of special circumstances, an agreement by a landlord not to enforce strictly an order for possession, whether conditional or unconditional, does not create a new secure tenancy or licence under Part IV of Act of 1985. Brent, by making the agreement of February 5 1992, did not grant a new tenancy or licence to Miss Burrows as from February 12 1992. It follows that the possession order of January 29 1992 was properly enforced. I would therefore reverse the decisions of the Court of H Appeal and the trial judge and dismiss Miss Burrows' action.

As Brent have succeeded on a point not taken in the courts below, I propose that there should be no order as to costs before your lordships' House and the order for costs made in the courts below should remain undisturbed.

Lord Keith of Kinkel and **Lord Griffths** agreed with Lord Browne-Wilkinson and did not add anything.

In his speech, **Lord Jauncey of Tullichettte** said: This appeal raises an important question of principle in relation to the effect of forbearance by a landlord to enforce an order for possession granted J under section 82(2) of the Housing Act 1985 ("the Act"). I have had the advantage of reading in draft the speech of my noble and learned friend Lord Browne-Wilkinson and I gratefully adopt his account of the factual background giving rise to this appeal.

Section 82(1) of the Act provides that a secure tenancy such as was enjoyed by the respondent cannot be brought to an end by the landlord without an order of the court for possession. Section 82(2) provides:

> (2) Where the landlord obtains an order for the possession of the dwelling-house, the tenancy ends on the date on which the tenant is to give up possession in pursuance of the order. K

This provision is mandatory with the result that occupation by the tenant after the date specified in the order ceases to be attributable to the original secure tenancy. In view of the terms of that subsection the Court of Appeal concluded that the effect of the agreement of February 5 1992 was to grant to the respondent a new lease or licence to occupy which enjoyed the same protection as the terminated tenancy. Before this House Mr Ashley Underwood for the appellants advanced an argument which had not been deployed before the Court of Appeal and which relied upon detailed consideration of section 85 which provides, *inter alia*, as follows:

> 85(1) Where proceedings are brought for possession of a dwelling-house L let under a secure tenancy on any of the grounds set out in Part I or Part III of Schedule 2 [grounds 1 to 8 and 12 to 16: cases in which the court must be satisfied that it is reasonable to make a possession order], the court may adjourn the proceedings for such period or periods as it thinks fit.
>
> (2) On the making of an order for possession of such a dwelling-house on any of those grounds, or at any time before the execution of the order, the court may —
>
> (a) stay or suspend the execution of the order, or
>
> (b) postpone the date of possession,
>
> for such period or periods as the court thinks fit.
>
> (3) On such an adjournment, stay, suspension or postponement the court — M
>
> (a) shall impose conditions with respect to the payment by the tenant of arrears of rent (if any) and rent or payments in respect of occupation after the termination of the tenancy (*mesne* profits), unless it considers that to do so would cause exceptional hardship to the tenant or would otherwise be unreasonable, and
>
> (b) may impose such other conditions as it thinks fit.
>
> (4) If the conditions are complied with, the court may, if it thinks fit, discharge or rescind the order for possession.

Subsections 1(1) and (2) cover three different situations which are in turn reflected in subsection (3). If the proceedings for a possession order are adjourned in circumstances covered by subsection (1) the secure tenancy will continue in force but the court is required by subsection (3)(a) to impose conditions as to payment of any arrears of rent and of future rent, subject always to questions of exceptional hardship and unreasonableness. If the tenant fails to comply with these conditions the landlord will be entitled to go back to the court and seek an order for possession. If during the initial proceedings the court makes an order but postpones the date of possession the tenancy will not be terminated under section 82(2) until any condition imposed under subsection (3) has been breached by the tenant: see *Thompson* v *Elmbridge Borough Council* [1987] 1 WLR 1425. However, the court's power to make an order postponing the date of possession is not restricted to exercise on the first application for an order for possession but may be exercised on the application of either party at any time prior to execution of that order and even after the secure tenancy has ended by reason of section 82(2). This is made clear by the words in secdon 85(2) "or at any time before the execution of the order" see also *Greenwich London Borough Council* v *Regan* (1996) 28 HLR 469, at p476, *per* Millett LJ. In such an event the secure tenancy is reinstated or revived subject to any conditions imposed under subsection (3).

However, whereas an order postponing the date of possession necessarily affects the operation of section 82(2), an order staying or suspending the execution of an order for possession on a stated date has no effect on the operation of that subsection but merely postpones execution so long as the conditions of suspension are complied with. The reference to *mesne* profits in subsection (3)(a) as contrasted with that to payment of rent is relevant to the period of occupation after such a suspension as well as to any period of occupation between the termination of a tenancy under section 82 and the subsequent postponement of the date of possession under section 85(2)(a). The words are, in my view, particularly significant since they show that the Act contemplates the court requiring post-termination payments to be made by an occupier which are of a nature wholly inconsistent with the existence of any tenancy in his or her favour. Subsection (4) is similarly inconsistent with the creation of a new tenancy because a discharge of an order for possession would only be effective if the tenancy to which the order related had not been superseded by a new tenancy. I therefore conclude that the court has power under section 85 not only to permit an individual to continue to occupy premises after a secure tenancy has determined without the creation of a new tenancy in his or her favour but also to revive the determined secure tenancy on compliance by the individual with any stipulated conditions. During the period between the termination of the secure tenancy and either its revival or the execution of the order for possession the occupation of the former secure tenant derives not from any new lease but from the provisions of the Act which cast him in the role of what my noble and learned friend aptly describes as "a tolerated trespasser". Can the same result be achieved by agreement between the parties without the need to invoke the power of the court under section 85?

In *Greenwich London Borough Council* v *Regan* (1996) 28 HLR 469 the Court of Appeal considered an order for possession which Millett LJ, at p476, construed as one "requiring the delivery of possession to be postponed, in the first instance for 28 days and thereafter until there was a breach of the conditions of the order". The tenant failed to comply with the terms of the order and the parties thereafter entered into a fresh agreement whereby the tenant agreed to make weekly payments of a lesser amount than those stipulated in the order. Some time later the tenant again defaulted on his payments and the landlords sought to execute the order for possession. The tenant argued that the agreement between the parties varying the order had created a new tenancy and that therefore a new order for possession was required. This argument was rejected. Millett LJ after pointing out the impracticability of a landlord applying to the court whenever he granted some indulgence to a tenant went on to say, at p477:

> If, after a breach occurs and the tenancy is determined, and the Local Authority subsequently reaches agreement with the tenant the effect of which is to waive the breaches, then in my opinion in accordance with ordinary principle such breaches must be treated as if they had not occurred. It would follow from this that the tenancy must thereafter be treated as not having been determined. This appears to me to be the effect of an order which leaves the determination of the tenancy to the occurrence of a breach of the conditions of the order, and which leaves the occurrence of a breach to be dependent upon whether the landlord treats it as such or agrees to waive the breach.
>
> If the landlord does more and agrees not only to waive a breach of the conditions on which the order has been suspended, but also to accept payment of the arrears by smaller instalments and over a longer period than those provided for by the order, he thereby consents to such payments being treated as sufficient compliance with the order. This must either prevent the tenancy from being brought to an end or, if it has already been determined, prevent it from being so regarded.

He summarised the position, at p478, in four paragraphs of which I repeat only two:

> 2. The Local Authority may waive the breach, in which case the tenancy continues as if there had been no breach. This is so whether the Local Authority agrees in advance not to treat the particular failure to comply with the terms of the order as bringing the tenancy to an end, or agrees to waive the breach after it has occurred. In either case the tenancy is treated as if it has not determined.
>
> 3. The Local Authority may agree a consensual variation in the terms of the order. It will be a question of fact in every case whether the agreement which the parties had entered into should be characterised as an agreement which creates a new tenancy, or merely as one which provides for a variation in the terms on which the order has been suspended, waiving any past breaches and treating the tenancy as if it has not been terminated.

Mr David Neuburger QC for the respondent, did not challenge, correctly in my view, the decision in *Greenwich London Borough Council* v *Regan* but sought to distinguish it upon the ground that whereas a landlord could waive a condition imposed on a tenant in a court order the parties could not by agreement alter the impact of section 82(2) on an unconditional order for possession. It followed that any agreement between the parties as to the respondent's continued occupation must constitute a new tenancy. Mr Underwood argued that such a result would be not only contrary to the intention of the parties but to the whole scheme of the Act.

My lords, I agree with Mr Neuburger that the impact of section 82(2) cannot be altered by agreement between the parties. However, for the reasons which I have already elaborated, that subsection cannot be looked at in isolation from the other sections, in particular section 85, of Part IV of the Act. The whole scheme of that Part is to afford protection to the secure tenant and that is achieved in section 85 by conferring on the court flexible powers to continue an existing secure tenancy, to revive a determined secure tenancy or to create a state of statutory limbo which will afford to a defaulting tenant an opportunity to have restored to him all the benefits of the secure tenancy when he has complied with stipulated conditions. Parliament cannot have intended to penalise a landlord who acted within the spirit of the Act by granting indulgences to defaulting tenants without going through time-wasting and expensive court proceedings. Furthermore, a tenant who has reached an agreement advantageous to himself is not thereby prevented from making an application to the court under section 85(2) or (4). In this case the judge found as a fact that neither party contemplated that the agreement of February 5 1992 created a new tenancy and I can therefore see no reason why it should not be given the same effect as an order of the court in similar terms suspending execution of the order for possession of January 29 1992. Such a result would accord entirely with the spirit of the relevant statutory provisions, would be consistent with the intention of the parties and would preserve all the respondent's rights under subsections (2), (3) and (4) of section 85. For these reasons and for those given by my noble and learned friend Lord Browne-Wilkinson I would therefore allow the appeal.

Lord Steyn also agreed that the appeal should be allowed and did not add anything.

Prudential Assurance Co Ltd *v* Mount Eden Land Ltd

COURT OF APPEAL

November 12 1996

BELDHAM and MORRITT LJJ and SIR JOHN BALCOMBE

Estates Gazette April 5 1997

[1997] 14 EG 130

Landlord and tenant — Covenant against alterations without landlord's written consent — Whether consent granted by "subject to licence" letter

By four leases the respondent tenant held 999-year terms from dates in 1917 and 1924 of commercial premises. The leases were subject to covenants that certain alterations could not be carried out without the landlord's "previous written consent". On March 1 1993 the tenant sought the written confirmation of the landlord, the predecessor in title to the appellant, for specified works to commence in April. On May 18 1993 the agents for the landlord wrote a letter marked "subject to licence" giving consent to the works subject to conditions that, *inter alia*, a formal licence was entered into. On September 6 1994 the then landlord served a notice under section 146 of the Law of Property Act 1925 alleging that the work had been carried out without landlord's consent. Judge Rich QC, sitting as a High Court judge, held that the letter of May 18 1993 constituted a written consent for the purposes of the leases. The landlord appealed.

***Held*: The appeal was dismissed. It was not legitimate to extend the principle illustrated by the "subject to contract" or "subject to lease" cases from the field of bilateral negotiations to that of a unilateral act. In cases requiring unilateral act the only question is whether that act has occurred. The letter of May 18 1993 expressed consent in the clearest terms and was a consent as required by the leases. The heading "subject to licence" added little to the condition requiring formal licence, and meant that it served to emphasise the degree of formality required so that the express condition for a formal licence should not be satisfied in correspondence. However, no such document was required by the terms of the leases.**

The following cases are referred to in this report.

Akiens v *Salomon* (1992) 65 P&CR 364; [1993] 1 EGLR 101; [1993] 14 EG 97

Consolidated Pool Carriers GmbH v *CTM CIA Transmediterranea SA: The CPC Gallia* [1994] 1 Lloyd's Rep 68

Longman v *Viscount Chelsea* (1989) 58 P&CR 189; [1989] 2 EGLR 242, CA

Star Steamship Society v *Beogradska Plovidba: The Junior K* [1988] 2 Lloyd's Rep 583

Venetian Glass Gallery Ltd v *Next Properties Ltd* [1989] 2 EGLR 42; [1989] 30 EG 92

This was an appeal by the defendant, Mount Eden Land Ltd, from a decision of Judge Rich QC, sitting as a High Court judge, who granted a declaration in an application by originating summons made by the plaintiff, Prudential Assurance Co Ltd.

Kim Lewison QC (instructed by Herbert Smith) appeared on for the appellant; Derek Wood QC (instructed by Lovell White Durrant) represented the respondent.

Giving judgment, **MORRITT LJ** said: This is an appeal of Mount Eden Land Ltd, the landlord, from the order of Judge Rich QC, sitting as additional judge of the Chancery Division, made on July 14 1995. By that order he declared in answer to a preliminary issue that the appellant's predecessor in title to the reversion, Grovewood (LE) Ltd, had granted consent to the tenants, Prudential Assurance Co Ltd, for the execution of works to properties in Oxford Street and Wells Street, London, which in the absence of such consent would have constituted breaches of covenants contained in the relevant leases.

The leases in question are four in number, three were dated January 17 1917 and the fourth June 24 1924. They were leases of 134/140 Oxford Street and 80/83 Wells Street for terms which amounted in total, that is to say under each lease, to the equivalent of 999 years from April 6 1915. The four leases were so far as relevant in the same terms. The term which is relevant is the expression "previous written consent" contained in clauses II(9) and II(8) in these terms:

> That (without the previous written consent which may be temporary or permanent revocable or irrevocable or otherwise howsoever framed or qualified of the Lessor)

The other material provision is the covenant in question, which is against the making of alterations. That is contained in clauses II(11) and II(10) in the respective leases. That provides:

> That without such previous written consent as aforesaid and except in accordance with plans previously approved by the Lessor and to his satisfaction no building or erection shall at any time be built or placed on the said demised premises and no addition or alteration affecting the elevation external structure or stability of the said premises or any parts thereof shall at any time be made to any building or erection for the time being on the said demised premises and that no machinery shall at any time be erected upon or affixed to the said demised premises except under the supervision and to the satisfaction of the Lessor and that the said Lessees will pay to the Lessor on demand and indemnify the Lessor against all reasonable Surveyors Fees and other charges and expenses which the Lessor may incur in connection with any matter or thing under this present sub-clause.

By October 1992 Grovewood (LE) Ltd, then in administrative receivership, held the reversions to each of the four leases, the terms were vested in Prudential Assurance. Prudential had ascertained that the outer cladding of the building needed to be replaced and that that would give rise to an alteration of the appearance of the outside. On October 5 1992 Prudential wrote to the agents for the landlord notifying them of their proposals to carry out such works.

On January 5 1993 Prudential sent to the agents for the landlord, "the latest plans being considered by the Planners". On March 1 1993 Prudential wrote to the agents of the landlord seeking, "written confirmation of your client's agreement" for the works due to commence in April of that year. On March 24 1993 the agents for the landlord wrote to Prudential in these terms:

> Please note that I have not yet received confirmation from my clients that consent will be given for these works.
>
> I will endeavour to keep you fully informed of any progress in this matter.

On March 29 1993 Prudential expressed concern at the delay in obtaining an answer, and on April 1 1993 the agents for the landlord again replied that they had still not received:

> authority to grant consent for your alteration to the elevations.

On the planning front, on April 6 1993 the City of Westminster, as the planning authority, gave consent, but subject to certain specified conditions, to the proposed works of recladding. Then, on May 18 1993 the agents for the landlord wrote the crucial letter. It is in these terms:

> *Subject to License* [*sic*]
>
> Dear Mr Pritchard,
>
> WELLS HOUSE, 134/140 OXFORD STREET, LONDON W1
>
> I refer to your request to consent to alter the external appearance of the above building during the course of maintenance to the defective cladding.
>
> I can confirm that the Freeholder, Grovewood (LE) Limited (In Administrative Receivership), gives consent for the works subject to the following conditions:
>
> 1. A formal licence is entered into by your client, The Prudential Assurance Company Ltd.
>
> 2. The Prudential will pay my client's costs in this matter, including any additional insurance premium levied as a result of the works.
>
> 3. The Prudential are to obtain necessary consents and statutory approvals.

If you can confirm acceptance of the above terms, I will arrange for my client's solicitor to prepare the necessary licence. When replying, I should be grateful if you would confirm the name of the solicitor who will be representing your client and the appropriate drawing numbers to be incorporated in the licence.

On June 30 and then again on August 5 1993 the City of Westminster, as the planning authority, approved the reserved matters.

On September 7 1993 Prudential wrote confirming their agreement to the conditions set out in the letter of May 18 1993. That letter reads as follows:

Reference your letter dated 18th May. I confirm my client's agreement to items 1-3 inclusive.

I have forwarded drawings and details of the licence to my client's solicitors:— [the name and address of Lovell White Durrant is there set out]

I should be grateful if you would inform your client's solicitors of the position.

Subsequently, on September 6 1994 section 146 notices were served on Prudential, by the then landlord, claiming among other things that the replacement of the cladding had been carried out without the landlord's consent. This prompted Prudential to issue an originating summons on November 23 1995 claiming, among other things, that previous written consent had been given by the then landlord's predecessor in title.

In March 1995 Master Dyson ordered that that issue be tried as a preliminary issue. Then, on July 14 1995 the matter came before Judge Rich.

The judge concluded that the letter of May 18 1993 did constitute such consent as the leases required. After referring to the facts and the terms of the letter he posed the issue for his decision in these terms:

The issue therefore, it appears to me, between the parties is whether or not the expression "subject to licence" has a similar magic to the expression "subject to contract" such that it may contradict the effect which would otherwise arise from the words of the letter.

I put it that way, conscious that the letter has got to be construed as a whole and of course, even if those words lack magic, they may so reinforce the implications of other words in the letter as to contradict the expression "I can confirm that the freeholder gives consent", if that expression may, within the body of the letter, be said to give with one hand what is withdrawn with another.

I think that Mr Neuberger has wished, at the same time as he relies upon the magic of "subject to licence", also to put the totality of the effect of the letter as being other than the granting of a licence because of, in effect, the withdrawal of what is said in "I give consent" by the conditions and requirements identified subsequently in the letter.

The judge then considered the four reported cases to which he was referred. The fourth was the decision of Harman J in *Venetian Glass Gallery Ltd* v *Next Properties Ltd* [1989] 2 EGLR 42. At p12 of his judgment Judge Rich continued:

If the importation of the words "subject to licence" in any letter automatically had the effect that no consent is to be granted until the execution of a formal licence, then of course the letter upon which the plaintiffs here rely would be nullified in its effect, but Harman J at p44H, in my judgment, completely destroys Mr Neuberger's concept that those words are of themselves magic words. He referred to the three cases to which I have also referred and he said:

"All three go to show that there is a distinction recognised by the law between relationships, such as those between landlord and tenant, where there is an existing set of legal obligations between the parties and there is sought within those obligations a consent, and relations between strangers in law, as between prospective purchaser and prospective vendor, where there is no present tie and the parties are in mere negotiation.

I accept that there is such a distinction and I agree that one does not regard the need for a formal licence, probably under seal, as being the essential step without which there can be no effective licence."

He contrasted to that the situation in relation to vendor and purchaser transactions where there is no pre-existing relationship between the parties.

In other words, Harman J accepted, as I think it is necessary to accept in the light of many decisions to which I do not need to refer, that the words "subject to contract" do have what Mr Neuberger I think very reasonably describes as a magic, but that magic arises from the particular incantation in the particular context of particular transaction. Because the words "subject to contract" create magic it does not mean that similar magic is created by different words in different contexts.

Harman J has rejected that concept. I entirely agree with him in rejecting it. Without magic, this letter granted consent. There is no magic. It is a consent and I so declare.

From that decision the landlord now appeals.

It contends that the judge was wrong. Its case is that the heading "Subject to License [*sic*]" prevented the letter having any legal consequence. It submits that there is a general principle that:

where parties contemplate completion of a formal document in due course their negotiations are subject to the completion of that document and preclude any legal consequences from anything that precedes the execution of that document.

It is submitted that such a proposition is a general application and makes for certainty in a transaction such as this.

In support of that proposition we have been referred to cases dealing with phrases such as "subject to lease" and "subject to details" in the field of carriage of goods by sea. In those cases the same effect was given to those phrases as to the familiar phrase "subject to contract" in the field of sales of land. Thus in *Longman* v *Viscount Chelsea* (1989) 58 P&CR 189* Nourse LJ said at p193-194:

Wherever parties intend to enter into the relationship of landlord and tenant without a preliminary contract for the grant and acceptance of a lease, and their negotiations are expressed to be "subject to the completion of a lease," "subject to lease," "subject to contract" or the like, then, so long as the qualification remains in force, the relationship does not become binding on them unless and until there is an exchange of lease and counter-part, before which either party can withdraw. The parties intend to be bound at one ascertainable point of time, conventionally at a completion meeting where, if there is to be a mortgage, the mortgagee will also be represented and where all necessary formalities and exchanges will take place. With advances in modern technology it may increasingly be possible to dispense with a physical encounter between the parties. But in the absence of evidence to the contrary it is not possible to dispense with the common intention that the parties shall become bound at one ascertainable point of time.

The judgment of Nourse LJ was quoted and adopted by Dillon LJ in *Akiens* v *Salomon* (1992) 65 P&CR 364†. At p369 he said:

The parties were negotiating throughout on the basis that neither would be bound until the terms of lease had been embodied in formal lease and counterpart and these had been exchanged in the usual way. They had been negotiating on a basis that accepted, as each of the solicitors must have known, that either could withdraw from the negotiations at any point before the exchange had taken place.

In *Consolidated Pool Carriers GmbH* v *CTM CIA Transmediterranea SA: The CPC Gallia* [1994] 1 Lloyd's Rep 68 Potter J adopted a similar approach to the phrase "subject to details" when used in negotiations for the carriage of goods by sea. At p73 he quoted with approval the *dictum* of Steyn J in *Star Steamship Society* v *Beogradska Plovidba: The Junior K* [1988] 2 Lloyd's Rep 583 where he said:

I would respectfully suggest that it is in the interests of the chartering business that the Courts should recognise the efficacy of the maritime variant of the well-known "subject to contract". The expression "subject to details" enables owners and charterers to know where they are in negotiations and regulate their business accordingly. It is a device which tends to avoid disputes and the assumption of those in the shipping trade that it is effective to make clear that there is no binding agreement at that stage ought to be respected.

I do not accept that it is legitimate to extend the principle illustrated by those cases from the field of bilateral negotiations to that of a unilateral act. There can be no doubt that the distinction drawn by Harman J in the *Venetian Glass* case, to which the judge referred is a

*Editor's note: Also reported at [1989] 2 EGLR 242

†Editor's note: Also reported at [1993] 1 EGLR 101

valid one. The purpose of the suspensory condition "subject to contract" in the context of negotiations is to avoid the other side seeking prematurely to conclude a contract by the acceptance of an offer so as to give rise to unintended legal consequences. In cases requiring a unilateral act the only question is whether that act has occurred. So in this case the only question is whether the letter of May 18 1993 was a consent as required by the leases. That is a question of the construction of the letter in the light of all the surrounding circumstances.

So regarded I have no doubt that the letter does express the consent required by the leases. It will be remembered that such consent may be temporary or revocable or qualified. This letter expresses consent in the clearest terms. The consent was qualified by the stipulation for a formal licence as stated in the body of the letter. For that document it would be necessary to have the drawing numbers referred to in the concluding passage. In truth the heading "subject to licence" added little to the condition expressed in the body of the letter and could not qualify the unambiguous expression of consent it contained. If it be necessary to attribute some meaning to the heading then it might serve to emphasise the degree of formality required so that the express condition for a formal licence should not be satisfied in correspondence or by some less formal method than a licence strictly so-called. But no such document was required by the terms of the leases. In my view, it was right for the reasons he gave and I would dismiss this appeal.

Sir John Balcombe and **Beldam LJ** agreed and did not add anything.

Appeal dismissed.

Brown & Root Technology Ltd and another *v* Sun Alliance & London Assurance Co Ltd

COURT OF APPEAL

December 19 1996

Butler-Sloss and Mummery LJJ and Sir Ralph Gibson

Estates Gazette May 3 1997

[1997] 18 EG 123

Landlord and tenant — Assignment — Option to determine lease — Option ceasing to have effect after assignment of lease — Lease assigned but assignee never registered as proprietor under Land Registration Act 1925 — Whether exercise of option effective

In July 1989 the respondent landlords granted the first appellants a lease of business premises for a term of 25 years at an initial rent of £1.3m pa, subject to five-year reviews. The lease contained a break clause allowing the first appellant to determine the lease at the end of the seventh year of the term; the provisions of the clause were expressed as personal, were not capable of assignment and would cease to have effect on assignment. In August 1989 the first appellants were registered as proprietors of the lease, and have remained so registered. In 1993 the first appellants became a wholly owned subsidiary of the second appellants, and it was decided to transfer the business and assets of the first appellants to the second. In December 1993, following a licence to assign granted by the landlords and the execution of the necessary documents, the first appellants gave up possession of the premises and the second appellant commenced paying the rent to the landlords. In September 1994 the first appellants served notice under the break clause. In the court below the judge held that the lease had been assigned and that therefore the break clause could no longer be implemented. The appellants appealed.

***Held*: The appeal was allowed. On the correct construction of the break clause the first appellants were entitled to serve the notice terminating the lease. Under section 22 of the Land Registration Act 1925 the assignment of the legal estate in the lease was not effective as the first appellants were never registered as the proprietors. Assignment in the break clause meant assignment of the legal estate, which remained with the first appellants: *Gentle* v *Faulkner* applied. There was no detriment relied on by the landlords such as to make it inequitable for the first appellants to assert that there had been no assignment.**

The following cases are referred to in this report.

Gentle v *Faulkner* [1900] 2 QB 267; 16 TLR 397
R v *Edwards, ex parte Joseph* [1947] KB 392; [1947] 1 All ER 314
Rodenhurst Estates Ltd v *Barnes Ltd* [1936] 2 All ER 3
Rose, In re [1949] Ch 78; [1952] Ch 499; [1952] 1 All ER 1217; [1952] 1 TLR 1577, CA

This was an appeal by the plaintiffs, Brown & Root Technology Ltd and Brown & Root Ltd, from a decision of Judge Paul Baker QC refusing the appellants relief claimed in an originating summons to which Sun Alliance & London Assurance Co Ltd were defendants.

Kim Lewison QC (instructed by Vanderpump & Sykes, of Enfield) appeared for the appellants; Nicholas Dowding (instructed by Bates & Partners) represented the respondents.

Giving judgment at the invitation of Butler-Sloss LJ, **Mummery LJ** said: This is an appeal from an order of Judge Paul Baker QC (sitting as a judge of the High Court) on June 16 1995 in proceedings brought by lessees of office premises against the lessors for a declaration that the lease would determine in consequence of the service of a notice by the lessees pursuant to the provisions of a break clause.

The judge made a declaration that:

> on the true construction of the lease dated the 24th July 1989 made between (1) the Defendant (2) the 1st Plaintiff (then known as Brown & Root Vickers Ltd) and relating to 150 The Broadway, Wimbledon, London SW19: —
>
> (1) The 1st plaintiff was not entitled to serve the notice of determination dated 19th September 1994;
> (2) The lease will not determine pursuant to that notice.

The lessees served a notice of appeal dated July 26 1995 on the ground that the judge had erred in law in his construction of the relevant provisions of the lease. The lessors served a respondents' notice dated August 3 1995 contending that the judge's decision should be affirmed on an alternative ground of estoppel.

Background facts

As appears from the affidavit evidence sworn in support of and in opposition to the lessees' originating summons, there has never been any dispute about the relevant facts.

On July 24 1989 the lessors, Sun Alliance & London Assurance Ltd (Sun Alliance) granted a lease for a term of 25 years from June 24 1989 at an initial rent of £1.3m pa, with five-year reviews, to the first appellants, Brown & Root Technology Ltd ("Technology"), known as Brown & Root Vickers Ltd until a change of name on June 1 1993. Technology were, at the time of the grant of the lease, a partly owned subsidiary of the second appellants, Brown & Root Ltd ("B&R"). The appellants are members of a group of engineering companies. The premises consisted of offices at 150 The Broadway, Wimbledon, London SW19. Technology were registered as the proprietors of the lease on August 17 1989 and have remained so registered.

Clause 8.1 of the lease entitled the lessees to give notice to terminate the lease by serving not less than 12 months' notice expiring at the end of the seventh year. That notice would be effective to determine the term upon what was defined as "the first determination date". Technology subsequently purported to serve such a notice, but its effectiveness is disputed by reason of the provisions of clause 8.4 of the lease:

A the provisions contained in this sub clause 8.1 of this clause shall be personal and of benefit only to Brown & Root Vickers Ltd and shall not be capable of being assigned or otherwise dealt with by the said Brown & Root Vickers Ltd with the intent that the rights vested in the lessee under sub clause 8.1 shall cease to have effect upon the assignment of the lease by the lessee.

In 1993 Technology became a wholly owned subsidiary of B&R. At the end of May 1993 it was decided that B&R would take a transfer from Technology of their business and assets, including the lease. Sun Alliance granted a licence to assign. The necessary conveyancing documents in proper form were executed on December 10 1993. Technology gave up the property on that date. The records of Sun
B Alliance were altered so that B&R replaced Technology as the recorded tenants. Rent was invoiced to and paid by B&R, who had already started to pay the rent to Sun Alliance. There was an exchange of letters on December 13. B&R's letter to Sun Alliance's solicitors stated:

I refer to our previous correspondence in the matter and confirm that completion of the licence and assignment occurred on the 10th December by way of telephone exchange. Please forward to me as soon as possible the tenant's signed counterpart of the licence to assign.

Sun Alliance's solicitors' letter stated:

C We now enclose the licence duly dated 10th December last as agreed and have dated our copy of the transfer with the same date. There is accordingly no need to register the transfer as we are now closing our file.

(That was a reference to the requirement in a clause of the lease to register particulars in the lease.)

The transfer of the lease, which was registrable, has never in fact been registered at HM Land Registry in the name of B&R. Owing to unresolved stamp duty problems B&R failed to submit the transfer for registration.

On September 19 1994 Technology served a notice on Sun Alliance
D purporting to terminate the lease at the expiry of the seventh year (June 24 1996). Sun Alliance refused to accept that Technology's notice was valid and effective to terminate the lease.

Judge's decision

The issue for decision by the judge was whether there was an "assignment of the lease by the lessee" within the meaning of clause 8.4 of the lease. If there was an assignment, Technology were not entitled to serve the break notice, as the right to serve such a notice under clause 8.1 ceased to have effect upon an assignment.

The judge upheld the contention of Sun Alliance that the lease was not effectively terminated by service of the notice, as there had been an
E assignment of the lease by Technology on December 10 1993. He accordingly refused to make the declaration sought by Technology and B&R that, on the true construction of the lease, the lease would determine on June 24 1996 in accordance with clause 8.1 and made the order already referred to.

The judge's reason for finding in favour of Sun Alliance, as explained in his judgment now reported in [1996] Ch 51, was that the assignment of the lease by Technology to B&R took place when it was completed; and it was completed on December 10 1993 when Technology gave up the property subject to the lease and, having done all in their power to vest the lease in B&R, became a bare trustee for
F B&R of the legal title to the lease. B&R were the assignees of the lease, even though the legal estate had not been effectively transferred by entering the name of B&R in the Land Register as the legal proprietor.

The judge referred to the relevant provisions of the Land Registration Act 1925, in particular section 22(1), which relates to the registration of dispositions of leaseholds:

A transfer of the registered estate in the land or part thereof shall be completed by the Registrar entering on the register the transferee as proprietor of the estate transferred, but until such entry is made the transferor shall be deemed to remain the proprietor of the registered estate: and where part only of the land is transferred notice thereof shall also be noted on the register.

After reviewing the authorities, on which the judge commented that none were "directly in point", he concluded at pp67H to 68A: G

I accept, in effect, the defendant's submissions, that, where it is necessary to fix a date on which the assignment takes place, that date, in my judgment, is the date on which it is completed. The tenant/assignor gives up the property on that date; he has no control over the stamping of the transfer, or its submission to the Land Registry. He becomes a bare trustee for the assignee as regards the legal title. The landlord, a fortiori, has no such control. There is no ground for suggesting that either the assignor or the landlord can compel the assignee to submit the transfer for registration.

The judge rejected the submission of Technology and B&R which started from the provisions of the Land Registration Act and was H
summarised as follows by the judge at p65B:

. . . it is said here that the assignment means the assignment of the legal estate, and this does not occur until the name of the assignee is entered on the register. Until that happens the transferor is deemed to remain the proprietor of the lease. It is that conclusion that I am invited to draw from those authorities and the statutory language.

Although, in view of that conclusion, it was not necessary for the judge to deal with an argument on estoppel by Sun Alliance, the judge briefly dealt with that point at p68D:

I would say that on this issue I would accept that a representation had been J
made to the landlord that an assignment had taken place, and the landlord could expect that it would be registered in due course quite shortly. But I think the landlord is in a much more difficult position in regard to establishing any reliance to his detriment. What he had actually done is to change his records and collect the rent. That did not amount to very much. It was submitted to me generally that the landlord had changed his overall position and that was sufficient ,that it accepted the assignment and changed its position by accepting the assignee as the new lessee. The particular instances of rent and records and the general position that the landlord had accepted the assignee are not sufficient in themselves for reliance, and I would not have accepted that the tenant was estopped as against the landlord. I would respectfully associate myself with the way that Hoffmann J dealt with it in *Banner Industrial & Commercial Properties Ltd* v *Clark Paterson Ltd* [1990] 2 EGLR 139. The K
differences are too trivial to satisfy the test of reliance.

In their respondents' notice Sun Alliance contend that the judge's decision should be affirmed on the alternative ground that, in all the circumstances, Technology and B&R are estopped from denying that the lease was assigned by Technology to B&R prior to service of the break notice in September 1994.

Appellants' submissions

A striking feature of the rival submissions before the judge and on this appeal is that they start from different points. L

As the judge observed at p65B of his judgment, the argument for Sun Alliance starts from clause 8.4 itself and concentrates on what is meant, in the provisions of that clause, by "an assignment of the lease by the lessee". In contrast, the submissions of Technology and B&R start, as the judge observed at p61G, from the provisions of the Land Registration Act concerning effective dispositions of registered leasehold interests. Those submissions, repeated and elaborated on this appeal, are that Technology were not prevented from exercising their right to terminate the lease by service of a notice on the ground that the lease had been assigned within the meaning of clause 8.4. On the true M
construction of section 22 of the Land Registration Act an assignment of the legal estate in the registered lease was not effective, as B&R were not registered as proprietors. The legal estate in the lease therefore remained at all times vested in Technology. B&R had an equitable interest at most. Clause 8.4 only caused the right conferred by clause 8.1 to cease to have effect if there was an assignment of the lease by Technology. In the absence of registration of B&R as proprietors, there had been no assignment of the lease. Reliance was placed on *Gentle* v *Faulkner* [1900] 2 QB 267, a case on the construction of a covenant against assignment. It was submitted that, contrary to the view of the judge, the case was indistinguishable from the present case. In brief, the essential point is that the assignment

referred to in clause 8.4 must be an assignment of the legal estate, which, in the absence of registration, remained with Technology.

The absence of control by Technology over the stamping of the transfer and the submission of it to the Land Registry were irrelevant considerations. Contrary to the view of the judge, there was in fact a duty on Technology to register imposed by the terms of section 22 (1) that the transfer of a registered lease "shall be completed".

The judge's reliance on authorities relating to bequests and *inter vivos* gifts of shares was criticised. The judge cited from *In re Rose* [1949] Ch 78 and [1952] Ch 499 which, in the view of the judge, showed that:

> . . . in a case where the transfers were handed over and the relevant share certificates (that is a point that is made in both cases, that the relevant share certificates were handed over), and the registration then occurred after the critical date, death in the one case and the turning point for tax in the other, in such circumstances the documents took effect from the date of handing over and not subsequent registration.

The judge observed at p67H that the cases were "not directly in point". It is submitted on behalf of Technology and B&R that those cases are irrelevant, as they were concerned with the transfer of shares in the context of the construction of a will or of an *inter vivos* gift. This case is concerned with the legal rights of the parties to a lease: Technology are the lessees. Have they made an assignment of the lease? In that context the relevant estate is not the equitable interest, which may have passed to the assignee as a matter of contract, but the legal estate and its effect on third parties, such as the lessors, who are concerned with the location of the legal estate. It is the owners of the legal estate to whom the lessors look for the payment of rent and performance of the lessees' covenants.

On the estoppel question, it was contended that the judge was right and that Technology were not estopped, as against Sun Alliance, from asserting that there had been no assignment and that clause 8.4 did not apply. The essence of the argument of Technology and B&R on this point was that, even assuming that there was a representation on their behalf that the assignment had been completed, no estoppel would arise, because no detriment had been suffered by Sun Alliance as a result of reliance and a change of position, such as might make it unconscionable or inequitable to allow Technology and B&R to go back on their representation. The most that Sun Alliance could say in relation to change of position was in connection with the records and invoicing for rent. That was not detrimental reliance sufficient to raise an estoppel.

Submissions of Sun Alliance

Sun Alliance's answer to those submissions is short and simple: this is not a case about the technical rules relating to Land Registration. It is a question of the true construction of a particular provision (clause 8.4) in a lease and of the intention of the parties. The fact was that, as far as Technology and Sun Alliance were concerned, there had been an effective assignment of the lease. Technology had done all in their power to divest themselves of their interest in the lease. As far as Sun Alliance were concerned that had been done and they had acted accordingly. Mr Nicholas Dowding supported the decision of the judge by citation of the same authorities (and the further case of *R* v *Edwards, ex parte Joseph* [1947] KB 392), and by detailed reference to other provisions of the lease (clauses 3.15, 3.16 and 3.17, in particular).

On the estoppel point, Mr Dowding cited *Rodenhurst Estates Ltd* v *Barnes Ltd* [1936] 2 All ER 3 at p14 in support of a claim of estoppel by representation. Sun Alliance were informed that the assignment had been completed and, in reliance on that statement, had altered their records, demanded rent from B&R and treated B&R as the tenants occupying the premises in their own right. By being kept in the dark about the true position, Sun Alliance had lost the opportunity to consider their position. A similar submission was made on estoppel by convention: both parties acted on the common, but mistaken, assumption engendered by Technology and B&R that there was an assignment of the lease by Technology on December 10 1993. Sun Alliance had been misled into thinking that they had a new tenant and had acted accordingly. It would be wrong to allow Technology and B&R to depart from the common assumption which they themselves had engendered: for example, by allowing B&R to raise the absence of registration as a defence to a direct claim for rent by Sun Alliance.

Conclusion

In my judgment, Technology were entitled, on the correct construction of clause 8.4 of the lease, to serve the notice terminating the lease. I am persuaded by the excellent submissions of Mr Kim Lewison QC that the judge adopted an approach which, though sound in its analysis of the position in equity as between Technology and B&R, was inappropriate to determine the respective rights in law of Sun Alliance, as lessors, and Technology, as lessees, under the lease. I would allow the appeal and, subject to further argument on the precise form of order, would make a declaration that the lease determined on June 24 1996 in accordance with clause 8.1 of the lease.

My reasons are:

(1) Clause 8.4 had two purposes: first, that the right to terminate should be unassignable: second, that, if there was an assignment of the lease, the personal right to terminate conferred on Technology should cease to have effect. The second purpose is relevant to the resolution of this appeal, which turns on the identification of the precise event occasioning Technology's cesser of the right to invoke the break clause. The critical question is: has there been an assignment of the lease by the lessee and, if so, when did that event occur?

(2) It is common ground that there has been no transfer (and therefore no assignment) of the legal title to the lease; that, as between Technology and B&R, the equitable title to the lease was capable of passing by virtue of a specifically enforceable contract to assign the lease; that, if this were unregistered land, the assignment would occur on the execution of the deed of assignment and the conveyance of the legal estate thereby, and not on the conclusion of the contract to assign; and that, depending on the context, the passing of the equitable or beneficial interest may amount to a transfer or assignment of the property in question, even though there has been no registration of the transfer, as required by statute, to perfect the legal title. The judge referred to the cases of *In re Rose* [1949] Ch 78 and *In re Rose* [1952] Ch 499 as instances of a bequest of shares and an *inter vivos* gift of shares which took effect as between donor and donee and in accordance with donative intent before the registration of the transfers of the shares pursuant to the provisions of the Companies Act. Until registration there was no transfer so far as the company was concerned, but that did not prevent the gift from being effective as between others.

(3) This case is not a matter of beneficial ownership between parties to the transfer of the lease: the issue of assignment or no assignment affects the legal position of a third party, the lessors, who have given their licence to assign but are not a party to the transfer. As was observed by Jenkins LJ in *In re Rose* [1952] Ch 499 at p508 it is necessary to keep clear and distinct the position between the transferor and the transferee and the position of a third party. Transfer of the beneficial title is not, in this context, relevant to the legal relationship between the lessees and the lessors. The issue is not what rights Technology and B&R have against each other, but what rights Technology and Sun Alliance have against each other. That is a question of legal, not equitable, rights.

(4) As between lessors and lessees, there is binding Court of Appeal authority in *Gentle* v *Faulkner* (*supra*) for the proposition that "assignment" means, in the absence of a context showing an extended meaning, an assignment of the legal estate, and not of the beneficial interest, eg by declaration of trust of the lease. It is not a matter of intention to assign, a point highly relevant to the passing of beneficial title, but of whether a defined event has occurred. That event is not "completion", as Mr Dowding contended; it is the transfer of the legal title to the lease, so as to create the legal relationship of lessor and lessee between B&R and Sun Alliance.

A On the estoppel point, I agree with the judge that there has been no detrimental reliance by Sun Alliance such as to make it inequitable for Technology to assert, contrary to any representation or shared assumption, that there had been no assignment of the lease within the meaning of clause 8.4. For these reasons I would allow the appeal.

BUTLER-SLOSS LJ and **SIR RALPH GIBSON** agreed and did not add anything.

Appeal allowed.

B

Johnson and another *v* Davies and another

CHANCERY DIVISION

C December 5 1996

JACOB J

Estates Gazette May 10 1997

[1997] 19 EG 157

Landlord and tenant — Sureties — Tenant company sold by sureties — Purchasing shareholders giving indemnification to sureties — Insolvency — Whether co-debtors to sureties released by insolvency voluntary arrangement of one of their number — Whether accord and satisfaction

D The plaintiffs were sureties under a lease of commercial premises granted to a company of which the plaintiffs were originally the principal shareholders. In 1989 the plaintiffs sold their shares in the company to three individuals, H and the two defendants, and ceased to have any connection with the company. The company having failed to pay the rent, the plaintiffs satisfied the landlord's claim, but sought reimbursement from the defendants under the terms of the indemnification obligation in the share sale agreement. In April 1994 H entered into an insolvency voluntary arrangement ("IVA"); this followed a meeting of H's creditors at which the plaintiffs voted in favour of the IVA. In an appeal by the plaintiffs from the decision of E the deputy district judge, in proceedings by them under Ords 14 and 14A of the RSC for summary judgment and the determination of certain legal points, the defendants contended, *inter alia*, that by the IVA entered into by H, they were released from their obligation to indemnify the plaintiffs for any sums paid to the landlord.

***Held*: The appeal was allowed and judgment given to the plaintiffs. When a creditor votes for or against an IVA, he does nothing more than vote for or against a statutory binding; the IVA binds only the persons entitled to vote and an outsider can get no assistance from the terms of the IVA at all. The creditor does not enter into a contract with the debtor or with his co-creditors, and F he does not purport to touch his rights against third parties. An IVA does not have the effect of releasing perfectly solvent co-debtors. Accord and satisfaction does not apply to an IVA; a co-debtor is not released by the debtor's IVA.**

The following cases are referred to in this report.

Bradley-Hole, Re [1995] 1 WLR 1097; [1995] 4 All ER 865
Burford Midland Properties Ltd v *Marley Extrusions Ltd* [1994] BCC 604; [1995] 2 EGLR 15; [1995] 30 EG 89
Deanplan Ltd v *Mahmoud* [1993] Ch 151; [1992] 3 WLR 467; [1992] 3 All ER 945; (1992) 64 P&CR 409; [1992] 1 EGLR 79; [1992] 16 EG 100
EWA Re, A Debtor [1901] 2 KB 642; 70 LJKB 810; 85 LT 31; 49 WR 642, CA
March Park Estates v *Gunmark* [1996] 2 EGLR 38; [1996] 32 EG 75 G
Mytre Investments Ltd v *Reynolds* [1995] 3 All ER 588; [1995] 2 EGLR 40; [1995] 43 EG 131
Pybus v *Gibb* (1856) 6 E&B 902
RA Securities Ltd v *Mercantile Credit Co Ltd* [1995] 3 All ER 581; [1994] 2 EGLR 70; [1994] 44 EG 242
Watts v *Aldington* unreported December 15 1993

This was an appeal by the plaintiffs, Robert Arthur Johnson and Anne Johnson, from a decision of Deputy District Judge Radcliffe, sitting in Brighton District Registry, whereby he refused to give the plaintiffs the relief they claimed under summonses issued under Ords 14 and 14A of the RSC against the defendants, Suzanne Davies and H Nicholas Cole.

Christopher Wilson (instructed by Aldrich Crowther & Wood, of Brighton) appeared for the plaintiffs; Clifford Darton (instructed by Judge Sykes & Harrison, for the first defendant, and Edward Harte & Co, of Brighton, for the second defendant) appeared for the defendants.

Giving judgment, **JACOB J** said: This is a plaintiffs' appeal from the decision dated May 14 1996 of Deputy District Judge Radcliffe sitting in Brighton District Registry. The application was for summary judgment under the provisions of RSC Ord 14 and determination of a number of points of law pursuant to Ord 14A. Although the sum involved is comparatively small the appeal raises a question of general J importance in relation to the effect of an individual voluntary arrangement ("IVA") under the provisions of the Insolvency Act 1986. It was extremely well argued by counsel on both sides and I am grateful to them both.

The facts

The plaintiffs are two individuals. Originally they owned practically all the shares in a company. The company took a lease from a landlord, L. The plaintiffs were sureties under that lease. In 1989 the plaintiffs sold their shares in the company to three individuals, namely K the two present defendants and a Mr Hopkins. They also resigned their directorships and ceased to have any connection with the company. The plaintiffs say that by virtue of the share sale agreement the purchasers (namely the defendants and Mr Hopkins) agreed to indemnify the plaintiffs against all claims arising out of the lease. The next they heard was in 1993 when they were informed that the company had failed and that a claim was being made against them on the indemnity they had given in the lease. They had to meet that claim and accordingly seek to rely on the indemnity from the defendants contained in the share sale agreement. The lease has now come to an end, so there is no problem about any on-going liability for rent. The L plaintiffs claim from the defendants what they had to pay, after appropriate adjustments with which I am not concerned.

On April 19 1994, following the relevant procedural steps pursuant to the Insolvency Act, Mr Hopkins entered into an IVA with his creditors. The plaintiffs were given notice of the meeting and voted for the IVA. As it happens if they had voted against the proposal for an IVA it would not have got the necessary 75% in value majority called for by r5.18 of the Insolvency Rules 1986 (SI 1986 no 1925). One of the defendants, but not the other, was also given notice of the meeting. Under the IVA the creditors were to be paid 11p in the pound. Two provisions of the IVA read as follows: M

> 4. When all monies to be made available under these proposals have been realised and distributed to creditors in accordance with the terms herein I [ie Mr Hopkins] will be released from any further liability to them relating to claims in respect of which they were entitled to participate in this Voluntary Arrangement . . .
>
> 15. When all assets to be made available under these proposals have been realised and distributed to the creditors in accordance with the terms herein, I will be released from any further liability to them relating to claims in respect of which they were entitled to participate in this Voluntary Arrangement.

In summarising these facts I have omitted what is now agreed to be an immaterial complication, namely that at one point L granted a

A concurrent lease to another company. This was the subject of a point taken by the defendants before the deputy district judge. He held against the defendants on this point. Mr Clifford Darton, for the defendants, sensibly concluded that the judge was right on this point and abandoned it. He likewise abandoned another point about whether or not the 1989 agreement was a guarantee or an indemnity.

The issues

That leaves two issues to be determined. First, have the plaintiffs proved to the standard of Ord 14 that the defendants gave the indemnity relied upon and second, does the IVA with Mr Hopkins B release the defendants?

Was there an indemnity?

The problem is this: the plaintiffs are unable to produce the original of the 1989 share sale agreement. The best they can do is rely upon the following matters: that a draft, signed by the first plaintiff for both plaintiffs was in the possession of the defendants, that the shares in the company were undoubtedly paid for and transferred to the defendants and Mr Hopkins, that the defendants and Mr Hopkins took over the company, and that the plaintiff's solicitor recalls that the defendants C did sign the agreement. Further there is contemporaneous correspondence supporting the existence of a contract. The solicitor deposes that he has been unable to find the signed agreement. The defendants do not say there was no contract or that the draft contract produced in fact contains the terms to which they agreed. They just say they do not remember.

The deputy district judge took the view that the evidence of the plaintiffs' solicitor was not a firm enough foundation for Ord 14 and that his recollection may be successfully challenged on cross-examination. However, the defendants are in this difficulty: there plainly was an agreement of some sort. They are unable to suggest any agreement other than that signed by the plaintiffs. It is wholly D improbable that the plaintiffs would not have taken an indemnity. And Mr Darton fairly and sensibly accepted that the evidence (apart from a possible challenge to the solicitor's recollection) would not change at trial. I think in the circumstances there is no realistic prospect of the plaintiffs being unable to prove that there was indeed an indemnity.

The effect of the IVA

This is a pure point of law. The defendants say that the effect of the IVA with Mr Hopkins is to extinguish their liability to the plaintiffs. They say first that any co-debtor is released by operation of law when E the debt is taken into account in a debtor's IVA. Recognising that this would include the case where the creditor had voted against the IVA (producing the monstrous result that a creditor would, against his will, lose a perfectly good claim against a solvent co-debtor) they say alternatively that at least this is the case where the creditor votes for the IVA.

The solution to the problem depends, on one view, on the correct juridical analysis of an IVA. Is the arrangement to be regarded as "quasi-contractual" (as Mr Darton put it) or is it an arrangement whose effect is simply provided for in the statute? Section 260(2) provides that:

F The approved voluntary arrangement—

(*a*) . . .

(*b*) binds every person who in accordance with the rules had notice of, and was entitled to vote at, that meeting (whether or not he was present or represented at it) as if he were a party to the arrangement.

In *RA Securities Ltd* v *Mercantile Credit Co Ltd* [1995] 3 All ER 581* I had to consider the effect of the corresponding parallel provision relating to company voluntary arrangements in the 1986 Act, namely section 5. I said at p585:

*Editor's note: Also reported at [1994] 2 EGLR 70

Turning back to s 5, does "binds every person" have any effect outside the G voluntary arrangement? I think not. The effect of the binding is solely as between the parties bound, those entitled to vote, whether they did nor not. An outsider, such as Tl, can get no assistance from the terms of the voluntary arrangement as such. Of course if something is actually done as a result of the arrangement (eg property transferred, or as might have but did not happen here, surrender of a lease) then an outsider can rely upon that. But his right to rely upon it must result from an actual act done, not the arrangement.

I remain of that view. It is unnecessary to spell out of the statute anything more than that. I see no reason to think that when a creditor votes for or against an IVA he is doing anything more than just voting for or against a statutory binding. He is not entering into a contract H with the debtor or with his co-creditors and he is not purporting (unless the document expressly says so) to touch his rights against third parties. Either the IVA is approved, when it has the legal consequences set out in the Act, or it is not, in which case the debtor will probably go into bankruptcy. And I remain of the view that an IVA (or a company voluntary arrangement) does not have the effect of releasing perfectly solvent co-debtors. The Act does not say it does and I do not see why in principle it should have that effect. As I said in *RA Securities*, at p584:

The purpose of voluntary arrangements is to enable a company (or individual, J for which case provision is made in Pt VIII) to come to a composition with creditors so that the more drastic step of liquidation or bankruptcy can be avoided, if possible. It is better to keep the show on the road than close it down even if creditors have to accept less than their nominal (but not achievable) entitlement. The whole scheme is not for the benefit of solvent parties who happen to owe debts also owed by the debtor. It would in my judgment be unfair if a solvent debtor escaped liability as a sidewind of the voluntary arrangement system.

Mr Darton accepted that an IVA did not constitute a contract. But, he said, that did not matter. It still operated to extinguish the debt. He relied on the principle that the release of one joint debtor (or covenantor) will release the other joint debtors (or covenantors) unless K the creditor's (covenantee's) rights against the latter are reserved, referring to *Re EWA, A Debtor* [1901] 2 KB 642 and *Deanplan Ltd* v *Mahmoud* [1993] Ch 151*. In *EWA* Collins LJ said:

If this legal consequence had been pointed out to the bank at the time, they might have said that that was not what they intended, but that is a factor common probably to all cases in which a release is given to one of two joint debtors. The person giving it may not realise the full legal consequences of it as regards the release of the co-debtor; but that is not, in my opinion, a sufficient ground for reading into the document something that is not expressed in it; and unless you find in it something qualifying the general words, it appears to me that the legal consequences of the general words of L discharge must follow, notwithstanding that those consequences may go beyond what the person giving the document would have intended if they had been pointed out to him at the time, and he had had an opportunity of addressing his mind to them.

And in *Deanplan* three reasons for the rule were identified by Judge Paul Baker QC:

(1) There is only one obligation;

(2) The covenantee cannot have both the promised performance and some other performance which he agrees to accept;

(3) Unless the co-covenantors are also released they could claim a M right of contribution or indemnity from the covenantor who entered into the release.

Mr Darton said that even if the arrangement was purely statutory, so reason (2) did not apply, reasons (1) and (3) still did. As to reason (1) I do not think that an IVA operates to destroy the obligation as such. It merely binds the creditor (by virtue of section 260(2)) to an alternative. In practice that alternative is merely payment of a sum less than that which is due. If he agreed to accept less directly with the debtor, it is worth noticing that his agreement would have no consideration and

*Editor's note: Also reported at [1992] 1 EGLR 79

A would not be binding — it could hardly then amount to accord and satisfaction releasing a co-debtor.

As to reason (3) there is some force in what Mr Darton says. If a co-debtor is not released then he may upset it by claiming contribution from the debtor. This may pose a problem where he is not bound by the IVA (eg because he was not given notice or had only a contingent claim). He could go against the debtor and possibly upset the IVA, though the possibility of a further IVA varying the first would exist. Mr Darton adds this would be particularly unfair on the co-debtor if, as Rimer J held in *Re Bradley-Hole* [1995] 4 All ER 865, the assets the subject of an IVA did not vest in a trustee in bankruptcy but remained
B in a trust of which the trustee was the nominee of the IVA. (I was told that this decision has been questioned on the basis that normally a debtor cannot put his assets beyond the reach of creditors by putting them into a trust. I do not think it is necessary for me to consider it further here.)

But although there is some force in what Mr Darton says, I do not think it is sufficient to hold that the accord and satisfaction rule applies to an IVA. Most significantly the Act does not say it does. Further if the rule did apply no creditor, sensibly advised, would ever vote for an IVA when he had a solvent co-debtor to go against. This in practice could frustrate many a proposed IVA. It is true that the co-debtor may
C attack the debtor if not himself bound, but in practice there may be little point. And the creditor will often get more recovery from the debtor via an IVA (thus indirectly benefiting the co-debtor) than via a bankruptcy. Here, for instance, Mr Hopkins paid 11p in the £, which, on the figures in the IVA proposal, is more than a bankruptcy would have been likely to yield.

Mr Darton suggested that the creditor could protect his right against co-debtors by insisting upon an express reservation of his right at the meeting. If he did not get it, then he could be protected by an appeal to the court under section 262. Implicit in this suggestion is that the court would never allow a creditor to lose his right against a co-debtor,
D which is not an explicit limitation on the court's powers. In practice the consequence of such an insistence would be that many an IVA would fail at the outset.

Before turning to the second way the plaintiffs put their case I should touch upon certain further matters concerning my decision in *RA*. First, I commented on a passage in the 1992 edition of Andrews and Millett's *Law of Guarantees*. I said at pp586-587:

> This does not adopt the *Chitty* view, suggesting as a better analysis, the notion that a voluntary arrangement is in effect a statutory variation of pre-existing contractual rights. It follows, the authors argue, that the common law effect of discharge of a surety by a consensual discharge of the debtor (without
> E reservation of a right against the surety) will apply in a voluntary arrangement. This analysis to my mind turns a statutory binding into a consent and is in error.

I remain of that view. The authors, in their 2nd ed (1995) say at p244:

> In *RA* Jacob J disapproved this argument on the ground that a statutory binding cannot be equated with a consent. However it is difficult to see why the fact that the satisfaction of the debt under the voluntary arrangement is statutory rather than consensual on a creditor by creditor basis should make any difference to the result. Indeed in *Finch* v *Jukes*, it was held that a surety for
> F performance of obligations under a building contract was discharged when the obligations of the principal were altered by an Act of Parliament.

The authors then, in a footnote, cite *Pybus* v *Gibb* (1856) 6 E&B 902 and observe that neither this case nor *Finch* were cited to me in *RA*. That is true, but I do not think either case relevant. In both what happened was that the defendant had agreed to act as surety for obligation X, but, by Act of Parliament, that obligation had been varied to obligation Y. The court asked in each case whether the position of the surety was so altered that he must be treated as being discharged — the inquiry being the extent of the difference between X and Y.

Next it may be observed that *RA* has not been questioned in three subsequent cases, namely *Burford Midland Properties Ltd* v *Marley Extrusions Ltd* [1994] BCC 604*, *Mytre Investments Ltd* v *Reynolds* G
[1995] 3 All ER 588† and *March Park Estates* v *Gunmark* [1996] 32 EG 75‡. Now it is true that in *Burford* Judge Cooke QC, sitting as a High Court judge, took the contractual route, holding that on the construction of the IVA concerned, third parties were not absolved. And in *March Park* Lightman J allowed for the possibility that an IVA may affect third party rights. He said:

> A voluntary arrangement may expressly or by necessary implication regulate the rights *inter se* of a creditor of the company and of third parties liable for the same debt: consider [*Burford*]. Exceptional circumstances may justify this course. But if this is to be the case, the intention to regulate such rights must be made plain on the face of the proposal. Of course notice of the meeting must H
> also be given to those affected. In this way, those affected will know what is at stake, can make representations and vote at the meeting accordingly, and if necessary can within the statutory 28-day period apply to the court under section 6 of the Act (as to which see below) to revoke or suspend the approval of the meeting if this provision in the proposal unfairly prejudices their interests. But in the absence of either of these statutory safeguards to the interests of those affected being complied with, any arrangement can only affect the company alone, leaving unaffected the rights of the creditor against third parties. The voluntary arrangement may statutorily absolve the company from liability without absolving or releasing from liability any other party and will ordinarily be construed as reserving all rights of creditors against other parties: see *RA Securities* (*supra*) and [*Mytre*]. J

I do not think this in any way conflicts with my view that an IVA is in itself a mere statutory binding. If it expressly or by necessary implication deals with the position as between a creditor and a co-debtor that binding may have an effect. If the co-debtor is a party to the IVA he may be bound directly by statute. If he is not a party then he cannot take advantage directly of the IVA. In some cases there may be machinery which could operate to prevent the creditor from going against the third party, for instance a plea of estoppel by convention (which I have not investigated) or, by the third party turning on the debtor who, by virtue of the IVA, would be able to prevent the creditor from proceeding against the third party. K

The significance of the conclusion that an IVA operates to bind the parties to it merely by statute as compared by contract is that it is for the statute to spell out the consequences. It does not absolve solvent third party co-debtors: it merely binds the parties to the IVA. This IVA only releases Mr Hopkins from liability to the plaintiffs and does not seek to deal with the position between the plaintiffs and third parties. It is not a necessary implication that it goes any further.

Further, however, even if an IVA is to be regarded as something like a contract it is a question of construction as to whether it releases third parties. This IVA does not say it does. Mr Wilson, for the plaintiffs, drew my attention to the unreported Court of Appeal case of L
Watts v *Aldington*, unreported December 15 1993. A large sum by way of damages and costs had been awarded to Lord Aldington against Count Tolstoy and a Mr Watts. Lord Aldington entered into a settlement with Mr Watts whereby a much lesser sum was accepted (along with other matters providing a consideration). A term of the settlement provided:

> That Lord Aldington undertakes to accept the said sum in full and final settlement of the judgment and orders referred to above and any liability howsoever arising before today's date which could involve any payment directly or indirectly to Lord Aldington. M

Even though that clause was a good deal more explicit than the clause here, the Court of Appeal held that it did not absolve Mr Watt's co-debtor, Count Tolstoy. Morritt J's solution to the problem — that the agreement was merely an agreement not to sue (the well-settled escape route from the rule in *EWA*) — was rejected. But each of the lord justices held that the rule in *EWA* (which they accepted, reluctantly in the case of Steyn LJ, was binding) did not apply where

*Editor's note: Also reported at [1995] 2 EGLR 15
†Editor's note: Also reported at [1995] 2 EGLR 40
‡Editor's note: Also reported at [1996] 2 EGLR 38

the presumed intention of the parties was not to absolve the co-debtor, such intention to be learned from the language used and the surrounding circumstances. They were willing, in the circumstances of that case, to imply such a term. I would do likewise in this case. In particular, since this is not one of actual contract, one must focus on what was the intention of the creditors. Clearly if they voted against the IVA they could not be consenting to a release of the co-debtor. If they voted for the IVA I see no reason why, unless that was expressly what was involved in what they were voting for, they should be held to have voted for a release of a co-debtor. On the contrary, if the voter was asked "are you voting for a release" the answer would clearly be a testy "no".

In the result the IVA defence fails. I am not sorry. I think it would be absurd if the two sureties were better off because there was a third surety who failed (although contributing something to payment of the debt), than if they had stood alone. Similarly it would be very bizarre if the creditors were worse off because they had a further surety in addition to the two defendants. The actual result means that instead of shouldering a third of the liability, each of the defendants will have to take on something like 45%. That does not seem particularly unfair.

Accordingly the appeal succeeds. I will hear counsel.

Appeal allowed.

Devonshire Reid Properties Ltd *v* Trenaman

LANDS TRIBUNAL

October 17 1995

JUDGE RICH QC, sitting as a member of the Lands Tribunal

Estates Gazette May 17 1997

[1997] 20 EG 148

Landlord and tenant — Whether landlord entitled to create additional flat in roof space — Whether creation of flat would breach landlords' covenant to repair roof — Whether implied term that landlords covenanted not to create further leases

On December 17 1994 the leasehold valuation tribunal for the south east assessed the reversionary value on four 999-year leases of a building of four flats in the sum of £600. The freeholder appealed contending, *inter alia*, that the tribunal had failed to take into account the value of the right to create a fifth flat in the roof space of the building for which it had planning permission. The leases each contained provisions enabling the freeholder to recover the costs of repair and maintenance of the building, including the roof. Each lessee paying a proportion of the costs which the rateable value of his flat bore to the total rateable value of all the flats.

***Decision*: The appeal was dismissed. The landlords' proposed work of creating a flat in the roof space would constitute a breach of the covenant to maintain, repair, redecorate and renew the roof. The intention of the parties was obvious from the form in which the leases had been drawn. The costs of maintenance were to be shared equally among the four lessees of the flats in the building, with the necessary implication that the parties did not contemplate any beneficial occupation of the remainder of the premises either by the lessors or anybody holding under them. Care had been taken to ensure mutuality of covenants in the leases; it was necessary in order that there should be mutuality among them that they should be restricted to the number recited at the beginning of each lease, namely four. It was necessary to imply a covenant into the leases to carry out the intention of the parties that the lessors would not create within the building more than four leases. That covenant would be breached if the works contemplated were carried out.**

The following case is referred to in this report.

Gange v *Lockwood* (1860) 2 F&F 115

Michael Daiches (instructed by Morgan Bruce) appeared for the appellants; Jonathan Small (instructed by Donne Mileham & Haddock, of Brighton) represented the respondent.

Giving his decision, **JUDGE RICH QC** said: There falls for decision the determination of a preliminary point which I have directed should be tried in the course of an appeal against the decision of a leasehold valuation tribunal given on December 17 1994 whereby the tribunal assessed the value of the reversionary interest on some leases of four flats at 1 Wilbury Gardens, Hove, East Sussex, in the sum of £600. The appellants maintained before the tribunal and would have wished to maintain before this tribunal that the sums capitalised in order to arrive at the figure of £600 did not take into account the totality of the receipts to which the reversioner was entitled under the leases and would also have wished to urge that the years' purchase adopted by the leasehold valuation tribunal was insufficient. But the main thrust of the appellants' case before the leasehold valuation tribunal and the main subject-matter of their appeal was that the leasehold valuation tribunal had included nothing in the price of the reversion for the value which the appellants maintain attaches to the reversion by reason of the opportunity to create a fifth flat within the roof space of the building in which the four flats, whose reversions are thus valued, is contained.

I came to make a direction for a hearing of a preliminary issue in the following circumstances. The appellants applied on June 25 1995 for an order that the hearing which had been fixed for the 27th and 28th should be adjourned. I do not think it appropriate to rehearse the reasons for my concluding that that was not an application which I ought to grant, but it did transpire in the course of the application that the main reason for it was that the appellants wished to reconsider the evidence which they would wish to place before this tribunal in regard to the valuation of this right which they claimed to create a fifth flat in the roof space of the building. That right was said by the respondent to be precluded by reason of the provisions of the leases. It was in those circumstances that I thought that the fears of the appellants of not having a fair hearing of the real issue in the case could be overcome if I determined first whether or not they did indeed, as a matter of the true construction of the leases, have a right to create a fifth flat in the roof space. If they did have such right then it would be appropriate to give them some time in which to ensure that proper evidence as to its proper value was placed before this tribunal; but if they did not have such right their appeal would be concerned with only a derisory issue concerning whether the £600 was a sufficient valuation of the ground rent or whether the sum should be varied by some modest number of pounds.

In those circumstances, upon the appellants undertaking that they would not pursue those other matters of appeal and would submit therefore to the appeal being dismissed without further hearing if the preliminary issue was determined against them, I ordered the hearing of this preliminary issue as to whether on a true construction of the leases subject to which the appellants hold the freehold, the development for which planning permission was granted on October 19 1991, as shown in plans placed by Mr Fridell before the leasehold valuation tribunal, can lawfully be carried out. Those plans, it is agreed, show the variation of the existing roof of the building by the insertion into it of dormer windows; apart from that they show no construction works upon any part of the building which has been included in the demise to any of the four existing tenants of the block and it is agreed that either there has been retained, as undemised property of the reversioner, a means of access to the roof space or the reversioner has an easement providing such access, and, for the purposes at least of the consideration of this issue, I will assume that that access will be sufficient, although I suspect it may be barely sufficient and difficult in order to enable the proposed works to be

A carried out. It is also agreed that the roof and the roof space itself are not included within any demise out of the freehold of which the appellants are a reversionary owner.

The building in which these works are proposed is let in four leases all for a term of 999 years from Christmas 1963 at ground rents. The terms of each of the four leases are identical at least so far as is here material. Each contains a covenant by the lessors at clause 5(4)(a) that:

> subject to contribution as is herein provided (and that is a reference to Clause 7 to which I will return shortly) and except to such extent as the lessee or the tenant or any other part of the building shall be liable in respect thereof under the terms of this lease or any other lease, the lessor will maintain, repair,
> B redecorate and renew (a) the structure of the building and in particular the roof.

Clause 7 provides that the reversioner can recover from each of the lessees a proportion equal to that ratio which the rateable value of each flat has to the total rateable value of the four flats, of the money actually spent "on maintaining and managing the building during the year". If the proposed works are to any extent works of maintaining or managing the building, then the cost of them to that extent is recoverable from the four lessees. That appears to me to be a proper matter to note in seeking to construe the words in clause 5(4)(a) by which the lessors bind themselves to maintain, repair, redecorate and renew. I take maintaining in clause 7 to embrace the four kinds of
C work described in clause 5(4)(a) as maintain, repair, redecorate and renew.

I have been referred by Mr Jonathan Small to a recent work concerning dilapidations in leasehold premises. There reference is usefully made to a general principle that an obligation to repair contains a duty not to destroy. The learned authors of this work say that a covenant to repair imposes not only a positive duty to repair but also a negative obligation not to destroy the demised premises in whole or in part. Thus, they say, in *Gange* v *Lockwood* (1860) 2 F&F 115 where the tenant of certain rooms in a house opened two doorways in the walls between the house and an adjoining house, Willis J
D directed the jury in the following terms:

> A covenant to repair, uphold and maintain or keep in good repair raises a duty not to destroy the demised premises and the pulling them down, wholly or partly, is a breach of such covenant.

I take that citation from the judgment of Willis J to be accurately recorded in this text book and I accept and adopt the direction of Willis J as thus formulated. In my judgment, a covenant to repair and maintain raises a duty not to destroy wholly or partly and a proposal to do so is a breach of such covenant. The proposed works involve such an intention and would, unless there were some grounds for construing
E this particular covenant differently, constitute therefore a breach of the covenant to maintain, repair, redecorate and renew.

Mr Michael Daiches urges on me entirely properly that the direction of Willis J appears to have been given in a case where a tenant's covenant was being construed and at least in terms of a restriction upon the ability of the lessors to deal with that which is their own property it may not be so appropriate to construe the lessors as having a negative obligation not to destroy as opposed to the positive duty to repair; but there are two reasons for rejecting that submission as to the construction of this covenant. The first is that the lessees in the particular circumstances of this covenant have a significantly greater
F interest in the ability of the lessors to deal with the premises retained by the lessors than has the lessors, first, because they have the residue of a 999-year term in respect of their flats and so far as the part not demised is concerned it appears to be, for reasons which I will explain in the context of the second reason for answering the preliminary issue in favour of the respondent which I will give, the sole concern and function of the parts retained by the lessor in practice is to serve the parts which have actually been demised. But, second, the reason for rejecting the distinction which Mr Daiches has made is that which I have already referred to, that if the covenant permits under the head of maintenance the cutting into the roof and the changing in its form, then the cost of undertaking such work can be charged to the lessees, thus they have the greater interest in the scope of the work than does the lessor. The four words which I have read must all be construed as G
having some separate force so far as possible even though the word "repair" may well include and overlap with the obligation to redecorate and renew and even to maintain. The obligation to maintain in its natural meaning appears to me to mean to keep in its existing form. The obligation to renew may contemplate not merely the replacement of that which exists but even its replacement by different materials but reading the two together I think that an obligation to maintain and renew permits only such renewal as is consistent with the obligation to maintain, that is to say the maintenance of the existing form even if not the existing material of that which is to be maintained, repaired, redecorated and renewed. That which is proposed in respect of the roof H
departs from that restriction imposed by the covenant, and therefore could not be carried out, in my judgment, other than in breach of the covenant, and for that reason I would answer the preliminary point in the sense which I have indicated, that is to say as contended for by the respondent.

But issue was joined before me at greater length and possibly directing attention to a more difficult issue as to whether or not the proposed works are not in any case precluded because the proper construction of these leases is that the lessor is under an obligation not to create within the building more than the four flats which are referred to in the lease. In order to address this point I must make some further J
references to the provisions of the leases. They recite at recital 2 that the lessors have previously granted leases of, or intends hereafter to grant leases of, the three flats in the building other than the premises hereby demised. The building as I have indicated is the whole building including those parts which are not the subject of any of the four demises including, as I have said, the roof space. The recital goes on:

> And the lessor has in each lease imposed and intends in every future lease to impose, the restrictions set forth in the First Schedule hereto to the intent that any owner or lessees for the time being of any flat in the building may be able to enforce the observance of the said restrictions by the lessees of the other flats. K

The restrictions referred to in the first schedule are restrictive covenants as to the proper conduct of the occupiers of each flat in their own premises such as to avoid disturbance to their neighbours and restrictions as to the use of the common parts in a way which may cause annoyance to their neighbours. Mr Daiches accepts, and indeed contends, that that recital prefaces the creation of a letting scheme among the four proposed and actual lessees such as to create mutual enforceability of those restrictive covenants.

Mr Small for the respondent has sought to persuade me that the recital is properly to be construed as itself being a covenant to create the number of leases there referred to and no more than that number. I L
do not think that it is material to consider whether properly it is a covenant to create the number of leases there referred to but I think that I would accept that submission. I do not think that it justifies a conclusion that it is of itself a covenant to create no more than that number of leases and even if it were, I do not think that it will answer the question which I have to determine under the preliminary issue, namely whether or not the leases prevent the building work which is proposed by the appellants.

The reason why I have doubted as to whether or not these recitals can properly be construed as themselves creating covenants, is because there is in fact an express covenant within the body of each lease at M
clause 5(2) whereby the lessors covenanted that the lessors will require every person to whom they shall hereafter grant a lease of any flat in the building to covenant to observe the restrictions set forth in the first schedule hereto. That covenant I construe, and I think Mr Daiches accepted, was a covenant to obtain from each lessee a covenant not merely with the lessors but also with the other lessees as contemplated by the recital and in the form that the lessee in each lease actually covenants in regard to the first schedule at clause 3 of the lease, which provides that:

> The lessee hereby covenants with the lessor and with the owners and lessees of the other flats comprised in the building, that the lessee and the persons

deriving title under the lessee will at all times hereafter observe the restrictions set forth in the First Schedule hereto.

Thus, the express obligation accepted by the lessor was to ensure the creation of that mutual arrangement of covenants with a right of each lessee to enforce against the other, contemplated by the recital and imposed by clause 3 of the lease.

Mr Daiches accepted and indeed contended that it is of the nature of a letting scheme that its scope has to be determined at its outset. The clauses to which I have referred create thus a mutuality of covenants and Mr Daiches accepted and indeed contended that the requirement that the scope of a letting scheme shall from its beginning be certain and determined precluded, at least on the form of words adopted in the recital, there being added to that system of mutual covenants the owners of any fifth flat that might be created in the building. Thus, the position is achieved under the leases that if the lessors have, as they contend, a right to create within that part not demised to any of the four tenants a further dwelling, that dwelling will of necessity be outside the letting scheme contemplated by each of the four leases and established by each of the four leases. If that is what is the effect of the leases then so be it, the lessors have the rights that they contend for and that is what Mr Daiches seeks to persuade me to conclude.

It would appear to me that that conclusion must follow unless I am satisfied that it is necessary to imply into each of these four leases a covenant that the lessors will not create within the building more than the four leases contemplated. I am entitled to imply such a covenant only if it is necessary to do so in order to give business efficacy to the leases and in order to carry out the intention of the parties to the leases as objectively ascertained from the leases.

In my judgment, the intention of the parties to the leases is obvious from the form in which those leases have been drawn. Two things are absolutely apparent; the first is that the costs of maintenance in so far as the obligations fall upon the lessor should be shared in their entirety among the four lessees of the flats within the building. That carries with it the necessary implication that the parties did not contemplate any beneficial occupation of the remainder of the premises either by the lessor or by anybody holding under them. The second thing that, in my judgment, is clearly apparent is that great care has been taken to ensure a mutuality of covenants among the lessees. It is necessary in order that there shall be mutuality among them all that they should be restricted to the number recited at the beginning of each lease, namely four. Mr Daiches has suggested that the intention was for a letting scheme for four flats and that therefore it is not necessary to assume an intention not to have a fifth flat, but I think that the correct construction of the intention of the parties was that it was the intention to create a letting scheme between all the lessees occupying the building and that therefore it is necessary to restrict them to the number recited upon the determination of the scope of the letting scheme, namely the four referred to in each lease. For that reason I am satisfied that it is a necessary implied term in order to carry out the intention of the parties objectively ascertained in regard to these leases, to imply a covenant that the lessors will not create within the building more than the four leases referred to in each of the four leases. That covenant would likewise be breached if the works contemplated by the appellants were carried out and for that reason also I will therefore answer the preliminary issue in favour of the respondent.

Appeal dismissed.

Berrycroft Management Co Ltd and others *v* Sinclair Gardens Investments (Kensington) Ltd

COURT OF APPEAL

July 29 1996

BELDAM, HENRY and HUTCHISON LJJ

Estates Gazette May 31 1997

[1997] 22 EG 141

Landlord and tenant — Covenant to insure — Landlord and Tenant Act 1987 — Whether implied term that sum charged by insurer should not be unreasonable — Whether cost of insurance exceeds what is reasonable

Following the development of some 14 purpose-built blocks of flats, the flats were sold on long leases and the reversions were sold to the respondent landlord. The original developers created the appellant management companies and the leases imposed a requirement that the management companies should insure the buildings for such sums and through such companies as the landlord may direct. The lessees covenanted both with the reversioner and the management company to pay as part of the management charge the appropriate proportion of the insurance cost. When the present landlords acquired the reversions, it required the management companies to insure with Commercial Union through an associated company, C Ltd. The premium rate was considerably higher than the rate obtained by the original developers. The landlord paid the premiums and sought to recover the sums from the management companies. The agents for the management companies contended that the previous insurance arrangements were acceptable to the lessees and there was no justification for a change of insurer. The applications of the management companies and an individual tenant, seeking an order under sections 19 and 30A of the Landlord and Tenant Act 1985 that expenditure on insurance were not relevant costs reasonably incurred, was dismissed in the court below. The management companies appealed.

Held: **The appeals were dismissed. (1) A term cannot be implied into the covenants between the management companies and the landlord, and between the lessees and the appropriate management company, that the sum charged by the nominated insurer should not be unreasonable, or that a tenant should not be required to pay a substantially higher sum than he could himself arrange with an insurance office of repute. The right of the landlord to nominate the company and the agency for insurance was unqualified. The insurance rates proposed by Commercial Union were not other than market rates and the fact that the management companies could have secured lower rates was beside the point. (2) Notwithstanding the covenants by the lessees to pay service charges to the management companies, the landlord was landlord for the purposes of the Landlord and Tenant Act 1985. However although the insurance charges are incurred by the lessees as part of the service charges, they are not to be regarded as incurred by or on behalf of the landlord and are therefore not "relevant costs" within the meaning of section 18 of the 1985 Act. The trial judge had decided that the amounts quoted by Commercial Union were neither unreasonable nor excessive. *Quarere* whether, if the terms of the leases had given the right to nominate the insurer, the fact that the management companies undertook to insure on behalf of the landlords and the lessees would take the provisions of the leases outside para 8 of the Schedule to the 1985 Act.**

The following cases are referred to in this report.

Bandar Property Holdings Ltd v *Darwen (JS) (Successors) Ltd* [1968] 2 All ER 305; (1968) 19 P&CR 785

A *Finchbourne Ltd* v *Rodrigues* [1976] 3 All ER 581; [1976] 1 EGLR 51; (1976) 238 EG 717, CA
Havenridge Ltd v *Boston Dyers Ltd* [1994] 2 EGLR 73; [1994] 49 EG 111
Tredegar (Viscount) v *Harwood* [1929] AC 72

This was an appeal by Berrycroft Management Co Ltd and others from a decision of Judge Paul Baker QC, who had dismissed the appellants' applications for the determination of certain questions arising under leases of flats in buildings owned by the respondent, Sinclair Gardens Investments (Kensington) Ltd.

David Halpern (instructed by Taylor Vinter, of Cambridge) appeared for the appellants; David Neuberger QC and Nicholas B Dowding (instructed by Malthouse Chevalier) represented the respondent.

Giving judgment, **BELDAM LJ** said: These 14 appeals require the court to consider the extent of the landlord's right under the relevant provisions of the applicants' leases to direct the insurance of property held by them and the liability of the tenant to pay charges for such insurance. Further to consider the effect on those rights and liabilities of the Landlord and Tenant Act 1985, as amended by the Landlord and Tenant Act 1987.

The appellants in each case are a management company and a C representative tenant from an estate consisting of a block of flats and its surrounding grounds.

Facts

The applications relate to the rights and obligations of parties to leases of flats in purpose built blocks.

When a block of flats or other large development was being built, completion of the sales of the individual properties did not normally take place until the properties were finished. Depending upon the size of the development and the number of properties, there was bound to be a period when only some of the properties were sold and occupied D and others were waiting to be finished. There are usually common parts of the development which have to be maintained by the developer until the development is complete. It is a common practice to provide for the maintenance of the common parts and the unoccupied portion of the development by the creation of a management company. Until the development is complete the management company is controlled by the developer, who provides its directors. As the individual properties are sold, the purchasers are required to subscribe for a share in the management company. When the development is complete, control of the management company is handed over to the purchasers who in effect then become responsible for the common parts of the development and the other obligations E which until then had been carried out by the management company on behalf of the developers. Where the properties comprising the development are held on long leases as in the present case, the ordinary relationship between the developer who is the lessor or landlord and the purchaser who is the lessee or tenant is complicated by the introduction of the management company to perform some of the covenants which would normally be entered into either by the landlord or by the tenant.

After completing a development of leasehold properties, the developer as a rule assigns the reversion to an investment company. It is not difficult to see that even with the most carefully drafted leases F the introduction of a management company charged with the performance of certain of the covenants could give rise to questions of some difficulty in adapting the rules of privity of contract and estate between the landlord, the management company and the tenant.

In the present case the applicants did not raise any questions of privity and the questions for the court depend upon the construction of the terms of the leases and upon the application to the relationship of landlord and tenant in these circumstances of the Landlord and Tenant Act 1985 as amended by the Landlord and Tenant Act 1987.

Eleven of the 14 developments to which the applications related, were carried out by Rialto Homes plc (Rialto) and three by Tarmac Ltd (Tarmac). On December 31 1992 Rialto sold the reversions of 10 of the developments to the respondent, Sinclair Gardens Investments G (Kensington) Ltd (Sinclair). The reversions to the development at the Vale, Cricklewood, the 11th of Rialto's developments, were sold to Sinclair on February 10 1993. Tarmac sold the reversions of its three developments to Sinclair on January 22 1993. In every case there was a management company which by the time of the sale of the reversions was under the control of the tenants of the developments.

Many tenants' associations and management companies employ agents to carry out the functions entrusted to the management company and one company which specialises in running management companies is Corporate Property Management Ltd (CPM), which administers 450 residents' management companies with a H responsibility for some 17,000 flats and houses. CPM in fact was engaged by 12 of the management companies in the present applications. The questions which have arisen in these applications arise from the requirements in the leases that the management companies shall insure the buildings comprised by the flats.

During the construction of the development, insurance of the building was arranged and maintained by the developers. Rialto insured the buildings in their developments with Sun Alliance, transacting the business through brokers, CE Heath. As the construction of each development was completed and the flats became ready for sale, the proposal for insurance was made by Rialto in the J joint names of itself and the relevant management company. The period of insurance was a calendar year and Rialto paid a block premium and invoiced each management company for the appropriate proportion. CPM as managing agents then discharged the amount of the invoices to Rialto and collected a proportionate part from the individual tenants as part of the service charges. Tarmac for its part insured its developments with Norwich Union through brokers called Markfield.

When Sinclair acquired the reversions it decided to make fresh arrangements for insurance of the buildings. It required the management companies to insure with Commercial Union Assurance K plc through an associated company, Cullenglow Ltd, which, under the name of Princess Insurance Agencies, carried on insurance business as an exclusive agent for Commercial Union. As from January 1 1993, Sinclair insured 11 blocks with Commercial Union. The amount of the cover was based on the same valuation as in the policies arranged by Rialto, but the premium rate was considerably higher. In one instance the insurance was arranged from February 10 1993. Similar arrangements were made in the case of the three Tarmac developments. In fact on January 18 1993 CPM had undertaken negotiations on behalf of the management companies of the three Tarmac blocks and had arranged insurance.

In February 1993 Sinclair gave notice to CPM and to the other L management companies not represented by CPM that it required the managers in accordance with the several covenants under their leases to insure the property with the nominated insurer of the landlord, Commercial Union Assurance plc, and that such insurance be placed through the agency of the landlord, Princess Insurance Agencies, who would place and keep on foot the insurance cover with Commercial Union. This letter was not well received by CPM who took the stand that as the original block buildings insurances had been maintained and the cover was perfectly acceptable to the flat owners and their mortgagees there was no justification for any change in insurer. M Sinclair maintained that the management companies were not entitled to arrange insurance cover in any way other than in accordance with the terms of the leases and the notice which had been served. As Sinclair had already paid out premiums for insurance on behalf of the management companies, it contended it was entitled to recover the sums so paid.

The premiums required by Commercial Union were considerably higher than the premiums which the management companies had been paying and to the tenants it seemed unreasonable that they should be required to pay more since the management companies were already insuring with insurers of repute. The difference in rates per £1,000 insured were substantial. So, for example, the development at

A Willenhall Drive, Hayes, Middlesex, carried out by Tarmac and managed by Berrycroft Management Co Ltd was insured by Tarmac with Norwich Union for a sum insured of £3m at a premium of £0.50 per £1,000, but the premium charged by Commercial Union through Cullenglow was £1 per £1,000 for a sum insured of £2.4m. The development in Harrow Road, Wembley, carried out by Rialto was insured through Sun Alliance at a rate of £0.50 per £1,000 but for the same cover Commercial Union required a premium of £1.05 per £1,000.

Although the total premium could thereby be increased by as much as £1,000 or £2,000, when the increase was distributed pro rata among
B the tenants, the increases considered in individual cases appeared less dramatic.

In these applications an individual tenant and a management company from each of the developments applied for an order against Sinclair claiming a declaration that the premium in respect of insurance was excessive and irrecoverable. Alternatively, they sought an order under the provisions of sections 19 and 30(A) of the Landlord and Tenant Act 1985 and para 8 of the Schedule thereto that Sinclair were not entitled to recover expenditure upon insurance effected in respect of the property of the applicants and that the amounts incurred by Sinclair or at its direction in respect of insurance premiums payable
C to Commercial Union were not relevant costs reasonably incurred within the meaning of sections 18 and 19 of the Landlord and Tenant Act 1985 in determining the amount of service charges payable pursuant to the lease. Consequential orders were sought under the provisions of the Landlord and Tenant Act 1985.

On December 22 1994 Judge Paul Baker QC gave judgment in favour of Sinclair rejecting the claims of the tenants and management companies. He held that under the terms of the leases Sinclair as assignee of the reversion and landlord was entitled to require the management companies to insure through Cullenglow with Commercial Union. He held that no term was to be implied in the
D leases that the landlord could only require the management companies to pay a reasonable premium, but in any event he held that the premium required by Commercial Union was not unreasonable or excessive in the circumstances. Further, after considering extensive evidence about the rates of premium available in the market and the reasons why the premiums quoted by Sun Alliance and Norwich Union were substantially lower than those of Commercial Union, he held that the costs of insurance effected through Cullenglow with Commercial Union were not unreasonably incurred and he declared that expenditure incurred by Sinclair in respect of insurance premiums when CPM had refused to accept the claims of Sinclair was
E recoverable from the management companies. He refused the applicants relief under the Landlord and Tenant Act 1985.

Terms of the leases

I now come to the relevant terms of the leases. I propose to take as typical the terms of a lease entered into on March 17 1992 between the landlord, Rialto, the management company Harrow Road Sudbury Management Co Ltd, and the tenants, Michael Legg and Tracey Thornton. For a purchase price of £64,100 the tenants were granted a lease for a term of 99 years from September 1 1991. The initial yearly rent was £60, but that increased after the first 33 years. The lease was
F of a ground-floor flat, 35 Marian Court, 665 Harrow Road, Wembley, together with a private parking place.

As previously indicated, the terms of the leases are complicated by the fact that, in addition to the covenants entered into by the landlord and tenant, obligations are imposed and covenants are given by a management company which has no estate in the land. At first sight the covenants given by the management company, referred to in the leases as "the company", have the characteristics of personal covenants creating privity of contract but no privity of estate. In relation to insurance of the building under the fifth schedule of the lease the company covenants with the landlord and with the tenant (provided that the tenant shall have paid the management charge):

> (3) To insure at all times of the term . . . and to keep insured the building G
> and such other areas as the landlord decides to insure for such sum as the landlord thinks fit against loss or damage by fire and such other normal household risks in some insurance office of repute and if directed by the landlord through a company nominated by the landlord and if required through any agency of the landlord in that company (such insurance to be effected in the joint names of the landlord, the company and the tenant for their respective interests).

By clause 5 of the lease the tenant covenants with the landlord and its successors in title and *as a separate covenant with the company* in the terms set out in the fourth schedule hereto. The fourth schedule entitled "Covenants by the tenant in relation to payment of a H
management charge", provides a covenant by the tenant:

> To pay to the company on 1st January in each year or such other date as may from time to time be specified: —
> The sum reasonably specified by the company in advance and on account of the management charge for the then current management expenditure year.

"Management charge" is defined in the lease to mean:

> The annual contribution payable by the tenant under the provisions of this lease which shall be such proportion of the management expenditure in respect of the matters specified in Part 1 of the 5th schedule hereto . . . as shall be certified annually as being just and equitable by the accountants and auditors to the company . . . J

Thus the effect of clause 5 of the lease, the fourth schedule and the fifth schedule is that the tenant covenants directly with the landlord and its successors in title to pay to the company the management charges as defined and as set out in Part 1 of the fifth schedule and thus to pay a proportion of the insurance premium effected in the joint names and therefore on behalf of the landlord, the company and the tenant.

The first question as identified by the judge in his judgment was:

> What are the rights and obligations of the parties as regards insurance under the terms of the leases? K

He later stated that the issues which had been canvassed before him as arising from the terms of the leases were:

> (a) The extent of the obligation of the parties to the leases to insure;
> (b) The circumstances in which the landlord can exercise his rights of nomination;
> (c) The limits, if any, on the terms of insurance required by the nominated company and agent;
> (d) The circumstances under which the default provisions can be brought into play;
> (e) The effect of insurance by the landlords on the terms of the leases. L

The judge concluded that the obligation to insure and keep the blocks of flats insured was placed on the management company and the tenant's obligation was only to pay a due proportion of the premium. He held that under para 3 of the fifth schedule the landlord had the right to nominate the insurance company and agency, if any, of the landlord in that company through which the insurance was to be placed. He held that an assignee of the landlord was entitled to make his own nomination, that where the landlord exercised its right of nomination the role of the management company in practice was reduced to paying the premiums required by the agent and passing
them on to the lessees as part of the service charges. Although the M
default provisions were in some respects not aptly worded, they were sufficiently widely worded to empower the landlord to enforce a covenant even though it merely required expenditure, as for example of the insurance premium. He thus held in circumstances in which managing agents acting on behalf of two of the management companies had refused to insure through Cullenglow, there had been default by the management companies of their covenants to insure under the fifth schedule and Sinclair was entitled under the default provisions to insure and recover the cost. With this analysis of the effect of the provisions of the lease, I do not in general disagree but I find difficulty in interpreting para 3 of the fifth schedule as giving the

A landlord the right to nominate the insurance office of repute as such for it seems to me that the language used draws a distinction between the insurance office of repute and the company nominated by the landlord and the words "that company" which follow must refer back to the company nominated by the landlord. If the landlord was requiring the right to nominate the insurance office, it would have been simple to refer to "an insurance office of repute nominated by the landlord".

In the case of Cullenglow which had an exclusive agency for Commercial Union, the effect of para 3 of the fifth schedule effectively resulted in nomination of the insurer but it seems to me that arose from the special facts of the case rather than from a construction of the
B covenant.

First question

The management company and the tenants argued that a term was to be implied into the covenants between the company and the landlord and between the tenant and the company that the sum charged by the nominated insurer should not be unreasonable. Alternatively that the tenant could not be required to pay a substantially higher sum than he could himself arrange with an insurance office of repute.

For the appellants it was argued that the court should follow the reasoning of this court in *Finchbourne Ltd* v *Rodrigues* [1976] 3 All
C ER 581*. The lessors in that case were the owners of a block of flats and the defendant was tenant under a lease which provided that the tenant was required to contribute a yearly sum equal to 4% of the amount which the landlord should from time to time have expended in maintaining and running the block of flats as a whole.

The amount of the contribution was to be ascertained by an independent agent. The landlord brought an action against the tenant to recover the contribution which had been certified by his own firm and on a preliminary issue the judge held that no valid certificate had been issued in accordance with the terms of the lease. In addition, however, he held that the landlord could only recover for costs and outgoings
D such sums as were fair and reasonable.

On appeal on the question whether it was necessary having regard to the terms of that lease to imply a term that the costs should be fair and reasonable, Cairns LJ said at p587:

> Taking the strictest of tests on that matter, I am of that opinion such an implication must be made here. It cannot be supposed the plaintiffs were entitled to be as extravagant as they chose in the standards of repair, the appointment of porters etc . . . In my opinion, the parties cannot have intended that the landlords should have an unfettered discretion to adopt the highest conceivable standard and to charge the tenant with it.

E Browne LJ agreed that the term must be implied in the lease and the costs should be fair and reasonable.

The recovery by the landlord of the charges for insurance in that case were but part of the charges, 4% of which the tenant was required to pay, and no express attention was paid by the court to the recovery of charges for insurance in isolation. It is significant that the court referred expressly to the other outgoings which would make up the sum payable by the tenant.

For Sinclair reliance was placed on *Bandar Property Holdings* Ltd v *JS Darwen (Successors) Ltd* [1968] 2 All ER 305 in which Roskill J declined to imply a term that the landlord should act reasonably in
F placing insurance, the cost of which he was entitled to recover from the tenant. The term which he was asked to imply was that the lessor should place the insurance so as not to impose an unnecessarily heavy burden on the lessee since the amount of such insurance was to be repaid to the lessor by the lessee.

At p307G Roskill J said:

> It is axiomatic that a court will not imply a term which has not been expressed merely because, had the parties thought of the possibility of expressing that term it would have been reasonable for them to have done so. Before a term which has not been expressed can be implied it has got to be shown not only that it would be reasonable to make that implication, but that it is necessary in G
> order to make the contract work that such a term should be implied.

Sinclair also placed reliance on the more recent decision of this court in *Havenridge Ltd* v *Boston Dyers Ltd* [1994] 49 EG 111*. The terms of the clause in that case required the lessor to insure with some insurance office of repute and the tenant to pay as insurance rent all sums properly expended.

Again the court declined to imply a term of reasonableness or to construe "properly" as if it meant "reasonably". At p75 Evans LJ said:

> But the question remains, what limit should be placed upon the tenant's obligation to indemnify the landlord, so as to preclude an exorbitant claim or H
> what Cairns LJ described in *Finchbourne Ltd* as an "outlandish" result? In my judgment, it matters not whether the limit is expressed as the meaning or true construction of "properly pay" or as an implied restriction on the landlord's right of recovery under clause 2(6)(a). The limitation, in my judgment, can best be expressed by saying that the landlord cannot recover in excess of the premium which he has paid and agreed to pay in the ordinary course of business as between the insurer and himself. If the transaction was arranged otherwise than in the normal course of business, for whatever reason, then it can be said that the premium was not properly paid, having regard to the commercial nature of the leases in question, or, equally, it can be supposed that both parties would have agreed with the officious bystander that the tenant should not be liable for a premium which had not been arranged in that way.
>
> If this is the correct test, as in my judgment it is, then the fact that the J
> landlord might have obtained a lower premium elsewhere does not prevent him from recovering the premium which he has paid. Nor does it permit the tenant to defend the claim by showing what other insurers might have charged. Nor is it necessary for the landlord to approach more than one insurer, or to "shop around". If he approaches only one insurer, being one insurer of "repute", and a premium is negotiated and paid in the normal course of business as between them, reflecting the insurer's usual rate for business of that kind then, in my judgment, the landlord is entitled to succeed. The safeguard for the tenant is that, if the rate appears to be high in comparison with other rates that are available in the insurance markets at the time, then the landlord can be called upon to prove that there was no special feature of the
> transaction which took it outside the normal course of business. K

Mr David Neuberger QC for Sinclair also drew our attention to the decision of the House of Lords in *Viscount Tredegar* v *Harwood* [1929] AC 72. The lease in that case was for 99 years. It contained a covenant by the lessee to insure against fire during the term in the joint names of the lessee and lessor in the Law Fire Office or in some other responsible insurance office to be approved by the lessor.

The tenant assignee of the lease discontinued the policy with the Law Fire Office and took out a policy with Atlas. The lessor refused to approve the Atlas company on the ground that for the purposes of estate management he required that all houses upon his estate should
be insured in the same office. The lessee refused to insure with the L
Law Fire Office and on an action brought by the lessor to obtain the decision of the court on the construction of the covenant the House held that upon the true construction of the covenant the primary obligation of the respondent was to insure in the Law Fire Office and that the appellant had an absolute right to withhold his approval of an alternative office without entering upon reasons.

In the Court of Appeal it had been held that there should be implied into the lease the term that the consent of the lessor to any responsible insurance office selected by the lessee was not to be unreasonably withheld.

In the course of his speech Lord Shaw of Dunfermline said: M

> In the present case no such restriction or condition upon the right of approval or disapproval by the landlord is imposed. The case is the simple one of one sound office being named, with the alternative given of another responsible office approved by the lessor. If the lessee will not insure in the Law Fire Office nor in any other responsible office which the lessor has approved, the covenant is broken. It is a condition precedent to the alternative being resorted to that the lessor's consent has been given to the other office suggested. I am of opinion in these circumstances that this condition precedent cannot be removed or held as satisfied, because in the opinion of a court of law

---

*Editor's note: Also reported at [1976] 1 EGLR 51

*Editor's note: Also reported at [1994] 2 EGLR 73

A the lessor's refusal was unreasonable. The Court's opinion upon that subject cannot be allowed to supply the want of the lessor's consent in fact.

Later he said:

> Were it necessary to inquire, then the facts of this case would show that there were sound business reasons behind this unfettered simplicity. The lessor was the owner of many hundreds of houses erected under building leases, and the development of a considerable estate required strict attention to, among other things, the insurance of the buildings erected. The difficulty in such cases is to check the failure in renewals. With properties so numerous, and insured, it may be, in a dozen different offices, this point of management becomes very complex, and, in this case, it is pointed out that the cost involved becomes
> B considerable. By a simple working arrangement with one office, such as the Law Fire, simplicity and accuracy are promptly secured.

It is elementary that the specific covenants in each case have to be construed in the context of the lease in which they are contained. Construction of a clause in one lease in a particular way is no guide to the construction of a clause in another lease couched in different terms and set in a different context. However the observations to which I have drawn attention are, it seems to me, of assistance as a guide to the approach which the court should take to the construction of the covenants in this case.

C I can see no reason to imply any term into the covenant by the management company that there should be a restriction, however defined, on the landlord's right to nominate either the company or the agency through which the insurance is to be placed. The management company and the tenant are protected by the qualification that the insurance office must be of repute. The right of the landlord to nominate the company and the agency is unqualified.

No question arises in the present case of the insurance having been arranged otherwise than in the normal course of business nor otherwise than through an office of repute. Nor on the judge's findings were the rates proposed to be charged by Commercial Union other than market rates. That they were higher than the management company could have
D secured is beside the point. The landlord was not required to give reasons for requiring the insurance to be effected through his agency or with the company chosen. In fact he did give reasons which bore close resemblance to those suggested by Lord Shaw in the case cited.

I am of opinion that the judge was correct on the facts of this case to refuse to imply into the covenants in question any additional terms.

Second question

What was the effect on the rights and liabilities of the landlord and tenant under the leases of the provisions of the Landlord and Tenant
E Act 1985 as amended by the Landlord and Tenant Act 1987?

Sections 18 to 30 refer to restrictions on "Service Charges". The relevant provisions are:

> 18. (1) In the following provisions of this Act "service charge" means an amount payable by a tenant of a [dwelling] as part of or in addition to the rent —
>
> (a) which is payable, directly or indirectly, for services . . . or insurance or the landlord's cost of management, and
>
> (b) the whole or part of which varies or may vary according to the relevant costs.
>
> (2) The relevant costs are the costs or estimated costs incurred or to be
> F incurred by or on behalf of the landlord, or a superior landlord, in connection with the matters for which the service charge is payable . . .
>
> 19. (1) Relevant costs shall be taken into account in determining the amount of a service charge payable for a period —
>
> (a) only to the extent that they are reasonably incurred, and
>
> (b) . . .
>
> and the amount payable shall be limited accordingly.
>
> . . .
>
> (4) A county court may make a declaration —
>
> (a) that any such costs were or were not reasonably incurred,
>
> . . .
>
> (c) that any such amount is or is not reasonable, notwithstanding that no other relief is sought in the proceedings . . .

Section 30: G

> In the provisions of this Act relating to service charges —
> "landlord" includes any person who has a right to enforce payment of a service charge;"

Section 30A provides:

> The Schedule to this Act (which confers on tenants certain rights with respect to the insurance of their dwellings) shall have effect.

Although the company under the leases in the present case is not a recognised tenants association as defined by section 29, in a subsequent section of the Act, as amended, section 30(B)(8), a landlord is specifically defined for the purposes of that section in these H terms:

> "landlord", in relation to a recognised tenants association, means the immediate landlord of the tenants represented by the association or a person who has a right to enforce payment of service charges payable by any of those tenants;

Upon the second question the judge said of the statutory provisions:

> The application of these provisions to the cases in hand gives rise to the following questions.
>
> (a) Does the cost of insurance included in the service charge payable by each of the tenants exceed what is reasonable and if so by what amount? J
>
> (b) Do the tenancies of the flats require the tenant to insure the dwelling with an insurer nominated by the landlord?
>
> (c) If so, is the insurance available from the nominated insurer unsatisfactory in any respect or are the premiums excessive?

He said:

> It will be convenient to take question (b) first. If the answer is in the negative, only question (a) will remain . . . None of the leases of the flats with which I am concerned require the tenant to insure at all much less to insure with a nominated insurer. On the contrary it is the management company which covenants with each tenant to insure his flat as part of the block and that company is or may be required to insure with an insurer nominated by the landlord. It is true that the management company is owned by all the tenants in K a block but that, in my judgment, is insufficient to bring the leases either individually or collectively within the scope of para 8. In the first place the management company in relation to the tenant for the purposes of the schedule is a landlord not a tenant. Second, what is necessary to bring this paragraph into operation is a requirement of the tenant to insure the dwelling ie the flat of which he is the tenant. The insurances in these cases are of entire blocks for obvious reasons. The Act of 1987 extended the scope of the service charge provisions of the Act of 1985 to leases of individual dwellings whereas they had previously been confined to leases of flats forming part of a building. The paragraph was introduced to give a remedy where dwellings which unlike flats might frequently be insured separately. Accordingly I now go to the remaining question, question (a). L

Having regard to the view which I have formed that the judge's conclusion under his question (a) cannot reasonably be challenged, it is perhaps unnecessary to consider the basis of his answer to question (b). Nevertheless I do not think it is possible in the circumstances of this case and having regard to the terms of the Act to regard the company as the tenants' landlord to the exclusion of the landlord as defined in the lease.

In the first place the definition in section 30 includes any person who has a right to enforce payment of a service charge. As the tenant has covenanted directly with the landlord and his assignees under the provisions of clause 5 of the lease to pay the service charge to the M company, it seems to me the landlord does have the right to enforce payment of the service charge and so is a landlord for the purpose of the provisions of the Act. Under section 18 a service charge means an amount payable by a tenant for insurance but it still has to be an amount the whole or part of which varies or may vary according to the relevant costs and relevant costs are defined to be costs incurred or to be incurred by or on behalf of the landlord.

Although the management company effects insurance in the joint names of the landlord and the tenant, in so far as the insurance charges are incurred by the tenant as part of the service charge, they would not, in my view, be regarded as incurred by or on behalf of the landlord and not therefore "relevant costs". Even if the landlord directs that the insurance should be effected through a company or agency nominated by the landlord, the management company will have paid in accordance with its obligations contained in a contract to which the tenant is a party and there would be no basis for reducing the sum recoverable from the tenant.

Turning to the requirements of the Schedule, para 8 provides:

> (1) This paragraph applies to a tenancy of a dwelling which requires the tenant to insure the dwelling with an insurer nominated by the landlord.
>
> (2) Where, on an application made by the tenant under any such tenancy, the court is satisfied —
>
> (a) that the insurance which is available from the nominated insurer for insuring the tenants' dwelling is unsatisfactory in any respect, or
>
> (b) that the premiums payable in respect of any such insurance are excessive, the court may make either an order requiring the landlord to nominate such other insurer as is specified in the order or an order requiring him to nominate another insurer . . .

In para (1) of the Schedule "landlord" is defined:

> . . . in relation to a tenant by whom a service charge is payable which includes an amount payable directly or indirectly for insurance, includes any person who has a right to enforce payment of that service charge.

For the reasons already expressed, I consider that the landlord and his assigns as specified in the lease come within this definition. In the present case however it seems to me arguable that it is the landlord's nomination of the company and of the particular agency which results in the tenant having to pay for the cost of insuring with Commercial Union rather than the terms of the tenancy. If the terms of para 3 of schedule 5 of the lease had given the landlord the right to nominate the insurer, I would like to reserve an opinion whether the fact that the management company undertook to insure on behalf of the landlord and the tenant would take the provisions of the lease outside para 8 of the Schedule.

As previously indicated, the judge concluded after a thorough review of the evidence that the quotations for insurance from Commercial Union were competitive compared with a quotation obtainable by a single management company acting alone and that the active and responsible management of the agency nominated by Sinclair was, taken overall, beneficial to the lessees. Consequently the costs of the insurance were not unreasonably incurred. The tenor of his judgment is that the amounts quoted by Commercial Union were neither unreasonable nor excessive and were negotiated in the ordinary course of business. In my view, therefore, these appeals must be dismissed.

HENRY and HUTCHISON LJJ agreed and did not add anything.

Appeal allowed with costs.

Co-operative Insurance Society Ltd *v* Argyll Stores (Holdings) Ltd

HOUSE OF LORDS

May 21 1997

LORD BROWNE-WILKINSON, LORD SLYNN of HADLEY, LORD HOFFMANN, LORD HOPE of CRAIGHEAD and LORD CLYDE

Estates Gazette June 7 1997

[1997] 23 EG 141

Landlord and tenant — Keep open covenant — Anchor store — Whether tenant can be ordered to comply with covenant — Whether mandatory injunction appropriate remedy to enjoin person to carry on business

The appellant tenants decided in May 1995 to close their Safeway supermarket, in a shopping centre owned by the respondent landlords, because it had been losing money. The supermarket premises was the largest unit in the shopping centre. This was a breach of clause 4(19) of the lease which contained a positive obligation to keep the premises open for retail trade during the usual hours of business. The tenants admitted the breach and, in an action commenced by the landlords, consented to an order for damages to be assessed. Reversing the trial judge, the Court of Appeal ordered that the covenant be specifically performed making a final injunction: [1996] 1 EGLR 71. The tenants appealed.

***Held*: The appeal was allowed. The trial judge was correct in referring to the settled practice of the courts that orders which require a defendant to run a business would not be made. The heavy-handed nature of the enforcement mechanism, the quasi-criminal procedure of punishment for contempt, to compel the running of a business is perhaps the paradigm case of its disadvantage. A distinction must be made between court orders which require a defendant to carry on an activity, such as running a business, and orders which require him to achieve a result. This explains why the courts have ordered specific performance of building contracts and repairing covenants. The settled practice is based upon sound sense and no criticism could be made of the trial judge who had refused to make an order to compel the tenants to continue trading.**

***Per Lord Clyde*: Reserved his opinion on the approach which might be adopted by civilian systems.**

The following cases are referred to in this report.

Attorney-General v *Colchester Corporation* [1955] 2 QB 207; [1955] 2 WLR 913; [1955] 2 All ER 124; (1955) 53 LGR 415
Braddon Towers Ltd v *International Stores Ltd* [1987] 1 EGLR 209
Dowty Boulton Paul Ltd v *Wolverhampton Corporation* [1971] 1 WLR 204; [1971] 2 All ER 277; (1971) 69 LGR 192
Giles (CH) & Co Ltd v *Morris* [1972] 1 WLR 307; [1972] 1 All ER 960
Greene v *West Cheshire Railway Co* (1871) LR 13 Eq 44
Isenberg v *East India House Estate Co Ltd* (1863) 3 De GJ&Sm 263
Jeune v *Queens Cross Properties Ltd* [1974] 1 Ch 97; [1973] 3 WLR 378; [1973] 3 All ER 97; (1973) 26 P&CR 98; [1973] EGD 976; 228 EG 143
Redland Bricks Ltd v *Morris* [1970] AC 652; [1969] 2 WLR 1437; [1969] 2 All ER 576, HL
Shiloh Spinners Ltd v *Harding* [1973] AC 691; [1973] 2 WLR 28; [1973] 1 All ER 90; (1973) 25 P&CR 48, HL
Tito v *Waddell (No 2)* [1977] Ch 106; [1977] 2 WLR 496; [1977] 3 All ER 129
Williamson (JC) Ltd v *Lukey and Mulholland* (1931) 45 CLR 282
Wilson v *Northampton & Banbury Junction Railway Co* (1874) LR 9 Ch App 279
Wolverhampton Corporation v *Emmons* [1901] 1 KB 515

This was an appeal by the tenants, Argyll Stores (Holdings) Ltd, from a decision of the Court of Appeal ([1996] 1 EGLR 71) which had allowed an appeal by the landlords, Co-operative Insurance Society Ltd, from an order of Judge Maddocks QC, sitting as a judge of the Chancery Division, who had given summary judgment to the landlords but refused an order requiring the tenants to comply with a keep open covenant.

Jonathan Gaunt QC and Martin Seaward (instructed by Titmus Sainer Dechert) appeared for the tenants; Peter Smith QC and Paul Chaisty (instructed by BKJ Lewis, of Manchester) represented the landlords.

Giving his opinion, **Lord Hoffmann** said:

1. *The issue*

In 1955 Lord Goddard CJ said:

> No authority has been quoted to show that an injunction will be granted enjoining a person to carry on a business, nor can I think that one ever would be, certainly not where the business is a losing concern.

Attorney-General v *Colchester Corporation* [1955] 2 QB 207, at p217). In this case his prediction has been falsified. The appellants Argyll Stores (Holdings) Ltd ("Argyll") decided in May 1995 to close their Safeway supermarket in the Hillsborough Shopping Centre in Sheffield because it was losing money. This was a breach of a covenant in their lease, which contained in clause 4(19) a positive obligation to keep the premises open for retail trade during the usual hours of business. Argyll admitted the breach and, in an action by the landlord, the Co-operative Insurance Society ("CIS"), consented to an order for damages to be assessed. But the Court of Appeal*, reversing the trial judge, ordered that the covenant be specifically performed. It made a final injunction ordering Argyll to trade on the premises during the remainder of the term (which will expire on August 3 2014) or until an earlier subletting or assignment. The Court of Appeal suspended its order for three months to allow time for Argyll to complete an assignment which by that time had been agreed. After a short agreed extension, the lease was assigned with the landlord's consent. In fact, therefore, the injunction never took effect. The appeal to your lordships is substantially about costs. But the issue remains of great importance to landlords and tenants under other commercial leases.

2. *The facts*

A decree of specific performance is of course a discretionary remedy and the question for your lordships is whether the Court of Appeal was entitled to set aside the exercise of the judge's discretion. There are well established principles which govern the exercise of the discretion but these, like all equitable principles, are flexible and adaptable to achieve the ends of equity, which is, as Lord Selborne LC once remarked, to "do more perfect and complete justice" than would be the result of leaving the parties to their remedies at common law: *Wilson* v *Northampton & Banbury Junction Railway Co* (1874) LR 9 Ch App 279, at p284. Much therefore depends upon the facts of the particular case and I shall begin by describing these in more detail.

The Hillsborough Shopping Centre consists of about 25 shops. Safeway was by far the largest shop and the greatest attraction. Its presence was a commercial benefit to the smaller shops nearby. The lease was for a term of 35 years from August 4 1979 with five-year rent reviews. Clause 4(12)(a) contained a negative covenant as to the user of the premises:

> Not to use or suffer to be used the demised premises other than as a retail store for the sale of food groceries provisions and goods normally sold from time to time by a retail grocer food supermarkets and food superstores . . .

Clause 4(19) was the positive covenant enforced in this case:

> To keep the demised premises open for retail trade during the usual hours of business in the locality and the display windows properly dressed in a suitable manner in keeping with a good class parade of shops.

Competition in the supermarket business is fierce and in 1994 Argyll undertook a major review of its business and decided to reduce the scale of its operations. The management was to be reorganised, 27 loss-making or less profitable supermarkets closed and thousands of employees made redundant. Hillsborough, which according to Argyll's management accounts had made a loss of about £70,000 in the previous year, was on the list for closure. For administrative reasons as well as to avoid the demoralising effect of successive closure announcements, it was decided to close all the supermarkets at once and try to negotiate the disposal of their sites as a package. In early April 1995 Argyll announced that Hillsborough and the other supermarkets would close on May 6 1995.

As soon as CIS heard of the impending closure, it protested. On April 12 1995 Mr Wightman, the regional surveyor of the investment department, wrote to Mr Jefferies of Safeway:

> Whilst obviously there is little point in trying to influence your corporate decision with regard to the closure of this unit I am dismayed at the short period of notice given which will undoubtedly have immediate impact on the Centre and all the other tenants trading therein.

He drew attention to the covenant to keep open, invited Safeway to agree to continue trading until a suitable assignee had been found, offered to negotiate a temporary rent concession and asked for a reply by return of post.

Unfortunately he received no answer. Mr Jefferies had himself fallen victim to the reorganisation; he had been made redundant. No one else dealt with the letter. On Saturday May 6 1995 the supermarket closed and over the next two weeks its fittings were stripped out. On May 22 1995 CIS issued a writ claiming specific performance of the covenant to keep open and damages.

3. *The trial*

CIS issued a summons for judgment under RSC Ord 14 but when the matter came before Judge Maddocks, sitting as a judge of the High Court, on August 1 1995 it was agreed that, since the material facts were not in dispute, the hearing should be treated as the trial of the action. The learned judge was therefore invited by CIS to make a final order that the covenant be performed for the remainder of the lease or until an earlier assignment or subletting. By this time Argyll were already in serious negotiation with another supermarket chain for an assignment but no contract had yet been signed.

The judge refused to order specific performance. He said that there was on the authorities a settled practice that orders which would require a defendant to run a business would not be made. He was not content, however, merely to follow authority. He gave reasons why he thought that specific performance would be inappropriate. Two such reasons were by way of justification for the general practice. An order to carry on a business, as opposed to an order to perform a "single and well defined act", was difficult to enforce by the sanction of committal. And where a business was being run at a loss, specific relief would be "too far reaching and beyond the scope of control which the court should seek to impose". The other two related to the particular case. A resumption of business would be expensive (refitting the shop was estimated to cost over £1m) and although Argyll had knowingly acted in breach of covenant, it had done so "in the light of the settled practice of the court to award damages". Finally, while the assessment of damages might be difficult, it was the kind of exercise which the courts had done in the past.

4. *The settled practice*

There is no dispute about the existence of the settled practice to which the judge referred. It sufficient for this purpose to refer to *Braddon Towers Ltd* v *International Stores Ltd* [1987] 1 EGLR 209, at p213, where Slade J said:

---

*Editor's note: Also reported at [1996] 1 EGLR 71

A Whether or not this may be properly described as a rule of law, I do not doubt that for many years practitioners have advised their clients that it is the settled and invariable practice of this court never to grant mandatory injunctions requiring persons to carry on business.

But the practice has never, so far as I know, been examined by this House and it is open to the respondents to say that it rests upon inadequate grounds or that it has been too inflexibly applied.

Specific performance is traditionally regarded in English law as an exceptional remedy, as opposed to the common law damages to which a successful plaintiff is entitled as of right. There may have been some element of later rationalisation of an untidier history, but by the 19th
B century it was orthodox doctrine that the power to decree specific performance was part of the discretionary jurisdiction of the Court of Chancery to do justice in cases in which the remedies available at common law were inadequate. This is the basis of the general principle that specific performance will not be ordered when damages are an adequate remedy. By contrast, in countries with legal systems based on civil law, such as France, Germany and Scotland, the plaintiff is *prima facie* entitled to specific performance. The cases in which he is confined to a claim for damages are regarded as the exceptions. In practice, however, there is less difference between common law and civilian systems than these general statements might lead one to
C suppose. The principles upon which English judges exercise the discretion to grant specific performance are reasonably well settled and depend upon a number of considerations, mostly of a practical nature, which are of very general application. I have made no investigation of civilian systems, but *a priori* I would expect that judges take much the same matters into account in deciding whether specific performance would be inappropriate in a particular case.

The practice of not ordering a defendant to carry on a business is not entirely dependent upon damages being an adequate remedy. In *Dowty Boulton Paul Ltd* v *Wolverhampton Corporation* [1971] 1 WLR 204, Pennycuick V-C refused to order the corporation to maintain an
D airfield as a going concern because (at p211), "It is very well established that the court will not order specific performance of an obligation to carry on a business". He added (at p212): "It is unnecessary in the circumstances to discuss whether damages would be an adequate remedy to the company". Thus the reasons which underlie the established practice may justify a refusal of specific performance even when damages are not an adequate remedy.

The most frequent reason given in the cases for declining to order someone to carry on a business is that it would require constant supervision by the court. In *JC Williamson Ltd* v *Lukey and Mulholland* (1931) 45 CLR 282, at pp297-298, Dixon J said flatly:
E "Specific performance is inapplicable when the continued supervision of the Court is necessary in order to ensure the fulfilment of the contract".

There has, I think, been some misunderstanding about what is meant by continued superintendence. It may at first sight suggest that the judge (or some other officer of the court) would literally have to supervise the execution of the order. In *CH Giles & Co Ltd* v *Morris* [1972] 1 WLR 307, at p318C Megarry J said that "difficulties of constant superintendence" were a "narrow consideration" because:

> there is normally no question of the court having to send its officers to
> F supervise the performance of the order . . . Performance . . . is normally secured by the realisation of the person enjoined that he is liable to be punished for contempt if evidence of his disobedience to the order is put before the court; . . .

This is, of course, true but does not really meet the point. The judges who have said that the need for constant supervision was an objection to such orders were no doubt well aware that supervision would in practice take the form of rulings by the court, on application made by the parties, as to whether there had been a breach of the order. It is the possibility of the court having to give an indefinite series of such rulings in order to ensure the execution of the order which has been regarded as undesirable.

Why should this be so? A principal reason is that, as Megarry J G
pointed out in the passage to which I have referred, the only means available to the court to enforce its order is the quasi-criminal procedure of punishment for contempt. This is a powerful weapon; so powerful, in fact, as often to be unsuitable as an instrument for adjudicating upon the disputes which may arise over whether a business is being run in accordance with the terms of the court's order. The heavy-handed nature of the enforcement mechanism is a consideration which may go to the exercise of the court's discretion in other cases as well, but its use to compel the running of a business is perhaps the paradigm case of its disadvantages and it is in this context that I shall discuss them. H

The prospect of committal or even a fine, with the damage to commercial reputation which will be caused by a finding of contempt of court, is likely to have at least two undesirable consequences. First, the defendant, who *ex hypothesi* did not think that it was in his economic interest to run the business at all, now has to make decisions under a sword of Damocles which may descend if the way the business is run does not conform to the terms of the order. This is, as one might say, no way to run a business. In this case the Court of Appeal made light of the point because it assumed that, once the defendant had been ordered to run the business, self-interest and compliance with the order would thereafter go hand in hand. But, as I shall explain, this is not J
necessarily true.

Second, the seriousness of a finding of contempt for the defendant means that any application to enforce the order is likely to be a heavy and expensive piece of litigation. The possibility of repeated applications over a period of time means that, in comparison with a once and for all inquiry as to damages, the enforcement of the remedy is likely to be expensive in terms of cost to the parties and the resources of the judicial system.

This is a convenient point at which to distinguish between orders which require a defendant to carry on an activity, such as running a business over or more or less extended period of time, and orders K
which require him to achieve a result. The possibility of repeated applications for rulings on compliance with the order which arises in the former case does not exist to anything like the same extent in the latter. Even if the achievement of the result is a complicated matter which will take some time, the court, if called upon to rule, only has to examine the finished work and say whether it complies with the order. This point was made in the context of relief against forfeiture in *Shiloh Spinners Ltd* v *Harding* [1973] AC 691. If it is a condition of relief that the tenant should have complied with a repairing covenant, difficulty of supervision need not be an objection. As Lord Wilberforce said (at p724D): L

> [w]hat the court has to do is to satisfy itself, *ex post facto*, that the covenanted work has been done, and it has ample machinery, through certificates, or by enquiry, to do precisely this.

This distinction between orders to carry on activities and to achieve results explains why the courts have in appropriate circumstances ordered specific performance of building contracts and repairing covenants: see *Wolverhampton Corporation* v *Emmons* [1901] 1 KB 515 (building contract) and *Jeune* v *Queens Cross Properties Ltd* [1974] 1 Ch 97 (repairing covenant). It by no means follows, however, that even obligations to achieve a result will always be enforced by M
specific performance. There may be other objections, to some of which I now turn.

One such objection, which applies to orders to achieve a result and *a fortiori* to orders to carry on an activity, is imprecision in the terms of the order. If the terms of the court's order, reflecting the terms of the obligation, cannot be precisely drawn, the possibility of wasteful litigation over compliance is increased. So is the oppression caused by the defendant having to do things under threat of proceedings for contempt. The less precise the order, the fewer the signposts to the forensic minefield which he has to traverse. The fact that the terms of a contractual obligation are sufficiently definite to escape being void for uncertainty, or to found a claim for damages, or to permit compliance

A to be made a condition of relief against forfeiture, does not necessarily mean that they will be sufficiently precise to be capable of being specifically performed. So in *Wolverhampton Corporation* v *Emmons* [1901] 1 KB 515, Romer LJ said that the first condition for specific enforcement of a building contract was that "the particulars of the work are so far definitely ascertained that the Court can sufficiently see what is the exact nature of the work of which it is asked to order the performance". Similarly in *Redland Bricks Ltd* v *Morris* [1970] AC 652, at p666G Lord Upjohn stated the following general principle for the grant of mandatory injunctions to carry out building works:

> B [t]he court must be careful to see that the defendant knows exactly in fact what he has to do and this means not as a matter of law but as a matter of fact, so that in carrying out an order he can give his contractors the proper instructions.

Precision is of course a question of degree and the courts have shown themselves willing to cope with a certain degree of imprecision in cases of orders requiring the achievement of a result in which the plaintiff's merits appeared strong; like all the reasons which I have been discussing, it is, taken alone, merely a discretionary matter to be taken into account: see Spry on *Equitable Remedies* 4th ed at p112. It is, however, a very important one.

I should at this point draw attention to what seems to me to have C been a misreading of certain remarks of Lord Wilberforce in *Shiloh Spinners Ltd* v *Harding* [1973] AC 691, at p724. He pointed out, as I have said, that to grant relief against forfeiture subject to compliance with a repairing covenant involves the court in no more than the possibility of a retrospective assessment of whether the covenanted work has been done. For this reason, he said:

> Where it is necessary, and, in my opinion, right, to move away from some 19th century authorities, is to reject as a reason against granting relief, the impossibility for the courts to supervise the doing of work.

This is plainly a remark about cases involving the achievement of a D result, such as doing repairs, and, within that class, about making compliance a condition of relief against forfeiture. But in *Tito v Waddell (No 2)* [1977] Ch 106, at p322 Sir Robert Megarry V-C took it to be a generalisation about specific performance and, in particular, a rejection of difficulty of supervision as an objection, even in cases of orders to carry on an activity. The Vice-Chancellor regarded it as an adoption of his own views (based, as I have said, on incomplete analysis of what was meant by difficulty of supervision) in *CH Giles & Co Ltd* v *Morris* [1972] 1 WLR 307, at p318. In the present case, in the Court of Appeal took this claim at face value. In fact, Lord Wilberforce went on to say that impossibility of supervision "is a E reality, no doubt, and explains why specific performance cannot be granted of agreements to this effect". Lord Wilberforce was, in my view, drawing attention to the fact that the collection of reasons which the courts have in mind when they speak of difficulty of supervision apply with much greater force to orders for specific performance, giving rise to the possibility of committal for contempt, than they do to conditions for relief against forfeiture. While the paradigm case to which such objections apply is the order to carry on an activity, they can also apply to an order requiring the achievement of a result.

There is a further objection to an order requiring the defendant to carry on a business, which was emphasised by Millett LJ in the Court F of Appeal. This is that it may cause injustice by allowing the plaintiff to enrich himself at the defendant's expense. The loss which the plaintiff may suffer through having to comply with the order (for example, by running a business at a loss for an indefinite period) may be far greater than the plaintiff would suffer from the contract being broken. As Professor RJ Sharpe explains (*Specific Remedies for Contract Breach* in *Studies in Contract Law* (ed Reiter and Swan) at p129):

> In such circumstances, a specific decree in favour of the plaintiff will put him in a bargaining position *vis à vis* the defendant whereby the measure of what he will receive will be the value to the defendant of being released from performance. If the plaintiff bargains effectively, the amount he will set will exceed the value to him of performance and will approach the cost to the G defendant to complete.

This was the reason given by Lord Westbury LC in *Isenberg* v *East India House Estate Co Ltd* (1863) 3 De GJ&S 263, at p273 for refusing a mandatory injunction to compel the defendant to pull down part of a new building which interfered with the plaintiff's light and exercising instead the Court of Chancery's recently-acquired jurisdiction under Lord Cairns's Act 1858 to order payment of damages:

> . . . I hold it . . . to be the duty of the Court in such a case as the present not, by granting a mandatory injunction, to deliver over the Defendants to the Plaintiff H bound hand and foot, in order to be made subject to any extortionate demand that he may by possibility make, but to substitute for such mandatory injunction an inquiry before itself, in order to ascertain the measure of damage that has been actually sustained.

It is true that the defendant has, by his own breach of contract, put himself in such an unfortunate position. But the purpose of the law of contract is not to punish wrongdoing but to satisfy the expectations of the party entitled to performance. A remedy which enables him to secure, in money terms, more than the performance due to him is unjust. From a wider perspective, it cannot be in the public interest for the courts to require someone to carry on business at a loss if there is J any plausible alternative by which the other party can be given compensation. It is not only a waste of resources but yokes the parties together in a continuing hostile relationship. The order for specific performance prolongs the battle. If the defendant is ordered to run a business, its conduct becomes the subject of a flow of complaints, solicitors' letters and affidavits. This is wasteful for both parties and the legal system. An award of damages, on the other hand, brings the litigation to an end. The defendant pays damages, the forensic link between them is severed, they go their separate ways and the wounds of conflict can heal.

The cumulative effect of these various reasons, none of which K would necessarily be sufficient on its own, seems to me to show that the settled practice is based upon sound sense. Of course the grant or refusal of specific performance remains a matter for the judge's discretion. There are no binding rules, but this does not mean that there cannot be settled principles, founded upon practical considerations of the kind which I have discussed, which do not have to be re-examined in every case, but which the courts will apply in all but exceptional circumstances. As Slade J said in the passage which I have quoted from *Braddon Towers Ltd* v *International Stores Ltd* [1987] 1 EGLR 209, at p213 lawyers have no doubt for many years advised their clients on this basis. In the present case, Leggatt LJ remarked that there was no evidence that such advice had been given. L In my view, if the law or practice on a point is settled, it should be assumed that persons entering into legal transactions will have been advised accordingly. I am sure that the learned lord justice would not wish to encourage litigants to adduce evidence of the particular advice which they received. Indeed, I doubt whether such evidence would be admissible.

5. The decision of the Court of Appeal

I must now examine the grounds upon which the majority of the Court of Appeal thought it right to reverse the judge. In the first place, they regarded the practice which he followed as outmoded and treated Lord Wilberforce's remarks about relief against forfeiture in *Shiloh* M *Spinners Ltd* v *Harding* [1973] AC 691, at p724 as justifying a rejection of the arguments based on the need for constant supervision. Even Millett LJ, who dissented on other grounds, said that such objections had little force today. I do not agree. As I have already said, I think that Lord Wilberforce's remarks do not support this proposition in relation to specific performance of an obligation to carry on an activity and that the arguments based on difficulty of supervision remain powerful.

The Court of Appeal said that it was enough if the contract defined the tenant's obligation with sufficient precision to enable him to know what was necessary to comply with the order. Even assuming that this to be right, I do not think that the obligation in clause 4(19) can possibly be regarded as sufficiently precise to be capable of specific performance. It is to "keep the demised premises open for retail trade". It says nothing about the level of trade, the area of the premises within which trade is to be conducted, or even the kind of trade, although no doubt the tenant's choice would be restricted by the need to comply with the negative covenant in clause 4(12)(a) not to use the premises "other than as a retail store for the sale of food groceries provisions and goods normally sold from time to time by a retail grocer food supermarkets and food superstores". This language seems to me to provide ample room for argument over whether the tenant is doing enough to comply with the covenant.

The Court of Appeal thought that once Argyll had been ordered to comply with the covenant, it was, as Roch LJ said, "inconceivable that they would not operate the business efficiently". Leggatt LJ said that the requirement "was quite intelligible to the respondents, while they were carrying on business there . . . If the premises are to be run as a business, it cannot be in the respondents ' interest to run it half-heartedly or inefficiently. . .". This treats the way the tenant previously conducted business as measuring the extent of his obligation to do so. In my view, this is a non sequitur: the obligation depends upon the language of the covenant and not upon what the tenant has previously chosen to do. No doubt it is true that it would not be in the interests of the tenant to run the business inefficiently. But running the business efficiently does not necessarily mean running in the way it was run before. Argyll had decided that, from its point of view, the most efficient thing to do was to close the business altogether and concentrate its resources on achieving better returns elsewhere. If ordered to keep the business open, it might well decide that the next best strategy was to reduce its costs as far as was consistent with compliance with its obligations, in the expectation that a lower level of return would be more than compensated by higher returns from additional expenditure on more profitable shops. It is, in my view, wrong for the courts to speculate about whether Argyll might voluntarily carry on business in a way which would relieve the court from having to construe its order. The question of certainty must be decided on the assumption that the court might have to enforce the order according to its terms.

The respondent argued that the court should not be concerned about future difficulties which might arise in connection with the enforcement of the order. It should simply make the order and see what happened. In practice Argyll would be likely to find a suitable assignee (as it in fact did) or conduct the business so as to keep well clear of any possible enforcement proceedings or otherwise come to terms with the CIS. This may well be true, but the likelihood of Argyll having to perform beyond the requirements of its covenant or buy its way out of its obligation to incur losses seems to me to be in principle an objection to such an order rather than to recommend it. I think that it is normally undesirable for judges to make orders *in terrorem*, carrying a threat of imprisonment, which work only if no one inquires too closely into what they mean.

The likelihood that the order would be effective only for a short time until an assignment is an equivocal argument. It would be burdensome to make Argyll resume business only to stop again after a short while if a short stoppage would not cause any substantial damage to the business of the shopping centre. On the other hand, what would happen if a suitable assignee could not be found? Would Argyll then have to carry on business until 2014? Mr Smith QC, who appeared for the CIS, said that if the order became oppressive (for example, because Argyll were being driven into bankruptcy) or difficult to enforce, they could apply for it to be varied or discharged. But the order would be a final order and there is no case in this jurisdiction in which such an order has been varied or discharged, except when the injuncted activity has been legalised by statute. Even assuming that there was such a jurisdiction if circumstances were radically changed, I find it difficult to see how this could be made to apply. Difficulties of enforcement would not be a change of circumstances. They would have been entirely predictable when the order was made. And so would the fact that Argyll would suffer unquantifiable loss if it was obliged to continue trading. I do not think that such expedients are an answer to the difficulties on which the objections to such orders are based.

Finally, all three judges in the Court of Appeal took a very poor view of Argyll's conduct. Leggatt LJ said that they had acted "with gross commercial cynicism"; Roch LJ began his judgment by saying that they had "behaved very badly" and Millett LJ said that they had no merits. The principles of equity have always had a strong ethical content and nothing which I say is intended to diminish the influence of moral values in their application. I can envisage cases of gross breach of personal faith, or attempts to use the threat of non-performance as blackmail, in which the needs of justice will override all the considerations which support the settled practice. But although any breach of covenant is regrettable, the exercise of the discretion as to whether or not to grant specific performance starts from the fact that the covenant has been broken. Both landlord and tenant in this case are large sophisticated commercial organisations and I have no doubt that both were perfectly aware that the remedy for breach of the covenant was likely to be limited to an award of damages. The interests of both were purely financial: there was no element of personal breach of faith, as in the Victorian cases of railway companies which refused to honour obligations to build stations for landowners whose property they had taken: compare *Greene* v *West Cheshire Railway Co* (1871) LR 13 Eq 44. No doubt there was an effect on the businesses of other traders in the Centre, but Argyll had made no promises to them and it is not suggested that CIS warranted to other tenants that Argyll would remain. Their departure, with or without the consent of CIS, was a commercial risk which the tenants were able to deploy in negotiations for the next rent review. On the scale of broken promises, I can think of worse cases, but the language of the Court of Appeal left them with few adjectives to spare.

It was no doubt discourteous not to have answered Mr Wightman's letter. But to say, as Roch LJ did, that they had acted "wantonly and quite unreasonably" by removing their fixtures seems to me an exaggeration. There was no question of stealing a march, or attempting to present CIS with a *fait accompli*, because Argyll had no reason to believe that CIS would have been able to obtain a mandatory injunction whether the fixtures had been removed or not. They had made it perfectly clear that they were closing the shop and given CIS ample time to apply for such an injunction if so advised.

6. *Conclusion*

I think that no criticism can be made of the way in which Judge Maddocks QC exercised his discretion. All the reasons which he gave were proper matters for him to take into account. In my view, the Court of Appeal should not have interfered and I would allow the appeal and restore the order which he made.

Lord Browne-Wilkinson, Lord Slynn of Hadley and **Lord Hope of Craighead** agreed and did not add anything.

Also agreeing, **Lord Clyde** said: While I should wish to reserve my opinion on the approach which might be adopted by civilian systems I agree that the appeal should be allowed for the reasons which he has given.

A

Mannai Investment Co Ltd *v* Eagle Star Life Assurance Co Ltd

HOUSE OF LORDS

May 21 1997

LORD GOFF OF CHIEVELEY, LORD JAUNCEY OF TULLICHETTLE, LORD STEYN, LORD HOFFMANN and LORD CLYDE

[1997] 24 EG 122; [1997] 25 EG 138

B

Landlord and tenant — Break clause — Notices determining fixed term lease one day earlier than date specified in break clause — Whether notices valid

By two leases dated March 11 1992 the appellant tenant held a 10-year term of premises from the respondent landlord. Clause 7(13) of each lease provided that the tenant might determine the lease by serving a notice of not less than six months expiring on the third anniversary of the term commencement date. The tenant served notices expiring on January 12 1995. The landlord's contentions, that as the term commencement date was January 13 1992 and the third C anniversary of the term under each lease was therefore January 13 1995, the notices were ineffective, were accepted by the Court of Appeal: [1996] 1 EGLR 69. The tenant appealed.

***Held*: The appeal was allowed. Clause 7(13) did not prescribe as an indispensable condition for its effective service that the notice must contain specific information. The construction of the notices must be approached objectively taking into account the relevant objective contextual scene. The purpose of a notice under a break clause must not be lost sight of. There is no justification for placing such notices in a unique category; such notices belong to the general class of unilateral notices. It was obvious to a reasonable D recipient with knowledge of the leases and the third anniversary date that the tenant wished to determine the leases on that date. The reasonable recipient would not have been perplexed by the minor error in the notices. One must substitute for the rigid rule in *Hankey* v *Clavering* [1942] 2 KB 326 the standard of commercial construction. The correct test for the validity of a notice is that in *Carradine Properties Ltd* v *Aslam* [1976] 1 WLR 442. The meaning of words, as they would appear in a dictionary, and the effect of their syntactical arrangement, as it would appear in grammar, is part of the material used to understand a speaker's utterance. Another part is knowledge of the background against which the E utterance was made. *Hankey* v *Clavering* and the earlier authorities should no longer be followed.**

Lord Goff of Chieveley and Lord Jauncey of Tullichettle dissenting.

The following cases are referred to in this report.

Antaios Compania Naviera SA v *Salen Rederierna AB* [1985] AC 191; [1984] 3 WLR 592; [1984] 3 All ER 229; [1984] 2 Lloyd's Rep 235, HL
Cadby v *Martinez* (1840) 11 A&E 720
Carradine Properties Ltd v *Aslam* [1976] 1 WLR 442; [1976] 1 All ER 573; (1975) 32 P&CR 12
Delta Vale Properties Ltd v *Mills* [1990] 1 WLR 445; [1990] 2 All ER 176
F *Doe d Cox* v *Roe* (1802) 4 Esp 185
Doe d Spicer v *Lea* (1809) 11 East 312
Fish, In re [1894] 2 Ch 83
Gardner v *Ingram* (1889) 61 LT 729; [1886-90] All ER Rep 258; 54 JP 311; 6 TLR 75
Garston v *Scottish Widows' Fund & Life Assurance Society* [1996] 1 WLR 834; [1996] 1 EGLR 113; [1996] 23 EG 131
Germax Securities Ltd v *Spiegal* (1978) 37 P&CR 204; [1979] 1 EGLR 84; 250 EG 449, CA
Hankey v *Clavering* [1942] 2 KB 326; [1942] 2 All ER 311, CA
Mannai Investment Co Ltd v *Eagle Star Life Assurance Co Ltd* [1995] 1 WLR 1508; [1996] 1 EGLR 69; [1996] 06 EG 140
Micrografix v *Woking 8 Ltd* [1995] 2 EGLR 32; [1995] 37 EG 179
National Society for the Prevention of Cruelty to Children v *Scottish National Society for the Prevention of Cruelty to Children* [1915] AC 207
Norwegian American Cruises A/S v *Paul Mundy Ltd; The Vistafjord* [1988] 2 Lloyd's Rep 343, CA G
Peel (RR), In the Goods of (1870) LR 2 P&D 46
Phipps (P) & Co Ltd v *Rogers* [1925] 1 KB 14
Prenn v *Simmonds* [1971] 1 WLR 1381; [1971] 3 All ER 237, HL
Price v *Mann* [1942] 1 All ER 453, CA
Reardon Smith Line Ltd v *Yngvar Hansen-Tanger* [1976] 1 WLR 989; [1976] 3 All ER 570; [1976] 2 Lloyd's Rep 621, HL
Sidebotham v *Holland* [1895] 1 QB 378
Sudbrook Trading Estate Ltd v *Eggleton* [1983] 1 AC 444; [1982] 3 WLR 315; [1982] 3 All ER 1; (1982) 44 P&CR 153; [1983] 1 EGLR 47; [1983] EGD 392; 265 EG 215, HL
Sunrose Ltd v *Gould* [1962] 1 WLR 20; [1961] 3 All ER 1142, CA H

This was an appeal by the tenant, Mannai Investment Co Ltd, from a decision of the Court of Appeal ([1996] 1 EGLR 69), which had allowed an appeal by the landlord, Eagle Star Life Assurance Co Ltd, from a decision of Judge Rich QC, sitting as a judge of the High Court, who determined the validity of notices served by the tenant.

John Cherryman QC and Kenneth Munro (instructed by Manches & Co) appeared for the tenant; Nicholas Patten QC and Thomas Leech (instructed by Nabarro Nathanson) appeared for the landlord.

Giving a dissenting opinion, LORD GOFF OF CHIEVELEY said: This appeal is concerned with the question whether a notice given by a J tenant pursuant to a break clause in a lease was an effective notice. In fact, there were two leases with identical break clauses, and two identical notices were given. For convenience, however, I will assume that there was only one.

The premises were in Jermyn Street, in London SW1. The lease was dated March 11 1992, and was for a term of 10 years from January 13 1992. The respondent company was the landlord, and the appellant company was the tenant. The relevant clause was clause 7(13), which provided as follows:

> The Tenant may by serving not less than six months notice in writing on the Landlord or its Solicitors such notice to expire on the third anniversary of the K term commencement date determine this Lease and upon expiry of such notice this Lease shall cease and determine and have no further effect . . .

So the clause gave the tenant a single opportunity to bring the lease to an end. It so happened that the market for rents in this area had fallen, and the tenant decided to take advantage of this opportunity. It served the following notice on the landlord:

> Pursuant to Clause 7(13) of the Lease we as Tenant hereby give notice to you to determine the lease on 12 January 1995 . . .

Unfortunately, however, the tenant made a mistake. The third L anniversary of the term commencement date was not January 12 1995 but January 13 1995. The landlord has claimed that in the result the notice was ineffective. The Court of Appeal*, reversing the decision of Judge Rich QC (sitting as an additional judge of the Chancery Division), upheld the landlord's contention. It is from that decision that the tenant now appeals to your lordships' House.

I should record at once that the judge held that the tenant was entitled to succeed on the basis that, because January 12 and 13 were contiguous dates, there was a moment of time at which they coincided; and from this it followed that a notice expressed to determine the lease on January 12 was effective to do so on January 13 as required by the M clause. In so holding, he invoked the authority of *Sidebotham* v *Holland* [1895] 1 QB 378. But, as Nourse LJ pointed out in the Court of Appeal in the present case [1995] 1 WLR 1508, at pp1514-1515, that case provided no authority for the judge's conclusion. It was concerned with a notice to quit and deliver up possession by midnight. It related therefore to a notice to quit at a point of time which was held to be common to both dates, and not, as in the present case, a notice to take effect on a certain date. Here, a notice taking effect on a different, though contiguous date, could not be rendered effective on the basis of

---

*Editor's note: Also reported at [1996] 1 EGLR 69

A *Sidebotham* v *Holland*. With that reasoning, I find myself in complete agreement. It follows that the central question in the appeal before your lordships' House is whether the Court of Appeal was right to hold that the notice was in any event not an effective notice under the clause. To that question I now turn.

At first sight it seems unreasonable that the notice should not have been effective. It was obvious that the tenant was trying to give an effective notice under the clause, and that it had mistakenly assumed that the anniversary of the term commencement date was not January 13 but January 12 1995. As Hobhouse LJ pointed out (see [1995] 1 WLR 1508, at p1516F), the tenant had mistakenly read the clause as if
B it said "to expire at the end of the third year of the term" when it did not do so. It is tempting therefore to assist the tenant who has made a mistake of this kind, when it must have been obvious to the landlord that the tenant intended to give an effective notice under the clause. But the difficulty in the way of adopting this approach is that, on the authorities, it is inconsistent with the agreement of the parties as expressed in the clause.

An early authority in this line is *Cadby* v *Martinez* (1840) 11 A&E 720. In that case a tenant was entitled under a clause in his lease to determine it by notice expiring on Michaelmas Day 1837. The tenant mistakenly gave notice to quit and deliver up the premises on June 24
C 1837. The notice was expressed to be "agreeably to the covenants of the lease". It was held that notice was not effective to determine the lease. Lord Denman CJ said, at p726:

> We have heard the case argued, and are of opinion that the covenant to pay rent during the whole term cannot be got rid of by any notice to quit which is not in accordance with the proviso introduced into the lease for the purpose.
>
> The cases that seemed to point the other way merely shew that, where there is no covenant, a notice describing the premises, so as to be perfectly understood between the parties, will be sufficient: but in none has a proviso or covenant in a deed been held to be satisfied by a notice inconsistent with the terms of it.

D

The reasoning in this brief judgment is clear and compelling. You start with the position that, under the lease, the tenant has covenanted to pay rent for the full term; but under a proviso in the lease the tenant may, by notice, rid himself of the obligation under that covenant. However, to be effective for that purpose, the notice must conform to the terms of the proviso. If on its true construction the notice does not do so, it will not be effective for its purpose, because the parties have agreed that only a notice conforming to the terms of the proviso will be effective. The fact that the landlord realises that the tenant intended to take advantage of his rights under the proviso, but has only failed through some mistake to give the required notice, is irrelevant. The
E simple fact is that the tenant has failed to use the right key which alone is capable of turning the lock.

Nowadays, the leading case is *Hankey* v *Clavering* [1942] 2 KB 326. The lease in question was for a term of 21 years from December 25 1934. The break clause conferred on either party the right to determine the lease at the expiration of the first seven years, by six calendar months' notice. The landlord gave notice to the tenant's solicitors in the following terms:

> As I may have to be away for some time in the near future, I will be obliged if you would accept the six months' notice to terminate your client's lease which
> F I am allowed to give on June 21, 1941. This would mean that he would have to give up the cottage on December 21, 1941.

This was obviously a mistake on the part of the landlord, because the six months' notice should have expired on December 25 not 21. At first instance, Asquith J held that the notice could be saved on the basis that it had been accepted as a good notice by the tenant's solicitors; but that solution was rejected by the Court of Appeal on the facts. The Court of Appeal, which consisted of two distinguished and very experienced Chancery lawyers, Lord Greene MR and Lord Clauson, held that the notice was ineffective. They regarded the point as so clear that they gave judgment *ex tempore*. Indeed, Lord Clauson concluded his brief judgment as follows, at p331:

> I should have thought that, as a matter of construction, an argument other than G
> that which leads to the result the Master of the Rolls has announced was quite untenable.

Lord Greene MR introduced his judgment with the following passage, at p328:

> This appeal raises a short point in connection with a break clause in a lease wherein the plaintiff was the lessor and the defendant was the lessee. By his letter of January 15, 1940, the plaintiff, on the face of it, was purporting to determine the lease by notice on December 21, 1941. The whole thing was obviously a slip on his part, and there is a natural temptation to put a strained construction on language in aid of people who have been unfortunate enough
> to make slips. That, however, is a temptation which must be resisted, because H
> documents are not to be strained and principles of construction are not to be outraged in order to do what may appear to be fair in an individual case.

He expressed his reasoning and conclusion as follows, at pp329-330:

> Notices of this kind are documents of a technical nature, technical because they are not consensual documents, but, if they are in proper form, they have of their own force without any assent by the recipient the effect of bringing the demise to an end. They must on their face and on a fair and reasonable construction do what the lease provides that they are to do. It is perfectly true that in construing such a document, as in construing all documents, the court in
> a case of ambiguity will lean in favour of reading the document in such a way J
> as to give it validity, but I dissent entirely from the proposition that, where a document is clear and specific, but inaccurate on some matter, such as that of date, it is possible to ignore the inaccuracy and substitute the correct date or other particular because it appears that the error was inserted by a slip. By the clear wording of this notice the plaintiff purported to bring the lease to an end on December 21, 1941. In so doing he was attempting to do something which he had no power to do, and, however much the recipient might guess, or however certain he might be, that it was a mere slip, that would not cure the defect because the document was never capable on its face of producing the necessary legal consequence.

The essential point made by Lord Greene MR therefore was that K
notices of this kind are documents of a technical nature because they are not consensual documents; but, if they are in proper form, ie if they comply with the specification in the clause, they have of their own force the effect of bringing the demise to an end. It is necessary, therefore, to turn to the lease to ascertain what is required. Here the landlord had to give six months' notice of his desire to determine the demise at the expiration of the first seven years, which in fact expired on December 25 1941. This he failed to do; and accordingly the notice was not effective. The facts that he had obviously meant to give an effective notice at the end of the first seven years, that he had simply made a mistake about the date, and that this may have been obvious to L
the other party, made no difference. The simple position was that the notice he gave did not conform to the agreed specification in the clause which gave the right to determine the lease, and so was not effective for that purpose.

It is, in my opinion, correct in principle that a notice under such a clause will only be effective if it conforms to the specification in the clause. The specification in the clause is contained in a document which has been agreed between the parties, and so prescribes the requirements with which the notice must comply if it is to achieve the desired effect. In the case before your lordships, the notice must: (1) be not less than six months' notice, (2) be in writing, (3) be served on the M
landlord or its solicitors, and (4) expire on the third anniversary of the term commencement date. It is plain that fulfilment of all four of these requirements is essential. It is with the fourth requirement that we are here concerned; and it is well established that this required that the notice should be so *expressed* as to expire on the relevant date. The position was made clear by another distinguished lawyer, Atkin LJ, in *P Phipps & Co Ltd* v *Rogers* [1925] 1 KB 14, at p27. He there cited Lord Coleridge CJ as saying in *Gardner* v *Ingram* (1889) 61 LT 729, at p730, that: "Although no particular form need be followed, there must be plain unambiguous words claiming to determine the existing tenancy at a certain time". Atkin LJ continued: "The date of determination must be the right date".

The principle is therefore clear. The agreement between the parties provides what notice has to be given to be effective to achieve the relevant result. The question in each case is: does the notice which was given, properly construed, comply with the agreed specification? If it does, it is effective for its purpose. If it does not, it is not so effective; and the mere fact that the person serving the notice plainly intended, and was trying, to give an effective notice under the clause, and that the recipient of the notice realised that he was doing so, makes no difference. This is because the notice, properly construed, did not comply with the agreement between the parties. The key does not fit the lock, and so the door will not open.

I have been careful to use the expression "properly construed". This is because, although at first sight the notice given may not appear to comply with the agreed specification in the clause, nevertheless on examination it may transpire that, properly construed, it does in fact do so. This may occur where there is a latent ambiguity in the notice. A striking example is to be found in *Doe d Cox* v *Roe* (1802) 4 Esp 185 (which was cited in argument in *Cadby* v *Martinez*). There the landlord of a public house in Limehouse gave notice to quit "the premises which you hold of me . . . commonly called or known by the name of the Waterman's Arms". On the evidence, the only property let by the landlord to the tenant was a public house called the Bricklayer's Arms; moreover there was no public house in Limehouse called the Waterman's Arms. The notice was held effective in respect of the tenancy of the Bricklayer's Arms, the case being treated as one of latent ambiguity. Another example occurs when a date is specified in the notice which, as is plain from the face of the notice, was obviously stated in error for the true date which the giver of the notice must have intended to specify. This will usually occur nowadays through a simple typing error. In such a case, the date so given can properly be construed as a reference to the true date. An example of the application of this principle is to be found in *Carradine Properties Ltd* v *Aslam* [1976] 1 WLR 442. In that case, which was (like the present) concerned with a break clause in a lease, the relevant date upon which a notice given by either party under the clause might take effect was a date in September 1975, but a notice given by the landlord in September 1974 specified a date in September 1973. It was plain from the face of the notice that a date in 1973, which had already passed, could not conceivably have been intended by him. It must have been a clerical error, and could properly be read as intended to refer to 1975. In these circumstances Goulding J, applying ordinary principles of construction, held the notice to be an effective notice to determine the lease on September 27 1975. He said, at p446C:

> In an option clause the requirement is that a party must strictly comply with the condition for its exercise. If the condition includes the giving of a particular notice, it seems to me that the logical first approach is to interpret the notice, looking at the words and applying legal principles to their construction, and then ask whether it complies with the strict requirements as to the exercise of the option. If that is right, I think that a benevolent approach could be applied in this case, as in the *Duke of Bedford's Case* (1796) 7 Term Rep 63, because reasonably read by a reasonable tenant the mistake is obvious on the face of it, and there is no doubt what the mistake was. *Therefore one interprets the notice as asserting an intention to determine in 1975*. It is true that if whoever made the mistake had typed 1976 instead of 1973, the error would probably have been incurable because although the tenant might suspect there was a slip, it might be that the landlord did intend 1976, not knowing or understanding his rights under the lease. In such a case the tenant would be entitled to disregard the notice but because a past date was given in the notice it is insensible and therefore an authority such as the *Duke of Bedford's Case* is in point.

[Emphasis supplied.]

The decision of Goulding J in that case was therefore made in accordance with the principles I have stated. I should however mention that, in the report of the case in [1976] 1 WLR 442, at p443G, it is stated that, although the date upon which a notice given by either party under the clause might take effect was September 27 1975, the notice given by the landlord specified the date as September 17 1973. Your lordships were however informed that comparison of this report with other reports of the case reveals that the date of determination specified in the landlord's notice was, in fact, September 27, not September 17, so that the only error in the landlord's notice consisted of specifying 1973 instead of 1975. I am satisfied that this must have been the case. If not, Goulding J would certainly have addressed the question of the effect of the error as to the day of the month; and, if he had done so, he would, consistently with the reasoning in his judgment, have held that on that ground the notice was ineffective.

It is however well settled that, under a clause such as the present which does not require that the date of expiry should be specified as such in the notice but merely that the notice should be expressed to expire on the relevant date, it is enough that the notice should be so expressed to expire in accordance with the terms of the clause. This was made clear by Lord Coleridge CJ in *Gardner* v *Ingram* (1889) 61 LT 729, at p730, in the passage I have already quoted. So if the tenant in the present case had given a notice expressed to determine the lease on the third anniversary of the commencement date, that would have been a sufficient compliance with the specification in the clause. Furthermore it is also well settled that, if the person giving the notice specifies the actual date but out of caution also specifies, in the alternative, the end of the period at which the notice is required under the clause to take effect, the alternative so given will be effective to save the notice if the actual date so given should prove to have been mistaken. For obvious reasons, therefore, it is a commonplace for notices to quit to take this form, which avoids the trap into which the tenant fell in the present case. The position was stated clearly by AL Smith LJ in *Sidebotham* v *Holland* [1895] 1 QB 378, at p389, a case to which I have already referred, when he said:

> I would point out that the plaintiff has only himself to blame for the difficulties he is in in this case. Had he added the words which are very ordinarily inserted in a notice to quit, "or at the expiration of the year of your tenancy, which shall expire next after the end of one half-year from the service of this notice", and which are inserted to avoid such a point as that now taken, all would have been in order; but the words are not there. If the notice to quit in this case had been for May 20 or 21 or any later day I should have had no doubt but that it was a bad notice; and I own that the inclination of my opinion is that the present notice is bad because it does not expire upon the last day of some year of the tenancy; but, as Lord Halsbury and Lindley LJ are of opinion that, inasmuch as this was a full six months' notice given to quit upon the anniversary of the day upon which the tenancy commenced, it is good, though the tenancy expired at midnight the day before, I yield to what they say, and will not differ from them, and hold that this unmeritorious technicality must prevail; and I content myself with expressing what I have said.

That what I have said sets out the long established law on this subject appears clearly from the judgments of Lindley LJ (with whom Lord Halsbury agreed) and AL Smith LJ in that case. This was the established law which was applied by the Court of Appeal in *Hankey* v *Clavering* [1942] 2 KB 326, and by the Court of Appeal in the present case. I wish to stress however that these principles have been evolved in the case of clauses in leases, such as notices under a break clause, options to purchase, and notices to quit, in which provision is regularly made as to the conditions to be fulfilled in respect of the relevant notice and, in particular, that the notice must be expressed so as to take effect on a certain date.

Before your lordships Mr John Cherryman QC, for the tenant, faced with the relevant authorities, sought in the present case to distinguish *Hankey* v *Clavering*, which was the authority held by the Court of Appeal to be decisive against his client. He submitted that the date of determination specified by the tenant in its notice was not an essential part of the notice and must give way to the tenant's obvious intention, having regard to the express invocation by the tenant of clause 7(13), that the notice should have the effect of expiring on the third anniversary. But, in my opinion, that submission is contrary to the judgment of Lord Greene MR in *Hankey* v *Clavering* itself. He treated it as necessary that the person giving the notice should claim to determine the tenancy at a certain time, and (as Atkin LJ said in *Phipps* [1925] 1 KB 14, at p27) that must be the right time. The fact that the tenant in the present case expressly identified the clause in question as opposed to implicitly referring to it (as in *Hankey* v *Clavering*) cannot,

A as I see it, make any difference. If that were so, it would be enough for the person giving the notice simply to invoke the clause without specifying any time at which the notice should take effect. That would, however, be contrary to the construction which has for a long time been placed upon clauses of this kind.

Mr Cherryman next relied upon a test stated by Goulding J in *Carradine Properties Ltd* v *Aslam* [1976] 1 WLR 442, at p444G, to be generally applicable, *viz*: "Is the notice quite clear to a reasonable tenant reading it? Is it plain that he cannot be misled by it?" These words have been fastened upon to suggest that a broader test should be applied than is recognised in the authorities, and have been cited in B textbooks as having that effect. Indeed, in *Micrografix* v *Woking 8 Ltd* [1995] 2 EGLR 32 Jacob J referred to Goulding J as having "distinguished the celebrated, much distinguished, case of *Hankey* v *Clavering* [1942] 2 KB 326" and stated, at p33, that "*Carradine* . . . is the modern approach . . .". I have to say, however, that this interpretation reveals a complete misunderstanding of Goulding J's judgment in *Carradine* which, read as a whole, demonstrates that he had no intention of departing from the law as established in the previous authorities, including *Hankey* v *Clavering*, by which he was bound. In his judgment, Goulding J quoted at length from the judgment of Lord Greene MR in that case, and said correctly, at p446:

C

> In an option clause the requirement is that a party must strictly comply with the condition for its exercise. If the condition includes the giving of a particular notice, it seems to me that the logical first approach is to interpret the notice, looking at the words and applying legal principles to their construction, and then ask whether it complies with the strict requirements as to exercise of the option.

Applying those principles, he was able to distinguish *Hankey* v *Clavering* on the basis that, if the notice was reasonably read by a reasonable tenant, the mistake (*viz*, specifying 1973) was obvious on the face of it and there was no doubt what that mistake was, ie that the D date 1973 must have been typed instead of 1975 which the landlord had intended to specify. It is in this sense that the test stated by Goulding J earlier in his judgment must be understood, as was made plain when he went on to say that if, instead of the earlier date of 1973, the later date of 1976 had been typed for 1975, such a mistake would probably have been incurable, because "although the tenant might suspect there was a slip, it might be that the landlord did intend 1976, not knowing or understanding his rights under the lease" (p446E). Having regard to the reasoning of Goulding J in his judgment, the suggestion that the test earlier stated by him constituted a departure by him from the well-established authorities by which he was bound is, E frankly, untenable. Moreover, it is also plain from his judgment that, in considering the question whether the notice, on its true construction, complied with the clause in the lease, he was entitled to have regard to the terms of the lease; indeed, it is difficult to see how he could otherwise have considered that question. In this respect, the decision is well summarised in the headnote to the report.

In *Micrografix* the tenants gave a notice determining the lease on March 23 1995 when under the relevant clause they could only have done so on June 23 1995. Jacob J held that, as the landlords knew that the date of determination could only be June 23 1995, they would not have been misled, and the notice should therefore be treated as an F effective notice to determine the lease on the correct date, *viz* June 23 1995. In my opinion that decision was contrary to the long established line of authority, including *Hankey* v *Clavering* [1942] 2 KB 326, by which Jacob J was bound. It failed to give effect to the underlying principle that the notice must, on its true construction, conform to the agreed specification in the applicable clause, failing which it will not under the contract be effective for its purpose. On its true construction it could not be read as a notice taking effect on June 23 1995, because there was no reason to believe that the landlords must have intended to give a notice on that date; on the contrary, in all probability they gave the wrong date as a result of a mistaken view of their rights. I wish to add that the invocation of Goulding J's test by the Court of Appeal in G *Germax Securities Ltd* v *Spiegal* (1978) 37 P&CR 204* reveals no such error, because (as Buckley LJ pointed out at p206) the mistaken date in that case was not in the operative part of the notice.

Mr Cherryman also relied upon the decision of the Court of Appeal in *Delta Vale Properties Ltd* v *Mills* [1990] 1 WLR 445, in which the test propounded by Goulding J in *Carradine* was referred to in the judgment of Slade LJ, at p454. That case was concerned with the meaning and effect of notices to complete served by a vendor on a purchaser of land. Under the contract it was provided that, upon service of such a notice, the transaction should "be completed within 15 working days of service and in respect of such period time shall be of the essence". The notices however substituted a period of 28 days H for the period of 15 days. The purchaser, lacking the necessary funds, failed to complete within the period of 28 days; then, having obtained the necessary funds a few days later, sought specific performance of the contract. The vendor claimed that he had rescinded the contract on the purchaser's failure to comply with the notice to complete. The question at issue was whether that notice was effective. The Court of Appeal held that it was. Slade LJ said, at p455, that there was only one sense in which any reasonable recipient would have read it, *viz* that the vendor would not exercise the rights conferred by the contract (to rescind if the purchaser did not complete within 15 days), provided that the purchaser completed within 28 days. Bingham LJ, who J reached the same conclusion, said, at p457E:

> The authorities show that a notice will be invalid and ineffective unless it gives the precise notice which the contract requires and leaves the recipient in no reasonable doubt as to the effect of the notice.

I can see nothing in the judgments in this case which detracts from the principles which I have already stated.

In the end Mr Cherryman was forced to submit in his reply that *Hankey* v *Clavering* should be overruled. In doing so, he was faced with the dual handicap, first, that no such argument had been foreshadowed in his printed case, and, second, that he was in truth K inviting your lordships to overrule not merely *Hankey* v *Clavering* itself but the long line of authority which had preceded it. Even so, I have considered with some care whether your lordships should accept this submission. In doing so I have, as previously stated, recognised that, on occasion, the recipient of a notice under, for example, a break clause in a lease may, as here, treat it as ineffective because it does not comply with the contractual specification, even though he must have been aware that the giver of the notice intended to comply with the clause and only failed to do so through a mistake on his part. I am, however, driven to state that there are formidable obstacles in the way of Mr Cherryman's submission.

L The suggestion is that a more relaxed approach be adopted to the construction of notices of this kind, so that, if it is clear that the giver of the notice intended to comply with the provisions of the clause, he should be held to have done so, despite an erroneous choice of the date on which the notice is to take effect. It was submitted that, provided that this test is expressed in sufficiently strict terms, so that the recipient is left in no reasonable doubt that the giver of the notice intended to comply with the clause, this approach should not lead to any undue uncertainty in its application. This latter argument I do not find persuasive. There is a wide range of possible errors, and there are bound to be cases on the borderline in which there is doubt whether the M intention is sufficiently clear. More fundamentally, however, it seems to me that the adoption of such a test in truth requires that a new meaning should be given to clauses of this kind, so that they are read as requiring no more than that the giver of the notice should express a clear intention to exercise his rights under the clause, instead of requiring that the date on which it is to take effect must be expressly identified, either as such or with reference to the terms of the clause.

I have no doubt that this may indeed be the meaning properly to be attributed to many contractual provisions conferring a right exercisable upon notice. But the difficulty in the way of so holding in the case of

---

*Editor's note: Also reported at [1979] 1 EGLR 84

A clauses in leases, such as the break clause in the present case, is that for well over a century a different construction has been placed upon them. Innumerable leases, many of them still in force, must have been drafted on the understanding that they bear the well-known meaning established in the authorities. It follows that, quite apart from the element of uncertainty to which I have referred, the step which your lordships are being invited to take would have a retrospective effect. In these circumstances I am most reluctant to depart from the established meaning unless I am persuaded that there is very good reason to do so.

In considering this question, I start from the position not only that the established meaning is clear and well-known, but also that the trap B which the tenant fell into in the present case is easily avoided by adopting the familiar stratagem of invoking, as an alternative to the specified date, a date identifiable by reference to the terms of the clause itself. In these circumstances I am driven to wonder how often it occurs that, among the numerous notices of this kind which are given each year, mistakes of this kind are made. I simply do not know; though I cannot help suspecting that such cases are very few. If they were more frequent than I suspect, it would be surprising if there was not a greater move for such clauses to be drafted in different terms, or even for the legislature to require that such clauses should be read as having a different effect. In these circumstances I find myself C responding to the submission made by Mr Nicholas Patten QC, in his admirable argument on behalf of the respondent landlords, that it is inadvisable for the judges to disturb this well-settled branch of the law. After all, the number of notices given each year under leases must be very great. Their effect, if challenged, will ordinarily fall to be considered in the county court. At present, the applicable law is clear and well-settled, and Mr Patten informed your lordships that disputes were rare. In these circumstances the change in the law now proposed would not, in my opinion, be justified.

For these reasons, I would dismiss the appeal.

D Also dissenting, **Lord Jauncey of Tullichettle** said: Clause 7(13) of each lease empowered the appellant as tenant to terminate the lease by serving not less than six months notice "to expire on the third anniversary of the term commencement date [to] determine this lease and *upon expiry of such notice* this notice shall cease and determine . . . ". The words emphasised (by me) refer to the expiry of the period of not less than six months which must be contemporaneous with the termination of the lease. It is agreed that the third anniversary expired on January 13 1995. However the notices, although stated to be pursuant to clause 7(13) bore to determine the leases on January 12 the day before a notice given in accordance with that provision could expire. The question is whether the reference to January 12 instead of E 13 is fatal to the validity of the notices. The Court of Appeal, reversing the decision of the judge, held that it was.

Notices terminating a tenancy are technical documents because they are effective without the consent of the receiver. It is therefore essential that they conform to the statutory or contractual provisions under which they are given. In *Cadby* v *Martinez* (1840) 11 A&E 720, Lord Denman CJ, at p726, observed that a covenant in a lease "cannot be got rid of by any notice to quit which is not in accordance with the proviso introduced into the lease for the purpose". This proposition was expressly approved more than a hundred years later in *Hankey* v *Clavering* [1942] 2 KB 326, at p330 by Lord Greene MR. Mr F Cherryman QC for the tenants in his opening speech accepted this proposition but argued that properly construed the notices did accord with clause 7(13) of the lease inasmuch as the specific reference thereto predominated over the subordinate and unnecessary reference to the date January 12. The notices would have been perfectly good if the words "12 January 1995" had been omitted. My lords, I reject this contention. Clause 7(13) required that the notice be expressed to expire in accordance with the provisions thereof. This could be achieved either by reference to the correct date or by repeating the formula "to expire on . . ." in the clause. This was the form of notice which the tenants had agreed to serve but which in the event they did not. The notice contains within itself a specific date upon which the tenancy is to determine. G Is that date to be construed as another date because of the reference to the empowering provisions of clause 7(13) or are those provisions to be construed as authorising notices which do not strictly conform to what has been agreed? In *Hankey* v *Clavering* the Court of Appeal had to determine the validity of a landlord's notice to quit which purported to terminate a tenancy some four days before the break clause in the lease permitted this to be done. In terms of the lease termination thereof could properly take place on December 25 1941 whereas the notice was in, *inter alia*, the following terms (p327):

> As I may have to be away for some time in the near future, I will be obliged if you would accept the six months' notice to terminate your client's lease which I am allowed to give on June 21, 1941. This would mean that he would have to H give up the cottage on December 21, 1941 . . .

In accepting that the date in the notice was obviously a slip Lord Greene MR rejected the temptation to put a strained construction on the document in order to aid someone who had made an unfortunate slip. At pp329-330 he said:

> Notices of this kind are documents of a technical nature, technical because they are not consensual documents, but, if they are in proper form, they have of their own force without any assent by the recipient the effect of bringing the demise to an end. They must on their face and on a fair and reasonable construction do what the lease provides that they are to do. J It is perfectly true that in construing such a document, as in construing all documents, the court in a case of ambiguity will lean in favour of reading the document in such a way as to give it validity, but I dissent entirely from the proposition that, where a document is clear and specific, but inaccurate on some matter, such as that of date, it is possible to ignore the inaccuracy and substitute the correct date, or other particular because it appears that the error was inserted by a slip. By the clear wording of this notice the plaintiff purported to bring the lease to an end on December 21, 1941. In so doing he was attempting to do something which he had no power to do, and, however much the recipient might guess, or however certain he might be, that it was a mere slip, that would not cure the defect because the document was never capable on its face of producing the necessary legal consequence. The authority for that proposition is *Cadby* v *Martinez*, 11 A&E 720 . . . K

I take from these observations the proposition that if a notice to quit is *ex facie* clear and specific as to a matter such as the subjects to which it relates or the date upon which it is to operate it matters not that consideration of the empowering provision demonstrates a discrepancy between it and the notice. The case of ambiguity to which Lord Greene MR referred was one appearing on the face of the notice and not one which was only apparent from a consideration of the lease. Indeed it was implicit in the notice in that case that it was given in pursuance of the landlord's contractual powers with which it did not accord. However, the conflicting terms of the lease did not save it. L Furthermore I can see no reason for construing clause 7(13) as permitting non-conforming notices.

It was argued that the specific reference to clause 7(13) distinguished this case from *Hankey* v *Clavering*. I disagree. If it were appropriate to read into an otherwise clear and specific notice a contractual power for the purpose of substituting a correct for an incorrect date Lord Greene MR would no doubt have done so, given his observation as to the court endeavouring to give validity to a document. I see no difference in principle between an implied reference to a contractual provision and an express one. If *Hankey* v *Clavering* is still good law then the notice given by this tenant was M ineffective.

I turn to consider whether the law has, as Mr Cherryman submitted, during the last 50 years developed a more flexible attitude to notices to quit which contained inaccuracies. In *Carradine Properties Ltd* v *Aslam* [1976] 1 WLR 442 Goulding J held that a notice by landlords correctly dated September 6 1974, to determine a lease on September 27 1973 would be interpreted as a notice to determine on September 27 1975, a date which accorded with the provisions of the break clause. The judge after referring to the need for a party to comply strictly with the requirement for the exercise of an option or break clause continued, at p446C:

> If the condition includes the giving of a particular notice, it seems to me that the logical first approach is to interpret the notice, looking at the words and applying legal principles to their construction, and then ask whether it complies with the strict requirements as to exercise of the option. If that is right, I think a benevolent approach could be applied in this case, as in the [*Duke of Bedford's Case* (1796) 7 Term Rep 63], because reasonably read by a reasonable tenant the mistake is obvious on the face of it, and there is no doubt what the mistake was. Therefore one interprets the notice as asserting an intention to determine in 1975. It is true that if whoever made the mistake had typed 1976 instead of 1973, the error would probably have been incurable because although the tenant might suspect there was a slip, it might be that the landlord did intend 1976, not knowing or understanding his rights under the lease.

In this passage the judge was contrasting the situation where it was apparent from the face of the notice that the date of determination must be mistaken because it had already passed with one where the date of determination had yet to come and where it could not therefore be seen from the notice itself that the date was inserted in error. Mr Cherryman argued that the judge was correct in relation to the first situation but wrong in relation to the second. In my view, the distinction drawn by Goulding J was entirely logical. In the first situation, which was the case before him, it only required a glance at the notice to see that the date of determination must be a mistake. This was not the position in *Hankey* v *Clavering* [1942] 2 KB 326. In the second situation there was nothing within the four corners of the notice which showed that 1976 was a mistake. This was the position in *Hankey* v *Clavering*.

In *Delta Vale Properties Ltd* v *Mills* [1990] 1 WLR 445 a contract for sale of property provided, *inter alia*, for service of a completion notice requiring the transaction to be completed within 15 days. The seller served such a notice requiring completion within 28 days and the buyers challenged its validity. The Court of Appeal held that the notice was valid since a reasonable reader thereof could only have concluded that the notice was intended to be one conforming to the empowering provision and that the sellers were giving him 28 days instead of 15 to complete. This was not a case of a party serving a notice which he was not empowered to serve but rather of a party exercising forbearance in the strict compliance with a contractual term in his favour by giving, as he was entitled, to do, a longer period of notice than was required. The case is no support for the tenant's argument that January 13 should be substituted for January 12. In *Micrografix* v *Woking 8 Ltd* [1995] 2 EGLR 32 the provisions of a lease entitled tenants to give not less than 12 months' notice of their intention to determine the lease on June 23 1995. In January 1994 they sent a letter stating that they were enclosing a notice determining the lease on March 23 1995, which notice stated the date of determination as March 23 1994. Jacob J in holding that the notice was valid because the landlords would not have been misled by the pair of wrong dates and would have seen exactly what the tenants intended, observed that *Carradine Properties Ltd* v *Aslam* [1976] 1 WLR 442 was the modern approach to construction of documents which was more in favour of making them work than in the past. After referring to Goulding J's second situation, he stated, at p33:

> That was an *obiter* observation at a point in time up to which the rather stricter view of Lord Greene had prevailed. In 1995 I rather think that a court faced with an obvious slip would construe the document without the slip.

My lords I consider that Jacob J was in error in placing so much reliance on the understanding of the landlord. Lord Greene MR in the passage from *Hankey* v *Clavering* [1942] 2 KB 326, at pp329-330, to which I have already referred emphasised that even certainty by a recipient that a date was a slip would not cure the defect and this view was echoed in the Court of Appeal in this case by Nourse LJ [1995] 1 WLR 1508, at p1513F-G. In *Delta Vale Properties Ltd* v *Mills* [1990] 1 WLR 445, at p455F, Slade LJ accepted at p455F that the absence of confusion or prejudice on the part of the recipient was irrelevant. Their lordships were referred to no other case in which this approach had been questioned nor the stricter approach of Lord Greene MR in *Hankey* v *Clavering* [1942] 2 KB 326, at pp329-330, doubted. *Micrografix* [1995] 2 EGLR 32 was the only case cited in argument in which the court had construed a specific date as other than that stated and in which the error did not, as in *Carradine*, proclaim itself from the face of the notice. I consider that it was wrongly decided.

Hankey v *Clavering* is a case of considerable authority, which has stood unchallenged and been applied for more than 50 years. Mr Cherryman only sought to challenge it with no great enthusiasm in the last speech. I am not persuaded that there is any good reason for departing from the principles enunciated by Lord Greene MR. It would be tempting to say that where a notice is disconform to that stipulated in the relevant empowering provision due to a slip which was obvious to any reasonable receiver thereof it may be construed so as to conform. However I can see considerable difficulties arising out of the application of such a principle. On one view it could be said that the giver of a notice empowered by a contractual provision always intends to comply with that provision thus however far such notice departed from what was contractually required it could be construed to conform with such requirements. On another view it could be argued that a notice which departed substantially from what was required cannot have been intended to conform but was intended to depart from contractual requirements or given in ignorance thereof. Where is the line to be drawn between defective notices which can and those which cannot be construed so as to conform: such possibilities would create inevitable uncertainties which as Mr Patten QC for the landlord submitted would be likely to lead to increased litigation. There will, of course, be cases where an unintended slip in the drafting of a notice will result in hardship to the giver thereof but he will only have himself to blame for not complying with the terms of the empowering provision. Against this, however, must be measured the advantage to the recipient of certainty inasmuch as a date in a notice is to be read as stated unless it is obvious from the face of the notice that it must be mistaken.

Applying Lord Greene MR's principles to this case, the tenant had no power under clause 7(13) to determine the lease on January 12 from which it follows that their notice was ineffective.

My lords, I have had the advantage of reading in draft the speech of my noble and learned friend, Lord Goff of Chieveley with whose reasoning I am in entire agreement. I would therefore dismiss the appeal.

Giving his opinion, **Lord Steyn** said: The short but not uncomplicated question is whether notices given by a tenant to a landlord purportedly under a right to determine reserved in terms of leases for fixed terms were effective to determine the leases.

The landlord is Eagle Star Life Assurance Co Ltd. The tenant is Mannai Investment Co Ltd. There are two leases both dated March 11 1992. The demised property consists of office premises and car parking space in London SW1. The two leases demised the office premises and the car parking space in each case "For the term of 10 years subject to the provisions of clause 7(13) hereof from and including 13 January 1992 . . .". Clause 7 (13) of each lease provided as follows:

> The Tenant may by serving not less than six months notice in writing on the Landlord or its solicitors such notice to expire on the third anniversary of the term commencement date determine this Lease and upon expiry of such notice this Lease shall cease and determine and have no further effect . . .

Under each lease the term granted included January 13 as the first day of the term. The "term commencement date" in clause 7(13) was not defined in either lease. But it is common ground that by reason of the words "from and including 13 January 1992" it was that date. Accordingly, it was agreed that the third anniversary of the "term commencement date" was January 13 1995.

By two letters dated June 24 1994 and served on the landlord more than six months before January 13 1995 the tenant gave notice to the landlord in respect of each lease as follows:

> Pursuant to Clause 7(13) of the Lease we as Tenant hereby give notice to you to determine the Lease on 12 January 1995.

A The tenant contended that the leases were determined by the notices. The landlord claimed that the notices did not comply with the right reserved under clause 7(13) and that the notices were ineffective in law.

By an originating summons the tenant sought a declaration that the notices effectively determined the leases. The matter came before Judge Rich QC, sitting as a judge of the High Court in the Chancery Division. Relying on *Sidebotham* v *Holland* [1895] 1 QB 378 Judge Rich held that the notices were good inasmuch as they did not take effect until that moment of time which was both the last moment of January 12 and the first of January 13, so that they did determine the **B** leases on January 13. The landlord appealed to the Court of Appeal.

Two issues were debated in the Court of Appeal. The first was whether, despite the erroneous reference in the notices to January 12 rather than 13, the notices were nevertheless as a matter of construction sufficiently clear to be effective: this was described as the construction point. The second issue was whether the notices were saved by what was called the special rule in *Sidebotham* v *Holland*. The Court of Appeal decided both issues against the tenant: *Mannai Investment Co Ltd* v *Eagle Star Life Assurance Co Ltd* [1995] 1 WLR 1508. The Court of Appeal held that the notices did not determine the leases.

C On the appeal to your lordship's House counsel for the tenant again advanced both arguments. The point of construction is one of substance. Despite a lucid argument by junior counsel for the tenant your lordships were of the view, which I shared, that the argument based on *Sidebotham* v *Holland* [1895] 1 QB 378 was plainly wrong and did not find it necessary to call on counsel for the landlord to reply on this issue. As Nourse LJ [1995] 1 WLR 1508, at p1515F trenchantly explained in the Court of Appeal *Sidebotham* v *Holland* is no authority for the proposition that in respect of a lease for a fixed term a notice served on one day can be treated as a notice served on the next. That disposed of this point. I turn therefore to the construction of **D** the notices.

On reflection I have come to the conclusion that the question of the construction of the notices should be answered by holding that the notices were effective to determine the leases. I will first summarise my analysis of the problem before I explain why I feel unable to accept the attractively presented arguments of counsel for the landlord.

The reasons for my conclusion can be stated in the form of numbered propositions:

(1) This is not a case of a contractual right to determine which prescribes as an indispensable condition for its effective exercise that the notice must contain specific information. After providing for the **E** form of the notice ("in writing"), its duration ("not less than six months") and service ("on the landlord or its solicitors"), the only words in clause 7(13) relevant to the content of the notice are the words "notice to expire on the third anniversary of the term commencement date determine this lease". Those words do not have any customary meaning in a technical sense. No terms of art are involved. And neither side has suggested that anything should be implied into the language. That is not surprising since the tests governing the implication of terms could not conceivably be satisfied. The language of clause 7(13) must be given its ordinary meaning. A notice simply expressed to determine the lease on third anniversary of **F** the commencement date would therefore have been effective. The principle is that that is certain which the context renders certain: *Sunrose Ltd* v *Gould* [1962] 1 WLR 20.

(2) The question is not how the landlord understood the notices. The construction of the notices must be approached objectively. The issue is how a reasonable recipient would have understood the notices. And in considering this question the notices must be construed taking into account the relevant objective contextual scene. The approach in *Reardon Smith Line Ltd* v *Yngvar Hansen-Tanger* [1976] 1 WLR 989, which deals with the construction of commercial contracts, is by analogy of assistance in respect of unilateral notices such as those under consideration in the present case. Relying on the reasoning in Lord Wilberforce's speech in *Reardon Smith*, at pp996D to 997D, three propositions can be formulated. **G** First, in respect of contracts and contractual notices the contextual scene is always relevant. Second, what is *admissible* as a matter of the rules of evidence under this heading is what is arguably relevant. But admissibility is not the decisive matter. The real question is what evidence of surrounding circumstances may ultimately be allowed to influence the question of interpretation. That depends on what meanings the language read against the objective contextual scene will let in. Third, the inquiry is objective: the question is what reasonable persons, circumstanced as the actual parties were, would have had in mind. It follows that one cannot ignore that a reasonable recipient of the notices would have had in the forefront of his mind the terms of the leases. **H** Given that the reasonable recipient must be credited with knowledge of the critical date and the terms of clause 7(13) the question is simply how the reasonable recipient would have understood such a notice. This proposition may in other cases require qualification. Depending on the circumstances a party may be precluded by an estoppel by convention from raising a contention contrary to a common assumption of fact or law (which could include the validity of a notice) upon which they have acted: *Norwegian American Cruises A/S* v *Paul Mundy Ltd* (*The "Vistafjord"*) [1988] 2 Lloyd's Rep 343. Such an issue may involve subjective questions. That is, however, a different issue and not one relevant to this appeal. I proceed therefore to examine the matter **J** objectively.

(3) It is important not to lose sight of the purpose of a notice under the break clause. It serves one purpose only: to inform the landlord that the tenant has decided to determine the lease in accordance with the right reserved. That purpose must be relevant to the construction and validity of the notice. *Prima facie* one would expect that if a notice unambiguously conveys a decision to determine a court may nowadays ignore immaterial errors which would not have misled a reasonable recipient.

(4) There is no justification for placing notices under a break clause in leases in a unique category. Making due allowance for contextual **K** differences, such notices belong to the general class of unilateral notices served under contractual rights reserved, eg notices to quit, notices to determine licences and notices to complete: *Delta Vale Properties Ltd* v *Mills* [1990] 1 WLR 445, at pp454E-G. To those examples may be added notices under charter parties, contracts of affreightment, and so forth. Even if such notices under contractual rights reserved contain errors they may be valid if they are "sufficiently clear and unambiguous to leave a reasonable recipient in no reasonable doubt as to how and when they are intended to operate": the *Delta* case, at p454E-G, *per* Slade LJ and adopted by Stocker LJ and Bingham LJ: see also *Carradine Properties Ltd* v *Aslam* [1976] 1 **L** WLR 442, at p444. That test postulates that the reasonable recipient is left in no doubt that the right reserved is being exercised. It acknowledges the importance of such notices. The application of that test is principled and cannot cause any injustice to a recipient of the notice. I would gratefully adopt it.

(5) That brings me to the application of this test. The facts are simple. Crediting a reasonable recipient with knowledge of the terms of the lease and third anniversary date (January 13), I venture to suggest that it is obvious that a reasonable recipient would have appreciated that the tenant wished to determine the leases on the third anniversary date of the leases but wrongly described it as the 12th **M** instead of the 13th. The reasonable recipient would not have been perplexed in any way by the minor error in the notices. The notices would have achieved their intended purpose.

That brings me to the contrary reasoning of the Court of Appeal in this case, and the submission of counsel for the landlord. Central to both was the proposition that *Hankey* v *Clavering* [1942] 2 KB 326 precluded such a conclusion. Given the importance ascribed to this decision I must embark on what Justice Cardozo once described as a gruesome autopsy. In *Hankey* v *Clavering* there was a lease for a term of 21 years from December 25 1934. The lease provided, *inter alia*, that if either party should desire to determine the lease after seven or 14 years he could do so by giving to the other party six months' notice.

A The landlord gave notice to the tenant as from June 1941, which purported to terminate the lease on December 21 1941. On March 8 1940 the landlord's solicitors acknowledged the notice as "properly served upon us". The tenant refused to give up possession of the premises. The landlord brought an action for possession. Mr Denning KC, submitted that the landlord attempted to terminate the tenancy on December 21 whereas he was only entitled to terminate it on December 25. In these circumstance, he said, that the defect could not be cured by showing that the other party understood what was meant. Mr Blanco White KC, argued that if a notice can be understood by a reasonable person it is a good notice. The judge decided in favour of
B the landlord on the effect of the correspondence. The matter came before a two judge Court of Appeal. It was disposed of by *ex tempore* judgments. Lord Greene MR gave the principal judgment; Lord Clauson added nothing of substance. The Court of Appeal ruled that the judge had misinterpreted the correspondence. Turning to the construction of the notice, Lord Greene MR said (at pp329-330):

> Notices of this kind are documents of a technical nature, technical because they are not consensual documents, but, if they are in proper form, they have of their own force without any assent by the recipient the effect of bringing the demise to an end. They must on their face and on a fair and reasonable construction do what the lease provides that they are to do. It is perfectly true
> C that in construing such a document, as in construing all documents, the court in a case of ambiguity will lean in favour of reading the documents in such a way as to give it validity, but I dissent entirely from the proposition that, where a document is clear and specific, but inaccurate on some matter, such as that of date, it is possible to ignore the inaccuracy and substitute the correct date or other particular because it appears that the error was inserted by a slip. By the clear wording of this notice the plaintiff purported to bring the lease to an end on 21 December 1941. In so doing he was attempting to do something which he had no power to do, and, however much the recipient might guess, or however certain he might be, that it was a mere slip, that would not cure the defect because the document was never capable on its face of producing the necessary legal consequence.

D Lord Greene MR said that his ruling was based on *Cadby* v *Martinez* (1840) 11 A&E 720, which he described of as "a case of the highest authority in these matters". Lord Denman CJ said in that case (at p726) that "the covenant to pay rent during the whole term cannot be got rid of by any notice to quit which is not in accordance with the proviso introduced into the lease for the purpose".

It goes without saying that any judgment of Lord Greene MR is entitled to great respect. But one must put the case in context. First, in relying *Cadby* v *Martinez* Lord Greene was founding his proposition on a case which with the benefit of hindsight seems far from
E conclusive on the point. Lord Denman CJ's judgment contains no reasoning at all; it is purely conclusionary. A century and a half later it is sometimes necessary to consider the force of reasoning in decided cases. *Cadby* v *Martinez* does not enable one to do so. In any event, Lord Denman's conclusion is in such absolute terms as to be of little value. It is not supported in this case by the judgment in the Court of Appeal or by the submissions of counsel for the landlord: it is conceded that some errors in a notice can be ignored. The real question is: What errors can be overlooked? Taking due account of what Lord Greene thought of *Cadby* v *Martinez* in 1942, I incline to the view that the persuasive force of this precedent must be regarded as slight. Second, it is noteworthy that Lord Greene does not expressly pose an
F objective test. He dismisses the reaction of the recipient ("however much the recipient might guess"). The law has moved on. The test is entirely objective. Third, Lord Greene did not expressly deal with the position where the notice contains an error which proves wholly immaterial and incapable of causing any confusion. The second and third matters detract from the force of reasoning in *Hankey* v *Clavering*. Lord Greene was not considering the issues as they have to be faced in this case. But there is a fourth point to be taken into account. *Hankey* v *Clavering* was decided more than half a century ago. Since then there has been a shift from strict construction of commercial instruments to what is sometimes called purposive construction of such documents. Lord Diplock deprecated the use of
G that phrase in regard to the construction of private contracts as opposed to the construction of statutes: *Antaios Compania Naviera SA* v *Salen Rederierna AB* [1985] AC 191, at p201D. That is understandable. There are obvious differences between the processes of interpretation in regard to private contracts and public statutes. For a perceptive exploration of the differences in the context of United States law, see Robert S Summers, "Statutes and Contracts as Founts of Formal Reasoning", in *Essays for Patrick Atiyah*, edited by Peter Cane and Jane Stapleton (1991) at p71 *et seq*. It is better to speak of a shift towards commercial interpretation. About the fact of the change in approach to construction there is no doubt. One illustration will be sufficient. In *Antaios Compania Naviera SA* v *Salen Rederierna AB* H
[1985] AC 191, at p201D, Lord Diplock in a speech concurred in by his fellow Law Lords observed that:

> . . . if detailed semantic and syntactical analysis of words in a commercial contract is going to lead to a conclusion that flouts business commonsense, it must be made to yield to business common sense.

In determining the meaning of the language of a commercial contract, and unilateral contractual notices, the law therefore generally favours a commercially sensible construction. The reason for this approach is that a commercial construction is more likely to give effect to the intention of the parties. Words are therefore interpreted in the J
way in which a reasonable commercial person would construe them. And the standard of the reasonable commercial person is hostile to technical interpretations and undue emphasis on niceties of language. In contradistinction to this modern approach Lord Greene's judgment in *Hankey* v *Clavering* is rigid and formalistic. Nowadays one expects a notice to determine under a commercial lease to be interpreted not as a "technical document" but in accordance with business common sense: see *Micrografix* v *Woking 8 Ltd* [1995] 2 EGLR 32. After all, there is no reason whatever why such a document must be drafted by a lawyer. Qualitatively, the notices are of the same type as notices under charter parties and contracts of affreightment. Such notices, even if they entail the exercise of important options, are habitually drafted by K
commercial men rather than lawyers. It would be a disservice to commercial practice to classify such notices as technical documents and to require them to be interpreted as such. Nowadays one must substitute for the rigid rule in *Hankey* v *Clavering* the standard of a commercial construction.

It is, however, also important to note that the decision in *Hankey* v *Clavering* caused surprise even in 1942. There was a case note on *Hankey* v *Clavering* bearing the initials REM: (1943) 59 LQR 17. The author of the note was a great authority on the field of landlord and tenant. He later became Sir Robert Megarry, Vice-Chancellor of the Chancery Division. He was clearly surprised at the decision. He drew L
attention to Lord Greene MR's decision in *Price* v *Mann* [1942] 1 All ER 453. In that case the question was whether under the Landlord and Tenant (War Damage) Act 1939 a notice to avoid disclaimer given by the landlord, requiring the tenant to retain the lease on the terms set out in section 10, was invalid because section 10 was irrelevant and by mistake inserted for section 11. The Court of Appeal held that the notice was good. Lord Greene MR said (at p454):

> Reading this document as a whole, it seems to me perfectly manifest that a person who received it, and who had that familiarity with the provisions of the Act which a recipient of such a document must be presumed to have, could not M
> possibly be under any illusion as to what it was intended to be and what its legal consequences were.

Mr Megarry observed, at p18:

> But for the line of authority governing notices to quit, it would have occasioned little surprise to some had the words used by the Master of the Rolls in the two cases been interchanged. Read literally, both notices purported to do something which they could not do, yet read as a whole, neither was at all likely to mislead the recipient . . . The distinction may be that in *Price* v *Mann* the reference to the precise section of the Act was unnecessary whereas in *Hankey* v *Clavering* the insertion of the date was essential to the validity of a notice cast in the form employed. If this is not the case, perhaps all that can be said is that in *Price* v *Mann* each member of the Court of Appeal could use

A the words of MacKinnon LJ, and say to himself "The law as it stands does permit me to give effect to common-sense and decency" (*Heap* v *Ind Coope & Allsopp Ltd* [1940] 2 KB 476, at p484), whereas in *Hankey* v *Clavering* the hand of *stare decisis* was writ large.

Lord Greene's observation in *Price* v *Mann* is much closer to the modern standard of commercial construction than *Hankey* v *Clavering*.

Counsel for the tenant invited the House in his case and in his reply to say that *Hankey* v *Clavering* was wrongly decided. I am content to say that it no longer represents the law. Like Lord Hoffmann I would hold that the correct test for the validity of a notice is that posed by
B Goulding J in *Carradine Properties Ltd* v *Aslam* [1976] 1 WLR 442, at p444, *viz* "Is the notice quite clear to a reasonable tenant reading it? Is it plain that he cannot be misled by it?"

It is necessary to turn briefly to other arguments advanced by counsel for the landlord. He argued that the tenant's construction argument must fail because the tenant's error may have been due to a mistaken legal view. He said that it may not have been a case of a mistaken insertion of a date but the tenant may have intended to refer to the 12th. This reveals a contradiction in the landlord's argument. Counsel for the landlord accepted that the test is an objective one: How would a reasonable recipient have understood the notice? But then he invited your lordships to speculate that the tenant's error was due to a
C mistake of law rather than a typing or clerical error. That argument, if accepted, would drive a juggernaut through the objective test. Speculation about the subjective intention of the tenant is irrelevant. The only question is how a reasonable recipient would have understood the notice.

Counsel also argued that as a matter of legal logic a process of interpretation can never permit one to substitute January 13 for January 12. Why should that be so? If a contract contains a termination date linked to an intended three-year period, which is variously expressed in the contract as January 12 and January 13, why should the court not as a matter of interpretation be able to select the date which best matches
D the contractual intent? The same reasoning must apply to unilateral documents such as contractual notices. It is surely permissible in all cases satisfying the test that no reasonable recipient of the notice could be misled. Counsel's argument is based on too formalistic a formulation of the question to be decided. The question is not whether January 12 can mean January 13: it self-evidently cannot. The real question is a different one: Does the notice construed against its contextual setting unambiguously inform a reasonable recipient how and when the notice is to operate *under the right reserved*? As Lord Hoffmann has observed we no longer confuse the meaning of words with the question of what meaning in a particular setting the use of
E words was intended to convey.

That brings me to counsel's argument that, if the notices are treated as valid, there will be great deal of confusion and unnecessary litigation. Experience teaches that "floodgates" arguments need to be examined with an initial scepticism. In this case the predictions of counsel are unrealistic. Those arguments must be judged on the basis that the test posed above is accepted. That test can only be satisfied where the reasonable recipient could be left in no doubt whatever. It is in accord with business commonsense that in cases where that simple and straightforward test is satisfied the notices should be treated as valid.

F That brings me to my conclusion. I do not accept the extreme argument of counsel for the tenant that whenever a notice to determine refers to a break clause, and whatever the other circumstances of the case, the notice must be valid. That goes too far. One can easily conceive of much weaker cases where the test posed above could not be satisfied. But in the present case it would have been obvious to a reasonable recipient that the notices contained a minor misdescription and that the notices conveyed that the tenant sought to determine the leases on "on third anniversary of the term commencement", ie January 13. I end this judgment with the words with which in 1903 Sir Leslie Stephen concluded a famous series of lectures: "I hope I have not said anything original".

It follows that I would allow the appeal. G

Agreeing, **Lord Hoffman** said: The appellant was tenant under two 10-year leases of offices in Jermyn Street, each of which contained in clause 7(13) a right to terminate at the end of the third year in the following terms:

> The tenant may by serving not less than six months' notice in writing on the Landlord or its Solicitors such notice to expire on the third anniversary of the term commencement date determine this Lease and upon the expiry of such notice this Lease shall cease and determine and have no further effect . . .

After the grant of the leases the market rents of offices in the West H
End fell sharply. On June 24 1994 the tenant served on the landlord two notices, each of which read as follows: "Pursuant to Clause 7(13) of the Lease we as Tenant hereby give notice to you to determine the lease on 12 January 1995". It is agreed that the third anniversary of the commencement date was actually 13 January 1995. The question is whether notwithstanding this mistake the notices were effective to terminate the leases.

This might seem a straightforward question, particularly when it is remembered that such notices, operating, as they do, unilaterally to alter the rights of the parties, must comply strictly with the terms of the lease. The Court of Appeal held that the notice was ineffective on the J
simple ground that "12 January" could not mean "13 January". In so doing, they followed (as in my view they were bound to do) the decision of the Court of Appeal in *Hankey* v *Clavering* [1942] 2 KB 326 which in turn had followed the decision of the Court of Queen's Bench in *Cadby* v *Martinez* (1840) 11 A&E 720. In that case, the notice said Midsummer instead of Lady Day. It seemed obvious to Lord Denman CJ that there was no way in which it could be construed to refer to Lady Day and he merely observed, at p726, that "in [no case] has a proviso or covenant in a deed been held to be satisfied by a notice inconsistent with the terms of it".

And yet, my lords, the case is by no means straightforward. The K
clause does not require the tenant to use any particular form of words. He must use words which unambiguously convey a particular meaning, namely an intention to terminate the lease on January 13. In *Hankey* v *Clavering* [1942] 2 KB 326, where the notice to quit said "21 December" instead of "25 December", Lord Greene MR said, at pp328, 330, "The whole thing was obviously a slip" on the part of the landlord but that the notice was invalid "however much the recipient might guess, or however certain he might be" that it was a mere slip. So even if the recipient was certain that the landlord actually wanted to terminate his tenancy on the right date, which was December 25, so that the necessary intention was unambiguously communicated, the notice was bad. One is bound to be left with a feeling that something L
has gone wrong here. Common sense cannot produce such a result; it must be the result of some rule of law. If so, what is that rule and is it correct?

I propose to begin by examining the way we interpret utterances in everyday life. It is a matter of constant experience that people can convey their meaning unambiguously although they have used the wrong words. We start with an assumption that people will use words and grammar in a conventional way but quite often it becomes obvious that, for one reason or another, they are not doing so and we adjust our interpretation of what they are saying accordingly. We do so in order M
to make sense of their utterance: so that the different parts of the sentence fit together in a coherent way and also to enable the sentence to fit the background of facts which plays an indispensable part in the way we interpret what anyone is saying. No one, for example, has any difficulty in understanding Mrs Malaprop. When she says: "She is as obstinate as an allegory on the banks of the Nile", we reject the conventional or literal meaning of allegory as making nonsense of the sentence and substitute "alligator" by using our background knowledge of the things likely to be found on the banks of the Nile and choosing one which sounds rather like "allegory".

Mrs Malaprop's problem was an imperfect understanding of the conventional meanings of English words. But the reason for the mistake does not really matter. We use the same process of adjustment when people have made mistakes about names or descriptions or days or times because they have forgotten or become mixed up. If one meets an acquaintance and he says "And how is Mary?" it may be obvious that he is referring to one's wife, even if she is in fact called Jane. One may even, to avoid embarrassment, answer "Very well, thank you" without drawing attention to his mistake. The message has been unambiguously received and understood.

If one applies that kind of interpretation to the notice in this case, there will also be no ambiguity. The reasonable recipient will see that in purporting to terminate pursuant to clause 7(13) but naming January 12 1995 as the day upon which he will do so, the tenant has made a mistake. He will reject as too improbable the possibility that the tenant meant that unless he could terminate on January 12, he did not want to terminate at all. He will therefore understand the notice to mean that the tenant wants to terminate on the date on which, in accordance with clause 7(13), he may do so, ie January 13.

Why, then, do cases like *Hankey* v *Clavering* [1942] 2 KB 326 arrive at a different answer? I want first to deal with two explanations which seem to me obviously inadequate. First, it is sometimes said that the examples which I have given from ordinary life are concerned with what the speaker meant to say. He may subjectively have intended to say something different from what he actually said and it may be possible, by the kind of reasoning which I have described, to divine what his subjective intentions were. But the law is not concerned with subjective intentions. All that matters is the objective meaning of the words which he has used.

It is of course true that the law is not concerned with the speaker's subjective intentions. But the notion that the law's concern is therefore with the "meaning of his words" conceals an important ambiguity. The ambiguity lies in a failure to distinguish between the meanings of words and the question of what would be understood as the meaning of a person who uses words. The meaning of words, as they would appear in a dictionary, and the effect of their syntactical arrangement, as it would appear in a grammar, is part of the material which we use to understand a speaker's utterance. But it is only a part; another part is our knowledge of the background against which the utterance was made. It is that background which enables us, not only to choose the intended meaning when a word has more than one dictionary meaning but also, in the ways I have explained, to understand a speaker's meaning, often without ambiguity, when he has used the wrong words.

When, therefore, lawyers say that they are concerned, not with subjective meaning but with the meaning of the language which the speaker has used, what they mean is that they are concerned with what he would objectively have been understood to mean. This involves examining not only the words and the grammar but the background as well. So, for example, in *Doe d Cox* v *Roe* (1802) 4 Esp 185 the landlord of a public house in Limehouse gave notice to quit "the premises which you hold of me . . . commonly called . . . the Waterman's Arms". The evidence showed that the tenant held no premises called the Waterman's Arms; indeed, there were no such premises in the parish of Limehouse. But the tenant did hold premises of the landlord called the Bricklayer's Arms. By reference to the background, the notice was construed as referring to the Bricklayer's Arms. The meaning was objectively clear to a reasonable recipient, even though the landlord had used the wrong name. We therefore will in due course have to answer the question: if, as long ago as 1803, the background could be used to show that a person who speaks of the Waterman's Arms means the Bricklayer's Arms, why can it not show that a person who speaks of January 12 means January 13?

The immediate point, however, is that the fact that the law does not have regard to subjective meaning is no explanation of the way *Hankey* v *Clavering* [1942] 2 KB 326 was decided. There was no need to resort to subjective meaning: the notice would objectively have been understood to mean that the landlord wanted to terminate the tenancy on the day on which he was entitled to do so.

I pass on to a second explanation which also seems to me inadequate. Lord Greene MR said at pp329-330, that because such notices have unilateral operation, the conditions under which they may be served must be strictly complied with. I have already said that this principle is accepted on both sides. But, as an explanation of the method of construction used in *Hankey* v *Clavering*, it begs the question. If the clause had said that the notice had to be on blue paper, it would have been no good serving a notice on pink paper, however clear it might have been that the tenant wanted to terminate the lease. But the condition in clause 7(13) related solely to the meaning which the notice had to communicate to the landlord. If compliance had to be judged by applying the ordinary techniques for interpreting communications, there was strict compliance. The notice clearly and unambiguously communicated the required message. To say that compliance must be strict does not explain why some other technique of interpretation is being used or what it is.

A variation of this explanation is to say that the language of the notice must be strictly construed. But what does it mean to say that a document must be "strictly" construed, as opposed to the normal process of ascertaining the intentions of the author? The expression does not explain itself. If it operates merely by way of intensification, so that the intention must be clear, unambiguous, incapable of misleading, then I think that the notice in this case satisfied the test at that level. Likewise, as Lord Greene MR acknowledged when he said that the whole thing was obviously a slip, did the notice in *Hankey* v *Clavering*. So the concept of strict construction does not explain the decision.

A more promising clue to the explanation is Lord Greene's statement, in two places, that the notice must "on its face" comply with the terms of the lease. What does "on its face" mean? Clearly, the face of the document is being contrasted with the background, in law sometimes called the "extrinsic evidence", against which the language is ordinarily construed. But Lord Greene cannot have meant that the document must always be read without any background, because (even if, which I doubt, it were conceptually possible to interpret the use of language without the aid of any background) cases like the *Cox* case, (1802) 4 Esp 185 show that some background, at least, can be used. It appears, therefore, that Lord Greene is referring to some principle whereby background can be used to show that a person who speaks of the Waterman's Arms means the Bricklayer's Arms, but not that a person who speaks of January 12 means January 13. What principle is this?

It is, I think, to be found in an old rule about the admissibility of extrinsic evidence to construe legal documents. In its pure form, the rule was said to be that if the words of the document were capable of referring unambiguously to a person or thing, no extrinsic evidence was admissible to show that the author was using them to refer to something or someone else. An extreme example is in *In the Goods of RR Peel* (1870) LR 2 P&D 46, in which the testator appointed "Francis Courtnay Thorpe, of Hampton . . . Middlesex" to be his executor. There was a Francis Courtenay Thorpe of Hampton, Middlesex. He was however only 12 years old and his father Francis Corbet Thorpe, of Hampton, Middlesex, was an old friend of the testator. Lord Penzance said, at p47, that these facts were inadmissible: "The testator makes use of a description which applies in fact to one person, and not to any other". A variation on this rule was *In re Fish* [1894] 2 Ch 83 in which the testator left his residuary estate to his "niece Eliza". He had no niece called Eliza but his wife had an illegitimate grandniece called Eliza, to whom the evidence of their relationship showed that he must have intended to refer, and also, as it happened, a legitimate grandniece called Eliza. The Court of Appeal said that the estate went to the legitimate grandniece and that evidence of the relationship between the testator and the illegitimate grandniece was inadmissible. Lindley LJ said, at p85:

> . . . where the person most nearly answering the description is the legitimate grandniece of the testator's wife . . . no evidence can be admitted to prove that her illegitimate grandniece was intended.

On the other hand, if there was no one to whom the description accurately applied, there was said to be a "latent ambiguity" and evidence of background facts which showed what the testator must have meant, notwithstanding that he had used the wrong words, was admitted.

Let us compare this rule with ordinary common-sense interpretation of what people say. If someone has gone to great pains, well in advance, to secure tickets for himself and a friend for a Beethoven concert at the Royal Festival Hall by a famous visiting orchestra on January 13 and says to the friend a week earlier: "I'll see you at the Festival Hall concert on January 12," it will be obvious that he is referring to the concert on January 13. According to the old rules of construction, the law will agree if there is no concert at the Festival Hall on January 12. In that case there is a latent ambiguity. But if there is a concert on that date (Stockhausen, say, played by a different orchestra) he will be taken to have referred to that concert.

This extraordinary rule of construction is, as it seems to me, the only explanation for the decisions in *Hankey* v *Clavering* [1942] 2 KB 326 and *Cadby* v *Martinez* (1840) 11 A&E 720. The *Cox* case, (1802) 4 Esp 185, was distinguished by counsel in *Cadby*, at p723 as involving a latent ambiguity: there was no Waterman's Arms in Limehouse, so evidence that the landlord would have been understood by a reasonable tenant as intending to refer to the Bricklayer's Arms was admissible. But Midsummer 1837, or December 21 1941 (in *Hankey* v *Clavering*) or January 12 1995 (in this case) are all real dates to which the notices could have referred. Therefore evidence of background which showed that a reasonable recipient would have understood the person giving the notice as having intended to refer to a different date had to be disregarded. The effect is that apart from the exceptional case in which the date is obviously impossible on the face of the notice (as in *Carradine Properties Ltd* v *Aslam* [1976] 1 WLR 442) the intention which the notice would convey as to date has to be determined without regard to the terms of the lease (or anything else) as background. There is an artificial assumption that the reasonable recipient does not know what would be the correct date. On this basis, the interpretation of the notices as referring to the wrong dates and therefore being invalid is, of course, inescapable.

It is clear that this rule of construction has been applied to the interpretation of notices for at least 200 years and it is hardly surprising that Lord Greene MR and Lord Clauson felt obliged to apply it in *Hankey* v *Clavering* and that the Court of Appeal applied it in this case. It is, however, highly artificial and capable of producing results which offend against common sense. Lord Penzance began his decision that the testator had appointed a 12-year old boy as his executor by saying:

> If I am at liberty to look at the facts stated in the affidavits, I may possibly have no difficulty in deciding that the person meant is the father, but the question is, whether I am at liberty to do so. (*In the Goods of RR Peel* (1870) LR 2 P&D 46)

In *In re Fish* [1894] 2 Ch 83, at p84, Lindley LJ began his judgment by saying: "This is one of those painful cases in which it is probable that the testator's intention will be defeated" and AL Smith LJ said, at p86, that if he could have admitted the evidence about the testator's relationship with his wife's illegitimate grandniece he would gladly have done so.

I think that the rule is not merely capricious but also, for reasons which I need not develop at length, incoherent. It is based upon an ancient fallacy which assumes that descriptions and proper names can somehow inherently refer to people or things. In fact, of course, words do not in themselves refer to anything, it is people who *use* words to refer to things. The word "allegory" does not mean a large scaly creature or anything like it, but it is absurd to conclude, as judges sometimes do, that this is not an "available meaning" of the word in the interpretation of what someone has said. This is simply a confusion of two different concepts; as we have seen, a person can use the word "allegory", successfully and unambiguously, to refer to such a creature.

Even in its natural habitat, the construction of wills, the rule has not been (and, I think, cannot be) applied with any consistency. In *National Society for the Prevention of Cruelty to Children* v *Scottish National Society for the Prevention of Cruelty to Children* [1915] AC 207 this House refused to accept that a gift to the "National Society for the Prevention of Cruelty to Children" should go to the society of that name, which had its head office in Leicester Square. It relied upon the background facts that, as Earl Loreburn said, at pp211-214, the testator was "a Scotsman living in Scotland" who had made a "Scotch will" to construe the will as intended to refer to the Scottish "National Society for the Prevention of Cruelty to Children". Earl Loreburn refused to accept that there was "a rigid rule" that "once a persona is accurately named in a will" there is not to be "any further inquiry or consideration in regard to the person who is to take the benefit". The true rule, he said, was that "the accurate use of a name in a will creates a strong presumption against any rival who is not the possessor of the name". This demotes the rule to the common-sense proposition that in a formal document such as a will, one does not lightly accept that people have used the wrong words. I doubt whether anyone would dissent from this principle, which would present no obstacle to a conclusion that the tenant in this case must have used the wrong words.

If your lordships are to follow this path, it will be necessary to say that *Hankey* v *Clavering* and the older cases which it followed are no longer good law. It would be wrong, I think, to distinguish them on narrow grounds and leave them as wrecks in the channel, causing uncertainty and litigation in the future. Furthermore, the old rule of construction has been applied not only to notices exercising break clauses but also to notices to terminate periodic tenancies: *Doe d Spicer* v *Lea* (1809) 11 East 312. In his admirable submissions on behalf of the landlord, Mr Patten QC warned that a departure from the old rule would cause great uncertainty in the daily construction of notices to quit in county courts throughout the land. I confess that this prospect has caused me some anxiety and I think that it must be given serious consideration.

The rule as applied to wills, which restricts the use of background in aid of construction, reflects a distrust of the use of oral evidence to prove the background facts. The people who could give evidence about the background to a will would in most cases be members of the family interested in the outcome of the case and until 1843, persons with an interest in the litigation were not even competent witnesses. No doubt the exclusion of background makes, in a somewhat arbitrary way, for greater certainty in the sense that there is less room for dispute about what the background was and the effect which it has upon the intention to be attributed to the testator. But, as the cases mournfully show, this certainty is bought at the price of interpretations which everyone knows to be contrary to the meaning which he intended.

There are documents in which the need for certainty is paramount and which admissible background is restricted to avoid the possibility that the same document may have different meanings for different people according to their knowledge of the background. Documents required by bankers' commercial credits fall within this category. Article 13(*a*) of the Uniform Customs and Practice for Commercial Credits (1993 revision) says (echoing Lord Greene MR's phrase in *Hankey* v *Clavering*) that the documents must "upon their face" appear to be in accordance with the terms and conditions of the credit. But the reasons of policy which require the restriction of background in this case do not apply to notices given pursuant to clauses in leases. In practice, the only relevant background will be, as in this case, the terms of the lease itself, which may show beyond any reasonable doubt what was the intention of the person who gave the notice. There will be no question of the parties not being privy to the same background — both of them will have the lease — and no room for dispute over what the relevant background is.

In the case of commercial contracts, the restriction on the use of background has been quietly dropped. There are certain special kinds of evidence, such as previous negotiations and express declarations of intent, which for practical reasons which it is unnecessary to analyse, are inadmissible in aid of construction. They can be used only in an action for rectification. But apart from these exceptions, commercial contracts are construed in the light of all the background which could

A reasonably have been expected to have been available to the parties in order to ascertain what would objectively have been understood to be their intention: *Prenn* v *Simmonds* [1971] 1 WLR 1381, at p1383. The fact that the words are capable of a literal application is no obstacle to evidence which demonstrates what a reasonable person with knowledge of the background would have understood the parties to mean, even if this compels one to say that they used the wrong words. In this area, we no longer confuse the meaning of words with the question of what meaning the use of the words was intended to convey. Why, therefore, should the rules for the construction of notices be different from those for the construction of contracts? There seems
B to me no answer to this question. All that can be said is that the rules for the construction of notices, like those for the construction of wills, have not yet caught up with the move to common sense interpretation of contracts which is marked by the speeches of Lord Wilberforce in *Prenn* v *Simmonds* [1971] 1 WLR 1381 and *Reardon Smith Line Ltd* v *Yngvar Hansen-Tanger* [1976] 1 WLR 989. The question is therefore whether there is any reason not to bring the rules for notices up to date by overruling the old cases.

There can, I think, be no question of anyone having acted in reliance on the principle of construction used in *Hankey* v *Clavering* [1942] 2 KB 326. The consequence of such a construction is only to allow one
C party to take an unmeritorious advantage of another's verbal error, an adventitious bonus upon which no one could have relied. In this respect, the case for rejecting the old authorities is at least as strong as it was in *Sudbrook Trading Estate Ltd* v *Eggleton* [1983] 1 AC 444, in which this House overruled cases going back to the early 19th century on the construction of contracts for sale at a valuation.

Nor do I think that a decision overruling the old cases will create uncertainty as to what the law is. In fact I think that the present law is uncertain and that only a decision of this House, either adopting or rejecting the *Hankey* v *Clavering* rule of construction, will make it certain. So, for example, in *Carradine Properties Ltd* v *Aslam* [1976] 1
D WLR 442, at p444G, Goulding J said that the test for the validity of a notice was: "Is the notice quite clear to a reasonable tenant reading it? Is it plain that he cannot be misled by it?" and he went on to say that the reasonable tenant must be taken to know the terms of the lease. This test was approved by the Court of Appeal in *Germax Securities Ltd* v *Spiegal* (1978) 37 P&CR 204, at p206 and, as will be apparent from what I have already said, I think that it was the right test to adopt. It is, however, absolutely impossible to reconcile the application of such a test with the decision in *Hankey* v *Clavering*, in which no reasonable tenant who knew the terms of the lease could possibly have mistaken the landlord's meaning. It is therefore not surprising that in
E *Micrografix* v *Woking 8 Ltd* [1995] 2 EGLR 32, Jacob J felt free to dismiss *Hankey* v *Clavering* as "much distinguished" and to ignore it, or that Rattee J in *Garston* v *Scottish Widows' Fund & Life Assurance Society* [1996] 1 WLR 834* should be puzzled as to why the Court of Appeal in this case considered, as I think rightly, that they were bound by *Hankey* v *Clavering*.

In my view, therefore, the House should say unequivocally that the test stated by Goulding J in *Carradine Properties Ltd* v *Aslam* [1976] 1 WLR 442 was right and that *Hankey* v *Clavering* and the earlier cases should no longer be followed. The notice should be construed against the background of the terms of the lease. Interpreted in this way, the
F notice in the present case was valid and I would therefore allow the appeal.

Agreeing, **Lord Clyde** said: The question in this appeal is whether the two letters dated June 24 1994 and sent by the appellant tenant to the respondent landlord qualify as effective notices to determine the leases to which each letter respectively referred. Their validity as notices has to be tested against the terms of the power under which they were served. It is accepted that the two cases are for present purposes indistinguishable and that the relevant terms of the power were set out in clause 7(13) and were as follows:

> The Tenant may by serving not less than six months' notice in writing on the Landlord or its solicitors such notice to expire on the third anniversary of the term commencement date determine this Lease and upon expiry of such notice this Lease shall cease and determine and have no further effect . . . G

The substance of the power is expressed by the words, "The Tenant may . . . determine this Lease". The method of its exercise is specified by the intervening words. The tenant must give six months' notice; the notice must be in writing; the notice must be served on the landlord or its solicitors. The subclause also states that the notice is to expire on the third anniversary of the term commencement date. The significance of that statement is that the period of six months is to terminate on that date. This regulates the time for the giving of the notice. The third H anniversary marks the end of the period prior to which a notice under clause 7(13) must be given. But it is not required that the notice should include mention of the date of the intended determination of the lease. That date is prescribed by clause 7(13) where it states that the lease shall determine on expiry of the notice.

Where a notice of termination complies precisely and unambiguously with the provision which empowers the sending of the notice then its validity should be unquestioned. Where the terms of the notice do not altogether accord with the provisions of the contract that may or may not render the notice unenforceable. The problem then may come to be one of finding a fair and reasonable construction of the J notice. But there can be cases where the validity of the notice cannot be saved by any construction and will have to be regarded as bad.

In some cases it may be obvious from the notice by itself that an error has been made. In *Carradine Properties Ltd* v *Aslam* [1976] 1 WLR 442 an expressed intention to determine the lease at a date in 1973 was obviously incorrect in a notice served in 1974. In other cases the discrepancy can only be seen from a study of the terms of the lease. One would need to be aware of the provisions of the lease in such a case to appreciate that the permitted date was inaccurately stated. I see no reason in principle why in each of these kinds of case, provided of course that the wording is not absolutely clear and unambiguous, a K notice should not be equally open to construction with a view to its possible validation.

In the present case the two letters, in my view, satisfy the formal and technical requirements of clause 7(13). But they go further and call for a determination of each lease one day before the day which the subclause identified as the date for the determination of the notice and for the determination of the lease. As I have mentioned that was not a formal requisite of the notice. Each notice proclaims at the outset that it is given "Pursuant to Clause 7(13)". This was a precise reference to the particular provision under which the notices were each being sent, as distinct from some general reference to the agreement between the parties. But it is evident from a consideration of that clause that there is L a discrepancy between the date there indicated for the termination of the lease and the date stated in the notice. Whether that inaccuracy in the notice is fatal or not depends on the proper construction of the notices. The formulation propounded by Goulding J in *Carradine* (p444) was "Is the notice quite clear to a reasonable tenant reading it? Is it plain that he cannot be misled by it?" *Delta Vale Properties Ltd* v *Mills* [1990] 1 WLR 445 concerned a vendor's notice to complete which was in condition 23 of the conditions of sale, but I see no reason why any different principle of construction should apply. Slade LJ (at p454G) observed: M

> In my judgment, notices to complete served under condition 23, if they are to be valid, must be sufficiently clear and unambiguous to leave a reasonable recipient in no reasonable doubt as to how and when they are intended to operate.

The standard of reference is that of the reasonable man exercising his commonsense in the context and in the circumstances of the particular case. It is not an absolute clarity or an absolute absence of any possible ambiguity which is desiderated. To demand a perfect precision in matters which are not within the formal requirements of the relevant power would, in my view, impose an unduly high standard in the framing of notices such as those in issue here. While careless

*Editor's note: Also reported at [1996] 1 EGLR 113

drafting is certainly to be discouraged the evident intention of a notice should not in matters of this kind be rejected in preference for a technical precision.

The test is an objective one. In circumstances where an estoppel might arise the actual understanding of the recipient may be relevant, but in general the actual understanding of the parties is beside the point. That the test is an objective one was recognised in *Micrografix* v *Woking 8 Ltd* [1995] 2 EGLR 32. It was held there that the landlords would not have been misled by the references to a wrong date both in the notice to terminate the lease and in the covering letter. Each document was expressly written pursuant to the particular break clause in the lease. The recipients would have observed the errors because they would be familiar with the terms of the lease and would have known that the only date of determination had to be June 23 1995. They would know that there was no requirement to specify any date in the notice. They would see that the tenant wanted to leave. It was held that the notice was valid.

In my opinion, a like view should be taken in the circumstances of the present case. The notices were expressed to be "Pursuant to Clause 7(13)". It is plain from that that the tenant intended to invoke that clause. It is also plain that the tenant wished to determine the tenancy and that clause is the only clause under which the tenant could achieve that result. The landlord would be expected to know the terms of the lease and the date on which the lease fell to be determined under that clause. He would also be expected to know that there was no formal requirement for the tenant to specify in the notice the date of termination of the lease. There was no evident reason why the tenant should specify January 12 rather than January 13. The close proximity of the 13th makes it the more evident that it was erroneous and that the date intended was the date which the parties had agreed for a determination of the tenancy under clause 7(13). While there is a discrepancy evident in the notices between the reference to the clause and the statement of the date it seems to me that the notices were sufficiently clear and unambiguous. No reasonable landlord would, in my view, be misled by the statement of a date which in the context of a clear intention to invoke clause 7(13) was inaccurate. The landlord would, in my view, recognise that in each case the reference to January 12 was to be read as a reference to January 13 and I would so construe the notices.

In *Hankey* v *Clavering* [1942] 2 KB 326 the court refused to disregard a slip even although the intention of the notice was sufficiently clear from its terms and the recipient could not reasonably misunderstand it. In my view, that was too strict and too technical an approach. Counsel for the tenant sought to restrict the decision in *Hankey* to circumstances where the insertion of the date of termination is an essential requirement of a notice. I am, however, not persuaded that the decision in *Hankey* rested on the understanding that the specification of the date was an essential. I note that the corresponding provisions in *Carradine* and in *Micrografix* were not dissimilar and indeed in the latter case one element in the decision was the consideration that there was no requirement to specify a date. While to a considerable extent the cases in this field may turn upon their own circumstances I do not consider that the decision reached in *Hankey* was sound and in my opinion it should be overruled. In the circumstances of the present case I take the view that the notices were valid and effective. I agree with your lordships that the argument based on *Sidebotham* v *Holland* [1895] 1 QB 378 is without merit. But for the reasons which I have explained I would allow the appeal.

Marshall *v* Rubypoint Ltd

COURT OF APPEAL

January 31 1997

KENNEDY and PETER GIBSON LJJ and BUCKLEY J

Estates Gazette June 21 1997

[1997] 25 EG 142

Landlord and tenant — Covenant to repair — Disrepair of front door — Whether landlord liable for loss following entry of burglar

In 1993 the plaintiff acquired a 99-year lease of a flat in a building converted into flats with internal front doors sharing the use of common parts including the main front door; the defendant held the reversion. Under the lease the landlord covenanted to maintain, repair, redecorate and renew, *inter alia*, the common parts. On August 17, September 10 and 23 1993 the flat was burgled. The plaintiff's claim against the defendant landlord, for breach of repair of the front door, was allowed in the court below and he was awarded damages and interest of £6,123.19. The defendant appealed.

***Held*: The appeal was dismissed. The disrepair of the front door, and the breach of covenant by the defendant, was causative of the burglary. It would have been in the contemplation of the parties at the date of the lease in 1983 that if the front door fell into disrepair and did not provide any real obstacle to a would-be intruder, a burglary was not unlikely. The damages for the burglary were not too remote. Where loss results partly from a breach of contract and partly from an intervening act of a third party, the party in breach will be liable for the loss if the intervening act was reasonably foreseeable by the parties at the time of the contract. The breach by the defendant was a substantial cause of the loss to the plaintiff. The doctrine of *novus actus interveniens* did not apply.**

The following cases are referred to in this report.

Iron & Steel Holding & Realisation Agency v *Compensation Appeal Tribunal* [1966] 1 WLR 480; [1966] 1 All ER 769, DC

London Joint Stock Bank Ltd v *Macmillan and Arthur* [1918] AC 777

Stansbie v *Troman* [1948] 2 KB 48; [1948] 1 All ER 599; (1948) 64 TLR 226, CA

Tucker v *Linger* [1882] 21 ChD 18

This was an appeal by the defendant, Rubypoint Ltd, from a decision of Judge Viner CBE QC in Brighton County Court, whereby she gave judgment to the plaintiff, David Lester Marshall, for breach of covenant of a lease.

David Dabbs (instructed by Sharpe Pritchard, London agents for Grant Bott & Co, of Worthing) appeared for the appellant; Darren Howe (instructed by Fitzhugh Gates, of Brighton) represented the respondent.

Giving the first judgment at the invitation of Kennedy LJ, **BUCKLEY J** said: In 1993 Mr Marshall, the respondent to this appeal, lived in a flat at 47C West Hill Street, Brighton. On August 17, September 10 and again on September 23 1993 the premises were burgled. On each occasion property was stolen, but, more seriously, Mr Marshall was assaulted during the burglary on September 10. Proceedings were instituted in Brighton County Court against Rubypoint Ltd, the lessors of the flat, claiming damages to compensate Mr Marshall for his personal injuries, consequential expenses and the items stolen. By her judgment dated January 10 1996 Judge Viner CBE QC awarded damages and interest totalling £6,123.19. It is against that judgment that Rubypoint now appeals.

The original lease in respect of Mr Marshall's flat was in 1983. By its terms, the original lessors demised the premises therein described to the lessee for 99 years in return for payment of rent, but reserved from the demise certain parts of the building. By clause 5 (4) the lessor covenanted to "maintain, repair, re-decorate and renew", *inter alia*, the common parts. It is agreed that the common parts included the front door of the building. It is apparent from the recitals in the lease that the premises as a whole, which were owned by the lessor, were or were to be divided into flats and let separately. Each flat had its own internal front door. Mr Marshall took an assignment of the remainder of the lease in respect of flat C on November 30 1984. Rubypoint acquired the reversion in 1992, so that by 1993 Mr Marshall occupied flat C as lessee of Rubypoint, the lessor.

In the county court, Mr Marshall contended that the front door was in a state of disrepair which amounted to a breach of the covenant to repair and was sufficiently causative of the burglaries. Rubypoint, in effect, denied everything. However, it seems from the judgment, that the main issues were whether Mr Marshall's loss and damage was too remote and or whether there was a break in the chain of causation or a *novus actus interveniens*. As to the state of the front door, the learned judge found that in April 1993 someone forced the door open, there was a hole in the glass top half of it and a piece of cardboard had been stuck on with packing tape. Further, that the lock was hanging loose and one only had to push the door to gain entrance. Mr Marshall was the only tenant at the time. Squatters had moved into the premises some time in April and subsequently refitted the lock, but apparently not securely. Mr David Dabbs, counsel for Rubypoint, told us that the door could be pushed open. As a more general indication, the learned judge found that the premises were dilapidated in 1992 when Rubypoint purchased the reversion. A specification for external decoration and repairs was prepared by Peter Overill & Associates, chartered building surveyors, and dated April 21 1993. It concluded at para 18 of Part 3:

> Strip out front entrance door complete with frame and cart away. Supply and fix new front entrance door and frame to match existing in all respects complete with glazing, letterbox 3 No 100mm steel hinges and Yale night latch complete with cylinder pull.

No repairs had been carried out by the date of the first burglary on August 17. There is a photograph of the front door taken on October 3 which, the learned judge found, showed its condition as from August 20 onwards. It is a sorry sight. Although a carpenter attended on August 20, after the first burglary, and fitted a new Chubb lock to the front door and replaced Mr Marshall's front door, the Chubb lock was gouged out within a couple of days. Thereafter, there was no lock and the door could only be pulled to. That was the state of affairs at the time of the second and third burglaries.

There can be no doubt that Rubypoint was in breach of clause 5(4). Mr Dabbs has accepted that. There can equally be no doubt to my mind that the breach of covenant was a substantial cause of the burglaries although Mr Dabbs did not accept that. He submitted that, looking fairly and objectively at the lease and at the circumstances at the time of lease, the tenant had a duty to protect himself, that the front door was only the first line of defence and the learned judge should have balanced the obligations of both parties. Mr Dabbs' difficulty was the finding of fact, which he could not challenge, to the effect that Mr Marshall's door was the same as at the time of the lease save that he had added a security lock and chain after the first burglary at the behest of his insurers. In other words, no sensible criticisms or breach of obligation could be laid at his door. He had improved it.

One of the main purposes of a front door is to provide security. A door which is obviously broken and dilapidated, as this one was, and which can easily be pushed open is an invitation to would-be burglars. Once inside the premises, they are likely to use such force as is necessary to enter the flats.

I conclude that the disrepair of the door, and hence the breach of covenant by Rubypoint, was causative of the burglary.

Mr Dabbs also submitted that the damages were too remote and the acts of the burglars broke the chain of causation.

I turn to those issues. Loss from burglary in this case is not too remote in law if at the time of contracting it was within the reasonable contemplation of the parties as a not unlikely result of the disrepair of the front door. Mr Dabbs submitted that the learned judge had applied that test and considered the question of *novus actus* at the wrong time, that is during the period April to September 1993 as opposed to the time of contracting. Indeed, he made the same submission on causation. With respect to the learned judge, for whose careful and full judgment I am grateful, it is not clear to me what time she had in mind when applying the relevant tests. Certainly, she made several findings touching the question of foreseeability based on the April to September period. Perhaps that was because those matters were the subject of evidence and argued forcefully by counsel.

However, even if Mr Dabbs could make good that submission it would not avail him because it is clear to me that if one goes back to the date of the original lease, which I accept is the correct approach, the answer would be the same. The relevant circumstances at that time may be ascertained from the lease itself:

(i) the lessor reserved from the demise certain parts of the premises including the front door;

(ii) by clause 5 (4) he covenanted to maintain and repair it;

(iii) the premises were divided into six flats each with its own internal front door;

(iv) the front door of flat C was only 10ft from the main front door;

(v) the date of the lease was 1983 and the premises in Brighton.

I would only add the obvious, that the main purpose of a front door, apart from affording access and egress, is to keep out uninvited visitors and that includes burglars.

In my judgment, it would have been within the contemplation of the parties that if the front door fell into disrepair and did not provide any real obstacle to a would-be intruder, a burglary was not unlikely. I do not consider the fact that flat C had its own front door alters the position. A burglar, having gained easy access through the main front door would be out of sight of passers-by and feel more secure. He would be unlikely to baulk at the prospect of forcing an internal door.

I conclude that damages for burglary are not too remote, even without drawing on the various findings by the learned judge which undoubtedly enhanced Rubypoint's culpability. Those findings included that actual notice was given by Miss Marshall, Mr Marshall's sister, to Rubypoint of the state of the front door and the first burglary. To put it euphemistically, the learned judge preferred the evidence of Mr and Miss Marshall to the witnesses called on behalf of the defendant.

Mr Dabbs' submissions aimed at establishing a break in the chain of causation run into insuperable obstacles in law and on the facts. Rubypoint's duty to repair the front door was, at least in part, so as to secure the premises against intruders. It was in breach of that duty and cannot set up the burglar's acts as breaking the chain of causation. Alternatively, if it be any different, it can be said that burglary was a foreseeable consequence of the breach with the same result.

Mr Dabbs recognised his difficulties and, quite properly, reminded us of several of the well known cases in that area of the law. The first was *London Joint Stock Bank Ltd* v *MacMillan and Arthur* [1918] AC 777. That was the case which dealt with the duty of a customer of a bank to draw cheques in a sensible manner. Two very short citations may suffice, the first from the speech of Lord Finlay LC (p789):

> It has been often said that no one is bound to anticipate the commission of a crime, and that to take advantage of blank spaces left in a cheque for the purpose of increasing the amount is forgery, which the customer is not bound to guard against. It has been suggested that prevention of forgery must be left to the criminal law. I am unable to accept any such proposition without very great qualification. Every-day experience shows that advantage is taken of negligence for the purpose of perpetrating frauds. A warehouseman is bound to take precautions against theft, and if he fails to do so he will be liable to the owner if the goods are stolen.

At p794 (also in the speech of Lord Finlay):

> The fact that a crime was necessary to bring about the loss does not prevent its being the natural consequence of the carelessness.

Stansbie v *Troman* [1948] 2 KB 48 was a case concerning a contractual duty to take care. There, the contractor was carrying out decorations to the house in question which was burgled. Tucker LJ said (p52):

> The reason why the decorator owed a duty to the householder to leave the premises in a reasonably secure state was because otherwise, thieves or dishonest persons might gain access to them; and it seems to me that if the decorator was, as I think he was, negligent in leaving the house in this condition it was as a direct result of negligence that the thief entered by the front door, which was left unlocked, and stole those valuable goods.

It is clear from those authorities that the fact that a crime is committed does not, of itself, break the chain of causation. *Iron & Steel Holding & Realisation Agency* v *Compensation Appeal Tribunal* [1966] 1 WLR 480 is a case which illustrates again that an intervening event, if foreseeable, is not or does not constitute a *novus actus interveniens*. Winn LJ said (p492):

> In my opinion, wherever any intervening factor was itself foreseen or reasonably foreseeable by the actor, the person responsible for the act which initiated the chain of causes leading to the final result, that intervening cause is not itself, in the legal sense, *novus actus interveniens* breaking the chain of causation and isolating the initial act from the final result.

Mr Dabbs also referred us to *Tucker* v *Linger* [1882] 21 ChD 18. I do not derive any assistance from that case. It was one that turned on its own special facts where a tenant's corn was damaged when rainwater came in through the roof. The tenant attempted to recover damages against the landlord for his breach of obligation to supply materials to the tenant. The court held that under the terms of the lease it was the tenant's obligation to carry out the repairs and that obligation was not conditional on the landlord's obligation to supply the materials. It is not surprising that the tenant failed.

The learned judge considered the cases to which she was referred carefully. She concluded, as have I, that burglary was a foreseeable consequence of an insecure front door. The chain of causation was not broken. In this day and age — and 1983 was not sufficiently different — assault in the course of burglary is, unhappily, not unusual. Mr Dabbs did not submit the contrary.

Finally, Mr Dabbs submitted that Mr Marshall had failed to mitigate his loss in the sense that after the first burglary, in particular, he should have realised the situation was unsafe and taken it upon himself to repair the front door. The learned judge's findings of fact upon that were that Mr Marshall did not have the necessary skills or money to do so. Having notified Rubypoint of the situation and been assured that something would be done, he could not reasonably be criticised. I entirely agree.

In the end the learned judge carefully weighed all the evidence before her. I can find no basis for criticising any of her findings. This was a blatant breach of the landlord's duty to maintain or repair the front door at 47 West Hill with the result that Mr Marshall was burgled on three occasions in a short space of time and seriously assaulted. That outcome was not unlikely in the circumstances. It was readily foreseeable by the parties to the lease at all times.

While I am grateful to Mr Dabbs, and sorry not to have heard from Mr Darren Howe, I would dismiss the appeal.

Agreeing, **Peter Gibson LJ** said: It is elementary in the law of contract that a defendant in breach of contract is not liable for loss that is too remote and is not caused by that breach. Whether a loss is too remote depends on whether the loss was within the reasonable contemplation of the parties to the contract at the time of the contract. In the present case the contract was the lease between the respective predecessors in title to the plaintiff as tenant and the defendant as landlord. Mr Dabbs, for the landlord, criticised the judge for failing to distinguish between what was foreseeable at the time of the lease and subsequent events which occurred. It seems to me that there is a good deal of force in that criticism. However, despite Mr Dabbs's well sustained arguments, I am not persuaded that any of the judge's conclusions were wrong.

It is not in dispute that there was a breach of contract by the defendant at all material times in failing to observe the covenant on the landlord's part to repair, redecorate and renew the common parts, including the outer door to the street. The questions that, therefore, arise are: (1) whether the loss suffered by the plaintiff through the burglaries was within the reasonable contemplation of the parties to the lease at the time of the lease; and (2) whether the breach of contract was causative of the loss suffered by the plaintiff.

On the first question Mr Dabbs accepted: (a) at the commencement of the lease, in the event of a breach by the landlord of the covenant to keep in repair the outer door, it was reasonably foreseeable that a trespasser might thereby gain access to the common parts; and (b) if a trespasser could gain access to the common parts, it was reasonably foreseeable that he might commit a burglary. Mr Dabbs went on to submit that it was also reasonably foreseeable by both landlord and tenant that a trespasser might gain access without a breach of the covenant by the landlord and that therefore the tenant was well advised to prepare for either eventuality by taking reasonable security precautions for the protection of himself and his own property. Those further submissions seem to me not to be of great materiality on the question of remoteness. Conceivably they may be relevant to causation. It seems to me obvious that it was, at the date of the lease, foreseeable by the parties to the lease that if the outer door was not kept in proper repair so that a trespasser such as a burglar might enter, loss, through burglary of the plaintiff's premises, was not unlikely.

On the second question it is sufficient that the breach of contract should be *a* (and I stress that indefinite article) substantial cause of the loss. But if there is a new intervening act by a third party, that may negative any liability on the defendant. I take the relevant principle to be that where loss results partly from a breach of contract and partly from an intervening act of a third party, the party in breach will be liable for the loss if, but only if, the intervening act was reasonably foreseeable by the parties at the time of the contract.

Mr Dabbs stressed that the outer door was only the first line of defence. He suggested that the plaintiff had failed to take sufficient security precautions to protect himself in his property. He said that the judge should have asked the question whether the level of security provided by the plaintiff tenant was reasonable. That seems to me to approach the matter from the wrong end. The question is whether the defendant, by his breach of contract, caused the loss. If the plaintiff acts in such a way as to leave the original breach of contract a matter of irrelevance to the loss, for example, if the plaintiff had left the outer door wide open, then the defendant would not be liable for the loss which in such a case would have been caused by the plaintiff. That is not this case. The court has always answered questions of causation by the test of common sense. In my judgment, by that test, the judge was entirely right to find that the breach of contract was a substantial cause of the loss to the plaintiff. This was not a case where the doctrine of *novus actus interveniens* applied because the loss, through burglary, was foreseeable.

For those reasons, as well as those given by Buckley J, I, too, would dismiss this appeal.

Kennedy LJ agreed and did not add anything.

Appeal dismissed.

For a further case on this subject see p229

LANLORD AND TENANT
BUSINESS TENANCIES

A
Esselte AB and another *v* Pearl Assurance plc

COURT OF APPEAL

November 8 1996

STUART-SMITH and MORRITT LJJ and SIR JOHN MAY

Estates Gazzette January 11 1997
B

[1997] 02 EG 124

Landlord and tenant — Landlord and Tenant Act 1954 — Tenant giving up occupation before expiration of contractual term — Whether tenancy continues — Whether section 27(2) notice required to determine tenancy

By a subunderlease dated September 18 1989 the second respondent subunderlet part of an office block to the appellant tenants for a term of five years from February 15 1988 at a rent payable in advance. On the same date the first respondent sublet the whole office
C block to the appellants, subject to and with the benefit of the subunderlease, also for a term of five years from February 15 1988. The appellants occupied the whole of the office block for the purposes of their business until December 5 1992; the contractual term of the sublease and subunderlease expired on February 14 1993. The respondents sought rent from the appellants up to June 24 1993 on the basis that no notices satisfying section 27(2) of the Landlord and Tenant Act 1954 were served until January 5 and 15 respectively expressed to expire on the June quarter day. In the court below the judge held that the tenancies continued until determined by the section 27(2) notices; the appellants appealed.
D ***Held*****: The appeal was allowed. The 1954 Act did not apply to the tenancies on the contractual terms dates, which expired by effluxion of time. On the true construction of the provisions in Part I of the 1954 Act, of which section 24(3)(*a*) is of particular significance, a tenancy is not one to which the Act applies where the tenant ceases to be in occupation. Two previous decisions of the Court of Appeal were in conflict: *Long Acre Securities Ltd* v *Electro Acoustic Industries Ltd* [1990] 1 EGLR 91 and *Morrisons Holdings Ltd* v *Manders Property (Wolverhampton) Ltd* [1976] 1 WLR 533. The court was entitled to follow the *Morrisons* case,
E rather than the *Long Acre* decision, because it was more consistent with the wording of the legislation and accords with the dicta found in other cases decided both before and after *Morrisons*.**

The following cases are referred to in this report.

Caplan (I&H) Ltd v *Caplan (No2)* [1963] 1 WLR 1247; [1963] 2 All ER 930
Cheryl Investments Ltd v *Saldanha* [1978] 1 WLR 1329; [1979] 1 All ER 5; (1978) 37 P&CR 349; [1978] 2 EGLR 54; 248 EG 591, CA
Cornish v *Brook Green Laundry Ltd* [1959] 1 QB 394; [1959] 2 WLR 215; [1959] 1 All ER 373; [1959] EGD 116; (1959) 173 EG 307, CA
Demetriou v *Poolaction Ltd* (1990) 63 P&CR 536; [1991] 1 EGLR 100; [1991] 25 EG 113, CA
Graysim Holdings Ltd v *P&O Property Holdings Ltd* [1996] 1 AC 329; [1995] 3 WLR 854; [1995] 4 All ER 831; [1996] 1 EGLR 109; [1996] 03 EG 124, HL F
Green v *Bowes-Lyon* [1963] AC 420; [1961] 3 WLR 1044; [1961] 3 All ER 843; [1961] EGD 301; (1961) 180 EG 429, HL
Green v *Bowes-Lyon* [1961] 1 WLR 503; [1961] 1 All ER 13, CA
Hancock & Willis v *GMS Syndicate Ltd* [1983] 1 EGLR 70; [1983] EGD 114; (1982) 265 EG 473, CA
Long Acre Securities Ltd v *Electro Acoustic Industries Ltd* [1990] 1 EGLR 91; [1990] 06 EG 103
Morrisons Holdings Ltd v *Manders Property (Wolverhampton) Ltd* [1976] 1 WLR 533; [1976] 2 All ER 205; (1975) 32 P&CR 218; [1976] 1 EGLR 70; 238 EG 715, CA
Phillips v *Mobil Oil Co Ltd* [1989] 1 WLR 888; [1989] 3 All ER 97; (1989) 59 P&CR 292; [1989] 2 EGLR 246, CA G
Provident Mutual Life Assurance Association v *Greater London Employers' Association Ltd* [1996] 1 EGLR 106; [1996] 23 EG 129
Young v *Bristol Aeroplane Co Ltd* [1944] KB 718; [1944] 2 All ER 293

This was an appeal by the defendants, Pearl Assurance plc, from a decision of Judge Colyer QC, sitting as a judge of the High Court, who had given judgment to the respondents, Esselte AB and British Sugar Corporation, on the hearing of a preliminary point of law in their actions for arrears of rent.

Paul Morgan QC and Wayne Clark (instructed by Geoffrey Delaney, of Peterborough) appeared for the appellants; Jonathan Brock H
(instructed by Theodore Goddard) represented the respondents.

Giving judgment, **MORRITT LJ** said: The question for decision on this appeal is whether a tenancy for a term certain to which Part II Landlord and Tenant Act 1954 once applied of property which the tenants, Pearl Assurance plc ("Pearl"), had ceased to occupy for the purpose of their business before the expiration of that term determined at the expiration of the contractual term or only on the later expiration of a notice given by Pearl to its immediate landlords, Esselte AB and British Sugar plc ("the landlords"), pursuant to section 27 of that Act. By his order made on May 9 1995 Judge Colyer QC, sitting as a judge J
of the High Court, declared that the tenancy determined on June 24 1993 at the expiration of the notice given by Pearl Assurance plc to the landlords on January 15 1993 pursuant to section 27 of the Landlord and Tenant Act 1954. This is an appeal of Pearl. Pearl contends that the judge should have declared that the tenancy determined on the earlier expiration of the term on February 15 1993.

On September 18 1989 Esselte AB as the tenant of part and British Sugar as subtenant of another part of Guild House, Oundle Road, Peterborough, executed a lease and an underlease respectively demising those parts to Pearl Assurance plc for a term of five years commencing on February 15 1989 and expiring on February 14 1993. K
In each case the substantial rent payable by Pearl was payable in advance. By notices in writing given to each of the landlords on November 16 and December 24 1992 Pearl purported to terminate the lease or underlease as the case might be on February 14 and March 31 1993 respectively. It is common ground that all such notices were ineffective. The judge held, and there is no appeal from such finding, that by December 6 1992 Pearl ceased to occupy Guild House for the purposes of its business within section 23 of the Landlord and Tenant Act 1954. Thereafter three notices were served. The first was a notice under section 25(1) served by British Sugar on Pearl on January 6

1993 to determine the subtenancy created by the subunderlease from British Sugar on January 3 1994. The second and third were notices under section 27 served by Pearl on each of the landlords on January 16 1993 to determine both tenancies on June 24 1993. It is common ground that the second and third were effective if, contrary to the submissions of Pearl, the tenancies had not determined earlier.

Pearl paid the rent due up to February 14 1993. The landlords claimed that it should also pay rent up to June 24 1993 and on June 25 1993 commenced proceedings for the purpose of enforcing such a liability. On June 20 1994 Master Moncaster ordered the trial of a preliminary issue whether the tenancies terminated on February 15 1993, June 24 1993 or on some other and if so what date.

The answer to that question depends on the true construction of the relevant provisions of the Landlord and Tenant Act 1954 as amended by the Law of Property Act 1969. They are as follows:

Tenancies to which Part II applies

23.— (1) Subject to the provisions of this Act, this Part of the Act applies to any tenancy where the property comprised in the tenancy is or includes premises which are occupied by the tenant and are so occupied for the purposes of a business carried on by him or for those and other purposes.

(2) In this Part of the Act the expression "business" includes a trade, profession or employment and includes any activity carried on by a body of persons, whether corporate or unincorporate.

(3) In the following provisions of this Part of this Act the expression "the holding", in relation to a tenancy to which this Part of this Act applies, means the property comprised in the tenancy, there being excluded any part thereof which is occupied neither by the tenant nor by person employed by the tenant and so employed for the purposes of a business by reason of which the tenancy is one to which this Part of this Act applies . . .

Continuation of tenancies to which Part II applies and grant of new tenancies

24.— (1) A tenancy to which this Part of this Act applies shall not come to an end unless terminated in accordance with the provisions of this Part of this Act; and, subject to the provisions of twenty-nine of this Act the tenant under such a tenancy may apply to the court for a new tenancy —

(*a*) if the landlord has given notice under section 25 of this Act to terminate the tenancy, or

(*b*) if the tenant has made a request for a new tenancy in accordance with section twenty-six of this Act.

(2) The last foregoing subsection shall not prevent the coming to an end of a tenancy by notice to quit given by the tenant, by surrender or forfeiture, or by the forfeiture of a superior tenancy, unless —

(*a*) in the case of a notice to quit, the notice was given before the tenant has been in occupation in right of the tenancy for one month; or

(b) in the case of an instrument for surrender, the instrument was executed before, or was executed in pursuance of an agreement made before, the tenant had been in occupation in right of the tenancy for one month.

(3) Notwithstanding anything in subsection (1) of this section —

(*a*) where a tenancy to which this Part of this Act applies ceases to be such a tenancy, it shall not come to an end by reason of the cesser, but if it was granted for a term of years certain and has been continued by subsection (1) of this section then (without prejudice to the termination thereof in accordance with any terms of the tenancy) it may be terminated by not less than three nor more than six months' notice in writing given by the landlord to the tenant;

(*b*) where, at a time when a tenancy is not one to which this Part of this Act applies, the landlord gives notice to quit, the operation of the notice shall not be affected by reason that the tenancy becomes one to which this Part of this Act applies after the giving of the notice . . .

Termination of tenancy by the landlord

25.— (1) The landlord may terminate a tenancy to which this Part of the Act applies by a notice given to the tenant in the prescribed form specifying the date at which the tenancy is to come to an end (hereinafter referred to as "the date of termination"):

Provided that this subsection has effect subject to the provisions of Part IV of this Act as to the interim continuation of tenancies pending the disposal of applications to the court . . .

Tenant's request for a new tenancy

26.— (1) A tenant's request for a new tenancy may be made where the tenancy under which he holds for the time being (hereinafter referred to as the "current tenancy") is a tenancy granted for a term of years certain exceeding one year, whether or not continued by section 24 of this Act, or granted for a term of years certain and thereinafter from year to year.

(2) A tenant's request for a new tenancy shall be for a tenancy beginning with such date, not more than twelve nor less than six months after the making of the request, as may be specified therein:

Provided that the said date shall not be earlier than the date on which apart from this Act the current tenancy would come to an end by effluxion of time or could be brought to an end by notice to quit given by the tenant . . .

Termination by tenant of tenancy for fixed term

27. — (1) Where the tenant under a tenancy to which this Part of this Act applies, being a tenancy granted for a term of years certain, gives to the immediate landlord, not later than three months before the date on which apart from this Act the tenancy would come to an end by effluxion of time, a notice in writing that the tenant does not desire the tenancy to be continued, section 24 of this Act shall not have effect in relation to the tenancy unless the notice is given before the tenant has been in occupation in right of the tenancy for one month.

(2) A tenancy granted for a term of years certain which is continuing by virtue of section 24 of this Act may be brought to an end on any quarter day by not less than three months' notice in writing given by the tenant to the immediate landlord, whether the notice is given . . . after the date on which apart from this Act the tenancy would come to an end or before that date, but not before that tenant has been in occupation in right of the tenancy for one month . . .

Meaning of "the landlord" in Part II, and provisions as to *mesne* landlords, etc

44.— (1) Subject to the next following subsection, in this Part of this Act the expression "the landlord", in relation to a tenancy (in this section referred to as the "relevant tenancy"), means the person (whether or not he is the immediate landlord) who is the owner of that interest in the property comprised in the relevant tenancy which for the time being fulfils the following conditions, that is to say —

(*a*) that it is an interest in reversion expectant (whether immediately or not) on the termination of the relevant tenancy, and

(*b*) that it is either the fee simple or a tenancy which will not come to an end within fourteen months by effluxion of time and, if it is such a tenancy, that no notice has been given by virtue of which will come to an end within fourteen months or any further time by which it may be continued under section 36(2) or section 64 of this Act,

and is not itself in reversion expectant (whether immediately or not) of an interest which fulfils those conditions . . .

Interim continuation of tenancies pending determination by court

64.— (1) In any case where —

(*a*) a notice to terminate has been given under Part I or Part II of this Act or a request for a new tenancy has been made under Part II thereof, and

(*b*) an application to the court has been made under the said Part I or the said Part II, as the case may be, and

(*c*) apart from this section the effect of the notice or request would be to terminate the tenancy before the expiration of the period of three months beginning with the date on which the application is finally disposed of,

the effect of the notice or request shall be to terminate the tenancy at the expiration of the said period of three months and not at any other time.

The judge concluded that the decision of this court in *Long Acre Securities Ltd* v *Electro Acoustic Industries Ltd* [1990] 1 EGLR 91 was binding authority for the conclusion that the tenancy did not determine until the expiration on June 24 1993 of the notice given pursuant to section 27 by Pearl on January 15 1993. On the footing that, as submitted for Pearl, that decision was distinguishable and therefore not binding he then considered whether the reasoning of that decision was inconsistent with the relevant sections of the Act or with certain earlier decisions in respect of them. The cases to which he was referred were *Cornish* v *Brook Green Laundry Ltd* [1959] 1 QB 394; *I&H Caplan Ltd* v *Caplan (No2)* [1963] 1 WLR 1247; *Green* v *Bowes-Lyon* [1963] AC 420 and *Cheryl Investments Ltd* v *Saldanha* [1978] 1 WLR 1329*. He determined that there was no such inconsistency. In his view:

> Once one sees that the critical provision of the whole of Part II of the Act is that in section 24(1) that "a tenancy to which this Part of the Act applies shall not come to an end unless terminated in accordance with the provisions of the Act, once one sees that and once the tenancy has been a business tenancy, that is to say once it has in the past satisfied section 23, then section 24(1) has caught it".

---

*Editor's note: Also reported at [1978] 2 EGLR 54

He concluded:

> In the circumstances therefore of this case this tenancy continued until determined pursuant to the provisions of the Act, that is pursuant to section 27(2) upon June 24 1993. So to hold imposes the very minimal obligation upon tenants of serving notice on their landlords if they intend to vacate. *Ex hypothesi*, they are business tenants and since the giving of such notice by a business tenant does not involve the use of prescribed forms or involve undue complications or ceremony. It is not surprising in a business relationship that this minimal burden of having to take the initiative should rest upon the party in whose knowledge are all the relevant facts and whose decision it is or will be whether the tenant purposes to carry on its business use or to resume it before the contractual expiry date. To hold the contrary would leave the landlord a sitting duck, powerless to do everything save in respect of periodic tenancies, since it is common ground and emphasised by Mr Clark that a section 25 notice can only be served when the tenancy remains one to which the Act applies, that is which still at the date of the giving of that notice satisfies section 23.
>
> I therefore answer the question directed to be tried by declaring that the date upon which the tenancy created by the Pearl underlease as defined by para 3 of the amended statement of claim and the Pearl subunderlease as defined by para 5 thereof was determined on June 24 1993.

Pearl submits that the judge was wrong. They rely on the terms of the Act and earlier authorities as demonstrating that the tenancies determined by effluxion of time on February 14. They submit that the decision of this court in *Long Acre Securities Ltd* v *Electro Acoustic Industries Ltd* [1990] 1 EGLR 91 is distinguishable or arrived at *per incuriam* and is wrong; either way they contend that neither the judge nor this court is bound by that decision and that the judge was wrong to have thought that he was.

With regard to the terms of the Act they rely on the fact that the headings to and wording of sections 23(1) and 24(1) are in the present tense. They submit that the provisions of section 23(3) envisage that what constitutes the holding may vary from time to time and those of section 24(3)(*b*) recognise that the application of the Act depends on the statutory conditions being satisfied at a particular point in time. They contend that the latter point is reinforced by sections 25(1) and 27 which require that the tenancy is one to which the Act applies when the notice is served and suggest that the same requirement is implicit in section 26. With regard to the earlier authorities they rely, in addition to those to which the judge was referred, on *Morrisons Holdings Ltd* v *Manders Property (Wolverhampton) Ltd* [1976] 1 WLR 533*; *Hancock & Willis* v *GMS Syndicate Ltd* [1983] 1 EGLR 70; *Demetriou* v *Poolaction Ltd* [1991] 1 EGLR 100. They also rely on a passage in the speech of Lord Nicholls of Birkenhead in the subsequent decision of the House of Lords in *Graysim Holdings Ltd* v *P&O Property Holdings Ltd* [1995] 3 WLR 854†.

The landlords support the decision of the judge for the reasons he gave. They emphasise the importance of section 27 rather than of section 24 on which Pearl placed greater weight in the argument in this court than they appear to have done before the judge. In addition they rely on the subsequent decision of Lightman J in *Provident Mutual Life Assurance Association* v *Greater London Employers' Association Ltd* [1996] 23 EG 129‡ to the effect that *Long Acre Securities Ltd* v *Electro Acoustic Industries Ltd* is binding and not distinguishable.

In these circumstances I propose to consider first the proper construction of the relevant provisions of the Act without regard to authority. If the result of that process is to arrive at a *prima facie* conclusion contrary to that reached by this court in *Long Acre Securities Ltd* v *Electro Acoustic Industries Ltd* it will then be necessary to consider the basis on which this court would be entitled not to follow it. Finally it will be necessary to decide whether the suggested grounds of distinction or the prior authorities relied on by Pearl should cause this court to refuse so to do.

There is no dispute that the issue is whether a tenant who has ceased to occupy the demised premises before the end of a fixed term tenancy none the less continues to be the tenant under the Landlord and Tenant Act 1954 Part II until either the landlord or the tenant gives to the other a statutory notice to determine the tenancy. But for the intervention of the legislative scheme introduced by the Act the answer would be in the negative. Pearl submits that the answer is in the negative notwithstanding that scheme because there is nothing in section 24 to continue the tenancy beyond its fixed term and section 24(3)(*a*) assumes that it will not. The landlords contend that the answer is in the affirmative because section 27, in particular subsection (2), is drafted on the assumption that the tenancy will continue unless brought to an end by service of the notice which it authorises. The root of the difference between the two approaches lies in whether the statutory formula "a tenancy to which this Part of this Act applies" refers only to the present or whether it may in all or some contexts embrace the past so that it should be construed as "a tenancy to which this Part of this Act applies or has applied".

The phrase "a tenancy to which this Part of this Act applies" appears in all the relevant sections except section 26. It appears to me to be an abbreviated reference to the provisions of section 23 which set out the conditions for the application of Part II. That section shows that occupation for the statutory purposes is essential; it recognises that such occupation is likely to change from time to time while the tenancy endures and provides that the identity of the holding shall change with it. That section is expressed in the present tense, and consistently with the purpose of the legislation, cannot be construed so as to include the past.

The introductory words of section 24 are also expressed in the present tense. The stipulation that "a tenancy to which this Part of this Act applies shall not come to an end unless terminated in accordance with the provisions of this Act" has no effect on a tenancy to which the Act does not apply at the time of its termination. The point is emphasised when reference is made to the provisions of section 25(1) for the method of determination provided for in that section requires that the tenancy should be one "to which this Part of this Act applies".

In my view, section 24(3)(*a*) is of particular significance for it assumes that a fixed term tenancy, to which the Act has ceased to apply, will come to an end unless continued by subsection (1). This could not have been the case if the formula "a tenancy to which this Part of this Act applies" should be read as including a tenancy to which this Act has applied. Thus the cases in which the tenancy has been continued by subsection (1) must be confined to those in which the Act ceased to apply after they had been so continued. In those cases the paragraph prescribes how the continuing tenancy may subsequently be determined. It makes no reference to the termination of a tenancy which has not been so continued; but there is no reason why it should for, by definition, such a tenancy has come to an end by effluxion of time. Likewise subsection (3)(*b*) deals with the converse case where the tenancy becomes one to which the Act applies after the landlord has given a notice to determine it. All these references are to the present and are not, in my view, susceptible of including references to the past.

The landlords relied on sections 24(2), 25(1), 26(5) and 27 as providing a code for the determination of tenancies. It was submitted that the code did not include the case of the tenant vacating the premises at or before the term date and the recognition of determination in such a case would be contrary to section 24(1). But this carries the point no further unless the phrase "tenancy to which this Part of this Act applies" can be construed as including tenancies to which the Act formerly applied for it is a condition for the application of the code that the Act does apply.

In truth the validity or otherwise of the argument for the landlords depends, as their counsel realised, on whether section 27 necessarily assumes that a tenancy to which the Act formerly applied is included. Subsection (1) is of no assistance because it repeats the same formula without throwing any further light on its meaning. If the Act applies to the tenancy at the time the tenant gives his notice then it enables a tenant to ensure that his liabilities cease at the term date. But no obligation is cast upon him to do so. It has no bearing on the question whether, if the tenant vacates the premises after the notice is served (which may be many months before the term date) but before the term date, the tenancy is continued by subsection (1).

---

*Editor's note: Also reported at [1976] 1 EGLR 70

†Editor's note: Also reported at [1996] 1 EGLR 109

‡Editor's note: Also reported at [1996] 1 EGLR 106

A It was submitted that subsection (2) envisages that a fixed term tenancy is capable of being continued before its termination for otherwise it would not be possible to serve a notice before that time for the purpose of bringing it to an end on the quarter day following that time. For my part, I do not accept that section 27(2) can be given that construction so as to override the emphatic use of the present tense in the formula "a tenancy to which this Part of this Act applies". In my view, subsection (2) is looking to the point in time when the notice which it enables takes effect. At that time the tenancy is by definition continuing. But it does not follow that it is also "continuing" when the notice is given before the term date. Indeed it is manifest that though the tenancy will be continuing at that time it does B so not by virtue of section 24, but because the term date has not arrived. In my judgment, it is placing more weight on section 27(2) than it will bear to find in it justification for construing the formula as including "a tenancy to which this Part of this Act has applied".

It is contended that the construction which I favour leaves a landlord at the mercy of his tenant for he cannot be sure whether at the term date he will be able to relet the property or not. I do not accept that a landlord must find himself in that position. If the tenant is in occupation for the statutory purposes the landlord may serve a notice under section 25(1) to determine the tenancy at the term date. If the landlord is doubtful whether or not the tenant is in occupation for the statutory purposes then I see no reason why, C as counsel for Pearl submitted, he should not serve a notice under section 25(1) without prejudice to his principal contention that as the tenant is not in occupation no such notice is required. In any event the argument for the landlords would, if accepted, lead to a comparable problem. If the formula had to be applied as if it stated "a tenancy to which this Part of this Act applies or has applied" then landlords seeking to exercise a break clause after say 50 years of a 99-year lease would have to investigate the use to which the property had been put at all times in the past as well as by the current tenant at the time.

There is no doubt that this conclusion as to the true construction of the Act, which I favour, would be contrary to that reached by this court D (Dillon, Lloyd, and Farquarson LJJ) in *Long Acre Securities Ltd* v *Electro Acoustic Industries Ltd* [1990] 1 EGLR 91. In that case the tenants occupied the property under a lease for a term expiring on March 25 1988. The landlord served notice under section 25(1) to determine the tenancy on March 1 1989 to which the tenants responded with a notice under section 27(2) to determine the tenancy on June 24 1988. In fact the tenants vacated the property on or immediately before March 25 1988 and refused to pay rent for any subsequent period. The landlords sued in the county court for unpaid rent. The county court judge determined that the tenants were not liable. The Court of Appeal disagreed. The leading judgment was E given by Dillon LJ with which the other lord justices agreed. After setting out the facts and quoting the relevant sections Dillon LJ proceeded, at p 92, to give his reasons, which, in view of the submission that we should depart from his conclusion, I should quote in full. They were:

> It is quite plain therefore, under subsection (1) [of section 27], that the tenant for a fixed term who does not want to continue his tenancy under the Act and will be ceasing to carry on business in the premises on the expiration of the fixed term is expected to give notice to his landlord not later than three months before the date on which, apart from the Act, the tenancy would come to an end by effluxion of time. That is in line with the need for a landlord to have F reasonable notice of his tenant's departure, which one can find recognised in other provisions in the Act, such as section 24(3)(*a*), to which I have already referred. It is inconsistent, in my judgment, with the view adopted by the judge that if the tenant ceased carrying on business on the contractual date and left the premises he could do so without any notice at all because the provisions of the Act would automatically fall away.
>
> To my mind the key provision is that in section 24(1) which provides that "a tenancy to which this Part of this Act applies shall not come to an end unless terminated in accordance with the provisions of this Part of this Act". It follows that the tenancy, which was a business tenancy that the tenant took under the compromise agreement of December 1987, was a tenancy which could be terminated only in accordance with the provisions of the Act, even during its fixed term. When one comes to section 27(2) one has the opening phrase "A tenancy granted for a term of years certain which is continuing by virtue of section twenty-four of this Act". It is submitted that that, in the G present case, could apply only after March 25 1988 if the tenant was still in occupation, but the tenancy by virtue of section 24 is continued by the Act and is from the outset a continuing tenancy. Moreover, section 27(2) envisages a notice under the subsection being given before, as well as after, the date on which apart from the Act the tenancy would have come to an end; that is to say, in the present case, before March 25 1988. That necessarily shows that you cannot look at the tenancy as merely continuing by virtue of the Act after the date on which apart from the Act the tenancy would have come to an end. Subsection (2) envisages the notice given before that date, yet the tenancy is still a tenancy which is said to be continuing by virtue of section 24. That to my mind picks up the opening words of section 24, to which I have already referred, in the manner which I have mentioned. H
>
> In the present case, therefore, my conclusion is that it was open to the tenant to serve the notice under section 27(2). The tenant was not bound to await the expiration of the landlord's much longer notice. There is no reason why the tenant should be regarded as locked in by the duration of the landlord's notice. But the tenant was not entitled to bring its liability to pay rent to an immediate end just by quitting the premises on the original contractual term date. That term date is subject to the provisions of section 24(1).

As is well known the general rule is that the Court of Appeal is bound by decisions of the House of Lords and by its own previous decisions. In each case the principle is applied to the essential J reasoning behind the decisions of the respective courts and not to statements which go beyond and are not necessary to the decision itself. The first submission for Pearl is that the decision in *Long Acre Securities Ltd* v *Electro Acoustic Industries Ltd* is distinguishable because of the section 25 notice which had been served by the landlord to determine the tenancy on March 1 1989 before the service by the tenant of his notice under section 27. I do not accept that submission. The existence of the prior notice served by the landlord forms no part of the reasoning of Dillon LJ. This was the view of Lightman J in *Provident Mutual Life Assurance Association* v *Greater London Employers' Association Ltd* [1996] 1 EGLR 106 and I agree with him.

Pearl then submitted (to this court but not to the judge) that this case K came within one or both of two exceptions to the principle that this court is bound by its own decisions and those of the House of Lords. The first is that the Court of Appeal is not only entitled but bound to choose between two conflicting decisions of the Court of Appeal; the second is that the Court of Appeal is not bound to follow a previous decision of the Court of Appeal if it is satisfied that the decision was given *per incuriam*: *Young* v *Bristol Aeroplane Co Ltd* [1944] KB 718. The circumstances in which the Court of Appeal is entitled to regard its own previous decision as given *per incuriam* are:

> not only where that decision is given in ignorance or forgetfulness of some L inconsistent statutory provision or some authority binding upon it, but also, in rare and exceptional cases, if it is satisfied that the decision involved a manifest slip or error. *Rickards* v *Rickards* [1990] Fam194, at p203.

The Court of Appeal has been careful not to define further what might be regarded as the rare and exceptional case, leaving it to be recognised when seen.

Pearl relies on the decision of this court in *Morrisons Holdings Ltd* v *Manders Property (Wolverhampton) Ltd* [1976] 1 WLR 533 as the previous decision of this court which should be followed in this case in preference to that of this court in *Long Acre Securities Ltd* v *Electro Acoustic Industries Ltd*. Pearl also relies on what it describes as a series of *dicta* of the highest persuasive authority as constituting the rare and exceptional case. It is clear that neither *Morrisons Holdings Ltd* v *Manders Property (Wolverhampton) Ltd* nor any of the *dicta* relied on was brought to the attention of the Court of Appeal in *Long Acre Securities Ltd* v *Electro Acoustic Industries Ltd*. None of them is referred to in the report or in the skeleton argument of counsel for the appellant and counsel concerned had no recollection to the contrary. In the light of the submissions for Pearl it is necessary to consider all those cases in order to ascertain whether this court is entitled to depart from the decision on *Long Acre*. If it is then they are also of relevance in deciding what decision this court should reach.

A The first is *Green* v *Bowes-Lyon* [1961] 1 WLR 503, CA; [1963] AC 420, HL. The facts are complicated but it is sufficient for present purposes to record that Mrs Green had a tenancy of the ground floor and basement of the property in question expiring on April 1 1959. She occupied the basement for the purposes of her business. Mrs Green sublet the ground floor to Mrs Bowes-Lyon, which she occupied for the purposes of her business, for a term due to expire on March 31 1959. The landlord of Mrs Green, a Mr Wells, had a tenancy of the whole building expiring on April 4 1959. His landlord, a Mr Rye, granted a reversionary lease of the ground floor to Mrs Bowes-Lyon for a term commencing on April 5 1959. The question was to whom B Mrs Bowes-Lyon was bound to pay the rent for the quarter ended June 24 1959. In analysing the position of the various parties in the Court of Appeal Holroyd-Pearce LJ said of the tenancy of Mr Wells ([1961] 1 WLR 503 at p506):

> It expired on that date [4th April 1959] because the representatives of Mr Wells were not using any part of the premises for business purposes, and therefore there was no continuance of the tenancy by the Landlord and Tenant Act.

In the House of Lords, [1963] AC 420 at p34, Lord Reid's analysis of the position was that:

C

> the common law tenancy subsists with a statutory variation of the mode of determination or continuance by way of statutory extension. When the words [in section 44] have to be applied to an unprotected tenancy — in the present case to the tenancy of Wells — then the tenancy does come to an end at the end of the contractual term. But when they have to be applied to a protected tenancy — in the present case to the tenancy of the respondent — then the tenancy does not come to an end at the end of the contractual term but remains in existence as the same tenancy for an indefinite time until terminated in one of the ways provided by the Act.

A statement to the like effect appears in the speech of Lord Hodson at p447. It is not suggested by Pearl that these statements formed any D part of the *ratio decidendi* but they are relied on as *dicta* of high persuasive value.

The next case, in chronological order, is that of *I&H Caplan* v *Caplan (No 2)* [1963] 1 WLR 1247. In that case the tenant had claimed to be entitled to a new tenancy under Part II of the Landlord and Tenant Act 1954 and on April 14 1959 commenced proceedings to enforce that entitlement. On April 9 1960 the tenant ceased to trade and its occupation of the premises was confined to the quantity of stock in trade it left there. The question for determination was whether in those circumstances the tenant was entitled to maintain the proceedings in which it sought to compel the grant of a new tenancy. E Cross J held that it was a continuing condition of the tenant's right to a new tenancy that he should throughout the proceedings remain a tenant under a tenancy to which the Act applied but that in the circumstances the tenant had not ceased to occupy the premises for the purpose of its business. At p1253 he said:

> The only sort of tenant who is given by section 24(1) a right to apply for a new tenancy is a tenant under a tenancy to which the Act applies. And if the tenant ceased to occupy the property for business purposes after the service by the landlords of a statutory notice but before he himself issued a summons asking for a new tenancy, then, as it seems to me, the landlord could clearly apply to have the summons dismissed on the ground that the tenant had not *locus standi*
> F to apply to the court.

He concluded at p1255:

> As I see it, it really is a continuing condition of the tenant's right to a new tenancy that he should be throughout the proceedings tenant under a tenancy to which the Act applies. If this condition ceases to be fulfilled at any time, then I think that the landlord can apply to have the tenant's summons dismissed just as he could if the tenancy was not protected by the Act when the application was made.

I consider that the statement as to the rights conferred by section 24(1) are part of the decision and not *obiter dictum* only. In relation to that decision the distinction does not matter because it was a decision at first instance and would not of itself entitle this court to depart from G the decision in *Long Acre*. But it is relevant in the consideration of the decision of this court in *Morrisons Holdings Ltd* v *Manders Property (Wolverhampton) Ltd* [1976] 1 WLR 533 which followed it.

In *Morrisons Holdings Ltd* v *Manders Property (Wolverhampton) Ltd* [1976] 1 WLR 533 the tenants' lease was due to expire on December 24 1977 subject to a break clause exerciseable by the landlord should the demised and other neighbouring premises be substantially destroyed by fire. Such a fire occurred on May 20 1974 doing such damage to the demised premises as to render them unfit for the occupation of the tenants for the purposes of their business. Two days later, on May 22 1974, the landlord served notice to operate the H break clause and on June 17 1974 demolished the premises to which the tenants had not returned. On July 1 1974 the landlord served a notice under section 25 to determine the tenancy. On October 17 1974 the tenants applied to the court for the grant of a new tenancy. The landlord took the preliminary objection that the tenants had no *locus standi* to apply for the grant of a new tenancy since before they made their application they had ceased to occupy the demised premises for the purposes of their business. The county court judge upheld the objection of the landlord on the basis that it was to be inferred that the tenants' absence from the premises was permanent. The tenants appealed. J

In the Court of Appeal Scarman LJ posed the question to be determined as whether the tenants had *locus standi* to make an application for a new tenancy under Part II of the Act. After setting out the facts of the case, at p539 he stated that:

> the point turns on the construction of sections 23 and 24 of the Landlord and Tenant Act 1954 and from the proper inference to be drawn as to the relationship of the tenants towards the property after the devastating fire.

He considered the terms of sections 23 and 24 and passages in the judgment of Cross J in *I&H Caplan* v *Caplan (No 2)* [1963] 1 WLR 1247 including the first of the passages from that judgment which I K have quoted. Then, at p540, he said:

> I respectfully agree with the view of the law expressed by Cross J in the two passages to which I have referred. I would put it in my own words as follows: in order to apply for a new tenancy under the Act a tenant must show either that he is continuing in occupation of the premises for the purposes of a business carried on by him, or, if events over which he has no control have led him to absent himself from the premises, if he continues to exert and claim his right to occupancy. This is a question which did not arise in terms in *Caplan (I & H) Ltd* v *Caplan (No 2)*, where there was a temporary absence of the tenant which the judge held did not interrupt the continuity of his occupation in the circumstances of that case. However, the temporary absence in *Caplan (I & H) Ltd* v *Caplan (No 2)* which did not destroy the continuity of occupation was L absence at the volition of the tenant.

Later, at p540, he said:

> Of course, if the tenants were out of occupation or had abandoned their right of occupation at the time the notice was served, it would determine the contract of tenancy, since the tenancy would no longer be one to which Part II of the Act applied. But if the tenants still occupied or claimed their right of occupation at the time the notice was served, then the notice would not avail because it was not a notice as required by the Act.

Scarman LJ then considered the facts and came to the conclusion that it was not to be inferred that the tenants' cesser of occupation M amounted to its abandonment so as to preclude any *locus standi* to apply for the new tenancy. Sir Gordon Willmer reached the same conclusion and Stephenson LJ agreed with both judgments.

It is true, as counsel for the landlords submitted, that the decision related to the *locus standi* of a tenant not in occupation to bring proceedings under section 24(1) seeking the grant of a new tenancy rather than the effect on the termination of the contractual tenancy of ceasing to occupy the demised premises. In my judgment, this feature cannot alter the fact that it was an essential part of the reasoning of the Court of Appeal in arriving at that conclusion that if a tenant ceased to occupy the premises for the purposes of his business then the

A protection which the Act would otherwise afford would be lost. It must follow from this proposition that a tenancy for a term which would determine by effluxion of time if it were not one to which the Act applies will so determine if at the expiration of the term the tenant has ceased to occupy it for the purposes of any business. It is also true, as counsel for the landlords observed, that there was no argument addressed to the court based on the provisions of section 27. In my view, this is not material to the question whether the decision is binding but is relevant in the consideration whether, if free to do so, this court should follow the decision of the Court of Appeal in *Morrisons* in preference to the decision of the Court of Appeal in *Long*
B *Acre*. In my judgment, the decision in *Morrisons* covered exactly the same point as the court decided in *Long Acre* and was a decision which the court in *Long Acre* would have been bound to follow had they been referred to it.

It follows that, in my view, this court is entitled and bound to decide whether to follow *Morrisons* in preference to *Long Acre*. In those circumstances it is unnecessary to consider the alternative argument of the landlords to the effect that this is one of the rare and exceptional cases in which the Court of Appeal is entitled to depart from one of its previous decisions. Nevertheless it is necessary to refer to the other decisions and *dicta* relied on by Pearl on the question
C which of the two decisions to follow.

Morrisons Holdings Ltd v *Manders Property (Wolverhampton) Ltd* was followed in two subsequent decisions of the Court of Appeal, namely *Hancock & Willis* v *GMS Syndicate Ltd* [1983] 1 EGLR 70 and *Demetriou* v *Poolaction Ltd* [1991] 1 EGLR 100. It is unnecessary to refer to the judgments in either of them. But reference should be made to *Cheryl Investments Ltd* v *Saldanha* [1978] 1 WLR 1329. The case concerned whether the Act applied where the relevant occupation was partly residential and partly for the purposes of a business. It is common ground that it was not a decision on the point so as to be binding in this case. At p1338 Geoffrey Lane LJ said:

> D If the tenant continues the "business occupation" (if it may be called that) thereafter, no further difficulty arises. What is the situation, however, if between the end of the contractual tenancy and the service of notice of determination by the landlord the business occupation ceases? Section 25(1) provides that the landlord
> "may terminate a tenancy to which this Part of this Act applies by a notice given to the tenant in the prescribed form specifying the date at which the tenancy is to come to an end . . ."
> If at the time the notice is served the business occupation has ceased, there is no "tenancy to which this Part of this Act applies", and nothing on which a section 25 notice by the landlord can bite. It seems therefore that the business occupation must exist both at the time the contractual tenancy comes to an end
> E and at the date of service of the notice of determination. It is necessary to point out, however, that in neither of the two cases with which we are concerned was there any change of purpose between the term date and the service of notice of determination. In each case the question is, was there or was there not a section 23(1) business occupation at the time the contractual tenancy ended?

These observations are entirely in line with those of Lord Reid in *Green*, Cross J in *Caplan* and Scarman LJ in *Morrisons*. Finally there is the passage in the speech of Lord Nicholls of Birkenhead in *Graysim Holdings Ltd* v *P&O Property Holdings Ltd* [1996] 1 AC 329. The case was concerned with the question whether the tenant occupied the
F demised premises for the purposes of his business when the premises were occupied by stallholders under licence from the tenant. Lord Nicholls of Birkenhead observed:

> The Act seeks to protect the tenant in his continuing use of the property for the purposes of that activity. Thus the word [occupied] carries a connotation of some physical use of the property by the tenant for the purposes of his business.

In my view, this *dictum* is consistent with all the others to which I have referred but does not carry the argument any further than those others do.

It was submitted by counsel for the landlords that this court should follow the decision in *Long Acre*, whatever its views as to its correctness, on the basis that it is a decision of some standing in the G field of property and conveyancing matters on the faith of which many arrangements are likely to have been made. Such a principle is well established and was most recently referred to in this court in *Phillips* v *Mobil Oil Co Ltd* [1989] 1 WLR 888*, pp893-894. For my part, I do not think that weight should be given in this case to any such considerations. The decision is relatively recent and these proceedings were commenced only three years after it had been decided. But, of more importance, the decision in *Long Acre* was not received without criticism. Thus the editors of *Woodfall's Law of Landlord and Tenant* wrote in para 22.046 that: "it would seem that a tenancy for a fixed term cannot be continued under s.24(1) unless, immediately before the H term would at common law expire by effluxion of time, the tenancy is one to which the Act applies". In a footnote they commented "this point appears to have been overlooked by the Court of Appeal in" *Long Acre*.

In all these circumstances, in my judgment, the decision of this court in *Morrisons* should be preferred to its decision in *Long Acre* because, in my respectful opinion, it is more consistent with the wording of the legislation and accords with *dicta* of the most persuasive sort to be found in other cases decided both before and after *Morrisons*. It is most unfortunate that the Court of Appeal in *Long Acre* was not referred to any of the cases which I have mentioned; I J cannot believe that their decision would have been the same if they had been. The consequence is that on the true construction of the Act section 24(1) did not apply to the tenancies of Pearl on February 14 1993 when, in accordance with the common law, they expired by effluxion of time.

I would allow this appeal.

Sir John May and **Stuart-Smith LJ** agreed and did not add anything.

Appeal allowed. K

Long *v* Tower Hamlets London Borough Council

CHANCERY DIVISION

March 20 1996 L

MR JAMES MUNBY QC, sitting as a deputy judge of the division

Estates Gazette February 1 1997

[1997] 05 EG 157

Adverse possession — Landlord and tenant — Tenant not paying rent — Whether time ran from expiration of notice to quit or date when rent stopped — Whether "lease in writing" under para 5, Schedule 1 to Limitation Act 1980 — Whether tenant can be in adverse possession under Landlord and Tenant Act 1954 continuation tenancy M

By a letter dated September 4 1975 the then owners of shop premises offered the plaintiff a quarterly tenancy to commence on September 29 upon certain terms; the plaintiff endorsed a copy of the letter on September 8 to the effect that he agreed to abide by those terms, and entered into possession on September 29 1975. On October 13 1976 the Greater London Council were registered as proprietors of the freehold title, and in September 1983 gave the plaintiff notice to quit on March 25 1984. In August 1988 the freehold was transferred to the defendant council as the successor authority. The plaintiff claimed

---

*Editor's note: Also reported at [1989] 2 EGLR 246

A to have stopped paying rent in 1977, and to have taken possession of a maisonette on the upper floors in 1977. Accordingly, he had been in adverse possession of the premises and the council's claim for possession was statute-barred by section 15 of the Limitation Act 1980. The council's application to strike out the plaintiff's claim was allowed by Deputy Master Wall. On appeal it was contended for the defendant council that: (1) the plaintiff's tenancy was a "lease in writing" within the meaning of para 5(1) of Schedule 1 to the Limitation Act 1980 and that therefore time did not begin to run until the tenancy expired by notice to quit on March 25 1984, less than 12 years before the commencement of proceedings; (2) it was impossible
B for the plaintiff to be in adverse possession because, by reason of Part II of the Landlord and Tenant Act 1954, the plaintiff's tenancy was continued after the expiration of the notice to quit; and (3) the plaintiff acknowledged the council's title by correspondence.

Held: **The appeal was allowed. (1) The tenancy document, which was a reversionary lease not made by deed, and not conferring on the plaintiff an immediate right to take possession, was incapable of taking effect as a parol lease under section 54(2) of the Law of Property Act 1925. The tenancy, which did come into existence, was not one created by the tenancy document, but rather arose by operation of law; it was therefore not a "lease in**
C writing" within the meaning of para 5(1) of Schedule 1 to the 1980 Act. (2) Whether or not a tenant can be in adverse possession where a tenancy is continuing under the 1954 Act, there was no evidence that the Act applied at the time. (3) There was no acknowledgement of the council's title in the correspondence.

The following cases are referred to in this report.

Bolton (Lord) v *Tomlin* (1836) 5 Ad & El 856
Buckinghamshire County Council v *Moran* [1990] Ch 623; [1989] 3 WLR 152; [1989] 2 All ER 225, CA
Doe d Landsell v *Gower* (1851) 17 QB 589
Edge v *Strafford* (1831) 1 Cr&J 391
Foster v *Reeves* [1892] 2 QB 255
D *Hand* v *Hall* (1877) 2 ExD 355
Inman v *Stamp* (1815) 1 Stark 12
Kingsmill v *Millard* (1855) 11 Exch 313
Kushner v *Law Society* [1952] 1 KB 264; [1952] 1 All ER 404; (1952) 159 EG 139
Martin v *Smith* (1874) LR 9 Exch 50
Moses v *Lovegrove* [1952] 2 QB 533; [1952] 1 All ER 1279; [1952] 1 TLR 1324, CA
Parker v *Briggs* (1893) 37 Sol Jo 452
Parker v *Taswell* (1858) 2 De G&J 559
Rawlins v *Turner* (1699) 1 Ld Raym 736
Rollason v *Leon* (1861) 7 H&N 73
Ryley v *Hicks* (1725) 1 Stranage 651
E *Smirk* v *Lyndale Developments Ltd* [1975] Ch 317; [1975] 2 WLR 495; [1975] 1 All ER 690; (1974) 29 P&CR 407; [1975] 2 EGLR 43; 235 EG 751, CA
Tress v *Savage* (1854) 4 E&B 36
Walsh v *Lonsdale* (1882) 21 ChD 9
Wright v *Macadam* [1949] 2 KB 744; [1949] 2 All ER 565, CA

This was an appeal by the plaintiff, Ronald James Long, against the decision of Deputy Master Wall, who had allowed an application by the defendants, Tower Hamlets London Borough Council, to strike out the plaintiff's claim on the ground that it disclosed no cause of action.

Philip Walter (instructed by Wallace Bogan) appeared for the plaintiff; Sarah Hargreaves (instructed by the solicitor to the Tower
F Hamlets London Borough Council) represented the defendants.

Giving judgment, MR JAMES MUNBY QC said: This case raises the question what is meant by the words "tenancy from year to year or other period, without a lease in writing" in para 5(1) of Schedule 1 to the Limitation Act 1980. It also, at least on one view of the matter, raises the question what is meant by the words "creation by parol of leases taking effect in possession for a term not exceeding three years" in section 54(2) of the Law of Property Act 1925 — a question which, in turn, involves consideration of the law relating to reversionary leases. Both questions are of some interest and importance.

The matter comes before me on an appeal by the plaintiff, Ronald James Long ("Mr Long"), from an order made by Deputy Master Wall on September 26 1995. The master had before him an application by G the defendants, Tower Hamlets London Borough Council (the "council"), by summons dated March 21 1995 to strike out the originating summons which the plaintiff had issued on January 13 1995. The master struck out the originating summons on the ground that Mr Long's claim was doomed to fail.

Facts

The facts which are relevant to the main question that I have to decide lie within a very narrow compass. The dispute relates to the ground-floor shop ("the shop") and a maisonette on the upper floors ("the maisonette") at 21 Turners Road, Tower Hamlets, London E3 ("the premises"). The premises are part of the property registered at H Her Majesty's Land Registry under title number 255264. On September 4 1975 agents acting for the then owner of the premises, a Mr Jacobs, wrote a letter to Mr Long in the following terms:

> Dear Sir,
> 21 Turners Road E3
> *Ground Floor*
> We confirm on behalf of our client, Mr H Jacobs, that we are prepared to grant you a quarterly tenancy of the above upon the following terms:
> 1 The rent to be £55 per quarter, payable on the usual quarter days in advance, exclusive of general and water rates payable in respect of this part of the property. J
> 2 The premises are to be held by R J Long, Esq, trading as Long Bros, Upholstery, in connection with his business as an upholsterer.
> 3 The tenant is not to sublet the whole or any part of the premises.
> 4 The tenant is to maintain the interior and shop front in good decorative repair and condition.
> 5 The tenant is to give four weeks notice in writing before vacating the premises.
> 6 The tenancy is to commence on Monday, the 29th September 1975 on a quarterly basis.
> Yours faithfully,

On September 8 1975 Mr Long made the following endorsement on a copy of the letter: K

> I, Ronald James Long, agree to abide by the terms and conditions of the tenancy set out herein
> Signed R J Long
> Date 8-9-75

The copy letter endorsed by Mr Long ("the tenancy document") was returned to the landlord. It was retained by the landlord and his successors in title, being eventually produced from their files by the council after the present litigation had been commenced.

On or about September 29 1975 Mr Long took possession of the shop. On September 24 1976 Mr Jacobs' successor in title completed L the sale of the freehold of the premises to the Greater London Council ("GLC"). On October 13 1976 GLC were registered as proprietors of the premises. In or about September 1983 GLC gave Mr Long notice to quit the shop; the notice expired on March 25 1984. On August 25 1988 the freehold of the premises was transferred under statutory powers to the council as the successor authority to GLC. The council have since been registered as proprietor of the premises.

Issues

It is common ground that Mr Long has paid no rent for the shop for many years. According to Mr Long he stopped paying rent in 1977. M The council accept that rent has been neither demanded nor paid since March 1984. When precisely rent was last paid is a matter which can only be resolved at trial. Mr Long claims to have taken possession of the maisonette in 1977 at a time when, he says, it was unoccupied. The council, on the other hand, have produced documents which *prima facie* establish that the maisonette was lawfully occupied by a Mr Taylor as tenant of GLC until February 20 1978, when he was rehoused by GLC elsewhere. Precisely when and to what extent Mr Long ever made use of the maisonette is in dispute between the parties, as is the extent to which he has, in recent years, been in occupation of the shop. These also, however, are matters which can only be resolved at trial.

A At trial the case will turn on the application to the facts, whatever they may be found to be, of section 15(1) and (6) of, and paras 5 and 8 of Schedule 1 to, the Limitation Act 1980. Section 15(1) and (6) is in the following terms:

> (1) No action shall be brought by any person to recover any land after the expiration of twelve years from the date on which the right of action accrued to him or, if it first accrued to some person through whom he claims, to that person. . . .
>
> (6) Part I of Schedule 1 to this Act contains provisions for determining the date of accrual of rights of action to recover land in the cases there mentioned.

B Para 5 of Schedule 1 provides as follows:

> (1) Subject to sub-paragraph (2) below, a tenancy from year to year or other period, without a lease in writing, shall for the purposes of this Act be treated as being determined at the expiration of the first year or other period; and accordingly the right of action of the person entitled to the land subject to the tenancy shall be treated as having accrued at the date on which in accordance with this subparagraph the tenancy is determined.
>
> (2) Where any rent has subsequently been received in respect of the tenancy, the right of action shall be treated as having accrued on the date of the last receipt of rent.

Para 8(1) of Schedule 1 provides that:

> C No right of action to recover land shall be treated as accruing unless the land is in the possession of some person in whose favour the period of limitation can run (referred to below in this paragraph as "adverse possession"); and where under the preceding provisions of this Schedule any such right of action is treated as accruing on a certain date and no person is in adverse possession on that date, the right of action shall not be treated as accruing unless and until adverse possession is taken of the land.

Mr Long's case is simple. He says that he is and has been ever since 1977 continuously in possession of the whole of the premises (that is, of both the shop and the maisonette); that he has paid no rent since 1977 for any part of the premises and has, ever since then, been
D in adverse possession of the premises within the meaning of para 8(1); that in accordance with para 5(2) the GLC's right of action to recover the premises is to be treated as having accrued at the latest by January 1 1978; and that, in consequence, any claim by the council to recover the premises became statute-barred, in accordance with section 15(1), at the latest by January 1 1990, some five years before the present proceedings were commenced. On this footing Mr Long seeks: (i) a declaration that he is entitled beneficially to the premises; and (ii) rectification of the proprietorship register of title no 255264 to remove the premises from that title and to open a new title for the premises under which he is the registered proprietor.

E The council deny that Mr Long can prove either the factual possession or the *animus possidendi* which are necessary to found adverse possession: *Buckinghamshire County Council* v *Moran* [1990] Ch 623. Moreover, the council say that Mr Long last paid rent in March 1984, less than 12 years before the proceedings were commenced on January 13 1995. Accordingly, say the council, their claim to recover the premises is not yet statute-barred.

As I have already said, the factual disputes between the parties can only be resolved at trial. On what basis, therefore, do the council seek to strike out Mr Long's claim? Three arguments are put forward by Miss Sara Hargreaves of counsel on behalf of the council. Each, in
F substance, proceeds on the footing that, whatever may be the eventual resolution of the factual disputes to which I have referred, there are other grounds on which the council can establish that time cannot on any view have begun to run in Mr Long's favour until 1984 at the earliest.

Before considering Miss Hargreaves' arguments there is one preliminary matter which I must address. It will be appreciated that the tenancy document related only to the shop. There is no suggestion that Mr Long was ever granted a tenancy of the maisonette or that he has ever paid any additional rent for the maisonette. On what basis, then, does Miss Hargreaves seek to resist Mr Long's claim in so far as it relates to the maisonette? She relies upon the principle that if a tenant occupies land belonging to the landlord but not included in the demise, G
that land is presumed to be an addition to the land demised to the tenant, so that it becomes subject to the terms of the tenancy and must therefore be given up to the landlord when the tenancy ends: *Smirk* v *Lyndale Developments Ltd* [1975] Ch 317*. Mr Philip Walter of counsel, on behalf of Mr Long, seeks to confine the principle, which otherwise he accepts, in two ways: first, he says it is limited to cases where the encroachment is on land which is waste; second, he says that it is limited to cases where the tenant is estopped by words or conduct from denying that the additional land is within the demise. Neither condition, he says is satisfied in the present case.

I cannot accept either of Mr Walter's submissions. In *Smirk* v H
Lyndale Developments Ltd [1975] Ch 317, Pennycuick V-C considered all the relevant authorities, which as he said at p323G had got into something of a tangle, and concluded at pp324G, 329C, 332G that the law had been correctly stated by Alderson and Parke BB in *Kingsmill* v *Millard* (1855) 11 Exch 313. His decision on the point was not challenged on appeal and was accepted as correct by the Court of Appeal: [1975] Ch 317, at pp337G, 340E, 341H. The law as laid down in those two cases I take to be clear and binding on me. They show that the principle applies whether or not the land encroached upon is waste. They show, moreover, that the principle is based on a presumption, albeit a rebuttable presumption, which, although it may be considered J
to operate in a manner akin to an estoppel, is not dependent upon proof, as Mr Walter I think would have it, of any active representation to the landlords by the tenant. On the contrary, the presumption is treated as applying unless the tenant, during the term, communicates with the landlords in such a way as to show that he is asserting his own title as against the landlords or, as Alderson B put it, that he is setting the landlords at defiance. There is nothing in the evidence before me in the present case which even begins to suggest that Mr Long, in his dealings with the GLC and the council, ever sought to distinguish between the shop and the maisonette or ever sought to assert that, while he occupied the shop as tenant, his occupation of the maisonette K
was in some different character. Accordingly, in my judgment, Miss Hargreaves is entitled to say that, on the evidence as it stands, the presumption is wholly unrebutted and that the principle as laid down in *Smirk* v *Lyndale Developments Ltd* applies. It follows that, for present purposes, no distinction is to be drawn between Mr Long's claim as it relates to the shop and his claim as it relates to the maisonette.

I turn therefore to consider the three issues identified by Miss Hargreaves as justifying the striking out of Mr Long's claim.

Issue 1 — "lease in writing"

In the first place, and this is her principal argument, Miss Hargreaves submits that the tenancy document is a "lease in writing" L
within the meaning of para 5(1) of Schedule 1 to the Limitation Act 1980; that time accordingly did not begin to run in favour of Mr Long when he stopped paying rent (whenever that was) but only when his tenancy was brought to an end by notice to quit; and that the only relevant notice to quit was that which expired on March 25 1984, less than 12 years before the commencement of the present proceedings.

Mr Walter denies that the tenancy document is, or as a matter of law is capable of being, a "lease in writing" within the meaning of para 5(1). Put shortly, his argument can be summarised as follows:

(a) A document cannot be a "lease in writing" within the meaning of para 5(1) unless it is dispositive, that is, as Mr Walter put it in M
argument, the document must be one which at law creates, of and by itself, a leasehold estate in land.

(b) As the tenancy document was not executed as a deed, it could take effect as a lease only if fell within the exception in section 54(2) of the Law of Property Act 1925, that is, if it "[took] effect in possession" for a term not exceeding three years. It did not, says Mr Walter, because it took effect, if at all, as a reversionary lease.

At first blush the use in para 5(1) of the words "tenancy from year to year or other period, without a lease in writing" is rather puzzling to

---

*Editor's note: Also reports at [1975] 2 EGLR 43

A anyone familiar with the scheme of the Law of Property Act 1925 and its distinction between those leases which are required by section 52 to be made by deed and those leases which section 54(2) permits to be created by parol, in other words without the need even for writing. The provisions of the Law of Property Act 1925 itself are also, at first, rather puzzling. Notwithstanding the reference in section 52(2)(*d*) to "leases or tenancies . . . not required by law to be made in writing", the Act does not itself, so far as I am aware, require any class of lease to be made in writing, though certain leases made under the Settled Land Act 1925 which would otherwise be required by section 42(1)(i) of that Act to be made by deed are permitted by section 42(5)(ii) to be
B made by writing under hand.

(1) *Historical background*

The explanation for the seeming oddities is historical. As Mr Walter pointed out, para 5 of Schedule 1 to the Limitation Act 1980 reproduces provisions first enacted as section 8 of the Real Property Limitation Act 1833 and later re-enacted as section 9(2) of the Limitation Act 1939. Section 8 of the Real Property Limitation Act 1833 provided as follows:

> And be it further enacted, That when any Person shall be in Possession or in Receipt of the Profits of any Land, or in Receipt of any Rent, as Tenant from Year to Year or other Period, without any Lease in Writing, the Right of
> C the Person entitled subject thereto, or of the Person through whom he claims, to make an Entry or Distress or to bring an Action to recover such Land or Rent shall be deemed to have first accrued at the Determination of the first of such Years or other Periods, or at the last Time when any Rent payable in respect of such Tenancy shall have been received (which shall last happen).

As Mr Walter also pointed out, the context in which the Real Property Limitation Act 1833 was enacted was that of the Statute of Frauds 1677. However, in order to understand the exceedingly interesting and instructive arguments that have been so skilfully deployed by Mr Walter and Miss Hargreaves it is necessary to go even further back than that and to consider, albeit briefly, the relevant
D principles of the common law.

At common law a lease could be granted in any way, even orally: Megarry and Wade, *The Law of Real Property*, 5th ed 1984, at p636. Moreover, there was at common law no restraint upon the grant of a reversionary lease, that is, a lease to take effect in reversion on some future day, however distant, and conferring no right to take possession in the meantime. Such a lease (to be contrasted with a concurrent lease, that is, a lease *of* the reversion: Megarry and Wade at pp664-665) gave the lessee an immediate vested legal *interest* in the land, that interest being known as an *interesse termini*, though until the date when the lease was due to take effect this interest was vested in interest and not
E in possession. On the other hand, the lessee under a reversionary lease acquired no *estate* in the land until he had actually entered, that is, taken possession in accordance with the lease; until then all he had was an *interesse termini*: Megarry and Wade at pp231, 637, 647-648.

In 1677 the Statute of Frauds was enacted "For Prevention of many fraudulent Practices, which are commonly endeavoured to be upheld by Perjury and Subornation of Perjury". Section 1 of the Statute of Frauds provided that:

> all Leases, Estates . . . or Terms of Years . . . made or created by Livery and Seisin only, or by Parol, and not put in Writing, and signed by the Parties so making or creating the same, or their Agents thereunto lawfully authorised by
> F Writing, shall have the Force and Effect of Leases or Estates at Will only, and shall not either in Law or Equity be deemed or taken to have any other or greater Force or Effect; any Consideration for making any such Parol Leases or Estates, or any former Law or Usage, to the contrary notwithstanding.

Section 2 of the Statute of Frauds excepted from the operation of section 1:

> all Leases not exceeding the Term of Three Years from the making thereof, whereupon the Rent reserved to the Landlord, during such Term, shall amount unto Two third Parts at the least of the full improved Value of the Thing demised.

Section 4 of the Statute of Frauds provided, so far as is material, that:

> no Action shall be brought . . . to charge any Person . . . upon any Contract or Sale of Lands, . . . or any Interest in or concerning them . . . unless the Agreement upon which such Action shall be brought, or some Memorandum or Note thereof, shall be in Writing, and signed by the Party to be charged therewith, or some other Person thereunto by him lawfully authorized. G

Accordingly, when the Real Property Limitation Act 1833 was enacted the distinction was not, as it now is, between those leases required to be made by deed and those which may be created by parol, but rather between those leases required by section 1 of the Statute of Frauds to be "put in Writing" and those which, by virtue of section 2 of the Statute of Frauds, could still be created at common law by parol.

The modern requirement of a deed was first imposed by section 4 of H the Transfer of Property Act 1844, which provided:

> That no Lease in Writing of any . . . Land . . . shall be valid as a Lease . . . unless the same shall be made by Deed; but any Agreement in Writing to let . . . any such Land shall be valid and take effect as an Agreement to execute a Lease . . .; and the Person who shall be in the Possession of the Land in pursuance of any Agreement to let may, from Payment of Rent or other Circumstances, be construed to be a Tenant from Year to Year.

The drafting of section 4 of the Transfer of Property Act 1844 was soon perceived to have been defective: see *Tress* v *Savage* (1854) 4 E&B 36, at p43, *per* Coleridge J. Section 4 was therefore repealed by section 1 of the Real Property Act 1845. Section 3 of the latter Act J provided:

> That . . . a Lease, required by Law to be in Writing, . . . shall . . . be void at Law, unless made by Deed:

Sections 1, 2 and 4 of the Statute of Frauds remained in force.

Thus from 1845 the combined effect of the Statute of Frauds and the Real Property Act 1845 was that those leases which, prior to the Transfer of Property Act 1844, had been required by section 1 of the Statute of Frauds to be in writing were now required by section 3 of the Real Property Act 1845 to be made by deed; but that those leases which had been excepted by section 2 of the Statute of Frauds from the K operation of section 1 of the Statute of Frauds remained excepted from the operation of section 3 of the Real Property Act 1845.

Although the effect of section 3 of the Real Property Act 1845 was to make a written lease which did not fall within section 2 of the Statute of Frauds "void", that did not deprive the written document of all legal effect. At *law*, as soon as the tenant took possession, and rent was paid and accepted, a yearly or other periodic tenancy arose. That periodic tenancy was a legal estate which was treated as being held on such of the terms set out in the written document as were applicable to and not inconsistent with a periodic tenancy: *Tress* v *Savage* (1854) 4 E&B 36; *Martin* v *Smith* (1874) LR 9 Exch 50. In *equity* the imperfect L lease was treated as a contract to grant a lease and, moreover, provided there was valuable consideration and a sufficient act of part performance, a contract of which specific performance would, in principle, be granted under the rule in *Parker* v *Taswell* (1858) 2 De G&J 559 and *Walsh* v *Lonsdale* (1882) 21 ChD 9.

Sections 1 and 2 of the Statute of Frauds and section 3 of the Real Property Act 1845 were replaced by sections 52 and 54 of the Law of Property Act 1925. Section 4 of the Statute of Frauds, so far as is material, was replaced by section 40 of the Law of Property Act, itself since repealed by section 2(8) of the Law of Property (Miscellaneous Provisions) Act 1989. Section 52 of the Law of Property Act 1925, M replacing section 3 of the Real Property Act 1845, provides so far as is material, as follows:

> (1) All conveyances of land or of any interest therein are void for the purpose of conveying or creating a legal estate unless made by deed.
> (2) This section does not apply to NL . . .
> (*d*) leases or tenancies or other assurances not required by law to be made in writing;

"Conveyance" is defined by section 205(1)(ii) of the Law of Property Act 1925 as including a lease. Section 54 of the Law of Property Act 1925, replacing sections 1 and 2 of the Statute of Frauds, is in the following terms:

(1) All interests in land created by parol and not put in writing and signed by the persons so creating the same, or by their agents thereunto lawfully authorised in writing, have, notwithstanding any consideration having been given for the same, the force and effect of interests at will only.

(2) Nothing in the foregoing provisions of this Part of this Act shall affect the creation by parol of leases taking effect in possession for a term not exceeding three years (whether or not the lessee is given power to extend the term) at the best rent which can be reasonably obtained without taking a fine.

The words "taking effect in possession" are not defined as such, though "Term of years absolute" is defined by section 205(1)(xxvii) of the Law of Property Act 1925 in such a way as to distinguish a "term of years taking effect in possession" from a "term of years taking effect in reversion":

"Term of years absolute" means a term of years (taking effect either in possession or in reversion whether or not at a rent) . . .

"Possession" is defined by section 205(1)(xix) of the Law of Property Act 1925 as including receipt of rents or profits or the right to receive the same.

It will be observed that sections 52 and 54 of the Law of Property Act 1925 follow very closely, and in large part perpetuate the language of, their statutory precursors. As is pointed out in Megarry and Wade at p637:

The provisions of the two previous Acts were repeated by the Law of Property Act 1925 but with certain alterations. No attempt has been made to state the combined effect of the earlier Acts: section 54 follows the Statute of Frauds 1677 and section 52 follows the Real Property Act 1845 so that their differing provisions still stand side by side. This is presumably because they had become so familiar that it was thought best to preserve them.

Thus sections 52(2)(*d*) and 54(1) preserve fossilised references to leases in writing which, though they may reflect the formal requirements imposed in 1677 by the Statute of Frauds (and still representing the law when the Real Property Limitation Act 1833 was enacted), had been overtaken by the formal requirement of a deed imposed by the Transfer of Property Act 1844.

However, as is also pointed out in Megarry and Wade at p637, and as appears from a comparison of section 54(2) of the Law of Property Act 1925 with section 2 of the Statute of Frauds, section 54(2) did make two changes. The first, on which nothing turns for present purposes, was to substitute for the reference in the Statute of Frauds to a rent not less than two-thirds of the full improved value reference to the best rent reasonably obtainable without taking a fine. The second, which is crucial to Mr Walter's argument, was to substitute for the reference in the Statute of Frauds to a lease not exceeding three years "from the making thereof" reference in section 54(2) to a lease "taking effect in possession" for a term not exceeding three years.

The Law of Property Act 1925 made two other changes to the law which are material for present purposes: first, section 149(1) and (2) abolished the doctrine of *interesse termini* and the rule that a lessee took no estate until he had actually entered; second, section 149(3) limits to 21 years the time that may elapse before the term under a reversionary lease begins; a reversionary lease postponed for more than 21 years is void. The effect of section 149, therefore, is to give the lessee under a reversionary lease, so long as it is not postponed for more than 21 years, an immediate estate, rather than a mere interest, in the land — though until the date when the lease is due to take effect that estate will be vested in interest only and not in possession.

(2) *Para 5(1) — "lease in writing"*

Before the master Mr Walter argued that, notwithstanding the use of the words "lease in writing", the exception under para 5(1) of Schedule 1 to the Limitation Act 1980 applies only if the lease has been made by deed. That submission, which, in my judgment, is clearly wrong, was not pursued before me. Mr Walter and Miss Hargreaves were at one in agreeing that the word "writing" in para 5(1) has the meaning given to it by section 5 of and Schedule 1 to the Interpretation Act 1978:

"Writing" includes typing, printing, lithography, photography and other modes of representing or reproducing words in a visible form, and expressions referring to writing are construed accordingly.

I agree. A written document does not have to be a deed in order to be a "writing" for the purposes of para 5(1), though on the other hand, a document which is a deed will also be a "writing" for the purposes of para 5(1).

As I have said, Mr Walter's argument is that a document is not a "lease in writing" unless it is dispositive, that is, a document which at law creates, of and by itself, a leasehold estate in land. A document which is merely an agreement for a lease, or merely evidential of the terms of a lease is not, he submits, a "lease in writing". Miss Hargreaves' submission is that this is much too subtle and technical an approach. She contends for a simple and practical approach, suggesting that the reason why para 5(1) distinguishes between those leases which are, and those leases which are not, "in writing" is evidential. She asserts, I have no doubt correctly, that Mr Long acquired, at the latest when he took possession and started paying rent on or about September 29 1975, a periodic tenancy at law and that that tenancy is to be treated at law as having been held on the terms of the tenancy document. She says, again no doubt correctly, that equity would in principle treat the tenancy document as a specifically enforceable agreement for a lease, so that Mr Long had what is often called an equitable lease. Thus far, I did not understand Mr Walter to challenge Miss Hargreaves' analysis of the effect of the tenancy document in the events which had happened. So, says Miss Hargreaves, there plainly was a lease, all the terms of the lease are to be found set out in a written document, and there is, therefore, a "lease in writing". It is, says Miss Hargreaves, as simple and straight forward as that. No, says Mr Walter, Miss Hargreaves' argument does not go far enough. She has to show that the tenancy agreement itself was dispositive and created a legal estate in land. An equitable lease is not, he says, a lease and does not of itself create an estate in land.

In my judgment, Mr Walter's submission on this point is correct.

In the light of the historical background, and bearing in mind the familiar distinction, drawn for example in sections 9 and 7 of the Statute of Frauds (now sections 53(1)(*c*) and 53(1)(*b*) of the Law of Property Act 1925), between those transactions required to be "in" writing, that is, where the writing required by the statute is dispositive, and those requiring only to be "manifested and proved by some writing", that is, where the writing required by the statute need be no more than evidential, one's initial impression is that the "writing" referred to in para 5(1) of Schedule 1 to the Limitation Act 1980 must be dispositive and not merely evidential. Para 5(1), after all, refers to a "lease *in* writing" (emphasis supplied). Indeed, as a matter of simple English language, that is what one would expect the words to mean.

But the matter does not rest there, because I have been referred to two authorities which, in my judgment, indicate that this is indeed the correct meaning of the words. In *Doe d Landsell* v *Gower* (1851) 17 QB 589 the question was whether an entry in a vestry-book, signed by the tenant and by one, but not by all, of the parish officers, was a "lease in writing" within the meaning of section 8 of the Real Property Limitation Act 1833. It was held by the Court of Queen's Bench (Patteson, Coleridge and Wightman JJ) that it was not, because only one of the parish officers had signed the vestry-book and he had not professed to sign on behalf of all. The facts of the case were very far removed indeed from the facts of the present case, but the reasons given by the judges for arriving at their decision are apposite. They are all to the same effect. Patteson J at p598 made it clear that, in his opinion, what section 8 required was an . . . "actual lease", that is

. . . not . . . a document, generally, but . . . a lease; . . . something which is to pass an interest.

Coleridge J at p599 said that what section 8 required was

. . . a lease in writing: that is, not merely an instrument which would be evidence of the conditions of holding, but one passing an interest.

Wightman J at p600 said that section 8

> ... requires an instrument in writing that may operate as a lease.

In my judgment, this is clear authority for the proposition that there was no "lease in writing" for the purposes of section 8 of the Real Property Limitation Act 1833, and thus there is no "lease in writing" for the purposes of para 5(1) of Schedule 1 to the Limitation Act 1980, if the "writing" is merely evidential. If there is to be a "lease in writing" the writing must itself at law "pass an interest" (to use the words of Patteson and Coleridge JJ) and "operate as a lease" (to use Wightman J's words).

Moses v *Lovegrove* [1952] 2 QB 533, the only other authority on the point to which I was referred, is to the like effect. One of the questions in that case was whether a rent book was a "lease in writing" within the meaning of section 9(2) of the Limitation Act 1939. The Court of Appeal held that it was not. Sir Raymond Evershed MR said at p536:

> The alleged lease in writing consisted of the rent book, which was put in evidence. The judge rejected the view that the rent book was such a lease in writing within the meaning of the Act, and I think he was entirely right in that conclusion.
>
> The rent book is, I think, what it purports to be, and what it is called, a rent book, that is, a book containing acknowledgements for payment of weekly sums of rent, and containing also, in pursuance of the terms of the legislation, a reference to the conditions on which the tenant was holding his tenancy. I think that on the face of it, it was not intended to be, and is not a contract for granting a tenancy, still less a lease creating an estate. It is, I think, at most what it was intended to be, and what it is on the face of it called, a book evidencing the terms on which the tenant held.

Romer LJ at p543 said that he entirely agreed with the Master of the Rolls.

These authorities make it clear, in my judgment, that a written document, whatever its terms, however clearly referable to the existence of a lease, and however comprehensive it may be in setting out the terms of the lease, is not a "lease in writing" for the purposes of para 5(1) of Schedule 1 to the Limitation Act 1980 unless at law the document itself operates to "pass an interest" (the words used by Patteson and Coleridge JJ), itself "operates as a lease" (the words used by Wightman J) or itself "creates an estate" (the words used by Sir Raymond Evershed MR).

(3) *Section 54 (2) — "taking effect in possession"*

That being so, the next question which arises is whether the tenancy document operated to create a leasehold estate. It plainly did not take effect as a deed. The question therefore is whether or not it operated in accordance with section 54(2) of the Law of Property Act 1925 as a "lease taking effect in possession for a term not exceeding three years".

Mr Walter accepts that, notwithstanding the reference in section 54(2) to the creation of leases "by parol", a lease which otherwise satisfies the requirements of section 54(2) can be created either by parol or in writing. This is plainly correct: see, for example, *Wright* v *Macadam* [1949] 2 KB 744, at p747 *per* Jenkins LJ; *Kushner* v *Law Society* [1952] 1 KB 264, at p272 *per* Lord Goddard CJ. Mr Walter concedes that the tenancy document, properly construed, purported to take effect as a tenancy and not as a mere agreement to grant a tenancy: *Rollason* v *Leon* (1861) 7 H&N 73; *Hand* v *Hall* (1877) 2 ExD 355. He also accepts that the tenancy document, since it stipulated for a quarterly tenancy, was for a "term not exceeding three years" as that phrase is used in section 54(2).

However, submits Mr Walter, the tenancy did not "tak[e] effect in possession" and the tenancy document was therefore not, as required by section 54(2), a lease taking "effect in possession". Although the tenancy document was executed on September 8 1975, the tenancy was not to commence immediately but rather on a future date, September 29 1975, and Mr Long had no right to take possession until the latter date. Therefore, says Mr Walter, it took effect, if at all, as a reversionary lease. Whatever may have been the position at common law or prior to 1925, he says, the inclusion in section 54(2) of the words "taking effect in possession" has the effect of excluding altogether from the ambit of section 54(2) reversionary leases conferring no immediate right to take possession. Thus, he submits, a reversionary lease can only take effect if made by deed. If that is right then it follows, says Mr Walter, that a reversionary lease cannot be a "lease in writing" for the purposes of para 5(1) of Schedule 1 to the Limitation Act 1925 unless it is made by deed. Mr Walter submits that the only tenancy which ever came into existence in the present case was not one created by the tenancy document. It arose, he says, by operation of law as a result of Mr Long taking possession on or about September 29 1975 and paying rent. The tenancy document, says Mr Walter, is not a lease. It is no more than a document whose terms, so far as they are consistent with a periodic tenancy, will be treated as the terms on which Mr Long held the periodic tenancy which arose by operation of law on or about September 29 1975.

Miss Hargreaves, on the other hand, disputes that reversionary leases were excluded from the exception in section 2 of the Statute of Frauds or that they are excluded from the ambit of section 54(2). She accepts that the three-year period, referred to both in section 2 of the Statute of Frauds and in section 54(2), provides what she calls the basic measure of computation, and that the computation of the three-year period is from the date of the parol grant. But that, according to Miss Hargreaves, is the only limit imposed by section 54(2). The fundamental principle, common both to section 2 of the Statute of Frauds and to section 54(2), she says, is: the term whether it is to commence immediately or at some date in the future, must take effect in possession within three years from the date of the parol grant and must not extend beyond three years from the date of the parol grant. So long as it satisfies this condition it matters not, she submits, whether the term takes effect in possession at the date of the parol grant. It is implicit in this argument that the words "taking effect in possession" in section 54(2) are descriptive only of the quality of the lease at the date when the term commences and not of the quality of the lease at the date of the parol grant. She suggests that the restrictions on the grant of reversionary leases are to be found in section 149 of the Law of Property Act 1925, and not in section 54(2).

(a) *Statute of Frauds*

Before turning to consider the question as it arises in relation to section 54(2) I must first examine certain older authorities on section 2 of the Statute of Frauds. The first case is *Rawlins* v *Turner* (1699) 1 Ld Raym 736. The report is very short.

> No lease by parol is good which imports to convey an interest for more than three years from the time of the making.
>
> It was ruled by Holt Chief Justice at Lent Assizes at Kingston 1699, that such lease for three years of land, as will be good without deed within the [Statute of Frauds] . . . s2, must be for three years, to be computed from the time of the agreement; and not for three years to be computed from any day after.

The second case is *Ryley* v *Hicks* (1725) 1 Stranage 651. Again the report is very short:

> In *Middlesex*, *coram* Raymond, Chief Justice.
> Leases by parol for less than three years from the making, to commence at a future day, are not within Statute of Frauds. . . .
>
> The plaintiff declares, that 24 February 1723, she demised to the defendant a chamber, a cellar, and half a shop, habendum from LadyDay then next for a quarter of a year, and so from quarter to quarter, so long as both parties shall please, at 51. per quarter.
>
> It was objected by Whitaker, that this being to commence at a future day, was but a lease at will since the Statute of Frauds. The Chief Justice at first thought it a good objection, but upon farther consideration he was of opinion, that the exception was not confined to leases that were to commence from the time of making, but was general as to all leases that were not to hold for above three years from the making. So the plaintiff had a verdict . . .

The case is also reported in Buller's *Nisi Prius* 6th ed at p177:

A A lease by parol for a year and an half, to commence after the expiration of a lease which wants a year of expiring, is a good lease within the statute of frauds, for it does not exceed three years from the making.

It will be seen that the two reports differ in their summary of the relevant facts. As observed in Selwyn, *An Abridgment of the Law of Nisi Prius*, 13th ed 1869, vol 2 at p759, the report in Strange is "probably from a different note". But the statement of principle is the same in both reports.

Ryley v *Hicks* was criticised by Dampier J in *Inman* v *Stamp* (1815) 1 Stark 12, reported on this point only in Selwyn at p759 (the earlier proceedings at *nisi prius* are reported, but on a different and for present B purposes irrelevant point, as *Inman* v *Stamp*:

In *Inman* v *Stamp* BR Trin 55 Geo III., Dampier, J, said the practice had been with the foregoing case of *Ryley* v *Hicks*, although he rather inclined to think that the 2nd section of the statute [of frauds], taken with section 4 [of the statute of frauds], was confined to leases executed by possession, on which two-thirds of the improved rent was reserved.

In *Edge* v *Strafford* (1831) 1 Cr&J 391 it was argued that *Ryley* v *Hicks* had in effect been over-ruled by *Inman* v *Stamp*. The argument was rejected at p396 by Bayley B, giving the judgment of the Court of Exchequer, who appears to have treated *Ryley* v *Hicks* as good C authority for the proposition that:

a lease, though it were to commence in futuro, would be within the exception in the statute of frauds, if it did not exceed three years from the making.

A similar argument in *Lord Bolton* v *Tomlin* (1836) 5 Ad & El 856 fared little better, though the court in that case managed to avoid deciding, or even expressing any opinion on, the point.

In *Hand* v *Hall* (1877) 2 ExD 355 an agreement was made on January 26 1876 for a tenancy until Midsummer 12 months from February 14 1876. It was held by the Court of Appeal, reversing the Exchequer Division, that the agreement operated as a devise within the exception in section 2 of the Statute of Frauds. As was pointed out in D Foa, *The Relationship of Landlord and Tenant*, 6th ed 1924, at p11, the decision of the Court of Appeal is consistent only with the view that *Ryley* v *Hicks* was correctly decided, although that case seems not to have been referred to and the contrary appears not to have been argued.

The next case to which I was referred is *Foster* v *Reeves* [1892] 2 QB 255. In that case an agreement was made on May 12 1890 for a tenancy for three years to commence on June 24 1890 and thereafter from year to year until determined. The landlord sued to recover rent after the tenant had given up possession. It seems to have been common ground that there was no lease at law, and the argument in the Court of Appeal was as to the power of the county court, under the E legislation then in force, to grant specific performance of the agreement under the rule in *Walsh* v *Lonsdale* (1882) 21 ChD 9. Each of the judges, however, referred, albeit briefly, to the position at law. Lord Esher MR said at p257:

When the agreement is looked at, it is found that it assumes to create a tenancy for three years to begin at a subsequent date. Such a tenancy can at common law only be created by deed . . . If, therefore, we look to the common law alone, this claim cannot be supported.

Fry LJ said at pp258-259:

. . . it was a sum which could be recovered in equity, provided the agreement F under which it was claimed was one that would be enforced by a court of equity. Independently of this, there was no right to sue for it.

Lopes LJ said at p259:

By the joint operation of the Statute of Frauds and . . . [the Real Property Act 1845], a lease for more than three years must be by deed. Here the agreement purports to create a term of more than three years, and it is clear that rent could not, at common law, have been recovered under that agreement.

At first, Mr Walter was minded to argue that Lord Esher MR's observations (though not, he accepted, those of Lopes LJ) were authority for the proposition that a tenancy to begin at a subsequent date could at law only be created by deed. On reflection, and having considered *Ryley* v *Hicks* (1725) 1 Stranage 651, Mr Walter accepted G that this was not what Lord Esher MR had meant. In my judgment, Mr Walter was entirely correct to make that concession. When Lord Esher MR said at p257 that "Such a tenancy can at common law only be created by deed", the tenancy he was referring to was, as the immediately preceding words make clear, "a tenancy *for three years*, to begin at a subsequent date" (emphasis supplied). Read thus, Lord Esher MR was making the same point as Lopes LJ, and indeed the very same point that had been made by Lord Holt CJ as long ago as 1699: a tenancy for three years to commence at a future date was not within section 2 of the Statute of Frauds, and thus at law could only be made in writing from 1677 or, from 1844, by deed. This was not H because the tenancy was to commence at a future date, for *Ryley* v *Hicks* shows that, even after the enactment of the Statute of Frauds, there was nothing objectionable in that. It was because the tenancy was to last for a period expiring more than three years from the date of the agreement, and it had been established by *Rawlins* v *Turner* (1699) 1 Ld Raym 736 that such a lease fell outside the exception in section 2. There is, in my judgment, nothing in *Foster* v *Reeves* to suggest that the Court of Appeal was seeking to overturn established authority, although in that case, as previously in *Hand* v *Hall* (1877) 2 ExD 355, *Ryley* v *Hicks* seems not to have been referred to *Foster* v *Reeves* is, therefore, in my judgment, no authority for the proposition that, after J the enactment of the Real Property Act 1845, a lease for less than three years to commence at a future date could at law only be created by deed.

Be that as it may, the matter appears to have been resolved the following year in *Parker* v *Briggs*, a decision of the Court of Appeal reported only, and there rather inadequately, in (1893) 37 Sol Jo 452:

the appellants contend . . . that a parol lease to commence at a future date was in point of law and in fact an agreement for a lease, and as such must be in writing under section 4 of the Statute of Frauds, even though the lease agreed to be granted were for a less period than three years (*Edge* v *Strafford* 1 Cr&J 391) . . . The court, however, decided against the appellants . . ., two of the K learned judges holding that it was possible to create a parol lease to commence in the future (*Ryley* v *Hicks* Stra 651) . . .

Accordingly, Mr Walter was, in my judgment, correct not to press the argument that, after the enactment of the Real Property Act 1845, a lease to commence at a future date, could at law only be created by deed, even if it would otherwise fall within the exception in section 2 of the Statute of Frauds. The contrary is, in my judgment, clear from the authorities to which I have referred. That *Ryley* v *Hicks* (1725) 1 Stranage 651 remained good law until the coming into force of the Law of Property Act 1925 appears, moreover, to have been the general L view of the commentators: see *Halsbury's Laws of England*, 1st ed 1911, vol 18, at p384, para 830; Foa, *The Relationship of Landlord and Tenant*, 6th ed 1924, at p11; Redman, *The Law of Landlord and Tenant*, 8th ed 1924, at pp121, 439-440.

It follows, in my judgment, that prior to the coming into force of the Law of Property Act 1925 Mr Walter's argument would have failed. So far as concerns the effect of the exception in section 2 of the Statute of Frauds, Miss Hargreaves correctly relies upon *Rawlins* v *Turner* (1699) 1 Ld Raym 736 and *Ryley* v *Hicks* (1725) 1 Stranage 651 to make good her submissions. Unless, therefore, the effect of section 54(2) of the Law of Property Act 1925 was, as Mr Walter asserts, to change the law as previously laid down in *Ryley* v *Hicks*, the tenancy M document will, as it seems to me, constitute a "lease in writing" for the purposes of para 5(1) of Schedule 1 to the Limitation Act 1980.

Before turning to consider the meaning of section 54(2) there is another aspect of the Statute of Frauds on which Mr Walter placed some reliance and to which I must briefly advert. Cases such as *Inman* v *Stamp* (1815) 1 Stark 12, *Edge* v *Strafford* (1831) 1 Cr&J 391 and *Lord Bolton* v *Tomlin* (1836) 5 Ad & El 856 show that if there was an oral letting *in futuro* for less than three years from the date of the agreement there was no remedy in contract, though an action could be brought on the lease once possession had been taken, so that what had previously been a mere *interesse termini* had become an estate in the land. The reason for the divergent treatment was, as Mr Walter points out, that such an oral transaction was caught, *qua* contract, by section 4 of the Statute of Frauds but, falling as it did within the exception in section 2, was not caught, *qua* lease, by section 1 of the Statute of Frauds. As Mr Walter observes, the law as formulated in particular in *Edge* v *Strafford* was unsatisfactory in treating very differently the enforceability of the same transaction *qua* contract and *qua* demise. Given that the exception in section 2 of the Statute of Frauds extended to oral demises, it could be a fine point whether a particular oral transaction was a mere agreement, and thus wholly unenforceable by virtue of section 4, or a demise, and thus effective *qua* lease under the exception in section 2.

(b) *Section 54(2)*

As I have already mentioned, the words "taking effect in possession" which appear in section 54(2) of the Law of Property Act 1925 were not to be found in section 2 of the Statute of Frauds. There appears to be no judicial authority as to the effect (if any) of this change: certainly I was referred to none. The question is canvassed by some of the commentators, though unhappily for Mr Long and the council, who as a result have been driven to litigate the matter, the commentators do not speak with one voice.

Foa, *The General Law of Landlord and Tenant*, 7th ed 1947, at p10, puts the point very clearly:

> A parol lease might be made under the exception in the Statute of Frauds to take effect in futuro, provided that it came to an end within three years from the making thereof (*Ryley* v *Hicks* (1725) 1 Stra 651). The present statute seems to except only those taking effect immediately in possession.

The same view is expressed in Megarry and Wade at p637, where it is said that:

> A lease taking effect in possession is to be contrasted with one taking effect in reversion, eg a lease granted today to take effect in 19 days' time . . .

Emmet on Title, 19th ed 1986, vol 2, para 26.006 expresses the same view as Megarry and Wade in very similar language. Hill and Redman's *Law of Landlord and Tenant*, 18th ed 1988, vol 1, at ppA264-265, paras A[631][639], is to the same effect, drawing a distinction between

> (*a*) reversionary leases, that is leases to take effect in reversion at the time of the grant, possession being postponed to a future date (which are outside the s 54(2) exception); and (*b*) leases which take effect in possession at once but where the term is calculated from a future date (which are within the exception).

These plainly support Mr Walter's argument. Miss Hargreaves referred me to the contrary view expressed in *Halsbury's Laws of England*, 4th ed 1994 reissue, vol 27(1), at p87, para 78, where it says of section 54(2):

> The three years are computed from the day of the making of the lease; and, if the term does not commence at once, it must expire, or be capable of expiring, within three years from that day. [*Rawlins* v *Turner* (1699) 1 Ld Raym 736; *Foster* v *Reeves* [1892] 2 QB 255, *per* Lord Esher MR.] The exception to the requirement of a deed is not confined to leases commencing at the time they are made. [*Ryley* v *Hicks* (1725) 1 Stranage 651]

One's confidence in that view might be somewhat greater were it not for the fact that the passage has remained unchanged since before 1925: compare *Halsbury's Laws*, 1st ed 1911, vol 18, at p384, para 830. Miss Hargreaves also referred me to Woodfall's *Law of Landlord and Tenant*, looseleaf ed 1994, vol 1, at p5/2, para 5.003, where *Ryley* v *Hicks* (1725) 1 Stranage 651 is still referred to as good authority, albeit not for the point at present in issue. Again, the passage in question has remained virtually unchanged since before 1925: compare Woodfall's *Law of Landlord and Tenant*, 21st ed 1924, at p158. Moreover, elsewhere in the current edition the view appears to be expressed that a lease takes effect in possession for the purposes of section 54(2) only if it is to commence immediately and not at a future date: Woodfall, vol 1, at p6/8, para 6.019.

The preponderance of the commentaries therefore supports Mr Walter. That fact though suggestive can hardly be decisive. There are, however, in my judgment, three reasons why Mr Walter's argument is indeed correct.

In the first place, the words "in possession", when used as part of the phrase "taking effect in possession for a term not exceeding three years", in my judgment, have their normal legal meaning. They connote an estate or interest in the land which is vested "in possession" rather than merely vested "in interest". This reading is powerfully reinforced by the distinction drawn in section 205(1)(xxvii) of the Law of Property Act 1925 between a "term of years taking effect in possession" and a "term of years taking effect in reversion". The words "taking effect in possession" in section 54(2) are, in my judgment, used in the same sense in which those words are used in section 205(1)(xxvii) and thus, and this is the critical point, in distinction to the words "taking effect in reversion". This, as it seems to me, demonstrates that, as Mr Walter submits, reversionary leases were not intended to come within the ambit of section 54(2). Putting the same point rather differently, the words "taking effect in possession" are used in section 54(2) in such a way as to achieve the same result as that contended for by Dampier J in *Inman* v *Stamp* when he sought to argue that the exception in section 2 of the Statute of Frauds applied only to leases "executed by possession".

Second, it is very difficult, if Miss Hargreaves' argument is correct, to see what was the purpose of including the words "in possession" in section 54(2) at all. It is implicit in her argument that the words "taking effect in possession" in section 54(2) are descriptive of the quality of the lease at the date when the term commences and not of the quality of the lease at the date of the parol grant. Given the definition of "possession" in section 205(1)(xix) of the Law of Property Act 1925, however, it would seem that any lease, even a concurrent lease, must necessarily take effect "in possession" at the date when the term commences. So, if Miss Hargreaves is correct, the words "in possession" add nothing to what would otherwise be the effect of section 54(2).

The third reason is, in my judgment, decisive. It will be remembered that those leases which fell within the exception in section 2 of the Statute of Frauds were defined, so far as is material for present purposes, as being

> Leases not exceeding the Term of Three Years from the making thereof.

In contrast, those leases which fall within the exception in section 54(2) are defined, so far as is material for present purposes, as being

> leases taking effect in possession for a term not exceeding three years.

It will be seen that the three-year period referred to in section 2 of the Statute of Frauds is not the same as the three-year period referred to in section 54(2). Under the Statute of Frauds the three-year period was the period "from the making" of the lease, in other words the period of three-years from the date of the parol grant. Thus, whether or not the language of section 2 implicitly required the term of the lease to commence on "the making thereof" (the point on which Dampier J in *Inman* v *Stamp* thought that *Ryley* v *Hicks* (1725) 1 Stranage 651 had been wrongly decided), the words "not exceeding the Term of Three Years from the making thereof" plainly meant that, as held by Holt CJ in *Rawlins* v *Turner* (1699) 1 Ld Raym 736, no lease by parol could be good which imported to convey an interest for more than three-years from the time of the making. Under section 4(2), on the other hand, the three-year period is the period of the term comprised in the lease. Moreover, there has been omitted from section 4(2) any express reference to the date of "the making" of the lease. Thus, if Miss Hargreaves' argument is correct, there is no limit expressed in section 54(2) to the period which may elapse before the lease "tak[es] effect in possession", the only requirement being that the lease, when eventually it does "tak[e] effect in possession", must be "for a term not exceeding three years". On this footing the only limit to the period which might elapse before the term "tak[es] effect in possession" would be the period of 21 years specified in section 149(3) of the Law of Property Act 1925.

Miss Hargreaves, of course, recoils from going this far. Her argument acknowledges that the term, whether it is to commence immediately or at some date in the future, must not extend beyond three years from the date of the parol grant. However it is impossible to find that limitation in the language of section 54(2) unless it is to be found, as I think it is to be found, in the crucial words "taking effect in possession", those words then being understood, as they have to be understood, in the sense which I have already indicated. Indeed, Miss Hargreaves comes very close to recognising how far the logic of her argument drives her when she suggests that the restrictions on the grant of reversionary leases are to be found in section 149 of the Law of Property Act 1925, and not in section 54(2). But this, in my judgment, is quite impossible. There is simply no basis for suggesting that what parliament envisaged when it repealed section 2 of the Statute of Frauds and enacted section 54(2) was the grant of parol leases which might not come into effect for 21 years and might not come to an end for 24 years. It makes sense to suggest, as is implicit in Mr Walter's argument, that parliament intended to clarify the law and, by adopting Dampier J's approach, to abolish the gloss to *Rawlins* v *Turner* (1699) 1 Ld Raym 736 represented by the somewhat controversial decision in *Ryley* v *Hicks* (1725) 1 Stranage 651. It is a very different thing indeed to suggest that parliament intended to sweep away the basic principle articulated in *Rawlins* v *Turner* and to widen so drastically the class of leases excepted from the operation of what are now sections 52(1) and 54(1) of the Law of Property Act 1925.

One final comment on the matter is perhaps appropriate. As Mr Walter points out, the effect of construing section 54(2) in the way for which he contends is to put an end to the distinction, based on sections 2 and 4 of the Statute of Frauds and illustrated by cases such as *Edge* v *Strafford* (1831) 1 Cr&J 391, of an oral letting *in futuro* being enforceable *qua* lease, but not *qua* contract. If his construction of section 54(2) is correct, the effect of the Law of Property Act 1925 was to make equally unenforceable both an oral executory agreement to grant a lease (section 40) and an oral attempt to grant a lease taking effect in possession in the future (section 54(2)), and, accordingly, to put an end to the need to distinguish between those oral transactions for a future tenancy which, as a matter of construction, took effect as mere agreements, and were thus unenforceable by virtue of section 4 of the Statute of Frauds, and those oral transactions which, as a matter of construction, took effect as leases, and thus fell within the exception in section 2 of the Statute of Frauds. It is tempting to think that this was, indeed, parliament's intention when it replaced section 2 of the Statute of Frauds with section 54(2) of the Law of Property Act 1925, but the temptation, which rests on mere speculation, must in my judgment be resisted. Be that as it may, it is of some comfort to note that, accepting Mr Walter's submission, as I do, this is indeed the effect of the legislation.

Accordingly Mr Walter is correct, in my judgment, when he submits that reversionary leases conferring no immediate right to take possession are altogether excluded from the ambit of section 54(2) of the Law of Property Act 1925 and that such reversionary leases can take effect only if made by deed. It follows that the tenancy document (which it is common ground was a reversionary lease not made by deed and not conferring on Mr Long an immediate right to take possession) was incapable of taking effect as a parol lease under section 54(2). Thus, as Mr Walter submits, the tenancy which undoubtedly came into existence was not one created by the tenancy document but rather one which arose by operation of law. From this it follows, in my judgment, that the tenancy document is not a "lease in writing" within the meaning of para 5(1) of Schedule 1 to the Limitation Act 1980. Accordingly, and as a matter of law, Miss Hargreaves fails in her first argument for striking out Mr Long's claim.

Issue 2 — the Landlord and Tenant Act 1954

Miss Hargreaves' second argument is based on the provisions of the Landlord and Tenant Act 1954. She submits that Mr Long's tenancy was a tenancy to which Part II of the Landlord and Tenant Act 1954 applied. Section 24(1) of that Act provides that:

> A tenancy to which this Part of this Act applies shall not come to an end unless terminated in accordance with the provisions of this Part of this Act;

an event which, she says, did not occur in the present case until the notice to quit given to Mr Long in or about September 1983 expired on March 25 1984. She points out that section 24(2) of the Landlord and Tenant Act 1954 does not include the operation of the Limitation Act 1980 among the list of matters which, notwithstanding section 24(1), are expressed not to prevent a tenancy to which the Act applies coming to an end. In short, she submits, the effect of the clear and mandatory language of section 24(1) is to make it impossible for the tenant under a tenancy to which Part II of the Act applies to be in adverse possession within the meaning of para 8(1) of Schedule 1 to the Limitation Act 1980 until such time as the tenancy has been terminated in accordance with the provisions of Part II of the Act. The argument is entirely novel: neither Miss Hargreaves nor Mr Walter was able to point me to any authority in which the point has been canvassed let alone decided.

A similar argument in the context of the Rent Acts was rejected in *Moses* v *Lovegrove* [1952] 2 QB 533. I have to confess that my initial reaction to Miss Hargreaves' argument reflected that of Sir Raymond Evershed MR when he said at p537:

> Such a view, if well founded is, I think, somewhat startling, for it would necessarily follow therefrom that in a case of premises controlled or covered by the Rent Restriction legislation, unless the statutory period under the Limitation Act, 1939, had expired before they came under the control, no Limitation Act could ever run in favour of the tenant, or could ever run at all, so long as the Rent Restriction legislation remains on the statute book.
>
> There is a further reason for thinking such a conclusion surprising, and it was a reason noticed by my brother Romer during the argument: the Rent Restriction legislation is intended to protect tenants. If . . . [counsel] is right, it would, in this instance, have the peculiar and oblique effect of preserving to the landlord a right in respect of this property, which she would indubitably, on the argument, otherwise have lost.

Miss Hargreaves points out, however, that there is a fundamental difference between the operation of the Rent Acts and the operation of Part II of the Landlord and Tenant Act 1954. Whereas the Rent Acts do not prevent the termination of the tenancy by a common law notice to quit, and thus confer on the protected tenant no more than what has been described as a "personal status or statutory right of irremovability", Part II of the Landlord and Tenant Act 1954 continues the common law tenancy, albeit with a statutory variation as to the mode of determination. What Miss Hargreaves says as to the distinction between the two statutory schemes of protection is, of course, entirely correct. For my part, however, I doubt that the distinction which Miss Hargreaves draws really bears upon the reasoning which underlay the decision of the Court of Appeal in *Moses* v *Lovegrove* [1952] 2 QB 533 or that it is sufficient, for the purposes of the proper application of the Limitation Act 1980, to justify distinguishing the two types of case in the way she suggests.

Be that as it may (and I wish to emphasise that I am expressing no final view on Miss Hargreaves' submissions), there is, in my judgment, a fundamental flaw in her argument which makes it impossible for her to succeed at this stage in the litigation. This is that her argument assumes that Mr Long's tenancy was a tenancy to which Part II of the Landlord and Tenant Act 1954 applied at the material times. However, Part II of the Act applies only to tenancies where the property comprised in the tenancy is or includes premises occupied by the tenant for the purposes of a business carried on by him or for those and other purposes: section 23(1). Given the factual disputes between the parties to which I have already referred, and the general paucity of the evidence at present before the court, it would not, in my judgment, be appropriate to assume for the purposes of this striking out application, as on Miss Hargreaves' case to have to, that Mr Long was at the material times occupying the whole or some part of the premises for the purposes of a business being carried on by him.

A Accordingly I reject Miss Hargreaves' second argument for striking out Mr Long's claim.

Issue 3 — acknowledgement

Miss Hargreaves' final argument is that Mr Long acknowledged the GLC's title to the premises in 1984, that is less than 12 years before the present proceedings were commenced. In this connection she relies upon sections 29 to 31 of the Limitation Act 1980. She accepts that the acknowledgement must be in writing. She puts her case in two ways.

First, she relies upon a letter dated January 12 1984 written to the GLC by Jones, Son & Day, a firm of estate agents and valuers who Mr B Walter accepts were acting as Mr Long's agents. The letter was in the following terms:

> Long Brothers
> Clemence Street Site
> *11/19 & 21 Turners Road*
>
> We refer to your letter of 14th September with enclosure. Since it would appear our clients are entitled to compensation for disturbance we shall be obliged if appropriate claim form be forwarded.
>
> As our clients would prefer alternative accommodation they ask if your Council have any lock-up type store in the area to let of 500/600 square feet.

C Neither the letter of September 14 1983 nor its enclosure was in evidence before me. Both seem to have been lost.

Deputy Master Wall thought that the reference to compensation might have been to statutory compensation following the service of a compulsory purchase order. Miss Hargreaves was able to demonstrate to my satisfaction that this cannot have been the case, and Mr Walter does not seek to argue the contrary. Miss Hargreaves and Mr Walter are at one in accepting that the enclosure to the missing letter of September 14 1983 was in all probability a notice to quit served in accordance with Part II of the Landlord and Tenant Act 1954, and in accepting that the reference to compensation in the letter of January 12 1984 was in all probability a reference to compensation payable under D that Act. Mr Walter, however, draws attention to the fact that, while the letter of January 12 1984 referred to two other properties in addition to the premises, namely 11 and 19 Turners Road, it referred to the letter of September 14 1983 as having apparently contained only a single enclosure. Therefore, he says, it is not shown with that degree of probability necessary to justify the striking out of Mr Long's claim, either that the letter of September 14 1983 contained a notice to quit the premises, as opposed to a notice to quit one or other or both of the other properties, or that the letter of January 12 1984 was written with reference to the premises.

Furthermore, says Mr Walter, the letter of January 12 1984 did not E constitute an acknowledgment of the GLC's title to anything because it was not making any definite assertion. The critical words "it would appear" are, he says, suggestive of no more than an inference derived from a document just received rather than an assertion or admission of fact. The letter of January 12 1984 merely assumes for the sake of argument that something said in the missing letter of September 14 1983 may be correct, but without acknowledging that it necessarily is. It amounts, he says, to nothing more than a comment to the effect that "on the basis of what you say in your letter, and assuming for the sake of argument that it is correct, it would appear . . . ". The request for a claim form cannot, he says, be read as an admission of anything. He F points out that it is difficult to construe the letter of January 12 1984 without knowing what had been said in the letter of September 14 1983 and without knowing the precise form of its enclosure. He relies upon the subjunctive use of the word "would" which, he says, is not consistent with an unequivocal acknowledgment.

I can see the force of Mr Walter's arguments. The point, in a sense, is a short one, but at the end of the day, and not having seen either the letter of September 14 1983 or its enclosure, I am not prepared to hold that the letter of January 12 1984 constituted an acknowledgment by Mr Long of GLC's title to the premises.

In the alternative, Miss Hargreaves relies upon the following letter written to the GLC by Mr Long on March 1 1984:

> As explained to your dept on many times we are exempt from water rates, we G have no toilets no running water at all. One of your men came round and had a look at the premises later part of last year to see for himself what I said was correct.

Miss Hargreaves says, on the evidence I have seen no doubt correctly, that the GLC's practice was to collect water rates on behalf of Thames Water only in respect of *tenanted* properties. Therefore, she says, the very fact that in 1984 the GLC were, as the evidence shows, trying to collect water rates from Mr Long indicates that he must still have been paying rent to GLC for the shop. That may or may not be so, but as I have already said the date when Mr Long last paid rent is a matter which can only be resolved at trial. More to the point, however, H Miss Hargreaves submits that there would have been no purpose in Mr Long approaching GLC with the complaint about the water supply set out in his letter of March 1 1984 unless he had thought that GLC was responsible for the water supply by reason of his tenancy. Therefore, she says, the letter of March 1 1984 constitutes an acknowledgment by Mr Long of the GLC's title to the premises.

Mr Walter says that the letter of March 1 1984 cannot possibly be construed as an acknowledgment of GLC's title as freeholder. The only admission, he says, is that the GLC were demanding payment of the water rate. I agree. It is, in my judgment, nothing to the point that the GLC's practice was to collect water rates only from tenanted J properties and that the GLC may therefore have thought that Mr Long was their tenant. There is nothing in the evidence I have seen to show that Mr Long either knew of the GLC's practice or had any reason to believe that the reason why the GLC were demanding payment of the water rates was because they were treating him as their tenant. There is, in my judgment, no warrant for Miss Hargreaves' assertion that Mr Long must have thought that the GLC were responsible for the water supply *by reason of his tenancy*. Absent knowledge of the GLC's practice, there is any number of reasons why someone in Mr Long's position may have thought that the GLC, a local authority exercising many public functions, was responsible for collecting water rates. Acknowledgment by Mr Long that the GLC had some form of K responsibility for the water supply to the premises cannot of itself and without more ado, in my judgment, constitute an acknowledgment by him of the GLC's title to the land. Someone like Mr Long is as likely to have thought that the GLC were exercising their functions as local authority as to have thought that their involvement was as owner of the land.

Conclusion

Accordingly, in my judgment, Miss Hargreaves has failed to make out any of the grounds she relies upon to justify striking out Mr Long's claim. It may be that, at trial, she will succeed in making good one or L other of the matters on which I have held there to be triable issues. But she has failed to show, as she must if it is to be struck out at this stage, that Mr Long's claim is doomed to fail. Mr Long is entitled to go to trial to have his claim determined.

It follows, in my judgment, that Mr Long's appeal must be allowed. I shall therefore set aside the order made by Deputy Master Wall on September 26 1995 and dismiss the summons issued by the council on March 21 1995.

Appeal allowed. M

A Keepers and Governors of the Possessions Revenues and Goods of the Free Grammar School of John Lyon *v* Mayhew

COURT OF APPEAL

December 12 1996

B LEGGATT, HENRY and HUTCHISON LJJ

Estates Gazette April 26 1997

[1997] 17 EG 163

Landlord and Tenant Act 1954 — Ineffective section 25 notice — Representation and estoppel — Whether tenant accepted validity of notice by unconditional counternotice

In 1981 the appellant acquired a lease of premises due to expire in June 1992. He lived and practised as a solicitor at the premises. In 1991 the appellant served a notice of leaseholder's claim under the
C Leasehold Reform Act 1967 ("the 1991 notice"). On September 2 1991, and following earlier proceedings, in which the Court of Appeal held that the 1991 notice was ineffective because when it was served no notional reduction in the rateable value of the premises to reflect tenant's improvements ([1991] 2 EGLR 89) had been obtained, the plaintiff landlords served notice on the appellant under section 25 of the Landlord and Tenant 1954. The section 25 notice was defective because it was not in the prescribed form and failed to refer to the tenant's rights under the Leasehold Reform Act 1967. The appellant served a counternotice and commenced proceedings for a new tenancy. On November 5 1993, after the Leasehold Reform, Housing and Urban
D Development Act 1993 came into effect removing the rateable value limit for enfranchisement under the 1967 Act, the appellant served a second enfranchisement notice ("the 1993 notice"). In the court below it was held that the appellant was estopped from disputing the validity of the section 25 notice, and that, because the 1993 notice was served more than two months after the section 25 notice, the 1993 notice was invalid. The appellant appealed.

Held: **The appeal was dismissed. The defect in the landlords' section 25 notice was plain on its face in view of the statement that it should not be used where the tenancy was one to which the 1967 Act may apply; the defect was known to the appellant at the time.**
E Although the landlords' section 25 notice was in the wrong form, the appellant's counternotice assumed the validity of the section 25 notice. By his actions in carrying on with the 1954 Act procedure, the appellant represented that the section 25 notice was a good notice. The landlords suffered detriment in lawyers' and surveyors' fees, and refrained from serving a second notice. It was unconscionable for the appellant to contend, in relation to the 1993 notice, that the section 25 notice was defective. Further, by the appellant's dealings with the landlords after the 1991 notice was held to be ineffective, he thereby represented that he had no continuing right to claim that the 1991 notice was valid. The
F principle, that a landlord and tenant may not contract out of statutory rights given to a tenant for his protection, did not apply because at the time the appellant knew he had no rights under the 1967 Act.

The following cases are referred to in this report.

Bremen v *Varden* [1978] 2 Lloyd's Rep 109

Collin v *Duke of Westminster* [1985] QB 581; [1985] 2 WLR 553; [1985] 1 All ER 463; (1984) 50 P&CR 380; [1985] 1 EGLR 109; 273 EG 881; 25 RVR 4, CA

Daejan Properties Ltd v *Mahoney* [1995] 2 EGLR 75; [1995] 45 EG 128, CA

Joseph v *Joseph* [1967] Ch 78; [1966] 3 WLR 631; [1966] 3 All ER 486, CA

Keen v *Holland* [1984] 1 WLR 251; [1984] 1 All ER 75; (1984) 47 P&CR 639; [1984] 1 EGLR 9; [1984] EGD 9; 269 EG 1043, CA

Tegerdine v *Brooks* (1977) 36 P&CR 261; [1978] 1 EGLR 33; [1978] EGD 49; 245 EG 51, CA G

Tennant v *London County Council* (1957) 121 JP 428; 55 LGR 421; 169 EG 689, CA

This was an appeal by the appellant, William George Mayhew, from the decision of Judge Zucker QC in Central London County Court, who on August 25 1995 dismissed the appellant's counterclaim and granted the respondents, the Keepers and Governors of the Possessions Revenues and Goods of the Free Grammar School of John Lyon (Harrow School), a declaration that the appellant was not entitled to acquire the freehold of 31 St John's Wood, London NW8.

Michael Driscoll QC, Julius Seal and Andrew McGuinness H
(instructed by Mayhew & Co) appeared for the appellant; Anthony Radevsky (instructed by Lee & Pemberton) represented the respondents.

Giving the first judgment, **LEGGATT LJ** said: William George Mayhew appeals against the order of Judge Zucker QC sitting in Central London County Court whereby on August 25 1995 he dismissed Mr Mayhew's counterclaim and granted to his landlords, Harrow School, a declaration that Mr Mayhew was not entitled to have the freehold of 31 St John's Wood, London NW8, pursuant to Leasehold Reform Act 1967 ("the 1967 Act"). On February 9 1981 Mr
Mayhew became the lessee under a long lease of those premises. The J
landlords own the freehold. The lease was for a term of 70 years from June 24 1922, expiring in June 1992. When Mr Mayhew took his assignment, it had 11 years to run. Mr Mayhew lives at the premises and, as is permitted by the lease, also practises there as a solicitor. It is common ground that his tenancy therefore fell within Part II of the Landlord and Tenant Act 1954 ("the 1954 Act"). Because the lease was a long lease at a low rent it would have fallen within the provisions of the 1967 Act, had the rateable value on April 1 1973 not exceeded £1,500. It was in fact £1,680.

By section 1(4A) of the 1967 Act, Mr Mayhew could have applied
under Schedule 8 to the Housing Act 1974 to have the rateable value K
notionally reduced in consequence of tenant's improvements so as to bring the premises within the Act. But in order to be able to serve a valid notice of leaseholder's claim it is first necessary to have obtained the notional reduction in accordance with the procedure prescribed by the 1974 Act. By para 2 of Schedule 3 to the 1967 Act a claim to acquire the freehold is of no effect if made more than two months after a landlord's notice to terminate the tenancy under section 25 of the 1954 Act; and, on the other hand, a landlord's notice terminating a tenancy under that section is of no effect if served during the currency of a claim to acquire the freehold. Mr Mayhew's first notice of
leaseholder's claim was, as this court held on May 21 1991*, not L
effective, because when it was served no notional reduction in rateable value had been obtained. Mr Mayhew's petition to the House of Lords for leave to appeal against that decision was dismissed on July 22 1991. Soon thereafter on September 2 1991 the landlords served a notice under section 25 of the 1954 Act. So far from challenging the validity of that notice Mr Mayhew served a counternotice and started proceedings for a new tenancy. The section 25 notice was served in form 1 in Schedule 2 to the Landlord and Tenant Act 1954 Part II (Notices) Regulations 1983 (as amended) ("the 1983 regulations"). There is no dispute that that notice ought to have been in form 13.

On November 1 1993 the Leasehold Reform, Housing and Urban M
Development Act 1993 took effect removing the limit on rateable value for the purpose of enfranchisement under the 1967 Act. Four days later Mr Mayhew gave to the landlords a second enfranchisement notice ("the 1993 notice").

The relationship between the notices under the Acts is this. A tenant's claim is of no effect if made more than two months after a section 25 notice has been given. Since the 1993 notice was given more than two months after service of the section 25 notice, the 1993 notice will have been of no effect if the section 25 notice was valid. A section

*Editor's note: Reported at [1991] 2 EGLR 89

25 notice is of no effect if served during the currency of a tenant's claim. Hence the need to consider whether the 1991 notice was ineffective or ceased to have effect before service of the section 25 notice. The critical question on this appeal is whether the section 25 notice has effect. If it does not, Mr Mayhew's 1993 notice is valid, but if it does, that notice is of no effect. On behalf of Mr Mayhew, Mr Driscoll QC submitted that for two reasons the section 25 notice did not have effect. First, it did not satisfy para 10 of Schedule 3 to the 1967 Act in that it did not contain the statements required by that paragraph, because it was in form 1 instead of being in form 13. Second, it was served during the currency of Mr Mayhew's 1991 notice under the 1967 Act and was therefore of no effect by force of para 2(2) of Schedule 3 to the 1967 Act.

Section 25(1) of the 1954 Act provides that:

> The landlord may terminate a tenancy to which this Part of this Act applies by a notice given to the tenant in the prescribed form . . .

Forms 1 and 13 are prescribed by the 1983 regulations. The forms are similar except that form 13 alerts the tenant to his rights under the 1967 Act, whereas form 1 does not.

The judge held at p23E of the transcript of his judgment that:

> As at the date of the [1991] notice Mr Mayhew had no right to acquire the freehold or to serve a notice under section 8 of the 1967 Act, a notice in form 13 was totally unnecessary. A notice in form 1 sufficed for all purposes.

He relied on *Tegerdine* v *Brooks* (1977) 36 P&CR 261*. In that case certain notes were omitted from a section 25 notice. The 1983 regulations permit the use of a form "substantially to the like effect" as the prescribed form. This court held that the omission of notes which in the circumstances were irrelevant did not invalidate the notice. But in this case the judge's attention was not drawn to Schedule 3 para 10(2) to the 1967 Act. That provides that a landlord's section 25 notice shall not have effect unless it states:

> (a) that, if the tenant has a right under Part I of this Act to acquire the freehold or an extended lease of property comprised in the tenancy, notice of his desire to have the freehold or an extended lease cannot be given more than two months after the service of the landlord's notice; . . .

For the landlords, Mr Anthony Radevsky bravely argued that because Mr Mayhew had no rights under the 1967 Act, the words omitted from form 1 were immaterial, and so for present purposes the notice was "substantially to the like effect" as form 13. This is, in my judgment, a hopeless argument, because it disregards the express language of para 10(2). That provides that the notice shall not have effect unless it states what was omitted from the form used. The omission was therefore fatal. The 1991 notice was of no effect.

Mr Radevsky, however, also succeeded before the judge on the ground of estoppel. The judge held at p26E that by serving a counternotice and issuing an originating application claiming a new tenancy Mr Mayhew "clearly and unequivocally represented that he would not exercise any right he might have had to claim that the section 25 notice was invalid, either by reason of the first enfranchisement notice or of any defects in the notice itself".

In this court Mr Driscoll QC for Mr Mayhew invoked section 23(1) of the 1967 Act, which provides that:

> (1) Except as provided by this section, any agreement relating to a tenancy (whether contained in the instrument creating the tenancy or not and whether made before the creation of the tenancy or not) shall be void in so far as it purports to exclude or modify any right to acquire the freehold or an extended lease or right to compensation under this Part of this Act . . .

He argued that where a tenant is prevented by statute from contracting out of a statutory right, he cannot achieve the same effect by an estoppel. That principle is well established: the question is whether it applies here.

Mr Driscoll argued that the landlords were here relying on an estoppel when parliament has said that the tenant cannot agree to give up the right. Mr Mayhew has a right to acquire the freehold unless he has been served with a valid section 25 notice. He cited *Keen* v *Holland* [1984] 1 WLR 251* at p261C for the proposition that the parties could not by estoppel confer on the court a jurisdiction which they could not confer on it by express agreement. Mr Mayhew could not have agreed with the landlords that the notice should be treated as in the correct form. In *Joseph* v *Joseph* [1967] Ch 78 this court held that the words of section 38(1) of the 1954 Act "purports to preclude the tenant from making an application or request" meant "would but for its avoidance by the sub-section have the effect of precluding the tenant from making an application or request".

Mr Driscoll referred also to *Daejan Properties Ltd* v *Mahoney* [1995] 2 EGLR 75 at p79L. This court was there concerned with para 13(2) of Schedule 1 to the Rent Act 1977, which provided that an agreement in writing between a statutory tenant and an incoming tenant, having the effect of deeming the incoming tenant to be the statutory tenant, shall not have effect unless the landlord is party to it. The landlord, without being party to the agreement between the statutory and incoming tenants, recognised both as joint statutory tenants. It was held that the landlord was estopped from denying that that was the position. At p79M Hoffmann LJ said:

> It is true that the effect of the estoppel is to allow the transfer of the statutory tenancy to take place without the written agreement between outgoing and incoming tenant required by para 13(1). It seems to me, however, that this is a formality for the protection of the individual parties rather than one imposed in the public interest. The parties are therefore entitled to waive it: . . .

The 1954 Act contains no provision comparable with para 10(2). There was therefore nothing to prevent the tenant from agreeing or representing that he would not dispute the validity of the section 25 notice. I can see an argument that such an agreement would not be an "agreement relating to a tenancy" within the meaning of section 23(1) of the 1967 Act. But in circumstances where, as the judge held, Mr Mayhew knew that the 1991 notice ceased to be effective after the House of Lords had dismissed his petition, I can see no reason why he should not have agreed not to dispute the validity of the section 25 notice. He was then in the same position as he would have been, had the 1954 Act stood alone. In any event, the agreement could not have had the effect of excluding a right to acquire the freehold, because by the time the notice was served he had no such right.

A statement that Mr Mayhew would not be willing to give up possession would have been purposeless unless the section 25 notice was valid. That it was given is explicable only as a response to the requirement of notification under section 25(5). Similarly, the counternotice assumed the validity of the section 25 notice: cf *Tennant* v *London County Council* (1957) 55 LGR 421 at pp434 and 436. The defect was plain on the face of form 13 in view of the statement at its foot that it should not be used where the tenancy is one to which the 1967 Act may apply. Since the representation was merely that Mr Mayhew was not disputing the validity of the section 25 notice, it matters not what grounds he would have had for doing so, or whether he knew what those grounds were. In any event, the judge was entitled to infer and find, as he did, that Mr Mayhew knew that the wrong form had been used.

Mr Driscoll did faintly argue that the landlords called no relevant evidence that they were affected by what Mr Mayhew did. But, in my judgment, it is indisputable that the landlords relied on the section 25 notice, and that they relied on nothing else. They do not have to go further and show what they would have relied on if there had been no representation that its validity was not disputed. By incurring lawyers' and surveyors' fees the landlords suffered obvious detriment, despite Mr Driscoll's ingenious argument that by an award of costs they might be reimbursed at least in part. They also refrained from serving another section 25 notice in the correct form. Since the landlords refrained

---

*Editor's note: Also reported at [1978] 1 EGLR 33

*Editor's note: Also reported at [1984] 1 EGLR 9

A from doing so on the faith of Mr Mayhew's representation that he was treating the section 25 notice as valid, it would plainly be unconscionable for Mr Mayhew to take the point thereafter that the section 25 notice was defective. Since Mr Mayhew knew that the wrong form had been served, while it is evident that the landlords did not, there is no unfairness in Mr Mayhew being bound by the consequence of treating it as valid.

Mr Driscoll also argued that the section 25 notice was invalid because it was served during the currency of his 1991 notice. That depends on the true construction of para 2(2) of Part I of Schedule 3 which provides that:

B

> A landlord's notice terminating a tenancy of any property under section . . . 25 of the Landlord and Tenant Act 1954 . . . shall be of no effect if given . . . during the currency of a claim made in respect of the tenancy to acquire the freehold . . .

By para 5(1)(c):

> references to the currency of a claim shall be taken as references to the period from the giving of a notice which has effect or would, if valid, have effect to the time when the notice is effective or ceases to have effect, or (not being a valid notice) is set aside by the court or withdrawn or would, if valid, cease to have effect, and those references shall include any period when the notice is
> C suspended.

The judge summarised his conclusions on this issue by saying:

> In sum, I hold firstly that the 1991 enfranchisement notice was from the outset invalid and ineffective; alternatively, that after the decision of the House of Lords on 22nd July 1991, the first enfranchisement notice ceased to have any effect and was both explicitly and implicitly withdrawn or abandoned.

Mr Driscoll argued that, because it was not, and never could have been, a valid notice, the hypothesis that it "would, if valid, cease to have effect" was inapplicable. Its period of currency could only be determined if it was set aside by the court or withdrawn. Since it has
D never been set aside by the court or withdrawn, it is still current. He cited *Collin* v *Duke of Westminster* [1985] QB 581* for the proposition that the right to rely on a notice cannot be abandoned by inactivity. But although the judge mentioned abandonment he relied also on withdrawal.

As Mr Radevsky pointed out, after the dismissal of his petition for leave Mr Mayhew, after corresponding with his Member of Parliament, dealt with the landlords' surveyors on the footing that, although he asserted no current right to enfranchise, the landlords should negotiate with him outside the scope of the 1967 Act in anticipation of a change in the law. It was to that conduct that the judge
E was referring when he declared that the notice "was both explicitly and implicitly withdrawn or abandoned". There is force in the argument that by that conduct it was by necessary implication withdrawn. Alternatively, if valid (which it was not), the notice would have ceased to have effect when leave to appeal was refused. The fact that it never could have been valid ought not to put it in a better position than it would have been if it could have been valid. But there are difficulties in the way of both arguments. The draftsmen did not directly cater for the position where, as here, it is definitively stated that a purported notice never could have been valid and so could not "cease to have effect". I prefer to rest on the judge's alternative conclusion at p27F of
F his judgment that Mr Mayhew is now estopped from challenging the validity of the section 25 notice "either on the ground that he had previously served the first enfranchisement notice or by reason of the use of the wrong form". His dealings with the landlords' surveyors after he had been refused leave necessarily presupposed, and he thereby represented, that he had no continuing right to claim that the 1991 notice was valid.

I would accordingly dismiss the appeal.

Agreeing, **Henry LJ** said: As the judge found, from the moment when Mr Mayhew took his assignment of the fag end of this lease, he

> a shrewd solicitor who is well acquainted with the provisions of the relevant G
> statutes and regulations . . . has been absolutely determined . . . to take any and every step to secure enfranchisement if he could.

Initially his problem was that the rateable value of the property was too high to be within the enfranchisement limit. But Schedule 8 to the Housing Act 1980 provided machinery whereby the rateable value might be adjusted downwards where the tenant (or previous tenants) had been responsible for qualifying improvements — and thus the property might be brought within the limit.

In 1983 Mr Mayhew served such a notice on his landlord. The statutory scheme gives the parties six weeks to agree, and failing agreement, another six weeks to apply to the court. Mr Mayhew did H
not apply in time. Nor was any extension granted to him. So he tried to revive his claim by making a fresh Schedule 8 application, followed by his first enfranchisement notice, the 1991 notice. But this court, on May 21 1991, allowed the landlords' appeal, and held that as the statutory time limits were mandatory, so there could be no second application after the first had failed. Parker LJ said [1991] 2 EGLR 89 at p91K that the respondent:

> having failed to apply to the county court within the time-limit, was not entitled to serve a second notice; that, although he has now done so and obtained a certificate which, if valid, would have entitled him to serve a section 8 notice, the factual rateable value remains the value for the purposes J
> of the Act; and that he has accordingly no right to acquire the freehold under the 1967 Act.

Leave to appeal to the House of Lords was sought, but refused on July 22 1991. At this time, questions in parliament had given rise to the hope that at some time in the future the enfranchisement limit would be done away with altogether (this eventually happened, but not until the passing of the Leasehold Reform, Housing and Urban Development Act 1993, with effect from November 1 1993). Returning to 1991, Mr Mayhew informed the landlords of the prospect of extending the right to enfranchisement, and invited them to negotiate with him in anticipation of a change in the law. K

This invitation the landlords declined. Instead they served him with a section 25 notice. As Leggatt LJ has already made clear, if that notice was effective when given (or if the tenant has waived his right to or is estopped from any challenge to its validity) then the tenant cannot succeed in his second enfranchisement notice, given on November 5 1993, immediately after the change in the law, and so is not entitled to have the freehold of the property, and accordingly this appeal would fail.

In my judgment, the live issue here is whether the appellant left it too late to take the point that the landlords' section 25 notice did not "have effect" because it did not inform the tenant of his 1967 Act L
rights. I agree with Leggatt LJ that, because of the clear language of para 10(2) of the third Schedule to that Act, the notice was ineffective. An ineffective notice clearly does not require a response, but the tenant would then risk losing his chance of a new tenancy by a court finding the notice to be effective. Accordingly, the prudent course for a tenant so placed is to serve the counternotice without prejudice to the contention that (on the basis of the appellant's submissions in this case) the section 25 notice was ineffective: (i) as to its form (for the reasons set out above); and (ii) because the section 25 notice had been given "during the currency of [the tenant's] 1991 enfranchisement claim" under the extended meaning given to "currency of a claim" by M
para 5(1)c of Schedule 3 to the 1967 Act. What the tenant in fact did was to take no point as to the efficacy of the section 25 notice, but to proceed in all respects as if that notice was valid and effective, and only to take the point after the property had come within the scope of the enfranchisement provisions (November 1 1993) and after the second enfranchisement notice had been served (November 5 1993). The point was taken by letter dated November 22 1993.

There is no shortage of authorities in many branches of the law to the effect that failure to object to a recognised defect or omission or to take an obvious defence may be objectively construed as that party's unequivocal indication of his intention not to insist on his strict legal

*Editor's note: Also reported at [1985] 1 EGLR 109

A rights: see for example, *Bremen* v *Varden* [1978] 2 Lloyd's Rep 109. In such cases, the main question is whether the party against whom the point has been taken has reasonably inferred that the notice in question has been accepted as a valid and effective notice. So in all such cases, it is ultimately a question of fact to be determined by the trial judge.

Mr Mayhew in the witness box pleaded a degree of ignorance in relation to the 1954 and 1967 Acts. The judge clearly did not accept his evidence as to that. I have already quoted his finding as to his knowledge and determination. The judge was also to find:

1. In relation to the submission that the section 25 notice was ineffective because the 1991 enfranchisement claim was still "current",
B he found that at the time of the service of the section 25 notice, Mr Mayhew realised there was no further legal life in his 1991 enfranchisement notice, and that he made his abandonment of the claim clear by inviting the landlords to negotiate in anticipation of a change of the law. Otherwise he would have said that realistically they had to deal with him, because until those proceedings were set aside by the court or withdrawn, the landlords' proceedings were frozen. Thus that point was both implicitly and explicitly abandoned.

2. In relation to the form of the notice, the judge crucially found that Mr Mayhew knew from the outset that the landlord had used form 1 when it should have used form 13. Mr Mayhew did not take the point
C at the time as: (i) it was then irrelevant, given the state of the law; and (ii) if the point were taken, the result would be that the correct form would be used — so to take the point would be futile.

Against the background of Mr Mayhew's knowledge, it is clear to me that all his actions, objectively viewed, would appear to a reasonable landlord as an acceptance of the section 25 notice as valid. D

First, it is clear that once the notice has served its statutory purpose of starting the proceedings, if no point on its validity is taken, it will not generally be critically examined again. Second, the tenant by his actions in taking no point on the notice but carrying on with the 1954 Act procedure as though it was a good notice represented that it was, and so discouraged any critical examination of it. He intended that there should be none, because the defect was easily curable up until: (1) the change in the law; and (2) the service of an enfranchisement notice. So the tenant clearly intended the landlords to act on his representation, and the landlords did so act, first, in not examining the notice critically and so not amending it, and, second, by incurring E
expense on activities only relevant to the 1954 proceedings for a new lease (ie predicated on acceptance of the notice as valid). And as Mr Mayhew was aware of the defect, and was hoping to take advantage of the landlords' mistake, he lulled them into a sense of false security by not taking this point. So there is nothing inequitable in holding him to the stance he presented to the world: namely that the notice was effective.

Last, while I entirely accept the general principle that landlord and tenant may not contract out of statutory rights given to the tenant for his protection, that principle does not apply here for the reasons given by Leggatt LJ. F

I, too, would dismiss this appeal.

HUTCHISON LJ agreed and did not add anything.

Appeal dismissed.

For a further case on this subject see p233

LANDLORD AND TENANT
LEASEHOLD REFORM

A
Mainwaring and another *v* Henry Smith's Charity Trustees (No 2)

COURT OF APPEAL

October 3 1996

HIRST, PETER GIBSON AND PILL LJJ

Estates Gazette March 22 1997

B [1997] 12 EG 152

Leasehold enfranchisement — Landlord and Tenant Act 1987 — Whether valid section 6 notice served — Whether a tenant entitled to withdraw from acquisition process notwithstanding irrevocable agreement with other tenants to proceed

By an agreement dated January 17 1996 the tenants of a block of flats entered into an agreement expressed to be irrevocable that, in the event of the landlords serving a notice under section 5 of the Landlord and Tenant Act 1987, they would accept the offer; the tenants, who included A, signed a draft notice of acceptance under section 6. The
C first respondent landlords gave a notice dated March 15 1996 under section 5 of the Landlord and Tenant Act 1987; they informed the tenants that they intended to dispose of their interest. The appellant served a photostat copy of the draft section 6 notice in response, which included the signature of A. In fact A had withdrawn his authority and support on February 5 1996. In the court below the judge held that A was entitled to withdraw his support. As there was no longer a qualifying majority of tenants, the tenants were not entitled to acquire the freehold. The appellant, one of the tenants, appealed.

***Held*: The appeal was dismissed. There was no valid section 6 notice. Before the appellant served the photostat copy which**
D purported to be a section 6 notice, A had to her knowledge crossed out his signature on the copy in his possession; he had made known that he did not accept the offer. In the context of the 1987 Act a tenant cannot by contract deprive himself of his right to withdraw his acceptance of the landlords' offer. The appointment by the tenants of one of their number as agent is not irrevocable in this context.

The following cases are referred to in this report.

Belvedere Court Management Ltd v *Frogmore Developments Ltd* [1996] 1 All ER 312; [1996] 1 EGLR 59; [1996] 05 EG 131
E *Mainwaring* v *Henry Smith's Charity Trustees* [1996] 2 All ER 220, [1996] 2 EGLR 25; [1996] 29 EG 110
Smart v *Sanders* (1848) 5CB 895
Southern Foundries (1926) Ltd v *Shirlaw* [1940] AC 707

This was an appeal by Zipporah Mainwaring from a decision of Judge Cowell in West London County Court whereby he dismissed the application of the appellant and the second plaintiff, Yeoman's Row Management Ltd, for an order that the respondents, Henry Smith's Charity Trustees, provide and enter into a contract to sell premises under the Landlord and Tenant Act 1987.

The appellant appeared in person; Timothy Fancourt (instructed by Denton Hall) represented the respondents. F

Giving judgment at the invitation of Hirst LJ, **PILL LJ** said: This is the second occasion this year when Miss Mainwaring and the trustees of Henry Smith's Charity have appeared before this court in a dispute arising from the sale of a building containing flats and known as 38 to 62 (even numbers) Yeoman's Row, London SW3 ("the building"), and the application of the Landlord and Tenant Act 1987 ("the Act"). Miss Mainwaring seeks a sale by the trustees to her and Yeoman's Row Management Ltd ("the company") of the building under the provisions of sections 6 and 19 of the Act. Miss Mainwaring is the statutory G
tenant of a flat in the building.

In the appeal heard in February 1996, *Mainwaring* v *Henry Smith's Charity Trustees* [1996] 2 All ER 220*, Sir Thomas Bingham MR, giving the judgment of the court, stated at p223c:

> The Act provides that if, in certain closely defined situations, a landlord proposes to dispose of his interest in premises of a certain kind, the tenants living in those premises shall have a priority right to acquire that interest on the same terms as those on which the landlord is willing to dispose of the interest to another. The long title of the 1987 Act describes it as "An Act to confer on tenants of flats rights with respect to the acquisiton by them of their landlord's reversion"... H

The court held, *inter alia*, and allowing the appeal of Miss Mainwaring, who is the appellant in the present appeal, that the trustees, when proposing to sell the building to the trustees of the Wellcome Trust, should have served notice under section 5 of the Act on the appellant and other qualifying tenants. That section provides for the service of notices by a landlord who proposes to make a relevant disposal. A suggestion that she is entitled to relief in this action, because of the delay caused by the need for her to pursue the earlier appeal, sensibly has not being pursued by the appellant.

Section 5 notices were served on March 15 1996. On March 25 the appellant served what purported to be a notice under section 6 of the J
Act accepting an offer to purchase the building for a sum of just over £1m, and stating in clause 4 that she and the company had:

> been nominated for the purposes of Section 6 of the Act as being the only persons to whom the disposal of your estate or interest in the property may be made during the relevant period as prescribed by the Act.

Section 6 provides in so far as is relevant:

> (1) Where —
> (*a*) the landlord has, in accordance with the provisions of section 5, served an offer notice on the qualifying tenants of the constituent flats, and
> (*b*) within the period specified in that notice under section 5(2)(*c*), a notice K
> is served on him by the requisite majority of qualifying tenants of the constituent flats informing him that the persons by whom it is served accept the offer contained in his notice,
> the landlord shall not during the relevant period dispose of the protected interest except to a person or persons nominated for the purposes of this section by the requisite majority of qualifying tenants of the constituent flats.
> (2) In subsection (1) "the relevant period" means —

---

*Editor's note: Also reported at [1996] 2 EGLR 25

A (*a*) in every case, the period beginning with the date of service of the acceptance notice and ending with the end of the period specified in the offer notice under section 5(2)(*d*), and

(*b*) if any person is nominated for the purposes of this section within that period, an additional period of three months beginning with the end of the period so specified.

(3) If no person has been nominated for the purposes of this section during the period so specified, the landlord may, during the period of 12 months beginning with the end of that period, dispose of the protected interest to such person as he thinks fit, but subject to the following restrictions, namely —

(*a*) that the consideration required by him for the disposal must not be less than that specified in the offer notice, and

B (*b*) that the other terms on which the disposal is made must, so far as relating to any matters covered by the terms specified in the offer notice, correspond to those terms . . .

(5) A person nominated for the purposes of this section by the requisite majority of qualifying tenants of the constituent flats may only be replaced by another person so nominated if he has (for any reason) ceased to be able to act as a person so nominated.

(6) Where two or more persons have been so nominated and any of them ceases to act as such a person without being replaced in accordance with subsection (5), any remaining person or persons so nominated shall be entitled to continue to act in his or their capacity as such.

What was served was a photocopy document containing the
C signature of five of the eight tenants of the building, including Mr James Addison. There is no majority of qualifying tenants for the purpose of section 6 unless he is included.

On April 9 1996 the appellant served on the trustees a notice under section 19 of the Act, which deals with enforcement of obligations under the Act, requiring them to provide her with a draft contract for the purchase of the building. Upon their failure to do so, she issued on June 16 an originating application for an order that the trustees provide and enter into a contract. The trustees served a counternotice under section 19, requiring the appellant to serve a notice under section 9(3) of the Act, and they sought a declaration that the section 5 offer had
D not been validly accepted, and in the alternative, an order requiring service of a section 9(3) notice.

The case came before Judge Cowell in West London County Court on September 16 1996. Several issues of fact were determined by the learned judge and are not the subject of appeal. The judge dismissed Miss Mainwaring's application and made a declaration that:

No valid acceptance of the offer contained in the respondent's section 5 notice was made by the requisite majority of qualifying tenants of the constituent flats in the property by May 15th 1996 or at all, and that accordingly the respondents are free to proceed with the sale of the property to the Wellcome Trust Limited.

E The case, so far as it is now relevant, turns upon the conduct of the appellant and Addison and the effect of that conduct in the statutory context.

Section 9 of the Act bears the side heading "Withdrawal of either party from transaction", and provides so far as is relevant:

(1) Where —

(*a*) section 6(1) applies to a landlord by virtue of any provision of sections 6 to 8, and

(*b*) any person has been nominated for the purposes of section 6 by the requisite majority of qualifying tenants of the constituent flats within the period specified by the landlord in his offer notice under section 5(2)(*d*)
F (taking into account any postponement of the commencement of that period effected by any of the preceding provisions of this Part), and

(*c*) the nominated person serves a notice on the landlord indicating an intention no longer to proceed with the acquisition of the protected interest,

the landlord may, during the period of 12 months beginning with the date of service of the nominated person's notice, dispose of the protected interest to such person as he thinks fit, but subject to the restrictions mentioned in subsection (2).

(2) The restrictions referred to in subsection (1) are —

(*a*) that the consideration required by him for the disposal must not be less than the amount which has been agreed to by the parties (subject to contract) for the disposal of the protected interest, and

(*b*) that the other terms on which the disposal is made must correspond to those so agreed to by the parties in relation to the disposal.

(3) If at any time the nominated person becomes aware that the number of G
the qualifying tenants of the constituent flats desiring to proceed with the acquisition of the protected interest is less than the requisite majority of qualifying tenants of those flats, he shall forthwith serve on the landlord such a notice as is mentioned in subsection (1)(*c*) . . .

(8) Nothing in this section applies where a binding contract for the disposal of the protected interest has been entered into by the landlord and the nominated person.

The section also deals with the question of costs following the service of notices.

It is also relevant to refer to section 20(2) of the Act, which provides that: H

(*a*) any reference to an offer or counter-offer is a reference to an offer or counter-offer made subject to contract, and

(*b*) any reference to the acceptance of an offer or counter-offer is a reference to its acceptance subject to contract.

The judge found that on January 17 1996 Addison, as a tenant, made an agreement in writing with the appellant and the company. It provides in clause 1:

In consideration of the undertaking contained in clause 2 hereof the Tenants agree irrevocably that in the event of a Notice being served pursuant to the J
provisions of Section 5 of the Act such of the Tenants as is then the qualifying Tenant will accept the offer contained in such Notice within the period stipulated in the Act and will nominate the Company and Ms Mainwaring in accordance with the provisions of Section 6 of the Act.

Clause 2 limited the rent which would be payable in the event of an acquisition by the appellant or the company.

Addison also signed in a separate document a section 6 notice prepared by the appellant. Clause 4 has already been mentioned. Clause 2 states:

A requisite majority of the qualifying tenants in the Property have decided K
that they wish to accept the terms set out in the Offer Notice.

Clause 3:

We are duly appointed by such majority to notify such acceptances to you and we give notice under Section 6(1)(b) of the Act that the requisite majority of the qualifying tenants, whose names and addresses appear in the schedule below and on whose behalf this notice is served, accept the offer contained in the Offer Notice.

Addison's name and signature appear in the schedule.

The judge found that on February 5 1996 Addison withdrew his L
"authority and support". Addison crossed out his signature on the section 6 notice still in his possession, and stated that he withdrew his agreement to clause 3. He had signed the draft section 6 notice at a date before the valid section 5 notice had been served by the landlord. The judge also found that the appellant knew of this withdrawal of the support and authority, although she believed it to be ineffective because of the terms of her agreement with Addison. The judge also found that Addison did not support the appellant at any time after February 5 1996 and did not desire to proceed with the project to acquire the freehold of the property.

It has to be said that Addison's state of mind did vacillate as he M
spoke to different people at the relevant times, and he does not appear to have made his position clear to the trustees' agents until much later. What the appellant served upon the trustees was a photostat copy of the section 6 agreement which did not show the deletion by Addison of his earlier signature. She inserted the date of the section 5 notice, March 15, in the copy served. The judge found:

I have come to the conclusion on this point of law that it is contrary to the policy of the Act to rely upon the agreement of Mr Addison on January 17 or to enforce that agreement in some way because he has, in my judgment, always been entitled to decide at every stage under the Act whether he wishes to exercise his rights of deciding whether or not to proceed.

The judge stated that, while it was unnecessary to consider further points, he concluded that Addison was entitled to revoke what in the January 17 agreement had been described as an irrevocable contract.

The appellant has appeared in person and has presented her arguments most persuasively. Her central submission is that Addison could not withdraw his acceptance of the offer contained in the section 5 notice when he had made an irrevocable agreement with her not to do so and had signed and had signified his acceptance of the offer by signing the draft section 6 notice. By signifying his acceptance and making his nomination Addison had contracted into the Act. It would be impracticable for tenants to exercise their right to purchase if individual tenants could "jump in and out", as the appellant put it. The Act conferred no express right in a tenant to withdraw an acceptance and nomination once made. Section 6(5) expressly limited the circumstances in which a person nominated could be replaced, and given the purpose of the Act, it is submitted, a tenant should not be permitted to withdraw an acceptance or nomination. The parties are obliged to work towards the performance of their contracts with other tenants. The appellant relied on no less an authority than Lord Atkin, who stated in *Southern Foundries (1926) Ltd* v *Shirlaw* [1940] AC 707, at p717:

> The arrangement between the parties appears to me to be exactly described by the words of Cockburn CJ in *Stirling* v *Maitland* (1864) 5 B&S 840, 852: "If a party enters into an arrangement which can only take effect by the continuance of an existing state of circumstances"; and in such a state of things the Lord Chief Justice said: "I look on the law to be that . . . there is an implied engagement on his part that he shall do nothing of his own motion to put an end to that state of circumstances, under which alone the arrangement can be operative." That proposition in my opinion is well established law. Personally I should not so much base the law on an implied term, as on a positive rule of the law of contract that conduct of either promiser or promisee which can be said to amount to himself "of his own motion" bringing about the impossibility of performance is in itself a breach.

Addison had created a state of affairs, it is submitted, which was irreversible and he is obliged to proceed. As to section 9(3), it is submitted that Addison has contracted himself out of the right to withdraw and not to proceed. Any right to withdraw which that subsection contemplates is qualified by arrangements such as the present and made between tenants.

Adopting the approach to the Act suggested by Hobhouse LJ in *Belvedere Court Management Ltd* v *Frogmore Developments Ltd* [1996] 1 All ER 312*, at p335, the appellant submits that section 9(3) is a secondary and not a substantive provision and should not be construed so as to defeat the purpose of the Act, which is the conferring upon tenants of the opportunity to purchase their landlords' interest.

I am unable to accept that there was a valid section 6 notice. I would have regarded the document served as a valid notice had it been what it purported to be. The document purported to establish that by their signatures the majority of five qualifying tenants accepted the landlords' offer of March 15 1996. However, before the appellant served the photostat copy which purported to establish that acceptance, Addison had, to her knowledge, crossed out the signature on the copy of the document in his possession and endorsed it "withdraw agreement to clause 3". He had thereby made it known that he did not accept the offer. I do not doubt that one qualifying tenant can serve an acceptance on behalf of others who signify their acceptance, but when to the knowledge of the person effecting service Addison had indicated that he did not want to accept, there is no service for the purposes of section 6.

I would say in mitigation of the appellant's conduct that she believed that Addison could not in law withdraw his consent. He had vacillated, and it appears that even in subsequent discussion with the trustees' agent, who understandably made inquiries, he did not make clear until much later that his acceptance had been withdrawn. However, a photostat signature cannot in these circumstances signify acceptance when the signature on the original document had previously been crossed out. Further, I do not accept that in the context of this Act a tenant can by contract deprive himself of his right to withdraw his acceptance of the landlord's offer. Until there is a binding contract for the purchase of the premises, the offers and acceptances mentioned in the Act are subject to contract: section 20(2). Time-limits are imposed in the statute at each stage of the procedure. Until there is a binding contract, a nominated person is under the duty set out in section 9(3). For the duty to inform the landlord to arise, the nominated person need only be aware that the number of tenants "desiring to proceed" is less than the requisite majority. A desire not to proceed need not be expressed in any formal way. The duty in section 9(3) has meaning only if a tenant has a freedom not to desire to proceed, which must necessarily be read into the statutory framework. It is conferred by statute, and cannot be surrendered by contract. The section 9(3) duty can be enforced by an application under section 19. Section 6(5), by limiting the circumstances in which the nominated person can be replaced, does limit the freedom of tenants in the acquisition to that extent, but it does not, in my judgment, bear upon the freedom to withdraw the "desire to proceed" with the acquisition.

I would dismiss the appeal on those grounds, and the case does not, in my view, turn upon the question of agency upon which both parties addressed us. I would only say that I cannot accept the submission that the appointment by tenants of one of their number as agent is irrevocable in this context. It is not one of those situations in which the authority is conferred as protecting or securing the agent's interest: see *Smart* v *Sandars* (1848) 5CB 895. Each of the parties is acting in his own interests, although the benefits sought to be obtained may vary. While the agent may be vulnerable to changes of mind, the appointment did not confer a property or security interest on him.

Neither can I accept the submission that arrangements made by parties seeking to utilise the provisions of the Act are necessarily void. Initiating a section 6 procedure may well have serious financial implication for the person doing so. Almost anyone embarking upon a procedure to acquire tenancies to which the Act applies will need to make some arrangements with other interested parties in an attempt to protect his position along the way. It is not necessary in the present case to consider what agreements or categories of agreement may be void or the effect of breaches of agreement. For present purposes it need only be said that the appellant's agreement with Addison did not have the effect of defeating his freedom to express a desire no longer to proceed with the acquisition.

The difficulties of a tenant who seeks to mobilise fellow tenants to exercise their rights under the Act are obvious, and there will inevitably be pitfalls along the path to success. The appellant made considerable efforts to achieve a purchase and feels let down by the change of mind of a fellow tenant who had made a contract with her. However, as Sir Thomas Bingham stated in *Mainwaring* v *Henry Smith's Charity Trustees* at p235:

> . . . the choice between the competing contentions . . . must be governed by the proper construction of the 1987 Act, and not by consideration of how the proper construction may in practice affect either one party or the other.

I would dismiss this appeal.

Peter Gibson and **Hirst LJJ** agreed and did not add anything.

Appeal dismissed.

---

*Editor's note: Also reported at [1996] 1 EGLR 59

A
Dugan-Chapman *v* Grosvenor Estates

WEST LONDON COUNTY COURT

April 25 1996

MR RECORDER RUSSELL

ESTATES GAZETTE MARCH 8 1997

B

[1997] 10 EG 152

Leasehold Reform Act 1967 — Enfranchisement — Mews house containing flat and garage — Whether mews house enfranchisable with main house — Whether garage part of "house and premises"

By a lease dated March 25 1960 the applicant tenant held a term of 57 years of premises consisting of two buildings; a substantial house on five floors and, to the rear, a mews house consisting of a garage and flat above. The tenant used the garage of the mews house for his motor car and the flat above for the purposes of an antiques' business. On
C
August 12 1994 the tenant gave notice under the Leasehold Reform Act 1967 to purchase the freehold of the premises. The respondent landlord contended that the tenant was not entitled to acquire the mews house.

***Held*: The tenant was entitled to acquire the main house together with the garage only of the mews house. The main house and the mews house taken together are not "a house" within the meaning of section 291 of the Act. However the main house and the garage in the mews house fall within the expression of "house and premises". The flat above the garage cannot be regarded as an outhouse or appurtenance so as to fall within the expression**
D
"premises".

The following cases are referred to in this report.

Methuen-Campbell v *Walters* [1979] QB 525; [1979] 2 WLR 113; [1979] 1 All ER 606; (1978) 38 P&CR 693; [1978] 2 EGLR 58; 247 EG 899, CA

St Thomas' Hospital (Governors of) v *Charing Cross Railway Co* (1861) 1 John & H 400; 30 LJ Ch 395; 25 JP 771; 7 Jur NS 256; 9 WR 411

Anthony Radevsky (instructed by Lee & Pembertons) appeared for the applicant; Frank Hinks (instructed by Boodle Hatfield) represented the respondent.

E

Giving judgment, MR RECORDER RUSSELL said: By this application dated May 18 1995 the applicant seeks an order that he is entitled by virtue of the provisions of Part I of the Leasehold Reform Act 1967 to acquire the freehold house and premises at 4 Wilton Crescent and 6 Kinnerton Street, London SW1.

The respondent's answer dated July 17 1995 denies that the applicant is entitled to the order sought on the grounds set out, namely that: (1) 6 Kinnerton Street does not constitute premises as referred to in subsection (3) of section 2 of the Leasehold Reform Act 1967; and (2) that on the date when the applicant's notice was received
F
he was not occupying 4 Wilton Crescent as his only or main residence and had not been occupying such property as his only or main residence for the requisite period.

I heard this matter on April 1, 2 and 3 1996 and at the end of counsel's argument reserved judgment. At the outset of the hearing (or at a very early stage) counsel for the respondent abandoned the second ground of opposition so far as it concerned 4 Wilton Crescent and accepted that (subject to formal requirements) the applicant would be entitled to enfranchise 4 Wilton Crescent and its garden, so that the arguments before me related entirely to 6 Kinnerton Street and whether the applicant was entitled (as he claimed and the respondent denied) to acquire the freehold thereof.

G
In this judgment, I propose to call the applicant "the tenant" and the respondent "the landlord"; and in referring to page numbers will be referring to the corresponding page of the agreed bundle of documents.

The landlord is the freeholder of a property known as 4 Wilton Crescent and 6 Kinnerton Street, London SW1, which by a lease dated March 25 1960 was demised for a term of 57 years, expiring on Midsummer Day 2015, which term has been vested in the tenant since September 1963.

The subject-matter of the demise is described as being:

> all that piece of land situate on the west side of Wilton Crescent fronting towards the east on Wilton Crescent which said piece of land with the dimensions thereof is delineated and coloured in the plan in the margin hereof, together with the messuage and buildings thereon erected and known as No 4 Wilton Crescent and No 6 Kinnerton Street.

H

As is apparent from the plan there are two buildings on the land demised, namely 4 Wilton Crescent which is a substantial house on five floors in a curved terrace of houses, and 6 Kinnerton Street which is a mews house on two floors in a terrace of mews houses. 4 Wilton Crescent and 6 Kinnerton Street are separated by the garden but joined on either side by garden walls, which are of a lesser height than the two buildings. The relation between these buildings to their surroundings is shown on the plan at p33, and there is a picture of the frontage of 4 Wilton Crescent at p31 and of the frontage of 6 J
Kinnerton Street at p25. 6 Kinnerton Street consists of a garage or garages on the ground floor and of a residential flat on the first floor. Access to 6 Kinnerton Street can be gained from Kinnerton Street and also from 4 Wilton Crescent via a door from the garage leading to the garden.

The lease in clause 2(x) contains a covenant as to "user" which provides that the demised premises shall be kept and used only as follows:

(a) 4 Wilton Crescent as a private dwelling-house in the occupation of one household only; K

(b) the garage at the rear of 4 Wilton Crescent as a private garage for the housing of private motor cars only;

(c) 6 Kinnerton Street as a private dwelling in the occupation of one household only, with a private garage on the ground floor for the housing of private motor cars only.

It is not disputed that the reference in (b) to "the garage at the rear of 4 Wilton Crescent" relates to either the whole of the ground floor of 6 Kinnerton Street or to the larger half of the garage which has the door into the garden. I have come to the conclusion that it refers to the whole of the ground floor. There never has been any other garage at the rear of 4 Wilton Crescent. L

Property as valuable as the property with which this case is concerned was only brought within the ambit of leasehold enfranchisement by an amending Act passed in 1993. On August 12 1994 the tenant gave notice of his claim to purchase the freehold to which the landlord's solicitors responded by a notice in reply dated November 3 1994, para 3 whereof set out the landlord's contentions. Thereafter in May of 1995 the present proceedings were commenced by the tenant.

For the tenant evidence was given by his wife and by Mr Simon Marr-Johnson [FRICS]. The tenant himself did not give evidence, on account of his poor state of health. M

For the landlord I heard the evidence of Mr John Lindgren [FRICS], a surveyor employed by the landlord, and of Mr George Pope [ARICS].

All these witnesses were cross-examined, but I do not think it necessary to refer to the answers given in cross-examination in any detail. I am satisfied that all the witnesses I heard were completely truthful. To the extent that the experts were expressing their own views of the law as it impinged on the facts of this case, I found this helpful; although I have always borne in mind that all decisions as to law are for me and not for them.

The landlord accepts that the tenant has at all material times occupied 4 Wilton Crescent as his main residence. But in relation to 6

A Kinnerton Street the identity of its occupant and the nature of its occupation has been an issue between the parties.

In relation to 6 Kinnerton Street I find that the tenant has proved, to my satisfaction, that he has at all the material times been in occupation of the whole of the mews house, and that he has occupied it together with 4 Wilton Crescent. I further find as proved that the ground floor of 6 Kinnerton Street was at the material date (that is to say, on August 12 1994) and at all other relevant times used by the tenant for the purposes of 4 Wilton Crescent, namely in order to garage one or more of the tenant's motor cars and to house a freezer which it was convenient to keep in the garage.

B However, in relation to the flat on the first floor of 6 Kinnerton Street, I find that at all material times it was being used for the purposes of the business of Wilton Antiques Ltd, a company controlled by the tenant and his wife. While I accept Mrs Dugan-Chapman's evidence that some furniture and a few pictures belonging to the tenant remained in the flat, I do not consider it sufficient to constitute "use for the purposes of 4 Wilton Crescent". If I am wrong in that view, then I consider that such use was *de minimis* and can be disregarded. In taking this view I have taken into account that the outgoings of the flat such as rates, electricity, telephone and gas have been met entirely by the company, although this has only been one of the factors which C have led me to the conclusion stated.

This brings me to an important decision which I have to make, which is either a matter of pure law or possibly mixed fact and law: namely, whether 4 Wilton Crescent and 6 Kinnerton Street can together be treated as "the house" within the meaning ascribed to that expression by the Leasehold Reform Act 1967. Under this Act the right to enfranchise is given to a tenant of a leasehold house who occupies it as his only or main residence. There are other preconditions which are not relevant in this case and to which I therefore need not refer. Such a tenant is given the right to acquire the freehold of *the house and premises* — (emphasis supplied).

The meaning of "house and premises" is defined by section 2 of the D Act. By section 2(1) "house" includes any building designed or adapted for living in and reasonably so called . . . and (b) where a building is divided vertically the building as a whole is not "a house", though any of the units into which it is divided may be. Flats in buildings of which a material part lies above or below another building are expressly excluded by section 2(l)(a) and section 2(2).

I am of the clear view that 4 Wilton Crescent and 6 Kinnerton Street, the buildings which form part of the property demised are, looked at objectively and in a common-sense way, two houses and not one.

This raises the question whether it is possible to construe the words "any building" where it occurs in section 2(1) as including the plural, E as if it read "any building or buildings". Whether one can do so depends on the context in which the words occur. I bear in mind that under other statutes several buildings, sometimes adjoining, sometimes detached from each other, have together been held to constitute a house, or a house let as a separate dwelling: see the cases of *Governors of St Thomas' Hospital* v *Charing Cross Railway Co* (1861) 7 Jur NS 256 and *Whitty* v *Scott-Russell* [1950] 2 KB 32.

But I also bear in mind that it has been held that decisions under other Acts (and in particular under the Rent Acts) are no reliable guide to the construction of the Leasehold Reform Act, because its objectives F are not the same.

It is also clear on the authorities cited to me that 6 Kinnerton Street could on its own be "a house" within the definition, but this, in my view, is by no means conclusive. What to my mind indicates the correct answer is the fact that the whole context of the relevant subsection indicates that the draftsman of the Act did not include the plural when using the words "any building". When one asks whether several freestanding buildings can be "a house reasonably so called", I think that the answer is "No". Equally, the provisions of section 2(1)(b) which relate to a building which is divided vertically and state that the whole building is not a house, suggests *a fortiori* that two separate buildings or houses are not "a house" within the definition of section 2(1). I do not overlook the fact that the word used in the G definition is "includes"; but I am satisfied nevertheless that anything that is outside the definition cannot be "a house".

For these reasons I am driven to conclude that 4 Wilton Crescent and 6 Kinnerton Street taken together are not "a house" within the meaning of the Act.

I therefore now turn to see whether 4 Wilton Crescent and 6 Kinnerton Street are "a house and premises" within the Act. By subsection (3) of section 2 of the Act:

> the reference to premises is to be taken as referring to any garage, outhouse, garden, yard and appurtenances which at the relevant time are let to (the H tenant) with the house and are occupied with and used for the purposes of the house or any part of it by him or by another occupant.

The landlord accepts that in the present case the garden falls within the above definition, but denies that the mews house (that is to say 6 Kinnerton Street) does so. The tenant argues that the ground floor of 6 Kinnerton Street is patently a garage, that it was let to him with the house, and that it has been occupied and used by him for the purposes of the house, namely to garage his motor car or cars and to store food in the freezer. The landlord concedes that the ground floor of 6 Kinnerton Street is a garage, but argues that it was not let to the tenant with the house, either because it was let to him with another house — J namely 4 Wilton Crescent — or alternatively that a garage which is part of a separate house and outside the curtilage of the relevant house cannot be a garage within the definition of "premises".

I think that on this issue the arguments of the tenant are to be preferred, and I hold that the ground floor of 6 Kinnerton Street is a garage let with the house, namely 4 Wilton Crescent, and falls fairly and squarely within the expression "house and premises" used by the Act. It was urged on me by the landlord that this conclusion could lead in some cases to absurd results, but I found this argument unconvincing.

Turning to the first-floor flat at 6 Kinnerton Street, the tenant argued K that it passed automatically with the freehold on the ground floor or, if it did not, the whole of 6 Kinnerton Street (or alternatively the flat alone) came within the words "outhouse or appurtenances" in subsection (3) of section 2. The landlord argued that if 6 Kinnerton Street was properly to be regarded as "a house" it could not be "an outhouse"; and that, as it was outside the curtilage of the main house (namely 4 Wilton Crescent) — which may be debatable — then it could not be "an appurtenance": see *Methuen-Campbell* v *Walters* [1979] QB 525*.

I reject the argument that the first floor passes with the ground floor L automatically. I do not think that can be right.

It seems to me that 150 years ago when this property was originally erected, the mews were probably for stabling on the ground floor, and that the first floor was used for the storage of hay or other stable equipment or for the accommodation of domestic staff, so that it would have been reasonable to regard the entity (and by that I mean the composite entity) of 4 Wilton Crescent and 6 Kinnerton Street as one domestic household or establishment. So that at that time 6 Kinnerton Street could have been an outhouse or appurtenance of 4 Wilton Crescent. But when this lease was granted and when the tenant's notice was served the social conditions in London were so different that M anyone describing the subject matter of this lease would have said that it demised two houses and a garden and not just one house with an outhouse or appurtenance of a mews.

I have therefore decided that the landlord's arguments are to be preferred on this issue, and that 6 Kinnerton Street (or the first-floor flat therein) cannot be properly regarded as an outhouse or appurtenance of 4 Wilton Crescent. Moreover, even if this were wrong, it is clear from my findings of fact that the user of the first-floor flat of 6 Kinnerton Street was not for the purposes of 4 Wilton

---

*Editor's note: Also reported at [1978] 2 EGLR 58

A Crescent or any part of it so as to come within the definition of "premises". For the reasons set out above, I have decided that "the house and premises" which the tenant is entitled to enfranchise under the provisions of the Act includes 4 Wilton Crescent, the garden and the ground floor of 6 Kinnerton Street, but does not include the first-floor flat of 6 Kinnerton Street. In coming to this conclusion I have not overlooked the fact that the Act clearly seeks to avoid the creation of "flying freeholds", with all the difficulties that these are known to involve. But it seems clear that the draftsman of the Act included subsections (4), (5), (6) and (7) of section 2 in order to overcome some B of these very problems.

In the result, having regard to the terms of the tenant's notice the tenant is not entitled to the order sought and this application fails. If there is an application for leave to amend the notice pursuant to para 6 subpara (3) of the Third Schedule to the Act, I am willing to consider it, but shall require further argument from the parties before making any decision on whether leave should be given, and, if so, on what terms. I also propose to hear the parties on the issue of costs.

LANDLORD AND TENANT
RENT REVIEW

A
Addin *v* Secretary of State for the Environment

CHANCERY DIVISION

November 27 1996

JACOB J

Estates Gazette April 5 1997

B [1997] 14 EG 132

Landlord and tenant — Rent review — Rent to be higher of current market rental value or initial rent — Current market rent below initial rent — Landlord only entitled to serve trigger notice — Whether rent must be reviewed — Whether landlord must serve trigger notice

By a lease dated September 8 1975 the defendant held a 42-year term of commercial premises at an initial rent of £148,500 subject to seven-year rent reviews; the plaintiff owns the reversion. The rent was reviewed in 1988 to £440,000. The *reddendum* provided that for each rent review period the rent "shall be the higher of the sum of £148,500
C . . . or such sum as shall be assessed as the current open market rent . . . such assessment being made in the following manner . . . Either (a) such assessment as shall be specified in . . . a notice by the lessors . . . ". There were further provisions concerning the assessment where the parties could not agree. In respect of the 1995 rent review, the landlords contended that these provisions gave the landlords an option to serve a trigger notice, and to have a review; in the absence of a review the tenant was required to continue paying the rent of £440,000 (which was above the rental value). (*Editor's note*: it is understood that the landlord served a trigger notice without prejudice to any contractual obligation to do so.)

D ***Held*****: The parties, assumed to be men of commercial substance, contracted to pay from a review date the higher of the sum of £148,500 and the current market rent. The remaining provisions of the rent review clause, including that providing for a trigger notice, were essentially machinery. The language of the *****reddendum***** contained a "shall" in providing that the rent for each succeeding rent review period shall be the higher of the two figures. The landlord is to serve a trigger notice and initiate the dispute procedure. The rent paid in the previous rent review period continues to be payable but, if the review is triggered, as it was, it then reverts to the formula in the *****reddendum*****.**

E

The following cases are referred to in this report.

Basingstoke and Deane Borough Council v *Host Group Ltd* [1988] 1 WLR 348; [1988] 1 All ER 824; (1987) 56 P&CR 31; [1987] 2 EGLR 147; 284 EG 1587, CA

Harben Style Ltd v *Rhodes Trust* [1995] 1 EGLR 118; [1995] 17 EG 125

Royal Bank of Scotland plc v *Jennings* (1995) 70 P&CR 459; [1995] 2 EGLR 87; [1995] 35 EG 140

Sudbrook Trading Estate Ltd v *Eggleton* [1983] 1 AC 444; [1982] 3 WLR 315; [1982] 3 All ER 1; (1982) 44 P&CR 153; [1983] 1 EGLR 47; [1983] EGD 392; 265 EG 215, HL

This was a hearing of an originating summons in which the plaintiffs, Addin Investments Ltd, sought declarations as to the meaning of a lease against the defendant, the Secretary of State for the Environment. F

Simon Berry QC (instructed by Denton Hall) appeared for the plaintiff; Nicholas Dowding (instructed by Berwin Leighton) represented the defendant.

Giving judgment, **JACOB J** said:This is yet another dispute over the construction of the rent review provisions of a lease. The plaintiff landlord holds the reversion in a lease granted on September 8 1975. The lessee is a minister of the Crown. The term of the lease is 42 years. G
So it still has more than 20 years to go. The lease contains a provision about rent reviews taking place every seven years. The review in 1988 fixed the open market rate at £440,000 with effect from July 18 1988. It is common ground that commercial rents have fallen since that date. The landlord says that, even though another review is possible under the terms of the lease, upon its true construction the lease gives him and him alone an option to initiate such a review and that, if there is no review, the tenant must continue to pay the rent fixed last time. That is agreed by both sides to be substantially more than the current open market rent.

I am not concerned with the details of what the current market rent ought to be. The parties apparently differ on this considerably. Even on H
the landlord's contention as to the current market rent the sums at stake are considerable, both in terms of the annual rent involved and the knock-on effect on the capital value of the property.

The tenant contends that the lease contains a provision requiring rent review and, if that be wrong, that the only rent it has to pay in the absence of rent review is the minimum rent set in the lease.

I begin by reading the *reddendum* concerned:

PAYING THEREFOR until the end of the seventh year of the said term the exclusive yearly rent of £148,500 such rent to be paid by equal quarterly instalments on the twenty-fifth day of March the twenty-fourth day of June the J
twenty-ninth day of September and the twenty-fifth day of December in every year to run from the eighteenth day of July 1974 the first payment (being a proportionate sum) to be made having become due on the eighteenth day of July 1974 PROVIDED ALWAYS AND IT IS HEREBY AGREED that the yearly rent payable by the Minister during the next and each succeeding period of seven years of the said term shall be the higher of the sum of £148,500 aforesaid or such sum as shall be assessed as the current open market rent of the demised premises for the appropriate period such assessment being made in the following manner, that is to say

Either

(a) such assessment as shall be specified in writing and in a notice by the lessors to the Minister given not more than twelve and not less than six months K
before the review date (which expression means the expiration of the seventh fourteenth twenty-first twenty-eighth or thirty-fifth years of the said terms as the context requires)

(b) as shall be agreed between the parties hereto in writing within three months after such notice or

(c) in the event of the parties hereto failing to reach agreement as aforesaid on or before the date appointed (in respect of which time is to be deemed to be of the essence) then the current open market rent of the demised premises ready for use as office premises for the next seven years of the said terms shall be fixed or assessed by an independent surveyor appointed for that purpose by the parties hereto or failing agreement as to such appointment three months before the review date (time in this respect also to be deemed to be of the essence) then by an independent surveyor appointed for that purpose by the President for the time being of the Royal Institution of Chartered Surveyors . . .

— I can leave out some immaterial words which follow —

The assessment fixed by the independent surveyor shall be communicated to the parties hereto in writing and immediately upon such communication the rent so assessed as the current open market rent for the next seven years of the said term or £148,000 whichever shall be the higher shall be the rent payable during that period under the terms hereof and the fees payable to the independent surveyor hereinbefore mentioned in respect of the assessment to be made herein shall be borne by the parties hereto in equal shares

(d) until the revised rent has been agreed or determined as aforesaid the Minister shall continue to pay the rent payable during the previous rent period PROVIDED THAT when such rent has been agreed or determined for such new rent period it shall become payable on the first quarter day after such agreement or determination and on such quarter day the Minister shall pay in addition to the quarter's rent then due a sum equal to the difference between the sum or sums in fact paid from the beginning of such period and the sum or sums which would in fact have been paid had the new rent been fixed before the beginning of the new rent period.

The issues before me turn on the true construction of the language which I have just read. Before turning to it I make some initial observations. The (a), (b), (c) and (d) of the language are a little odd. Normally the word "either" is followed by just two alternatives. Here there appear to be four. It is common ground, however, that (a) and (b) are to be read as one; (c) is the other; and (d) is something else altogether. In the end I do not think it matters very much.

It is also common ground that (d), which provides for payments being made by the minister in the event that a rent is raised above that which was being paid during an interim period, has a corresponding implied term that money would be paid back if the assessed rent came to be less than was being paid during that interim period.

Again before turning to consider the language in detail, I remind myself that I am concerned with a lease. In the well known case of *Basingstoke and Deane Borough Council* v *Host Group Ltd* [1988] 1 WLR 348*, at p353 Nicholls LJ said:

We recognise, therefore, that the particular language used will always be of paramount importance. Nonetheless it is proper and only sensible, when construing a rent review clause, to have in mind what normally is the commercial purpose of such a clause.

That purpose has been referred to in several recent cases, and is not in doubt. Sir Nicolas Browne-Wilkinson V-C expressed it in these terms in *British Gas Corporation* v *Universities Superannuation Scheme Ltd* [1986] 1 WLR 398, p401:

"There is really no dispute that the general purpose of a provision for rent review is to enable the landlord to obtain from time to time the market rental which the premises would command if let on the same terms on the open market at the review dates. The purpose is to reflect the changes in the value of money and real increases in the value of the property during a long term."

To the same effect Dillon LJ said in *Equity & Law Life Assurance Society plc* v *Bodfield Ltd* [1987] 1 EGLR 124, 125:

"There is no doubt that the general object of a rent review clause, which provides that the rent cannot be reduced on a review, is to provide the landlord with some measure of relief where, by increases in property values or falls in the real value of money in an inflationary period, a fixed rent has become out of date and unduly favourable to the tenant. The exact measure of relief depends on the true construction of the particular rent review clause."

I mention this because Mr Simon Berry QC, for the landlord suggested that these passages were confined only to the cases of protecting the landlord against the effects of inflation and the upward movement of market rents. I do not agree. The general background is that landlords and tenants expect to receive and pay respectively the market rents. One can normally expect that that is what either side would want and expect from the other party, and I bear that in mind in construing this document, although of course the question of paramount importance is the language used here.

*Editor's note: Also reported at [1987] 2 EGLR 147

Mr Berry says that (a) of the *reddendum* gives the landlord an option to serve a trigger notice. If he does then either the parties agree or they do not. If they do not then the machinery of (c), going to president, may be invoked by either party. The reference in (c), he says, to a date appointed, albeit slightly clumsy, means the date three months after the date of the notice under (a) because (b) gives a three-month period for negotiation. He accepts it is a slightly odd way to describe that date as a "date appointed", but he says that is the only thing that can make sense.

He then says that if no trigger notice is given none of the machinery in (a) to (c) applies and then, he says, it follows as night follows day that (d) applies and that the tenant must accordingly pay the rent payable during the previous rental period — in this case the £440,000.

In support of his arguments he referred me to *Harben Style Ltd* v *Rhodes Trust* [1995] 1 EGLR 118*. In that case the lease contained a provision expressly dealing with what happened if there was no rent review — the actual language was:

In the event that the yearly rent hereinbefore reserved and payable in respect of the review period is not ascertained at all . . . the tenants shall during the same review period . . . pay rent at the rate of the rent payable hereunder for the year immediately preceding the beginning of the same review period . . .

In that case it was possible for the landlords not to initiate a rent review and the lease itself had expressly contemplated what would happen if that option was, in effect, exercised. I do not think it assists Mr Berry.

Mr Berry also took me to the recent case of the *Royal Bank of Scotland plc* v *Jennings* in the Court of Appeal [1996] EGCS 168. It was held that the *reddendum* concerned did provide for a rent review. In the words of Sir Richard Scott, V-C that "there will be a rent review for each of the review periods". The express machinery for such review could only be initiated by the landlord, but refusing to initiate a review the landlord was frustrating the provisions of the contract that there should be a rent review. Accordingly, the Court of Appeal held that the court could substitute its own machinery following the decision of the House of Lords in *Sudbrook Trading Estate Ltd* v *Eggleton* [1983] 1 AC 444†.

Mr Berry said that the present case was distinguishable because on the true construction of the present *reddendum* there is no provision that there will be a rent review. The true construction here is that there is a rent review only if the landlord decides he wants one. If he decides he does not want one the agreement is perfectly complete in itself because it provides that the old rent shall continue to be payable pursuant to the provisions of para (d).

I do not agree. I think the correct way to approach the present case is to recognise that the key commercial provision is contained in the opening words. I think provisions (a) to (d) are essentially machinery. In the neat phrase used by Evans-Lombe J in *Royal Bank of Scotland plc* v *Jennings* at first instance [1995] 2 EGLR 87‡, at p89, they are "administrative and accounting provisions".

I think the way to read the agreement is that the parties, assumed to be men of commercial substance and alive to the ways of landlords and tenants, contracted that the rent was to bethe higher of £148,500 and the current market rent. It is true that they use the expression "assessed current market rent" and the agreement goes on to deal with how there should be assessment, but the substance of the agreement is that the higher of the current market rent and the £148,500 is to be paid. I think that is the most likely construction. It accords with my view of how commercial men behave and it accords with the principles which I have stated from the *Basingstoke* case and others.

It is unlikely that a tenant would agree to pay more than a current market rent. Why should he? It is unlikely that a landlord could get a tenant to pay that.

*Editor's note: Also reported at [1995] 17 EG 125
†Editor's note: Also reported at [1983] 1 EGLR 47
‡Editor's note: Also reported at [1995] 35 EG 140

A I am reinforced in view by the language of the opening part of the agreement. It says that the yearly rent payable shall be — and I emphasise the word "shall" — the higher of the two sums, and such sum as shall be assessed as the current open market rent. I think Mr Berry's argument amounts to this: that the second "shall" means "may be assessed if the landlord chooses". I do not think the agreement contemplates that there will be no assessment. I think one finds exactly the same conclusion to be derived from (a):

> such assessment as shall be specified in writing and in notice by the lessors . . .

Mr Berry says that means that the "shall" there was referring to what B assessment actually happens to be in the written notice if the landlord decides to serve such a notice. I think a fairer reading is that he shall specify his opening shot as to what he wants.

Finally, the language of (d) clearly contemplates that there will be a revised rent agreed or determined. It says:

> until the revised rent has been agreed or determined as aforesaid the Minister shall continue to pay the rent payable during rent period . . .

It does not say "unless and until". It contemplates that "the rent is going to be agreed or determined as aforesaid".

Mr Berry rather accepted that his construction meant "unless and C until", and he said in the context I should read the "until" as being only if the "aforesaid" referred to the machinery of (a), (b) and (c) and that it, therefore, contemplated that only the landlord could trigger the review machinery — that being his construction of (a), (b) (c). But that is circular and ignores the key "shalls" in the main operative portion of the *reddendum*. Very much the similar approach was adopted in the *Royal Bank of Scotland* case in relation to a clause very similar to clause (d).

I think, therefore, overall the agreement contemplates that the landlord is to serve a notice and initiate the procedure.

Suppose that is wrong, however. Then, says the tenant, the rent is to D be £148,500 and that is all. The argument goes like this: the rent shall be the higher of the sum of £148,500 or the assessed current open market rent. Well, if no current open market rent is assessed then it is the higher of the £148,500 and nothing, to wit, £148,500. The tenant then says that (d) only applies when a rent revision is under way in accordance with (a) or (b) and, while there is no such procedure in being, the only rent that is payable under the agreement is this minimum £148,500. If the procedure is initiated the tenant says, "Well, then you pay the last passing rent". Mr Berry says that is all very odd, two rents, first of all £148,500, then the last passing rent and finally the actual rent with adjustment between them all.

E Mr Nicholas Dowding for the tenant accepted it was somewhat odd, but he says he wins on, using the happy phrase, "the balance of oddities". He says it would be even odder that the tenant should have agreed to have paid above the market rent, and I agree. I think the oddity referred to by Mr Berry is more abstract than real. In reality, if no notice is served and the tenant starts paying £148,500 when, in fact, the landlord can get more, the immediate response will be for the landlord to initiate procedure by a specification in writing, or agreement, or going to the president.

I believe that disposes of the case, but I should deal with the final F argument of the tenant. Mr Dowding suggested that, on its true construction, clause (c) enabled the tenant to go to the president of the RICS after expiry of the date appointed — as clause (c) calls it. That depends on what is meant by "date appointed". He suggested that the date appointed meant as follows: if the landlord serves a notice then it is three months after the date of notice; if he does not then it is three months after the six months referred in (a).

I have to say that I find the argument perhaps too strained to work, but I do not think that that matters. Overall, I think the agreement does provide for an assessment which both parties expected to take place and ought to take place.

I will hear counsel as to the appropriate form of order.

Royal Bank of Scotland plc *v* Jennings and others G

COURT OF APPEAL

October 24 1996

SIR RICHARD SCOTT V-C, SAVILLE and POTTER LJJ

Estates Gazette May 10 1997

[1997] 19 EG 152 H

Landlord and tenant — Rent review — Expert appointment only upon application of landlords — Whether landlords could be compelled to seek appointment and have rent review

By an underlease dated March 31 1988 the plaintiff tenant holds a term of warehouse premises at a rent subject to five-year reviews. The rent review provisions directed that if the parties could not agree the rent at review, the matter would be referred to an expert to be agreed between the parties and, if not agreed, to be appointed by the president of the Royal Institution of Chartered Surveyors upon application of the landlords. The tenant, anticipating that current rental values would be less than the passing rent, sought to initiate a rent review; the J defendant landlords declined to agree an expert or to apply to the president to nominate one. In the court below Evans-Lombe J granted the tenant a declaration that the landlords are bound to apply to the president to determine the proper rent payable since the rent review date. The landlords appealed, conceding that the rent was reviewable upwards and downwards.

***Held*: The appeal was dismissed. The implication to be gained from the lease as a whole, and in particular the *reddendum* and rent review provisions, is that there will be a rent review for each of the rental periods. The provision relating to the nomination by the president being made upon the application of the landlord is a** K **matter of machinery and not of substance. If that machinery breaks down, the court would supplement it: see *Sudbrook Trading Estate Ltd* v *Eggleton* [1983] 1 EGLR 47**

The following cases are referred to in this report.

Sudbrook Trading Estate Ltd v *Eggleton* [1983] 1 AC 444; [1982] 3 WLR 315; [1982] 3 All ER 1; (1982) 44 P&CR 153; [1983] 1 EGLR 47; [1983] EGD 392; 265 EG 215, HL

United Scientific Holdings Ltd v *Burnley Borough Council* [1978] AC 904; [1977] 2 WLR 806; [1977] 2 All ER 62; (1977) 33 P&CR 220; [1977] 2 EGLR 61; [1977] EGD 195; 243 EG 43 & 127, HL L

This was an appeal by the defendant landlords, Francis Edward Jennings, Robert Joseph Pezaro and Circuitport (Brewery Road) Ltd, against a decision of Evans-Lombe J, [1995] 2 EGLR 87, who had given judgment to the plaintiff tenant, Royal Bank of Scotland plc, on its application for declaratory and injunctive relief in respect of the rent review provisions in a lease it held from the landlords.

Kim Lewison QC and Janet Bignall (instructed by John Summers & Co) appeared for the landlords; Robert Powell-Jones (instructed by Stephenson Harwood) represented the tenant.

Giving judgment, **SIR RICHARD SCOTT V-C** said: This is yet M another appeal concerned with the true construction of a rent review clause. The rent review clause, with which we are concerned, is contained in a lease dated March 3 1988. The appeal is from the judgment of Evans-Lombe J given on December 9 1994*. The judge decided that on the true construction of the relevant rent review provisions it was not open to the landlord unilaterally to prevent the operation of the rent review provisions. I will refer in more detail later to the exact order that the learned judge made to give effect to his conclusion.

---

*Editor's note: Reported at [1995] 2 EGLR 87

A It is necessary for me to refer to some of the contents of the lease. I describe it as a lease; it is in fact an underlease but nothing turns on that. The "landlord" under the lease consisted of a Mr Jennings, a Mr Pezaro and a company, Circuitpoint (Brewery Road) Ltd. They are the appellants before us; they were defendants below. The tenant, the plaintiff in the action and respondent before us, is the Royal Bank of Scotland plc. The term granted by the lease was one of 20 years from December 25 1987. The demised premises consists of Unit B, 15 Brewery Road, Islington. The initial yearly rent reserved under the lease was £50,000 pa exclusive. The lease commences with a number of definitions. Of these the "initial rent" is defined as "the sum so
B specified in the particulars" — I have already referred to that. The "rent commencement date" is defined as March 31 1988. The "rent review dates" are "the dates so specified in the particulars"; they are December 25 1992, December 25 1997 and December 25 2002. The "rental period" is defined as "each period of five years immediately following a rent review date" and the "first rental period" is defined as "the period from the term commencement date until the first review date".

I can now pass on to the *reddendum* to the lease under which the demised premises were demised to the tenant for the specified term:

C YIELDING AND PAYING

(a) Firstly the initial yearly rent from [and there then appears a blank] until the 24th day of December 1992. Thereafter during the remainder of the term the yearly rent subject to review as hereinafter mentioned such review to be calculated in accordance with the provisions of the Fourth Schedule such rent to be paid by equal quarterly payments in advance without any deduction or set off whether demanded or not on the usual Quarter Days every year . . .

Subpara (b) I need not read. That related to additional sums to be paid by way of further rent on various accounts.

I can now turn to the fourth schedule which, it will be recalled, contains the details as to how the reviewed rent is to be calculated. Para 1 of the fourth schedule reads as follows:

D

At the commencement of each Rental Period hereinbefore referred to, the revised rent may be agreed between the Landlord and the Tenant or in the absence of agreement determined by a specialist valuer (acting as an expert and not as an Arbitrator) such valuer to be agreed between the parties or nominated by the President for the time being of the Royal Institution of Chartered Surveyors upon application of the Landlord made at any time after the commencement of the relevant rental period so that in case of such valuation the revised rent to be determined by the valuer (who shall give written reasons for his decision) shall be as in his opinion represents the best yearly rent reasonably obtainable for the Demised Premises and on the supposition (if not a fact) . . .

E

And there then follow five suppositions. I need not refer to them all, but it is perhaps helpful for me to refer to the second of them which is in these terms:

Of a new letting of the Demised Premises as a whole or in the permitted parts in the open market at the commencement of the relevant Rental Period by a willing Landlord to a willing Tenant without a premium with full vacant possession for a term equal to the term unexpired or a term of ten years whichever is longer as at the commencement of the relevant Rental Period and on similar terms to this lease (other than the rent hereby reserved but including the provisions for the review of rent herein contained).

F

There are also some specified disregards to which I need not refer.

Para 2 of the fourth schedule is, in my opinion, of considerable relevance for the light it casts on the contractual intentions of the parties underlying the first paragraph. Para 2 reads:

If and so often as any revised rent in respect of any Rental Period has not been ascertained pursuant to the foregoing provisions before the first day hereby appointed for payment of rent for the relevant Rental Period, rent shall continue to be payable during the Rental Period at a rate equal to the highest rent previously payable hereunder until the first day for payment of rent after the revised rent has been ascertained or until the expiration of that Rental Period (whichever shall first happen) and on the first day for payment of rent after the revised rent has been ascertained there shall be payable by way of rent (in addition to the amount of the rent otherwise due on that day) the aggregate G of the amounts by which the instalments of rent payable in respect of that Rental Period fell short of the amounts which would have been payable if the revised rent had been ascertained before the first day for payment of rent for the relevant Rental Period and in addition that the tenant shall pay to the Landlord interest on such sum calculated at the Base Rate from time to time of National Westminster Bank PLC calculated on a daily basis from the relevant Review Date until the date of payment.

There are two comments which I would at this stage make on the contents of para 2. One is that the drafting is deficient. A minor point is that the reference to interest to be paid from the relevant review date until the date of payment must be an error. Interest would need to be H paid from the date on which the revised rent would, if it had been ascertained in time, have been paid until the date of payment. Bearing in mind that each successive rental period is a period of five years, and that each year's rent is paid in quarterly instalments on the usual quarter days, it could not possibly be right for the revised rent when ascertained to bear interest from the relevant review date which would be at the beginning of the five-year period. That is the first respect in which the contents of this paragraph appear to have been defectively drafted.

The other respect is in the failure of the paragraph to provide for what should happen if the revised rent, when ascertained, should turn J out to be less than the rent that had previously been paid. Up until the time the revised rent was ascertained, the tenant would have been continuing to pay quarterly rent on the old basis. The amount of the new rent will have been based upon the best yearly rent reasonably obtainable for the premises as at the commencement of the relevant rental period and it may be that that will be less than the rent previously payable. In the heady days when this clause was first drafted, and perhaps still in 1988 when this lease was granted, it may have occurred to nobody, or at any rate not to those who were concerned with this property, that there might come a time when rental values of property such as this would fall. But it is the case that from, I am told, about 1990 commercial rents, at least for properties such as K the property comprised in this lease, did fall. The draftsman of these rent review provisions did not take that possibility into account and, accordingly, there is no provision in para 2 for the repayment by the landlord to the tenant of the excess amount of rent that will have been paid if a revised rent is fixed for the review date at a lower figure than the rent that had previously been paid.

In the course of the hearing below, junior counsel for the landlord conceded that an appropriate implied term providing for repayment in the eventuality that I have described needed to be added to para 2. I have no doubt that that concession was rightly made.

L Mr Lewison conceded also, and this concession too was, in my judgment, unquestionably correct, that in the context of the lease construed as a whole, it is clear that the revised rent may be an upward revision or a downward revision of the previous rent. It may be that in the euphoria of the 1988 property market nobody thought the time might come when rent would have to be revised down in order to accord with best yearly rents at the time; but none the less as a matter of construction the rent to be fixed for the new rental periods as they arrive is clearly to be the best yearly rent whatever that may be and whether above or below the current rent.

Those are the rent review provisions. But I should refer also to one M other clause of this lease, clause 5. I refer to it because it bears upon some of the arguments that have been addressed to us regarding the correct construction of the rent review provisions. Clause 5 of the lease is headed "Rent Cesser" and deals with what is to happen regarding payment of rent during the period when the premises are unfit for occupation or use having been destroyed or damaged by some insured risk. The clause provides that:

. . . the rents hereby reserved or a fair proportion thereof according to the nature and extent of the damage sustained shall be suspended until the Demised Premises shall have again been rendered fit for occupation or use by the Tenant or until the expiration of three years from the date of the damage or destruction whichever shall be earlier and in any dispute shall be referred to the award of a single arbitrator to be appointed in default of agreement upon the application of either party to the President for the time being of the Royal Institution of Chartered Surveyors in accordance with the provisions of the Arbitration Act 1980 or any statutory modifications thereof.

Attention has been drawn to the clear difference between this provision in clause 5 and the rent review provisions. The clause 5 provision contemplates an arbitration in accordance with the provisions of the Arbitration Act, whereas the rent review provisions contemplate a specialist valuer acting as an expert not as an arbitrator in fixing the revised rent of the premises. But, none the less, there clause 5 provides expressly that the application for the president of the RICS to appoint an arbitrator may be an application made by either party, whereas the rent review provisions provide that the appointment by the president of the RICS of a specialist valuer is to be "upon application of the landlord".

In the present case correspondence took place between the tenant and the landlord as to the ascertaining of a revised rent for the rent period commencing December 25 1992. It appears from the evidence that the tactic of the landlord was simply to leave letters on this subject from the tenant unanswered. In her affidavit sworn on May 25 1994 on behalf of the tenant, Nicola Jane Seagar said that the chartered surveyors instructed by the plaintiff had put forward the names of three proposed valuers with a view to a valuer being agreed pursuant to the provisions of para 1 of the fourth schedule to the underlease.

She went on:

The defendants have not responded in open correspondence. The valuer has thus not been agreed. The plaintiff's surveyors have then requested the defendants to apply to the President of the Royal Institution of Chartered Surveyors to make a nomination. There has been no response in open correspondence. The defendants have not applied to the President to make a nomination.

She then said:

It seems that the Defendants' position is that it is a matter for them whether or not to apply to the President for a nomination, or, in other words, it is a matter for them whether or not any review takes place.

That indeed is the crux of the matter. It is the landlord's contention, put very cogently by Mr Kim Lewison QC on their behalf, that on its true construction para 1 of the fourth schedule places the question of whether there will or will not be a rent review for any particular review period at the option of the landlord. It is a fairly common feature of rent review provisions in leases that the landlord is given the option whether or not to invoke the rent review machinery. This is commonly done by providing that the rent review machinery may be invoked by a notice in writing served by the landlord not later than some specified date. Clauses of that character have given rise to a number of cases and gave rise, in particular, to the leading case, *United Scientific Holdings Ltd* v *Burnley Borough Council* [1978] AC 904, in which the House of Lords ruled that the time limitations in rent review clauses were normally not to be treated as of the essence of the agreement so that a failure by the landlord to comply with the requisite time-limits did not necessarily preclude the service out of time of an effective notice invoking the rent review machinery. But in those cases the lessor had expressly been given the option whether or not to serve a notice invoking the rent review machinery. The present lease contains no such express option. Indeed, in my view, the implication to be gained from the lease as a whole, in particular the *reddendum* which I have read and paras 1 and 2 of the fourth schedule, is that there will be a rent review for each of the rental periods. It will be recalled that para 2 provides that if the revised rent in respect of a new rental period has not been ascertained by the first day for payment of rent during that rental period, rent shall continue to be paid during that rental period at a rate equal to the highest rent payable previously "until the first day for payment of rent after the revised rent has been ascertained or until the expiration of that Rental Period (whichever shall first happen)". The paragraph then continues:

. . . and on the first day for payment of rent after the revised rent has been ascertained there shall be payment by way of rent [of the additional sums necessary to bring the previous payments up to the requisite level].

That language, to my mind, shows that the parties contemplated that for each new rental period there would be a rent review and a revised rent.

The contention that the landlord has an option whether or not to invoke the rent review machinery is based on the provision in para 1 of the fourth schedule that the nomination by the president of the requisite specialist valuer shall be "upon the application of the landlord".

Evans-Lombe J took the view that that part of the rent revision provisions was no more than machinery and was not intended to vest in the landlord an option as to whether there should or should not be a rent review. Mr Lewison has argued that that is wrong and that that provision does enable the landlord, by declining to apply to the president for the nomination of the specialist valuer, to prevent a rent review from being carried out.

In my judgment, the issue depends upon whether construing the lease as a whole, the conclusion is justified that the landlord was intended to have that option. If the landlord was intended to have that option, the landlord was entitled to exercise it and to decide whether or not there should not be a rent review. But if the judge below was right in concluding that the provision in question was no more than mere machinery for the carrying out of rent reviews which were intended to happen in any event, then, on authority, there is no reason why the landlord's failure to make the application should be allowed to frustrate the contractual intention discerned from the lease as a whole. The court will in that event if necessary supply machinery to prevent that frustrating refusal from achieving its purpose.

In his judgment Evans-Lombe J ([1995] 2 EGLR 87, at p89H) concluded that:

. . . the provisions of para 1 of the fourth schedule are to be construed as placing an enforceable requirement on the defendants as landlords to apply to the president in the event that the plaintiff has sought a review of rent and the parties have failed to agree the level of such rent or the identity of a valuer to determine that level in default of agreement.

I am not sure that I would support that reasoning. I agree with the judge that the provision relating to the nomination by the president being made upon the application of the landlord is a matter of machinery and not a matter of substance. But it does not follow that the provision should be read as placing an enforceable contractual requirement on the landlord to apply. I would prefer, for my part, the alternative ground on which Evans-Lombe J placed his conclusion, namely that if the machinery provided by para 1 should break down the court would supplement it.

Authority that that can and should be done is to be found in the judgment of the House of Lords in *Sudbrook Trading Estate Ltd* v *Eggleton* [1983] 1 AC 444* which indeed was the authority relied upon by the judge. The case did not involve a rent review clause in a lease. It involved an option to purchase contained in a lease. The option was exercised by the lessee. The option provision in the lease provided that the price at which the property would, if the option were exercised, be sold, would be "such a price not being less than £12,000 as may be agreed upon by two valuers, one to be nominated by the lessor and the other by the lessee and in default of such agreement by umpire appointed by the valuers". The problem in the case was that the lessor had refused to appoint a valuer and so there were no two valuers either to settle the price or to appoint the umpire. The House of Lords held that the case was one in which the agreement was for a sale of the premises at a fair and reasonable price to be ascertained by the application of the objective standards that would have been applied by the valuers and that the provisions relating to the appointment of the valuers was no more than machinery for the ascertaining of that fair and reasonable price so that if the machinery broke down for any

*Editor's note: Also reported at [1983] 1 EGLR 47

A reason the court would substitute its own machinery in order to enable the fair and reasonable price to be ascertained.

Mr Lewison distinguishes that case on the ground that there, by reason of the exercise of the option, there was an enforceable contract for sale and purchase. But, on my construction of the rent review provisions in the present case, there is a contract between lessor and lessee for a revised rent to be ascertained in accordance with the best yearly rent formula to be found in para 1 and to be substituted for the previous rent. The contents of para 1 regarding the appointment of valuers and the application for the president to nominate a valuer is, in my view, as much machinery as was the corresponding machinery
B referred to in the *Sudbrook Trading Estate Ltd* v *Eggleton* case.

In my judgment, the judge below came to the correct conclusion in regarding the case as one in which the landlord could not choose to frustrate the rent review provisions and in which the court would, if necessary, provide the requisite machinery. Whether, if I had been deciding the case at trial I would have chosen the route adopted by the judge, namely imposing a mandatory obligation on the landlord to make the requisite application to the president, I am not sure. That was the route chosen by the judge and perhaps it followed from his conclusion that an implied term with mandatory effect could be read into the rent review provisions. The mandatory order has been
C complied with by the landlord. In obedience to the order the landlord has applied to the president for the appointment of a specialist valuer. The specialist valuer has fixed a revised rent on the best yearly rent formula. The revised rent is, by some considerable sum, lower than the previous rent. In the circumstances, therefore, it is a matter of no materiality whether the judge was correct in choosing the mandatory obligation route as the appropriate form of relief or whether he should have devised some form of machinery to be substituted for the machinery in para 1 of the fourth schedule that the landlord had refused to operate. It may make a difference when subsequent rent periods arrive. But the decision in this case that the landlord does not
D have an option as to whether the rent review procedure is to be invoked or not will, I imagine, prevent any silly dispute arising in the future as to what should be done. I think the judge came to the correct conclusion for the reasons I have given and I would dismiss this appeal.

SAVILLE and POTTER LJJ agreed and did not add anything.

Appeal dismissed.

E

Ashworth Frazer Ltd *v* Gloucester City Council

COURT OF APPEAL

January 20 1997

McCOWAN, MILLETT and AULD LJJ

F Estates Gazette June 28 1997

[1997] 26 EG 150

Landlord and tenant — Rent review — Rent geared to 8% of rack-rents receivable from sublessees — Whether any increase where premises not sublet

By a development lease dated April 28 1969 the appellant city council grant a term of an old cattle market for 114 years. The site covered 14½ acres. The tenant covenanted to develop the site. A peppercorn rent was reserved for the first year, a rent of £9,425 from April 1970 to December 1989, and thereafter a rent to be reviewed each 21 years until the last review. The *reddendum* provided that the G rent at review is to be a fair market rent and by clause 4 of the lease the reviewed rent is to be the higher of £9,425 or a rent geared to 8% of "the rack rents receivable by the Lessees in respect of the demised premises on the relevant date". The tenant did not sublet the site, but disposed of parts of the site by six successive assignments to a variety of tenants each holding directly from the city council, the lease having been severed and the rent apportioned. The respondent held one such part of the site at an apportioned rent of £702. In the court below the judge held that as no part of the site was sublet, there were no rack-rents receivable and therefore no figure to which the 8% could attach; the rent remained the passing rent of £702. The city council appealed. H

***Held*: The appeal was allowed.**

***Per* McCowan and Auld LJJ: the scheme of the rent review provisions was to identify a fair ground market rent calculated by reference to 8% of a fair market rent of developed property if sublet. The use of the word "receivable" was in the sense of capable of being received if sublet. The fact that the property is not sublet is irrelevant.**

Millett LJ *dissenting:* Clause 4 requires the rent to be ascertained by a simple process of aggregating the actual rents receivable from the subtenants. That mechanism could not work because the lease did not impose an obligation on the tenant to use J its best endeavours to sublet.

The following cases are referred to in this report.

Basingstoke and Deane Borough Council v *Host Group Ltd* [1988] 1 WLR 348; [1988] 1 All ER 824; (1987) 56 P&CR 31; [1987] 2 EGLR 147; 284 EG 1587, CA

Compton Group Ltd v *Estates Gazette Ltd* (1977) 36 P&CR 148; [1977] 2 EGLR 73; [1977] EGD 237; 244 EG 799, CA

Equity & Law Life Assurance Society plc v *Bodfield Ltd* (1987) 54 P&CR 290; [1987] 1 EGLR 124; 281 EG 1448, CA; affirming [1985] 2 EGLR 144; (1985) 276 EG 1157

Philips Electronic Grant Public SA v *British Sky Broadcasting* [1995] EMLR 472 K

Prenn v *Simmonds* [1971] 1 WLR 1381; [1971] 3 All ER 237, HL

R&A Millett (Shops) Ltd v *Leon Allan International Fashions Ltd* [1989] 1 EGLR 138; [1989] 18 EG 107, CA

Reardon Smith Line Ltd v *Yngvar Hansen-Tanger* [1976] 1 WLR 989; [1976] 3 All ER 570; [1976] 2 Lloyd's Rep 621, HL

Sudbrook Trading Estate Ltd v *Eggleton* [1983] 1 AC 444; [1982] 3 WLR 315; [1982] 3 All ER 1; (1982) 44 P&CR 153; [1983] 1 EGLR 47; [1983] EGD 392; 265 EG 215, HL

This was an appeal by the defendants, Gloucester City Council, from a decision of Judge Roger Cooke QC, sitting as a judge of the L Chancery Division, in proceedings by the plaintiff, Ashworth Frazer Ltd, as to the meaning and effect of a lease.

Nigel Davis QC and Stephen Bate (instructed by the solicitor to Gloucester City Council) appeared for the appellants; Paul Morgan QC and Phillip Rainey (instructed by Rowe & Maw) represented the respondent.

Giving the first judgment at the invitation of McCowan LJ, **MILLETT LJ** said: This is an appeal by Gloucester City Council from the judgment of Judge Roger Cooke sitting as a judge of the High Court on February 3 1996. It raises a short but not altogether easy question of M construction of a few words in a badly drafted rent review provision in a lease. The clause requires the rent to be calculated by reference to "the rack rents receivable by the Lessee in respect of the demised premises". No express provision is made for voids or other cases where for any reason all or part of the demised premises is not sublet at a rack-rent. In the events which have happened part of the premises is in occupation of the respondent as lessee. The question is whether "the rack rents receivable by the Lessee" means, as the respondent contends and the judge found, "the rents which the lessee is entitled to receive from actual subtenants" or, as the appellant lessors contend, "the rents which the lessee would be entitled to receive if the premises were fully sublet at full open market rents".

A The lease is dated April 28 1969. It is a development lease made between (1) the appellant as lessors and (2) MacKenzie Hill Ltd ("the developer") as lessee. By the lease the developer was granted a lease of the site of the old cattle market in Gloucester for a term of 114 years from December 25 1969. The site covered $14^{1}/_{2}$ acres. No premium was payable on the grant of the lease and no rent was payable for the first 12 months of the term. The developer covenanted to develop the site in accordance with agreed plans and specifications for light industrial use and so as to be fit for immediate occupation. It was evidently contemplated that the developer would sublet the completed development to a variety of industrial subtenants and would then B dispose of the leasehold reversion, probably to an institution.

The *reddendum* reserved a peppercorn rent until April 28 1970, followed by a yearly rent of £9,425 from April 28 1970 to December 25 1989. Thereafter, the *reddendum* provided as follows:

> and thereafter such sum as may be agreed or determined as the fair market rent in accordance with the provisions of Clause 4 hereof . . .

Clause 4 of the lease reads as follows:

> It is mutually agreed that at the Twenty Fifth day of December One Thousand Nine Hundred and Eighty Nine and thereafter at the end of every Twenty-One C year period (hereinafter called "the relevant renewal date") the yearly rent payable by the Lessee shall be reviewed in manner following:
>
> (a) the yearly rent payable by the Lessee in respect of the said periods of Twenty One years and the last Nine years of the said term shall be either the sum of Nine Thousand Four Hundred and Twenty Five Pounds aforesaid or the rent payable during the previous rent period or such sum as shall be equivalent to Eight per cent of the rack rents receivable by the Lessees in respect of the demised premises on the relevant renewal date such figure of Eight per cent to be certified by the Lessee's accountants whichever figure shall be the higher.

I have read clause 4(a) as the parties agree it should be read. In fact, the word "on" in the expression "on the relevant renewal date" towards D the end of the clause appears as "or". But it is plain that that is a typographical error.

Thus the council's reward for contributing the undeveloped site was to consist not of a premium or ground rent (except during an initial period), but a share, fixed at 8%, of "the rack rents receivable by the Lessee" ascertained at 21-year intervals.

The lease contained a covenant on the part of the lessee against assigning the whole or any part of the premises without the lessors' consent, such consent not to be unreasonably withheld. It contained no prohibition on subletting the whole or any part of the demised E premises without the lessors' consent. More surprisingly (if the construction placed on clause 4 by the judge is correct) it contained no covenant by the lessee to use its best endeavours to sublet the various parts of the premises at the best rents reasonably obtainable without taking a premium and with appropriate provisions for rent review. The lease contained a full arbitration clause.

As expected, the developer completed the development of the site and disposed of the premises in parcels to various industrial occupiers. It did not do so, however, by subletting, but by successive assignments each made with the consent of the appellants. There have altogether been six severances of the lease, and the entirety of the demised F premises is now let to a variety of tenants each holding directly from the appellants under the terms of the lease, the rent having been apportioned. The respondent is the successor in title to Kentron Plastics Ltd, which took an assignment of part of the demised premises from the developer in September 1973. The rent payable in respect of the part of the premises thereby assigned for the period to December 25 1989 was apportioned by the assignment at £702, and it was agreed that the covenants and conditions in the lease were to be severed as if the lease had contained a demise at apportioned rent of the land comprised in the assignment. The assignment was at a premium of £21,600, which shows that even as early as 1973 the apportioned rent of £702 was already well below the open market rent for the part of the demised premises assigned to Kentron Ltd.

At the dates of the most recent rent review, the whole of the G premises let to the respondent was occupied by the respondent for the purposes of its own business. No part thereof being sublet, the judge held that there were no "rack rents receivable by [the respondent] in respect of the demised premises", and so nothing to which the 8% figure mentioned in clause 4 could attach. Accordingly, this being an upwards-only rent review clause, he held that the yearly rent payable by the respondent during the current period was the current passing rent of £702. The appellants submit that the judge's construction of the lease was wrong; alternatively they contend for an implied term which would have the effect of substituting the current open market value of the premises for the passing rent; in the further alternative, they submit H that the machinery for determining the open market rent of the property having broken down, the court should substitute other machinery.

Construction

Preliminary

Before dealing with the detailed arguments on the proper construction of the lease, it is convenient to dispose of some preliminary matters. In the first place, it is common ground that the wording of the assignment to Kentron Plastics Ltd does not affect the question now in dispute, and that accordingly that question must be J determined by construing the lease as at the date in 1969 when it was entered into and without reference to the later severances. The manner in which the premises demised by the lease were in fact disposed of is unlikely to have been foreseen and — while the parties ought perhaps to have foreseen the possibility of voids, for example, occurring at the review date — the possibility of the entirety of the premises being void, so that there would be no rents receivable from subtenants, could reasonably have been dismissed as too remote. Indeed, there is no suggestion that this has happened or that, if the judge's decision in the instant case is upheld, the appellants will be confined to the total passing yearly rent of £9,425. K

The appellants place great stress on what they claim is the commercial purpose of a rent review clause. This, they rightly observe, is to protect the landlord from the devaluation of the rent in real terms by reason of inflation or a rise in rental values since the date on which the lease was granted. As Dillon LJ said in *Equity & Law Life Assurance Society plc* v *Bodfield Ltd* [1987] 1 EGLR 124, at p125B:

> There is no doubt that the general object of the rent review clause, which provides that the rent cannot be reduced on a review, is to provide the landlord with some measure of relief where, by increases in property values . . . records the real value of money in an inflationary period, a fixed rent has become out of date and unduly favourable to the tenant. The exact measure of relief L depends on the true construction of the particular rent review clause.

There are, however, several different techniques for achieving this. The most common is to provide for a periodic revision of the rent by reference to an expert revaluation in order to ascertain the open market rent or the fair rent of the demised premises (these concepts are not the same, and the expression "fair market rent" is a confused amalgam of two contradictory ideas). But this is not the only technique available. It may be reviewed by reference to the Retail Price Index, or to the tenant's turnover, or to the subrents which the tenant is entitled to receive from subtenants. Ordinarily, it is obvious which technique has been adopted, and if there is any difficulty it arises in connection with M the application of the chosen technique to the facts of the particular case. Unusually, the question in the present case is whether, as the *reddendum* indicates, the technique which was adopted was that of an expert valuation of "the fair market rent" or, as clause 4 indicates, that of a mathematical computation based on the aggregate of the underlying subrents.

The appellants stress the commercial purpose of a rent review clause as the foundation for their contention that this is frustrated by the construction placed upon the clause by the judge. The result, it is submitted, is surprising and not one likely to have been intended by the parties. Unless compelled to do so by clear and unambiguous language, the court should not adopt a construction which would have this result.

The respondent counters this argument by submitting that clause 4 is not a rent review clause at all, but a rent-sharing clause. I cannot accept this as a complete description of the effect of the clause, even on the construction for which the respondent contends. A true rent-sharing clause would require continuous, or at least annual, adjustment, certainly not adjustment at 21-year intervals. And the concept of rent-sharing does not sit entirely comfortably with upwards-only revision. At the same time, it is quite clear that the appellant was not to receive the full rental value of the premises, however ascertained, but only 8% thereof. On any view, it seems to me, the lease combines elements of both rent review and rent-sharing. The question is whether, as the appellant contends, the parties are to share "the fair market rent" (whatever that means) ascertained by an expert or, as the judge found and the respondent contends, the actual subrents, if any, certified by the lessee's accountants.

The reddendum

There is an issue between the parties as to the true relationship between the *reddendum* and clause 4. The appellants submit that the *reddendum* is the controlling provision. It reserves "the fair market rent", and directs attention to clause 4 only for the purpose of enabling "the fair market rent" to be determined in default of agreement. Clause 4 should, if at all possible, be construed so as to give effect to this direction.

Like the judge, I have no doubt that, taken in isolation, this is the more natural way to read the *reddendum*. The reader is given to expect that clause 4 will contain appropriate provisions for the assessment by an expert valuer of the full "fair market" rental value (whatever that may mean) of the demised premises and that the rent reserved by the *reddendum* is the amount so determined. But on any footing the *reddendum* is inaccurately drafted, and not just because of the use of the expression "fair market rent". Even on the construction of clause 4 contended for by the appellants, the lease does not entitle the lessors to the "fair market rent" of the demised premises as the *reddendum* indicates, but only to 8% thereof.

The *reddendum* reserves, not "the fair market rent" simpliciter, but "the fair market rent in accordance with the provisions of clause 4 hereof". Moreover, on any view, clause 4 does not provide machinery for determining "the fair market rent", but a definition of the rent reserved. When, in addition, it is found that the figure of 8% appears in clause 4 only and not in the *reddendum*, I find it impossible to accept the appellant's submission that the *reddendum* is the controlling provision. In my opinion, the *reddendum* must be read with clause 4 and the two provisions reconciled if possible. But it does not control the meaning of clause 4, and if the result of performing the exercise directed by clause 4 does not produce 8% of what could properly be described as either a fair or a market rent, then, in my judgment, the parties must be taken to have made their own dictionary when using the expression "fair market rent" in the *reddendum*. If they used that expression, as the respondent contends, to mean 8% of the rents receivable from such tenants, that is not an inappropriate description of what the parties had agreed to be a fair rent as between the lessee, on the one hand, and the lessors, on the other, in respect of the demised premises.

Clause 4

If the *reddendum* leads a reader to expect that clause 4 will contain machinery for the ascertainment by an expert valuer of either the fair or market rental value of the demised premises, he will be disappointed. The clause contains no reference to the rental value of the premises and no provision for its determination by expert valuation. Moreover, as I have already pointed out, it does not provide machinery for the determination of the rent but a self-contained definition of the rent payable in each rental period.

Rents receivable

The appellant submits that the expression "rents receivable" is ambiguous. It could mean either (i) the rents which the lessee is entitled to receive from subtenants under existing subleases or (ii) the rents which the lessee would be entitled to receive from subtenants if the demised premises were fully sublet. I agree that, given an appropriate context, the words could bear either meaning; the use of the word "receivable" rather than "received" denotes an intention to avoid the need to satisfy some unexpressed contingency. But I have no doubt that the former meaning is the ordinary and natural meaning of the words, and that it would require a special context to give them the meaning for which the appellant contends. Rent due from a tenant but not paid is a receivable of the landlord; rent which the landlord could obtain only by letting unlet property is not. The contingency to which the word "receivable" is ordinarily directed to avoid is non-receipt, not non-entitlement.

At various times in the argument, the appellants submitted that "receivable" means capable of being received. That may be so. But that expression itself involves the same ambiguity. It may mean rents which the lessor is capable of receiving under existing subleases or rent which he is capable of receiving by subletting unlet premises.

The appellants have placed much reliance on the decision of the court in *R&A Millett (Shops) Ltd* v *Leon Allan International Fashions Ltd* [1989] 1 EGLR 138, but, in my judgment, that case does not help it. The rent review clause there provided that:

> the rent payable by the Lessee shall be seventy-eight eighty fifth parts of the rent payable by the Lessor . . . as fixed in manner provided by the said Superior Lease . . .

The superior lease referred to was surrendered by the lessor and replaced by a new headlease. The trial judge held that the underlessor was still entitled to a rent review, but, the machinery for the ascertainment of the new rent having broken down, the court could substitute new machinery, adopting the approach of the House of Lords in *Sudbrook Trading Estates Ltd* v *Eggleton* [1983] 1 AC 444*. His decision was upheld by a majority of this court. Counsel for the underlessor conceded that he could not succeed by relying on the *Sudbrook* decision unless he could show that the rent review clause, on its true construction, required the underlessee to pay the appropriate portion of a fair market rent as determined in accordance with the original headlease; but that, the majority held, was exactly what the rent review clause provided.

The appellants submit that the court must implicitly have held that the words "payable by the lessor" in the clause there under consideration meant "which would have been payable under the headlease if it had not been surrendered"; and that it follows that the words "receivable by the lessee" are capable of meaning "which would have been receivable if the premises were sublet". I do not agree. Once the conclusion was reached that the words in dispute merely had the effect of incorporating the rent review clause in the headlease by reference, it necessarily followed that the continued subsistence of the headlease was not required. The words "payable by the lessor" did not, therefore, mean "which the lessor was still contractually liable to pay under the headlease" but "expressed to be payable by the lessor" or "reserved by the headlease". I am not myself convinced that in these circumstances it was necessary to invoke the doctrine enunciated in *Sudbrook Trading Estates* v *Eggleton* for I cannot see that the machinery for ascertaining the rent had broken down at all.

Rack-rents

The appellants, however, place their strongest reliance on the words "rack-rents". The term "rack-rent" is defined in *Blackstone's Commentaries* 4th ed vol 2 at p43 as "a rent of the full value of the tenement or near it". In *Compton Group Ltd* v *Estates Gazettes Ltd* (1977) 36 P&CR 148† the court had to decide whether the words

---

*Editor's note: Also reported at [1983] 1 EGLR 47

†Editor's note: Also reported at [1977] 2 EGLR 73

A "rack-rent" meant the rent which represents the full annual value of the holding or the maximum rent which is permitted by law. Sir John Pennycuick said, after referring to the definition of *Blackstone's Commentaries*, (p154):

> ... in a judgment of Holmes LJ in *Ex parte Connolly to Sheridan and Russell*, it is defined thus: "a rack rent in legal language means a rent that represents the full annual value of the holding".
>
> Sargent J [in *Re C R Sawyer and Withal*, solicitors] in terms approved of Holmes LJ's definition.
>
> However, the latter meaning has been adopted by the courts, including the Court of Appeal, in certain recent cases where the premises were subject to
> B statutory control; see in particular *Rawlence* v *Croydon Corporation* [1952] 2 QB 803 ...

Sir John Pennycuick went on to hold that the word "payable" could mean either "reserved" or "recoverable". The case is of limited assistance to the appellant, since, as Browne LJ pointed out (at p159), the expression "rack-rent" is not a phrase which has a precise meaning which it always bears, but rather a meaning which varies according to the context. In the present case the respondent submits that it is used to exclude insurance rents and service charges reserved as rent. I agree that this is a possible meaning, though not the primary meaning.

C *Method of ascertainment*

If the clause had ended with the words "on the relevant renewal date" it would, I think, have been a nice question whether it required a professional valuation of the open market or rack-rent of the demised premises or merely an arithmetical calculation of the aggregate rents at which they were sublet. The *reddendum* and the use of the words "rack-rents", as well as the absence of any provision for voids or own-occupation, would have pointed in one direction; the words "receivable by the lessee", the absence of any provision for valuation or of any directions to the valuer as to the basis of valuation (eg should the rental value of the demised premises as a whole be taken, or the aggregate of
D the rental values of the separate parts?), and the fact that the lessor is to receive only 8% of the resultant figure (which may provide a reason for seeking to avoid the expense of a professional valuation) in the other.

But, in my judgment, the matter is put beyond doubt by the concluding words of the clause. The figure of 8% is to be certified by the lessee's accountants. On the construction contended for by the respondent, this makes good sense. The accountant's function is to examine the books and records of the lessee and ascertain what rents are currently receivable from subtenants. They then certify the figure which represents 8% of the total. Their intervention is necessary to
E protect the lessor from the suppression by the lessee of a recent sublease or increase in subrent. On the construction of the clause contended for by the appellants, however, they serve no useful function. They cannot certify that any rent due from a subtenant represents the full rental value of the premises, unless perhaps the sublease is recent or the lessor is demanding an increase in the subrent at a pending rent review. The appellants' construction of clause 4 makes a professional valuation, for which the clause does not in terms provide, essential, and limits the function of the accountants, for which the clause makes express provision, to confirming the correctness of a simple arithmetical calculation.

F In my judgment, the judge was right to hold that clause 4 requires the rent to be ascertained by a simple process of aggregating the actual rents receivable by the lessor from subtenants, subject to the confirmation of the accountants' certificate and does not require any process of professional valuation. It follows that the use of the words "fair market rent" in the *reddendum* may not be an altogether accurate description of the result of the process required by clause 4, but rather a shorthand label chosen by the parties and sufficient for their purposes of what was provided in detail by clause 4.

The appellants submit that this result is clearly uncommercial and cannot have been what the parties intended; and that this fact alone requires the court to re-examine its decision. I do not agree. If the position is analysed, it can be seen that the lessor will fail to recover 8% of the full open market rent of a part of the demised premises in G
four situations: (1) where there is a void; (2) where the relevant part of the demised premises is occupied by the lessee itself; (3) where the sublease was granted at a premium so that the passing rent is not a rack-rent; and (4) where the sublease was granted at a rack-rent, but changes in property values since mean that the passing rent is below (or even above) the current market rent.

Taking these four cases in reverse order, it is, in my judgment, past belief that the parties would have contemplated substituting the market rent for the passing subrents in case (4). Quite apart from the fact that the lessor insisted on an upwards only rent review, there is no reason why the lessors should have been prepared to accept the substitution of H
a rent lower than that which the lessors were entitled to receive (and probably actually receiving). But what goes for changes in market value in one direction applies equally to changes in the other. There is no reason why the lessee should accept a liability to pay the lessors a sum greater than 8% for rents which it was in a position to receive. I agree that case (3) produces a nonsense, but what has gone wrong is not the rent review provisions, but the provisions with regard to subletting. Once the parties opted for rent reviews based on subrents, it was necessary to place the lessors under a positive obligation to use their best endeavours to sublet the premises at the best rents reasonably obtainable without taking a premium. This was not done. It would be J
quite wrong, in my opinion, to distort the meaning of clause 4 to provide a remedy for any failure in the drafting elsewhere. Case (2) (which is the case which has occurred) cannot have been in the reasonable contemplation of the parties. The developer plainly had no use itself for the premises. What actually happened was clearly not foreseen. Case (1) — and case (2) if foreseen — should no doubt have been provided for, but there were several ways of dealing with the problem. Given that the lessor's interest is only 8%, it may well have been thought inappropriate to incur the expense of a professional valuation merely to deal with a void in one of several parts of the premises. The problem could have been adequately dealt with, for K
example, by averaging.

This analysis shows, in my opinion, that the lease is defective, but that what has gone wrong is only partly due to a *casus omissus* in the rent review provisions. It does not throw doubt on the construction of clause 4 for which the respondents contend, that is to say, that the review is to be conducted by aggregating the subrents and not by valuing the premises.

Implied term

If clause 4 of the lease is not to be construed in the manner contended for by the appellant, then the appellant submits that an appropriate term should be implied. There is, it submits, a gap which it L
is necessary to fill in order to give commercial efficacy to the lease.

The appellant's first choice is for a term to be implied which would have the effect of substituting "the fair market rent" of the demised premises for the aggregate of the subrents provided for by clause 4. This is plainly impossible. Where the relevant part of the demised premises is sublet at the review date, that is to say, in cases (3) and (4), there is no gap to be filled, and the suggested implied term would be inconsistent with the express terms of the lease. And in case (4), as I have already indicated, it is most unlikely that the parties would have agreed to the substitution of the basis of assessment which is proposed. M

With somewhat greater realism, the appellant contends in the alternative for the implication of a term which would require the "fair market rent" to be adopted in any case where there was no existing subtenancy at the review date, that is to say, cases (1) and (2). A persuasive case for this can be made out, but in the end I am not persuaded. The lessee was entitled to retain 92% of any subrents, and this was a sufficient incentive to deter it from deliberately creating any voids. It is not at all obvious that it would have agreed to bear the whole (as distinct from 92%) of the loss due to difficulty in finding a tenant. There is more force in the submission that a reasonable lessee would have agreed that the rent should take account of any part of the premises in its own occupation. The difficulty here, however, is that it is not at all clear what provision they would have agreed. It is far from obvious that they would have agreed to incur the expense of a professional valuation in order to provide the lessor with 8% of the assessed value.

It is not enough to show that, had the parties foreseen the eventuality which in fact occurred, they would have wished to make provision for it, unless it can also be shown either that there was only one contractual solution or that one of several solutions would without doubt have been preferred: see *Philips Electronic Grant Public SA* v *British Sky Broadcasting* [1995] EMLR 472 at p482 *per* Sir Thomas Bingham MR. In my judgment, even the more limited implied term contended for by the appellant fails this test.

Substitution of new machinery

All else failing, the appellants invoke the doctrine enunciated in *Sudbrook Trading Estate* v *Eggleton*. On the construction of clause 4 which I consider to be correct, clause 4 is not subordinate machinery for the determination of "the fair market rent" of the property, but the essential definition of the rent reserved by the lease. The only part of clause 4 which could properly be described as subordinate machinery is the provision for the figure to be confirmed by an accountants' certificate. If they refused to co-operate, the court could substitute an alternative means of ascertaining the amount of the subrents receivable by the lessee at the review date, for example, by ordering an inquiry. But the rest of clause 4 is essential; the court could not provide a substitute without rewriting the bargain.

But in any case the machinery established by clause 4 has not broken down. The fact that it yields a result which is unsatisfactory to one party does not mean that it has broken down, but only that it was ill-chosen.

Conclusion

I would dismiss the appeal.

Disagreeing, **AULD LJ** said: I differ from Millett LJ. I do so with great diffidence, particularly in a matter such as this.

The point is one of construction of a rent review clause in a development lease of 114 years entered into in 1969. The lease clearly contemplated that the lessee would, after developing the demised property, sublease it. It provided for an upwards-only review of the rent at 21-yearly intervals, upwards, that is, only in a rising property market by reference to the return on any sublease or subleases. I say "the return". The issue before the court is whether, in a rising property market, the rent is to be calculated at the review date by reference to the potential return or to the actual return, if any, at that date. 8% of whichever is the appropriate figure is the rent payable. It is thus a rent review mechanism designed to achieve a constant income sharing ratio between the lessor and lessee, albeit at lengthy review intervals.

As my lord, Millett LJ, has indicated, clauses 1 and 4 are not well drawn. However, in my view, their intent and effect, when read together and with other relevant provisions of the lease, are plain. No implication of a term is necessary to give effect to that meaning. It is that the review rent is to be the fair market (ground) rent calculated by reference to a percentage of the rack-rent, namely the full annual rental value, but that for a notional sublease of the property as developed. Clause 1 provides expressly, and by reference to clause 4, that the rent is to be reviewed and that it is to be "the fair market rent" as agreed or determined in accordance with that clause. The scheme in clause 1 allows for some tinkering with the normal meaning of fair market rent because it permits of its agreement or, in the absence of agreement, subjects it to determination in accordance with whatever constraints clause 4 may impose. Nevertheless, the clause has at its core a familiar notion, that of the fair market rent, a notion which has no function if all that was intended was a reference to rent, if any, actually due and payable under a sublease or subleases at each review date.

Equally, I would not expect there to be a reference in the *reddendum* to "agreement" or "determination" of the review rent if all that is required is an identification of sublease rents, if any, actually due and receivable at each review date.

Turning to clause 4, it identifies, without specifying the machinery for its agreement or determination, "the fair market rent" referred to in clause 1. It does so in the third alternative circumstance by reference to another fair market rent, "the rack-rents receivable by the lessee . . . at the renewal date". In my view, there is no reason to give any other meaning to the term "rack-rent" in this context than that which it has ordinarily, namely a rent representing the full annual valuation of the holding: see *Compton Group Ltd* v *Estates Gazette Ltd* (1977) 36 P&CR 148, CA, *per* Sir John Pennycuick, at p154. In particular, I see no basis for treating the use of the word "rack" as a means for distinguishing the reference to rent receivable under the sublease from a ground rent, since the contemplated sublease or subleases were of developed property, or from an insurance or service charge. Nor can I see any basis for treating it as a reference to rent, if any, actually due and receivable under a sublease at the renewal date. If the latter had been intended there would be no reason to use the word "rack" at all in the provision. It would have been sufficient to refer only to rents due and receivable by the lessee at the review date.

It seems to me that the scheme of clauses 1 and 4, when read together, is to identify for the purpose of rent review a fair ground market rent for the lease calculated by reference to 8% of a fair market rent of developed property if subleased. The use of the word "receivable" in the sense of "capable of being received if sublet" is consistent with that conclusion, though not determinative of it. It is plain that words such as "payable" and "receivable" in leasehold documents and the like may mean "due to be paid and received" or "capable of being paid and received" according to their context: see, for example, *Compton Group Ltd* v *Estates Gazette Ltd per* Sir John Pennycuick at p154 and *per* Browne LJ at p160; and *R&A Millett (Shops) Ltd* v *Leon Allan International Fashions Ltd* [1989] 1 EGLR 138, CA.

As to clause 4's lack of provision of machinery for agreement or determination or other valuation of the rack-rent of a notional sublease, if there is agreement no machinery is necessary. If there is no agreement clause 5(c) provides, in its general provision for arbitration, a means of determination.

As to the provision in clause 4 for the certification by the lessee's accountant of the 8% of the rack-rents receivable by the lessee, on which the judge relied so heavily in favour of the lessee, it is not, in my view, a significant pointer either way. Whether viewed as an ascertainment of the rents due under any existing subleases and the calculation of 8% of them (the lessee's contention) or the calculation of 8% of agreed or determined rack or fair market rents (the lessor's contention), the exercise is essentially a mechanical one. On neither interpretation is it the accountant's function to certify the correctness of the rack-rent for the sublease. His certification can only be of the 8% following agreement or determination of that rack-rent as provided for in clauses 1 and 4.

For those reasons, I am of the view that the review rent under the lease is to be determined by reference, where it produces the highest figure, to the rack or fair market rent which could be obtained under a sublease. In the unlikely event of the property market not having risen significantly over the previous 21-year renewal period, the actual rent or rents payable may equal or be an acceptable indicator for agreement by the parties, or for determination in an arbitration, of the rack-rent. But history has shown that that would not ordinarily be the case. The fact that the property is not in fact sublet is irrelevant, save possibly as an indicator of its marketability if the lessor has attempted to sublet it.

I add that, even by the standard of drafting of this lease, it could not sensibly have been intended or contemplated by the parties to it that, with such long rent review intervals and in the event of a rising property market, the lessor would be confined to a rent calculable by reference to what had been paid in the previous rent review period or to the rental for the first 20 years, if, for whatever reason, the lessee sublet the property at well below the market rate or, as here, the property was not sublet at all at the review date.

Such a construction would, in my view, defeat the whole purpose of an upwards-only rent review clause which is to protect the lessor from property inflation, enabling him, subject to the length of the rent review periods, to recover the same rent in real terms throughout: see *Basingstoke and Deane Borough Council* v *Host Group Ltd* [1988] 1 WLR 348*, CA, *per* Nicholls LJ (as he then was) at pp353D-356C. As the deputy judge observed in his judgment in this case, and Mr Paul Morgan QC, on behalf of the lessee, effectively acknowledged in his submission to this court, the interpretation for which the lessee contended was unfair to the lessor. Its clear lack of commercial good sense reinforces me in my construction in favour of the lessor: see *Prenn* v *Simmonds* [1971] 1 WLR 1381, HL; and *Reardon Smith Line Ltd* v *Yngvar Hansen-Tanger* [1976] 1 WLR 989, HL, re Lord Wilberforce at pp995G-997E. The interpretation for which the lessor contends, on the other hand, does accord with the clear commercial purpose of the rent review provisions in clauses 1 and 4.

Accordingly, I would allow the appeal.

Agreeing that the appeal should be allowed, **McCowan LJ** said: the judge himself recognised in his judgment that his decision would not do substantive justice between the parties. Mr Morgan accepted before this court that it was "a bad bargain for the appellants". Whether in fact it is a bad bargain for the appellants depends on the construction of the lease.

Additional to not doing justice, it seems to me that the judge's construction is not in accordance with the commercial purpose of a rent review clause. Mr Morgan's answer to that was to say that it is not actually a rent review clause at all, but a rent sharing agreement. I cannot accept that. The words of clause 4 are themselves against this.

In my judgment, Mr Davies is clearly right in saying that the judge gave insufficient attention to the words of clause 1 and in particular to the words "the fair market rent". This is shown by his dismissive reference to those words as mere protestations. I noted that Mr Morgan separated himself from the judge on that description.

Again the judge said:

> It is hard to see how a draftsman could have both incorporated the closing words of clause 1 and the words of clause 4 in the same document.

Mr Morgan did not seek to support this view of the judge. Mr Morgan said that the two clauses are perfectly reconcilable. Clause 4, he said, is the measure of clause 1.

As to the words "rack-rents", the judge largely ignored them and treated them as superfluous. Mr Morgan, on the other hand, candidly said that his case would be more "straightforward" if it were not for the presence of the word "rack". Calling a spade a spade, I would say that he was profoundly embarrassed by it, because "rack-rent" means the full annual value of the holding.

In my judgment, the commercial purpose of a rent review provision (viewed objectively) is to enable the landlord to obtain from time to time the market rent which the premises would command if let on the open market on the same terms as at the review date. That object can be achieved in this case by interpreting "receivable" as "capable of being received" which I consider a wholly appropriate interpretation in the circumstances of the case.

By that means it is possible, failing agreement, to determine the yearly rent payable by the lessee pursuant to clause 4. In the event of a dispute there is in clause 5 a provision for arbitration.

I cannot, therefore, accept the judge's construction of the agreement and I would allow the appeal.

Appeal allowed.

---

*Editor's note: Also reported at [1987] 2 EGLR 147

For a further case on this subject see p239

NEGLIGENCE

A

Gardner and another *v* Marsh & Parsons and another

COURT OF APPEAL

November 13 1996

HIRST, PETER GIBSON and PILL LJJ

Estates Gazette April 12 1997

B **[1997] 15 EG 137**

Negligence — Structural survey — Damages — Purchase of long lease of maisonette — Landlord remedying defect at plaintiffs' instigation — Whether damages payable on Philips v Ward principles — Whether plaintiffs suffered any loss

In June 1985 the plaintiffs instructed the defendant firm to make a structural survey of a maisonette. The survey report was negligent in failing to disclose a serious structural fault. Following the report, the plaintiffs acquired a long lease of the maisonette paying a premium of C £114,000 with the assistance of a building society loan of £50,000. The structural defect was discovered in 1988, when the plaintiffs were trying to sell the maisonette; it was remedied by, and at the expense of, the landlord in 1990 following negotiations and the threat of proceedings. In the court below the judge awarded £29,000 damages as the difference between the value of the property without the defects and its value with the defects at the date of the purchase. The defendants appealed against the award of damages on the grounds that: (1) had the defect been discovered in 1985 the landlord would have remedied the defect before purchase; and (2) the landlord did in fact remedy it at the plaintiffs' instigation and therefore they suffered no D loss.

***Held*: The appeal was dismissed.**

***Per Hirst and Pill LJJ*: (1) It could not be inferred that had the defect been discovered at the time of purchase, the landlord would have remedied it before the plaintiffs purchased the maisonette. The plaintiffs did purchase it in its defective state as a result of the defendants' negligence. It is intrinsic to the principles in the *Philips* v *Ward* [1956] 1 WLR 471 line of cases that, in assessing the market price of the property in its defective condition, a hypothetical sale of the property in that state is assumed to have taken place. (2) The repairs executed in 1990 by the landlord were E not part of a continuous transaction of which the purchase of the lease as a result of the negligence was the inception. The repairs were *res inter alios acta* and therefore collateral to the negligence.**

***Per Peter Gibson LJ dissenting* on the second ground of appeal: If the plaintiff in fact avoids or mitigates his loss, he cannot recover for the loss thereby avoided even though the steps he took were more than could reasonably be required of him under the duty to mitigate his loss. When the plaintiffs realised they had purchased a defective property, they took steps to eliminate their loss by requiring the landlord to remedy the defect; that was an act of mitigation which eliminated the plaintiffs' loss: see *British Westinghouse Electric & Manufacturing Co Ltd* v *Underground Electric Railways Co of London Ltd* [1912] AC 673.** F

The following cases are referred to in this report.

Admiralty Commissioners v *SS Chekiang (The Chekiang)* [1926] AC 637
Bradburn v *Great Western Railway* (1874) LR 10 Exch 1
British Transport Commission v *Gourley* [1956] AC 185; [1956] 2 WLR 41; [1955] 3 All ER 796, HL
British Westinghouse Electric & Manufacturing Co Ltd v *Underground Electric Railways Co of London Ltd* [1912] AC 673; (1912) 81 LJKB 1132; 107 LT 325
Daisley v *BS Hall & Co* (1972) 225 EG 1553
Hodge v *Cowling (Clifford) & Co* [1990] 2 EGLR 89; [1990] 46 EG 120, CA G
Hussey v *Eels* [1990] 2 QB 227; [1990] 2 WLR 234; [1990] 1 All ER 449; [1990] 1 EGLR 215; [1990] 19 EG 77, CA
Jones v *Just* (1868) LR 3 QB 197; 9 B&S 141; 37 LJQB 89; 18 LT 208; 16 WR 643
Oswald v *Countrywide Surveyors Ltd* [1996] 2 EGLR 104; [1996] 37 EG 140
Pagnan (R) & Fratelli v *Corbisa Industrial Agropacuaria Limitada* [1970] 1 WLR 1306; [1971] 1 All ER 165; [1970] 2 Lloyd's Rep 14, CA
Parry v *Cleaver* [1970] AC 1; [1969] 2 WLR 821; [1969] 1 All ER 555; [1969] 1 Lloyd's Rep 183, HL
Perry v *Sidney Phillips & Son* [1982] 1 WLR 1297; [1982] 3 All ER 705; [1982] 2 EGLR 135; [1982] EGD 412; (1982) 263 EG 888, CA
Philips v *Ward* [1956] 1 WLR 471: [1956] 1 All ER 874, CA
Redpath v *Belfast & County Down Railway* [1947] NI 167 H
Walton v *Inland Revenue Commissioners* [1996] 1 EGLR 159; [1996] 21 EG 144; [1996] STC 68
Watts v *Morrow* [1991] 1 WLR 1421; [1991] 4 All ER 937; [1991] 2 EGLR 152; [1991] 43 EG 121, CA

This was an appeal by the defendants, Marsh & Parsons and Sean Dyson ARICS, from a decision of Judge Byrt QC in Mayor's and City of London Court on February 27 1995, whereby he had given judgment to the plaintiffs, James Piers Gardner and Penelope Helen Gardner, in their action for damages for negligence against the defendants.

Giving the first judgment, **HIRST LJ** said: This is an appeal against J the decision of Judge Byrt QC in the Mayor's and City of London Court on February 27 1995.

In or about June 1985 the plaintiffs Mr James Piers Gardner and Mrs Penelope Helen Gardner were interested in buying the maisonette on the third and fourth floors of 8 Royal Crescent, London W11, at the price £114,000 subject to a satisfactory survey. They instructed the first defendant, Marsh & Parsons, to carry out a full structural survey of the property, and this was undertaken by the second defendant, Mr Sean Dyson ARICS. Unfortunately Mr Dyson carried out this survey negligently, as the judge held and as is no longer disputed, in that he failed to spot a serious structural defect of which there were tell-tale K signs in the decoration of one of the rooms. The judge assessed the damages under this head at £29,000 plus interest, and it is against this award that the defendants presently appeal. An additional small amount of damages also awarded is not challenged.

8 Royal Crescent is a Grade II (Starred) listed building on the east side of Royal Crescent. In 1985 it was converted by Guidedale Ltd, who carry on business as residential and commercial property developers, into four dwellings, namely a maisonette on the basement and ground floors, two flats on the first and second floors, and the maisonette with which we are presently concerned, on the third and

fourth floors. Originally the building was a typical five-storey terrace house. Guidedale had employed a full professional team to carry out and market the development, including consulting constructional engineers, Glasspool & Thaiss.

In 1985 the plaintiffs were living in Strasbourg but expected to return to the UK in or about 1988. At this time the property market was rising rapidly and they were anxious to gain a foothold in the UK market by purchasing a property in London to let with a view to selling it in due course, and then buying another property with the combined proceeds of their English and French properties. They viewed the maisonette on June 4 1985, and put in an offer for £114,000 on June 7 1985 which was accepted by Guidedale, whereupon they instructed the first defendant to carry out the structural survey.

Mr Dyson's report was dated June 15 1985, contracts were exchanged on July 5 1985 and the purchase of the long lease was completed on September 11 1985, the plaintiffs having obtained a mortgage from a building society in the sum of £50,000. The plaintiffs' lease, and also those of the other tenants, contained a covenant under which Guidedale were responsible for structural repairs. It is common ground that at the time of Mr Dyson's survey the maisonette suffered from the structural defect, which was not discovered until three years later in 1988, when the plaintiffs were attempting unsuccessfully to sell the maisonette.

Originally the floors, which are of timber joists, were supported centrally on a trussed timber pine wall, which transferred the load from the floors to the party walls. The alterations, which included the adding of a floor at roof level, caused the spine wall to carry additional loads, and also affected its trussed structure; in consequence the first-floor joists had to support the full load of the spine wall above the first floor, and became overstressed by a factor of about two and a half, and therefore sagged, creating cracking in the maisonette and in the first and second floor flats. The necessary remedial works, which were carried out by Guidedale in response to demands by the plaintiffs and the other tenants in 1990, consisted of the insertion of a steel beam below the spine wall in the ceiling space of the first-floor flat, which belonged to a Miss Solomonides, who had exchanged contracts with Guidedale for the purchase of a long leasehold interest on July 1 1985 with completion scheduled for July 29 1985.

The judge's findings on negligence were set out in his judgment as follows:

> We know in this case, because of what is accepted by both parties, that the structural engineers did not get it right. The purpose of the plaintiffs instructing Mr Dyson was precisely to check out the opinions, the calculations and the recommendations of any structural engineer the builder might have engaged. I accept that Mr Dyson would not have had the experience effectively to challenge the details of his calculations. But the obligation was upon him to press upon such a structural engineer and/or those others to whom he would have reported and advised the tell-tale signs he would have seen from the rucking of the wallpaper in the drawing room, to have put his point of view, as that of an experienced structural surveyor, to the structural engineer so as to persuade him to re-evaluate his assessments. There is no evidence that Mr Dyson did that, and in his failure to do it I think he fell down on his duty to these clients, the plaintiffs, who were relying upon him to take a line, independent from that of the developer from whom they were buying the lease . . .
>
> If he had seen the rucking of the wallpaper and linked it with the other structural danger spots — namely, the sagging floor, the breaking into the spine wall and the alterations carried out to the fourth floor — if those had been put together with the rucking wallpaper, I am satisfied that he would have discharged his responsibilities to the plaintiffs and the probabilities are that the structural engineer would have uncovered or discovered the defects in his own calculations.
>
> If in fact he had not been able to persuade the structural engineer as to the faults in his calculations, I am satisfied that, within that situation, Mr Dyson would have been under an obligation to have entered a caution in his report to the plaintiffs and have afforded them the opportunity of instructing a structural engineer or of looking elsewhere for a property. Accordingly, on that basis, I find that the first and second defendants are liable to the plaintiffs in negligence for Mr Dyson's failure to warn of the structural defect.

On *quantum* the judge assessed the measure of damages as the difference between the value of the property without the defects and its value with the defects at the date of purchase, in accordance with a long line of authority, starting with *Philips* v *Ward* [1956] 1 WLR 471, and culminating in the very recent Court of Appeal decision in *Oswald* v *Countrywide Surveyors Ltd* [1996] 37 EG 140*. He accepted the evidence of the plaintiff's expert, Mr ID Taylor ARICS, who is a director of Chesterton, and head of their Residential Professional, Investment and Development Department in the UK, that the open market value of the flat in its structurally defective condition in July 1985 was £80,000, which the judge adjusted to £85,000 to allow for some small concessions made by Mr Taylor during cross-examination.

The defendants had contended that this conventional approach was fundamentally flawed on its special facts of the present case, on the footing, supported by both their experts Mr MH Fawcett FRICA and Mr RWC Horner FRICS, that, if the defect had been discovered in 1985 and had been drawn to the attention of Guidedale, they would forthwith have remedied it at their own expense prior to selling any of the units into which the building had been divided, including of course the plaintiffs' maisonette; consequently there would have been no diminution in the maisonette's value and no entitlement to more than nominal damages.

The judge dealt with this contention as follows:

> I am satisfied that Mr Taylor's approach is a perfectly reasonable means of calculating a rather difficult calculation, namely the market value of this property in the condition that it was. Market value is an objective assessment, that is the price paid by a willing purchaser to a willing vendor negotiating at arm's length. In my judgment, the willing vendor is not the developer in this or any particular case but a hypothetical vendor with a property on his hands which he wishes to sell. Mr Palmer's submission predicates that the plaintiffs were in a negotiating position in which they never had a chance to be. Second, if Mr Palmer is correct when he says that in each instance when you seek a market valuation you have to explore all the surrounding circumstances in order to ascertain the valuation figure, it would form a major exception to the general rule adumbrated in the cases I am referred to, and there is no reference to any such exception and no like case quoted. The fact is that the plaintiffs in this case paid for the property more than they should have done having regard to what they acquired for their price. The sole question is: how much more did they pay?

By their first ground of appeal the defendants reiterate this point and submit the judge was wrong.

The defendants' second ground of appeal is that, even if they are wrong on the first point, the plaintiffs avoided their loss by reason of the repair undertaken by Guidedale in 1990.

The judge rejected this submission as follows:

> The court is constrained to make an assessment of the loss recorded at the date of the breach. In this case, the damages to be assessed have to be assessed as at July 1985, and, in my judgment, it is not relevant that the risk is subsequently eliminated at the cost of someone else.

I shall deal with each of these grounds in turn, but first it is convenient to sketch in the subsequent history between 1985 and 1990.

As already noted, it was not until 1988 the structural defect was discovered. It came to light as a result of two very adverse reports prepared by the surveyors retained by two prospective purchasers, both of whom backed out in consequence. The remedial work was carried out almost two years later, having been begun in September and completed in October 1990. Meantime the plaintiffs, who co-ordinated the exercise, entered into tortuous and prolonged quadripartite negotiations with the other tenants, with Glasspool & Thaiss, and with Guidedale, who originally denied responsibility, and only agreed to carry out the work after the threat of legal proceedings. Meantime party wall awards were required in respect of both adjoining occupiers, the notices being served on May 16 1990, and the awards being forthcoming respectively in June and July 1990. Furthermore Miss Solomonides required an indemnity, which the plaintiffs granted, for

*Editor's note: Also reported at [1996] 2 EGLR 104

A any expense arising from the structural work, and her removal from her flat, together with her rehousing and storage expenses. I shall in future refer to these matters collectively as "the intervening events".

Turning now to the first ground of appeal, it is not in dispute that the applicable legal principles are laid down in the line of authorities mentioned above.

In *Philips* v *Ward* (*supra*) a surveyor negligently valued a property which the plaintiff had purchased. Denning LJ stated as follows at p473:

> I take it to be clear law that the proper measure of damage is the amount of
> B money which will put Mr Philips into as good a position as if the surveying contract had been properly fulfilled: see *British Westinghouse Electric and Manufacturing Co Ltd* v *Underground Electric Railways Co* [1912] AC 673, at 689 per Lord Haldane LC. Now if Mr Ward had carried out his contract, he would have reported the bad state of the timbers. On receiving that report, Mr Philips would either have refused to have anything to do with the house — in which case he would have suffered no damage — or he would have bought it for a sum which represented its fair value in its bad condition — in which case he would pay so much less on that account. The proper measure of damages is therefore the difference between the value in its assumed good condition and the value in the bad condition which should have been reported to the client.

C Morris LJ delivered a concurring judgment. Romer LJ. stated at p477:

> It appears to me that in order to arrive at a correct solution of the problem in this case one has to compare the position into which the plaintiff was put by the defendant's failure to perform his duty with the position in which he would have been had the defendant performed it; and in so far as the first position is more unfavourable to the plaintiff than the second, and the difference can be assessed in terms of money, then prima facie that assessment is the measure of the defendant's liability to the plaintiff.

In *Perry* v *Sidney Phillips & Son* [1982] 1 WLR 1297, where the facts were similar, Lord Denning MR said at p1301:

> D . . . where there is a contract by a prospective buyer with a surveyor under which the surveyor agrees to survey a house and make a report on it — and he makes it negligently — and the client buys the house on the faith of the report, then the damages are to be assessed at the time of the breach, according to the difference in price which the buyer would have given if the report had been carefully made from that which he in fact gave owing to the negligence of the surveyor. The surveyor gives no warranty that there are no defects other than those in his report. There is no question of specific performance. The contract has already been performed, albeit negligently. The buyer is not entitled to remedy the defects and charge the cost to the surveyor. He is only entitled to damages for the breach of contract or for negligence. It was so decided by this
> E court in *Philips* v *Ward* [1956] 1 WLR 471, followed in *Simple Simon Catering Ltd* v *Binstock Miller & Co* (1973) 117 SJ 529 . . .

Oliver LJ and Kerr LJ delivered concurring judgments.

In *Watts* v *Morrow* [1991] 4 All ER 937*, again on similar facts, Ralph Gibson LJ stated at p950c:

> The task of the court is to award to the plaintiffs that sum of money which will, so far as possible, put the plaintiff into as good a position as if the contract for the survey had been properly fulfilled: see Denning LJ in *Philips* v *Ward* [1956] 1 All ER 874 at 875, [1956] 1 WLR 471 at 473. It is important to note that the contract in this case, as in *Philips* v *Ward*, was the usual contract
> F for the survey of a house for occupation with no special terms beyond the undertaking of the surveyor to use proper care and skill in reporting upon the condition of the house.
>
> The decision in *Philips* v *Ward* was based upon that principle; in particular, if the contract had been properly performed the plaintiff either would not have bought, in which case he would have avoided any loss, or, after negotiation, he would have paid the reduced price. In the absence of evidence to show that any other or additional recoverable benefit would have been obtained as a result of proper performance, the price will be taken to have been reduced to the market price of the house in its true condition because it cannot be assumed that the vendor would have taken less.
>
> The cost of doing repairs to put right defects negligently not reported may be G relevant to the proof of the market price of the house in its true condition: see *Steward* v *Rapley* [1989] 1 EGLR 159; and the cost of doing repairs and the diminution in value may be shown to be the same. If, however, the cost of repairs would exceed the diminution in value, then the ruling in *Philips* v *Ward*, where it is applicable, prohibits recovery of the excess because it would give to the plaintiff more than his loss. It would put the plaintiff in the position of recovering damages for breach of a warranty that the condition of the house was correctly described by the surveyor and, in the ordinary case, as here, no such warranty has been given.

Bingham LJ as he then was, and Sir Stephen Brown P agreed.

In his submission on the first ground, Mr Adrian Brunner QC invited us to interpret the judge's findings on negligence quoted above H as follows:

> The Judge held:
>
> (i) that Mr Dyson should have advised Glasspool & Thaiss of the tell-tale signs of the rucking of the wallpaper and attempted to persuade them to re-evaluate their assessments relating to the spine wall;
>
> (ii) had Mr Dyson done so, the probabilities were that Glasspool & Thaiss would have discovered the defects in their calculations and in that event Mr Dyson would owe no further duty to the Gardners. It is implicit from the finding that the Judge considered that once the engineers were aware of the problem then such remedial work as was necessary would be carried out;
>
> (iii) that it was only in circumstances where Mr Dyson had failed to persuade the engineers of their error that he had a duty to report further to the J Gardners by way of a caution which would have enabled them to instruct their own structural engineer or to look elsewhere.

Let me say at once that I am unable to accept this interpretation. The judge himself referred to Mr Dyson's "failure to warn" and I have no doubt that the plaintiffs' retainer of Marsh & Parsons was to undertake the survey and to report to the plaintiffs themselves, as stated by the former in their letter dated June 14 1981 accepting the retainer:

> Thank you for your instructions to undertake a structural survey of the above property and we will let you have a formal report as soon as possible.

I am sure that the judge did not intend to state otherwise, though no K doubt Mr Dyson would have been in touch with Glasspool & Thaiss as well.

The crux of Mr Brunner's main submission was that it is an inevitable inference that, had the structural defect been drawn to Guidedale's attention in June 1985, they would forthwith have carried out the necessary repairs, being faced with the choice between undertaking a comparatively modest repair costing no more than £4,000, and facing a heavy loss; and that thereafter, before the conclusion of the plaintiffs' purchase, negotiations would have ensued which, having regard to the property boom of which both were well aware, would have resulted in an agreement for no more than a L nominal discount, it being unthinkable that Guidedale would have agreed to a reduction of the magnitude put forward by Mr Taylor.

Mr Brunner accepted that there was no evidence from Guidedale or from Glasspool & Thaiss to support his submission, but invited the court to draw the appropriate inference. He submitted that by contrast Mr Taylor's approach was artificial and unsound, and that the judge should not have adopted it.

I am unable to accept this submission, substantially for the reasons given by the judge as quoted above, and supported by Mr Edwin Johnson on behalf of the respondents.

In my judgment, it is intrinsic to the principles laid down in the M *Philips* v *Ward* line of cases that, in assessing the market price of the property in its defective condition, a hypothetical sale of the property in that state is assumed to have taken place. Mr Brunner proceeds on the opposite assumption, namely that in the particular circumstances of this case no sale would have taken place until after the defect had been remedied, leading to the negotiations he envisages and the resultant nominal discount which he foreshadows. This, to my mind, is basically unsound.

I should mention that in support of this contention Mr Brunner sought to draw comfort from the passage quoted above from Ralph Gibson LJ's judgment in *Watts* v *Morrow*. But that passage must be read in its context, and I am satisfied that Ralph Gibson LJ's sole

---

*Editor's note: Also reported at [1991] 2 EGLR 152

purpose was to make clear that, in the absence of specific evidence of any additional recoverable benefit, the purchaser is not entitled to recover more than the difference between the price paid and the market price of the property in its defective condition, and in particular that he is not entitled (as the appellant sought to do in that case) to recover the cost of the repairs if they exceed that difference.

A further flaw in Mr Brunner's argument is that his scenario does not square with the actual facts of the case, seeing that the plaintiffs did in fact purchase the maisonette while still in its defective state, which was a natural consequence flowing from Mr Dyson's negligence.

In my judgment, this is a straightforward *Philips* v *Ward* type of case from which it follows that the judge's approach was soundly based.

I should add that, quite apart from this decision in principle, I do not think that the inference which Mr Brunner invited us to draw was supported by the evidence. Following the judgment the appellants requested the judge to elaborate his findings, and on April 11 1995 he responded as follows:

> I have been asked to find as a question of fact whether the hypothetical proposal relating to valuations, advanced in argument by Mr Palmer in his submissions and through his various valuation witnesses in the course of the evidence was realistic and practical. If I were to attempt such a finding I would have to hedge it round with so many qualifications and caveats, some of which were dependent upon evidence which was not in fact given, that I do not think it really would be of any value to the Court of Appeal.
>
> There were certain parts of that proposal which I do not think were challenged by Mr Johnson; and in that they were not challenged, certainly I would happily endorse them as findings of fact. Here I am thinking of the fact that the works concerned could have been carried out for approximately £4,000, and further, as I think I mentioned in my judgment, if the defendants had had vacant possession when they carried out the works, the works could have been done much more simply and easily.
>
> But as I made plain in argument this morning, difficulties arise because we do not know the developer's approach to this matter, as to whether he would have been willing to carry out the rectifications for the plaintiffs in advance of completion or as to whether he would have just said, "No, I am not prepared to do any of this work. If you don't like it go elsewhere. I am certainly not prepared to drop my price". We do not know what his reaction would be.
>
> Equally we do not know what the attitude of Miss Solomonides would have been. She was the occupier of the first floor flat. She had exchanged contracts on 1st July and her completion date was on July 29. Had this scenario started off on July 5, at a date when the plaintiffs exchanged their contract, one just does not know how much of the remedied works the developer could have done in the time she made available. In all the circumstances, I do not think I can usefully make any specific findings of fact about any of those matters and I must leave Mr Palmer to argue his interesting point of law on the basis of the evidence which has been adduced and the findings that I have already made in my judgment.

This was an assessment of the evidence on the primary facts made by a judge who heard the witnesses, and I for my part do not think this court can properly interfere, let alone interpose findings or inferences which the judge himself was not prepared to make or draw.

Finally on this part of the appeal I turn to Mr Taylor's valuation, which the judge accepted subject to his small adjustments, but which the appellants criticise as basically unsound. The judge dealt with this as follows:

> The plaintiffs relied upon the evidence of Mr Taylor, another senior director with Chestertons, as to valuation. He carried out his calculations on the basis of a residual valuation, using a computer programme well-recognised by developers, as the plaintiffs say, for both large and small residential developments. It is based on construction costs, other remedial works, takes into account the legal fees, the structural engineer's fees, party wall awards, sale, agents' and legal fees. When first Mr Taylor calculated the residual valuation, he did so on the basis of a profit of 15% on costs. The result of his calculation was a market valuation, on the July 5, of the property in the condition that it was, not of £114,000 as paid by the plaintiffs, but only £80,014.

The judge then explained the small adjustments and proceeded:

> Making those corresponding adjustments, my assessment of the market valuation, in accordance with the evidence that Mr Taylor gave, is approximately £85,000. Subtracting that sum from £114,000, one has a difference amounting to £29,000. The plaintiffs say that this represents the damages they are entitled to. If the figure seems to be high, Mr Taylor points out that the property in effect was unmortgageable in the condition that it was in. As a result of that, when looking for a purchaser of property in that condition, one was looking for a purchaser who did not need a mortgage but had cash and who was prepared to make a speculative buy. He said that that necessarily directed you to a very limited market.

The defendants' valuers approached the problem from an entirely different angle, as shown for example by Mr Fawcett's report:

> With regard to the lack of internal structural support, I consider it reasonable to assume that if this point had been brought to the attention of the plaintiff by either Mr Dyson, the plaintiff's solicitors or through any other source, then the outcome would have been the same in 1985 as happened in 1988. That is, the problem would have been rectified at the expense of others. No works were required to the plaintiffs' accommodation and the structural work was undertaken in another part of the building.
>
> In reality sales are concluded over a period of time within which problems are resolved. It would not be necessary to have the works completed but only that there was an agreement to do so for the price to be unaffected. It is in my opinion reasonable to assume that if the defendant had recommended the services of a structural engineer who in turn recommended further structural work, this would have been considered favourably by the developer notwithstanding that no objection had been made by the local authority building inspector. The cost of undertaking this work would have been in the order of £2,500, a relatively small sum in relation to the total sale price of the flats.

As already demonstrated, this basis of valuation was incorrect, so the judge had before him no rival valuation prepared on the correct basis. In these circumstances it was perfectly proper for him to accept Mr Taylor's valuation.

The first ground of appeal therefore fails.

On the second ground (avoidance of loss) it is convenient first to summarise the relevant authorities on which both sides rely.

In the leading case of *British Westinghouse Electric & Manufacturing Co Ltd* v *Underground Electric Railways of London Ltd* [1912] AC 673 the House of Lords held unanimously that where a railway company had installed improved turbines, in replacement of defective turbines which had been supplied to them in breach of contract, the pecuniary advantage derived therefrom was a relevant consideration in the assessment of damages, since the purchase of the improved turbines was a reasonable and prudent course for them to have taken in mitigation of damages.

Giving the leading speech, with which the other members of the Appellate Committee agreed, Viscount Haldane LC stated:

> The fundamental basis is thus compensation for pecuniary loss naturally flowing from the breach; but this first principle is qualified by a second, which imposes on a plaintiff the duty of taking all reasonable steps to mitigate the loss consequent on the breach, and debars him from claiming any part of the damage which is due to his neglect to take such steps. In the words of James LJ in *Dunkirk Colliery Co* v *Lever* (1878) 9 Ch D 20, at p25, "The person who has broken the contract is not to be exposed to additional cost by reason of the plaintiffs not doing what they ought to have done as reasonable men, and the plaintiffs not being under any obligation to do anything otherwise than in the ordinary course of business."
>
> As James LJ indicates, this second principle does not impose on the plaintiff an obligation to take any step which a reasonable and prudent man would not ordinarily take in the course of his business. But when in the course of his business he has taken action arising out of the transaction, which action has diminished his loss, the effect in actual diminution of the loss he has suffered may be taken into account even though there was no duty on him to act.

Having cited a number of earlier authorities, he proceeded:

> I think the principle which applies here is that which makes it right for the jury or arbitrator to look at what actually happened, and to balance loss and gain. The transaction was not res inter alios acta, but one in which the person

A whose contract was broken took a reasonable and prudent course quite naturally arising out of the circumstances in which he was placed by the breach. Apart from the breach of contract, the lapse of time had rendered the appellants' machines obsolete, and men of business would be doing the only thing they could properly do in replacing them with new and up-to-date machines.

> The arbitrator does not in his finding of fact lay any stress on the increase in kilowatt power of the new machines, and I think that the proper inference is that such increase was regarded by him as a natural and prudent course followed by those whose object was to avoid further loss, and that it formed part of a continuous dealing with the situation in which they found themselves, and was not an independent or disconnected transaction.

B In *Hussey* v *Eels* [1990] 2 QB 227*, the plaintiffs purchased a bungalow in reliance on answers to precontractual inquiries which included a misrepresentation that the building had not been the subject of subsidence. The plaintiffs could not afford the necessary repairs to the foundations, and ultimately obtained planning permission to redevelop the land, which they then sold to developers for a substantial sum. The question arose whether what, if any, profit they made on the resale could be brought into account in assessing the damages.

The Court of Appeal (Mustill and Farquharson LJJ and Sir Michael Kerr) held that it could not.

C Giving the leading judgment, with which Farquharson LJ and Sir Michael Kerr agreed, Mustill LJ expressed his conclusion as follows:

> I have dealt with the authorities at some length, because it was said that in one direction or another they provided a direct solution to the present problem. For the reasons already stated, I do not see them in this light. Ultimately, as with so many disputes about damages, the issue is primarily one of fact. Did the negligence which caused the damage also cause the profit — if profit there was? I do not think so. It is true that in one sense there was a causal link between the inducement of the purchase by misrepresentation and the sale 2½ years later, for the sale represented a choice of one of the options with which the plaintiffs had been presented by the defendants' wrongful act. But only in that sense. To my mind the reality of the situation is that the plaintiffs bought
> D the house to live in, and did live in it for a substantial period. It was only after two years that the possibility of selling the land and moving elsewhere was explored, and six months later still that this possibility came to fruition. It seems to me that when the plaintiffs unlocked the development value of their land they did so for their own benefit, and not as part of a continuous transaction of which the purchase of land and bungalow was the inception.

Mustill LJ's review of the earlier authorities, which began with a discussion of the *Westinghouse* case, included two cases which he described as follows:

> I now turn to a pair of contrasting decisions in a different line of authority. The first is *Jamal* v *Moolla Dawood Sons & Co* [1916] 1 AC 175 where the
> E claim was for a failure by a buyer to accept shares under a contract of sale for delivery on a specified date. Two months after that date the sellers began to re-sell the shares on a rising market. It was held that the profit thus accruing should not be deducted from the damages for non-acceptance, which were to be ascertained as at the date of the breach. Delivering the opinion of the Board, Lord Wrenbury said, at pp179, 180:
>
> "The question therefore is the general question and may be stated thus: In a contract for sale of negotiable securities, is the measure of damages for breach the difference between the contract price and the market price at the date of the breach — with an obligation on the part of the seller to mitigate the damages by getting the best price he can at the date of the breach — or is the seller bound to reduce the damages, if he can, by subsequent sales at better prices? If
> F he is, and if the purchaser is entitled to the benefit of subsequent sales, it must also be true that he must bear the burden of subsequent losses. The latter proposition is in their Lordships' opinion impossible, and the former is equally unsound. If the seller retains the shares after the breach, the speculation as to the way the market will subsequently go is the speculation of the seller, not of the buyer; the seller cannot recover from the buyer the loss below the market price at the date of the breach if the market falls, nor is he liable to the purchaser for the profit if the market rises . . . The seller's loss at the date of the breach was and remained the difference between contract price and market price at that date. When the buyer committed this breach the seller remained entitled to the shares, and became entitled to damages such as the law allows.
> The first of these two properties, namely, the shares, he kept for a time and G
> subsequently sold them in a rising market. His pocket received benefit, but his loss at the date of the breach remained unaffected."
>
> This case was discussed in the *R Pagnan & Fratelli* case [1970] 1 WLR 1306, the facts of which were as follows. Corbisa sold to Pagnan a quantity of maize on cif terms; with extensions, the shipment period ended on 22 August 1965. The sellers failed to ship in time. On 21 September 1965 the parties met and the buyers agreed to accept a consignment on a named vessel if satisfied with its condition on arrival at Venice. Upon arrival part was found to be in bad condition, and the buyers rejected it. Meanwhile they had obtained on 13 October a decree sequestrating part of the cargo for the recovery of freight and premiums advanced and for reimbursement of damages for non-fulfilment.
> The sellers repaid the advances, and the sequestration was lifted pro tanto, H
> leaving 700 metric tons under sequestration in relation to the claim for damages. On 13 November the parties agreed that the buyers would purchase the rejected goods ex silo Trieste, at a price which the arbitrators found was unduly depressed by reason of the sequestration: so much so that it was below the market price. The arbitrators also found:
>
> "The purchase of 13 November 1965 formed part of a continuous dealing with the situation in which the buyers found themselves, and was not an independent or disconnected transaction. By such purchase the buyers diminished and mitigated any loss which they might have suffered:" see p1309D.
>
> On this basis the arbitrators dismissed the claim. The buyers appealed by case stated. Roskill J upheld the award, as did the Court of Appeal. After J
> referring to *Jamal* v *Moolla Dawood, Sons & Co* [1916] 1 AC 175 and to a similar case, Salmon LJ said at pp 1314-1315:
>
> "The principle of law is that where a buyer wrongfully neglects or refuses to accept and pay for the goods or a seller wrongfully neglects or refuses to deliver the goods to the buyer, the innocent buyer or seller as the case may be may maintain an action for damages for breach of contract. The measure of damage in each case is the estimated loss directly and naturally resulting in the normal course of events from the breach of contract. Where there is an available market for the goods, the measure of damage is prima facie to be ascertained by the difference between the contract price and the market price at the date of the breach: see section 50 and 51 of the Sale of Goods Act, 1893.
> The two authorities relied on by Mr Goff do no more than illustrate instances K
> in which the prima facie rule relating to the measure of damage applies. In such cases the innocent party is not bound to go on the market and buy or sell at the date of the breach. Nor is he bound to gamble on the market changing in his favour. He may wait, if he chooses; and if the market turns against him this cannot increase the liability of the party in default; similarly if the market turns in his favour, the liability of the party in default is not diminished. Normally if the innocent party goes on to the market and buys or sells after the date of the breach, this is *res inter alios acta* so far as the party in default is concerned. The present case, however, is quite different. The purchase of 13 November 1965 was certainly not *inter alios acta*; it was between the self-same buyers and sellers who were parties to the contract of 20 May 1965 and it related to
> the self-same goods that were the subject-matter of that contract. Moreover, as L
> already stated, the tribunal found that it was not an independent or disconnected transaction but formed part of a continuous dealing between the parties; and these findings of fact cannot be challenged in this court. Accordingly the prima facie rule for ascertaining the measure of damages cannot apply because the buyers suffered no loss or damage but instead made a handsome profit in spite of the sellers' breach."
>
> Salmon LJ concluded, at p1316:
>
> But the buyer cannot have his cake and eat it, as these buyers are seeking to do. They went through the motions of rejecting the goods in October 1965. Indeed
> they did, in law, reject them. They did so, however, in the confident expectation M
> that, as a result of their rejection and the sequestration order, they would be able to negotiate a new agreement under which they would acquire the goods at a price favourable to themselves. This they did by their purchase of November 13. The price was substantially below the market price and their resulting profit certainly exceeded the difference between the May contract price as varied and the prevailing market price at all relevant times. Damages for breach of contract are awarded for loss suffered. Here the buyers suffered no loss. It is only by looking in isolation at the sellers' failure to deliver sound goods that the buyers' claim is even arguable. This failure cannot in my view properly be looked at in isolation because together with the purchase of November 13 which arose out of the situation in which the buyers found themselves, it formed one continuous dealing between the same parties in respect of the same goods. As a result of this dealing, looked at as a whole, the buyers, notwithstanding the sellers' breach, made a

---

*Editor's note: Also reported at [1990] 1 EGLR 215

> A profit and no loss. To allow the buyers' claim would in my view be contrary alike to justice, common sense and authority. I would accordingly dismiss the appeal.
>
> It seems to me that the *Pagnan* case is as far away from the present case in one direction as the *Jamal* case is in the other. In *Pagnan*, not only had the tribunal made a finding of continuity, but the bare narrative shows that such a finding was inevitable. From the moment of breach the sellers were at a disadvantage which the buyers were able to exploit by successful measures leading to the ultimate repurchase. In no sense could the buyers be said to have purchased the same cargo from the same seller as a bargain for their own account quite separate from anything that had gone before.
>
> In reality I believe that neither *Pagnan* nor any of the other cases cited in argument presents a true analogy here. The plaintiffs are not claiming a
> B conventional prima facie measure of damages, as with a sale of goods or shares: they really suffered the loss claimed, for they would have had to pay the cost of repairs if they had remained permanently in residence. On the other hand the later transaction did not flow inexorably from the first, as was the case in *Pagnan*.

Earlier in his judgment, with reference to the question whether the plaintiffs had been under a duty to mitigate, Mustill LJ left open the question whether there could be any question of mitigation in cases where the loss has already crystallised, in contrast to cases where there is a continuing loss (as for example in the *Westinghouse* case), stating that he would not be prepared without a very full review of the
C authorities to underwrite any generalisation, "especially in the field of damages where broad statements of principle tend to be unreliable".

Finally in *Jones* v *Just* (1868) LR 3 QB 197, it was held that a normal measure of damages applied where the plaintiff bought first quality hemp and second quality hemp was delivered, although the plaintiff had resold the delivered hemp at substantially higher than the price at which the first quality hemp had stood at the time of delivery, the market price having meantime risen. This was the judgment of a strong court (Lord Cockburn, CJ, Blackburn and Mellor JJ) delivered by Mellor J, in which at p200 they approved the direction of the trial judge to the jury that if they found for the plaintiffs, "the damages
D should be measured by the rate which the hemp was worth when it arrived compared with the rate which the same hemp would have realised had it been shipped in the state in which it ought to have been shipped" thus, (in Mellor J's words) "in effect, giving the plaintiffs the benefit of the rise in the market".

Mr Brunner submitted that, the plaintiffs having themselves adopted the reasonable and prudent course of instigating the repairs, and those repairs having been undertaken with the result that the defect was rectified, the plaintiffs had suffered no loss, and were no worse off, seeing that the premises had been restored to their full value well before the date of trial. He stressed that these repairs remedied the very
E defect which Mr Dyson had originally failed to spot, and submitted that there was therefore a direct connection between Mr Dyson's negligence in 1985 and the remedial work undertaken in 1990. He further submitted that, as a general rule, where the plaintiff has in fact avoided his loss in whole or in part, whether or not under a duty to mitigate, by reasonable and prudent action, the resultant benefit must be taken into account. He recognised at the conclusion of his argument that this formulation is difficult to reconcile with the cases referred to above dealing with sale of goods or shares, but submitted that they formed a special class outside the general rule.

Mr Johnson, founding his argument on *Hussey* v *Eels*, submitted
F that where as a result of the defendant's negligence a plaintiff suffers loss in the form of diminution of value of the property, that loss is not avoided by the subsequent conduct of the plaintiff unless such conduct flows inexorably from the original transaction, and can properly be seen as part of a continuous course of dealing with the situation in which the plaintiff originally found himself.

Here, he submitted, the action of the landlords in repairing the property was collateral, and *res inter alios acta*; moreover, it did not flow inexorably from the original transaction (ie Mr Dyson's negligent valuation) and was in no sense part of a continuous course of dealing, in view of the long lapse of time and of the nature and magnitude of the intervening events.

In evaluating these arguments I bear very much in mind Mustill LJ's G
salutary warning against laying down potentially unreliable statements of principle in the field of damages, and I respectfully adopt his approach, namely that the issue is primarily one of fact, and that the relevant considerations are *mutatis mutandis* those cited by him in his conclusion, which seem to me in line with *Westinghouse*: see especially the final two paragraphs quoted above from Lord Haldane's speech. It follows that I accept Mr Johnson's analysis and reject Mr Brunner's broad brush formulation, not least because of its inconsistency with the cases dealing with sale of goods or shares cited above, which cannot, in my view, be segregated from the main stream of authority. H

In my judgment, having regard to the intervening events and to the long interval of time, the repairs executed in 1990 were not part of a continuous transaction of which the purchase of the lease as a result of Mr Dyson's negligence was the inception. Furthermore, these repairs undertaken by Guidedale at the plaintiff's insistence were *res inter alios acta* and therefore collateral to Mr Dyson's negligence.

I should add by way of postscript that I do not think this conclusion is affected by the decision of this court in *Hodge* v *Clifford Cowling & Co* [1990] 46 EG 120*, which came to our attention after the conclusion of the hearing, and in which *Hussey* v *Eels* was applied: see *per* Glidewell LJ at p92. J

For these reasons the appellants' second ground also fails, and consequently I would dismiss this appeal.

Disagreeing, **Peter Gibson LJ** said: Two points were taken by Mr Brunner QC on this appeal: (1) Judge Byrt QC erred in accepting the valuation approach by Mr Taylor, the plaintiffs' expert, it being inconceivable that any properly advised vendor would have sold the maisonette in 1985 for £85,000; (2) the judge erred in not finding that the loss incurred by the plaintiffs in purchasing a property for a price in excess of its true value by reason of the unreported structural defect was avoided by the remedying of the defect by the freeholder and its structural engineers at the instigation of the plaintiffs. K

On the first point Mr Brunner, while accepting that the value of the property is the price paid on a hypothetical sale between a willing vendor and a willing purchaser, submitted that the vendor developer would not have been a willing vendor at the price of £85,000. That submission proceeds on a mistaken appreciation of the valuation exercise. In my opinion the same principles apply, whether the valuation is, as in this case, for the purpose of assessing damages or for some statutory or other purpose (in the absence of specific statutory provisions requiring a different hypothesis). I venture to repeat what I said in *Walton* v *Inland Revenue Commissioners* [1996] STC 68† at pp
85f, 86 (in the context of a valuation for capital transfer tax purposes): L

> Second, it is agreed that the valuation required by s 38 [Finance Act 1975] is on the basis of a hypothetical sale in the open market. Although the statute says nothing about a willing seller or a willing buyer, the concept of the open market automatically implies a willing seller and a willing buyer, each of whom is a hypothetical abstraction. However, the willing buyer 'reflects reality in that he embodies whatever was actually the demand for that property at that time' (see *IRC* v *Gray* . . . [1994] STC 360 at 372 *per* Hoffmann LJ). Whilst both the seller and the buyer are assumed to be willing, neither is to be taken to be over-eager . . . The statute assumes a sale. That means that however improbable it is that there would ever be a sale of the property in the real world . . . nevertheless the sale must be treated as capable of being completed . . . It
> also means that the vendor, if he is offered the best price reasonably obtainable in the market, cannot be assumed to say that he will not sell because the price is too low as inadequately reflecting some feature of the property nor can the purchaser be assumed to say that he will not buy because the price is too high. M

It is therefore, in my view, impermissible to postulate that on the hypothetical sale by the hypothetical vendor of the property he would have refused to sell for £85,000 because he thought the price should be higher. If that was the best price obtainable, that is what he must be taken to accept. No less impermissible is it to change the hypothesis by

*Editor's note: Also reported at [1990] 2 EGLR 89

†Editor's note: Also reported at [1996] 1 EGLR 159

postulating that the property is in a different condition by reason of the vendor or the freeholder being prepared to do the repairs first.

For these and the reasons given by Hirst LJ the first point taken by the appellants must be rejected.

The second point taken by Mr Brunner raises a question of some difficulty in circumstances unlike those in any case to which we have been referred. I shall first recite the essential circumstances:

1. The plaintiffs, acting on the negligent advice of the defendants, in 1985 purchased property for £114,000, when its true value, had the structural defect been known was £85,000.

2. The structural defect became known in 1988, as a result of two abortive attempts by the plaintiffs in June and October 1988 to sell the property.

3. The plaintiffs were instrumental in procuring the remedying in 1990 of the structural defect by the freeholder by threatening in 1989 an action under the Defective Premises Act 1972, as a result of which the freeholder's structural engineers paid for the remedying of the defect.

4. Accordingly, at the time this action commenced in 1991, the defect, the existence of which had caused the property to be worth £29,000 less in 1985 than it would have been if the defect had been remedied before the contract to sell the property to the plaintiffs, had been remedied at no cost to themselves and the plaintiffs had the structurally sound property which in 1985 the defendants had represented it was and which they thought they had bought. In the judge's words, "the plaintiffs have a rectified building worth the equivalent of what they had paid for it without any extra cost".

The question is whether in such circumstances the law allows the plaintiffs to recover the £29,000 loss which they initially sustained on the purchase. On this aspect of the appeal we are not concerned with any other loss which the plaintiffs may have suffered consequent upon the negligence. In appropriate circumstances it may be that a plaintiff would recover both the costs and expenses of an abortive sale and also a sum equal to the profit on a lost bargain. I have considerable sympathy with the plaintiffs for the distress and inconvenience they are likely to have suffered, but that is beside the point in issue.

Although the claims of the plaintiffs are both in contract and tort against the first defendants and in tort alone against the second defendant, it has not been suggested that in this case there is in the result any difference between the contractual and the tortious measure of damages. The basic principle applicable in assessing damages is that there must be measured the sum of money which would put the innocent party in the same position as he would have been in if the contract had been performed or if he had not sustained the wrong. In the light of the decisions of this court in *Philips* v *Ward* [1956] 1 WLR 471 and *Perry* v *Sidney Phillips & Son* [1982] 1 WLR 1297* *prima facie* the proper measure is the difference between the price paid by the plaintiffs and the market value at the time of purchase of the property, as it should have been described. However at the date of trial the court may have to consider whether in the light of the circumstances then existing that represents what the plaintiff has actually lost. The law does not permit the plaintiff to recover more than is seen to be his actual loss and the rules of mitigation may deprive the plaintiff of all or part of the damages for loss which otherwise he might have recovered.

The first rule of mitigation is that the plaintiff must take all reasonable steps to mitigate his loss. It follows that he cannot recover for avoidable loss (though he may recover for loss incurred in reasonable attempts to avoid loss). The plaintiff also cannot recover for avoided loss at any rate where certain conditions are satisfied. This rule of mitigation is summarised in *McGregor on Damages*, 15th ed (1988) para 277 in this way:

> (3) The third rule is that, where the plaintiff does take steps to mitigate the loss to him consequent upon the defendant's wrong and these steps are successful, the defendant is entitled to the benefit accruing from the plaintiff's action and is liable only for the loss as lessened; this is so even though the plaintiff would not have been debarred under the first rule from recovering the whole loss, which would have accrued in the absence of his successful mitigating steps, by reason of these steps not being ones which were required of him under the first rule.

Thus if the plaintiff in fact avoids or mitigates his loss, he cannot recover for the loss thereby avoided even though the steps he took were more than could reasonably be required of him under the duty to mitigate his loss.

The leading authority is *British Westinghouse Electric Co Ltd* v *Underground Electric Railways of London Ltd* [1912] AC 673. In that case turbines were supplied by a manufacturer to a railway company. They did not accord with the contract for their supply. Four years after taking delivery of the first turbine the railway company purchased replacement turbines of greater efficiency and power from another supplier which caused the railway company to make a profit. The House of Lords held that even though the replacement turbines were superior to those which should have been supplied in accordance with the contract the profit should have been taken into account in assessing the damages. Viscount Haldane LC expressed the relevant principle as being : ". . . when in the course of business [the plaintiff] has taken action arising out of the transaction, which action has diminished his loss, the effect in actual diminution of the loss he has suffered may be taken into account even though there was no duty on him to act" (p689). At p690 he made it clear that the mitigating action must be "one which a reasonable and prudent person might in the ordinary conduct of business properly have taken" and must be "one arising out of the consequences of the breach and in the ordinary course of business". He described (at p691) the purchase from the new supplier of the replacement turbines as "not *res inter alios acta*, but one in which the person whose contract was broken took a reasonable and prudent course quite naturally arising out of the circumstances in which he was placed by the breach", and he said (at p692) that the increase in the power of the replacement turbines "formed part of a continuous dealing with the situation in which they found themselves, and was not an independent or disconnected transaction."

Viscount Haldane's remarks with the repeated references to the course of business must be read and understood in their context of a case relating to a breach of a contract between trading companies. They do not entail that the rule of mitigation relating to avoided loss cannot apply in other contexts such as where a tort has occurred and where the innocent party is not carrying on business. I would draw attention to the fact that neither a lapse of time of several years before the turbines were replaced, nor the purchase by the railway company of the replacement turbines from a third party prevented the advantages derived from the use of the replacement turbines from being taken into account and that the replacement with more powerful turbines was still described as part of a continuous dealing with the situation in which the innocent party found itself. That decision can therefore be seen to establish the principle that where an advantage accrues to a plaintiff from taking action to mitigate his loss, that advantage must be taken into account in assessing the loss, if any, to be recovered.

The principle thereby established has frequently been applied. It was expressly applied by this court in *R Pagnan & Fratelli* v *Corbisa Industrial Agropacuaria Limitada* [1970] 1 WLR 1306. In that case buyers justifiably rejected goods on the ground of their defective quality. Negotiations between the parties ensued and after some months the buyers finally accepted the same goods from the sellers at a reduced price. It was held that the *prima facie* rule for ascertaining the measure of damage in section 51(3) Sale of Goods Act 1893 could not apply in the circumstances because the buyers had in reality suffered no loss or damage. Salmon LJ (at p1315) said that the buyers in accepting the same goods at a reduced price did not do so with the motive of mitigating damages. He continued:

> But their motive or intention does not matter. No one can doubt that they took a reasonable and prudent and, I would add, an astute course, quite naturally arising out of the circumstances in which they were placed by the sellers'

---

*Editor's note: Also reported at [1982] 2 EGLR 135

A breach. It was a course, as the tribunal found, which formed part of a continuous dealing with the situation in which they found themselves and was not an independent or disconnected transaction. And it in fact extinguished the loss which they would otherwise have suffered . . . As a result of this dealing, looked at as a whole, the buyers, notwithstanding the sellers' breach, made a profit and no loss. To allow the buyers' claim would in my view be contrary alike to justice, common sense and authority.

In reaching that conclusion in a case involving the sale of goods where there was an available market, this court distinguished cases where the *prima facie* rule of section 51(3) (*viz* the measure of damages is to be ascertained by the difference between the contract
B price and the market price of the goods when they ought to have been delivered) and the corresponding rule in section 53(3) *ibid*, where the buyers are at fault, are applied. It also distinguished the cases where the same principles are by analogy applied to the sale of shares for which there is an available market. The assumption underlying the *prima facie* rule is that the innocent party can and should act immediately upon the breach and buy or sell the goods (or shares) in the available market. The same assumption cannot be made in the case of negligence or some other breach of contract or tort affecting the sale or purchase of land, where the right of action may not be known until long after the arising of the cause of action and there is unlikely to be
C an available market in the Sale of Goods Act sense.

The reason why the *British Westinghouse* principle does not apply to profits subsequently received by the innocent buyers or sellers is that the profits are independent of any act of mitigation: see *McGregor op cit* paras 336-340, *Benjamin on Sale of Goods* 4th ed (1992) para 16-046 and the similar statement by the same author in *Chitty on Contracts* 27th ed (1994) para 26-056. Thus where buyers in breach of contract do not accept goods but the sellers choose not to resell the goods on the date of the buyers' breach and retain the goods until a later sale, just as the sellers could not make the buyers liable for additional loss had the market fallen after the date of the breach, so the
D sellers are entitled not to account for any profit they make on the later sale. The decisions to retain and subsequently sell the goods are independent of the breach. The Sale of Goods Act rules were based on cases decided prior to the 1893 Act, such as *Jones* v *Just* (1868) LR 3 QB 197, to which Hirst LJ referred in his judgment. That case is silent on the reasoning underlying the acceptance by the Court of Queen's Bench (which included Blackburn J) of the correctness of the direction of Blackburn J as the trial judge to the jury which in effect gave the innocent buyers, to whom damaged hemp had been delivered, the benefit of the rise in the market between the date of breach and the subsequent sale by the buyers. But it may be inferred that similar
E reasoning applies, *viz* that the benefit does not arise from any act of mitigation and so is irrelevant in assessing damages: see *McGregor op cit* para 379, *Benjamin op cit* para16-046 and *Chitty* op cit para 26-056. Similarly advantages gained by a plaintiff from another independent transaction, such as an insurance contract entered into before the breach or wrong, cannot be taken into account: see, for example, *Bradburn* v *Great Western Railway* (1874) LR 10 Exch 1. As is said in *McGregor op cit* para 328, in the nature of things, actions taken before breach cannot be within the *British Westinghouse* principle, since the action must arise out of the consequences of the breach of duty.

Two cases are relied on in particular by Mr Johnson for the
F plaintiffs. One is *Daisley v BS Hall & Co* (1972) 225 EG 1553, on the basis of which the Judge decided the present case. In that case a surveyor in 1968 negligently failed to warn the plaintiff, the intending purchaser of a house, of the risk to the house caused by a shrinkable subsoil and the presence of poplars close to the house. Bristow J found that applying the *Philips* v *Ward* measure of damages the loss was £1,750. The plaintiff had, on advice in 1969, removed the poplars and the risk of damage became negligible by the date of the trial in 1972. Bristow J refused to reduce the damages to take account of that fact. He said (at p1557):

What is sauce for the goose is sauce for the gander, and if [the plaintiff] would never be entitled to recover more than £1,750, as *Philips* v *Ward* appears to me clearly to establish, it must follow that he will never be entitled to recover G
less . . .

While *Philips* v *Ward* establishes that the measure of damages is the price paid less the market value of the property at the date of the breach, even though the cost of repairing the property may be greater or smaller than that, it does not follow that the rules of mitigation can never apply to such a case. That would be contrary to the *British Westinghouse* principle. Indeed as Mustill LJ pointed out in *Hussey* v *Eels* [1990] 2 QB 227, at p233, any generalisation that where a loss has crystallised in terms of there being a conventional measure of damages at the date of breach, there can be no mitigation is shown by H
the *Pagnan* case to be unsound. For my part I cannot see why the advantage accruing from the action of the plaintiff in that case to mitigate his loss, *viz* the elimination of the risk to the house by the felling of the poplars, should be left out of account in arriving at the award of damages and there is nothing in *Philips* v *Ward* to compel such a result. In my judgment, the *Daisley* case was wrongly decided.

The second case relied on by Mr Johnson is *Hussey* v *Eels*. In that case the defendant vendors of a property on which there was a bungalow made a misrepresentation to the plaintiffs that the building had not been the subject of subsidence. In reliance thereon the plaintiffs purchased the property in February 1984. They lacked the J
means to pay the cost of repairs. They applied for planning permission to build two buildings elsewhere on the land and in October 1986 sold the property with planning permission to developers at a profit of £25,000. In their action for damages for misrepresentation the trial judge dismissed their claim on the ground that the profit had wiped out the initial loss. But this court allowed the appeal of the plaintiffs. The argument of the defendants to which most attention was paid by Mustill LJ (giving the only reasoned judgment of the court) was that when the plaintiffs' dealings were regarded as a whole it could be seen that they had suffered no loss. He considered the authorities from *British Westinghouse* to *Pagnan* and said that none presented a true
analogy. He continued (at pp239, 240): K

The plaintiffs are not claiming a conventional prima facie measure of damages, as with a sale of goods or shares: they really suffered the loss claimed, for they would have had to pay the cost of repairs if they had remained permanently in residence. On the other hand the later transaction did not flow inexorably from the first, as was the case in *Pagnan*.

After further explanation why none of the authorities provided a direct solution to the problem in the case before him, Mustill LJ said (at p241):

Ultimately, as with so many disputes about damages, the issue is primarily one L
of fact. Did the negligence which caused the damage also cause the profit — if profit there was? I do not think so. It is true that in one sense there was a causal link between the inducement of the purchase by misrepresentation and the sale $2^1/2$ years later, for the sale represented a choice of one of the options with which the plaintiffs had been presented by the defendants' wrongful act. But only in that sense. To my mind the reality of the situation is that the plaintiffs bought the house to live in, and did live in it for a substantial period. It was only after two years that the possibility of selling the land and moving elsewhere was explored, and six months later still that this possibility came to fruition. It seems to me that when the plaintiffs unlocked the development value of their land they did so for their own benefit, and not as part of a continuous transaction of which the purchase of land and bungalow was the M
inception.

Mr Johnson, basing himself on Mustill LJ's words, submitted that when the plaintiff suffers loss in the form of a diminution in value in consequence of the purchase of a property in reliance on the negligent advice of a surveyor, that loss is not subsequently avoided by a subsequent action of the plaintiff affecting the amount of the diminution in value or remedying the defect unless that subsequent action of the plaintiff: (1) flows inexorably from the original transaction; (2) can properly be seen as part of a continuous course of dealing with the situation in which the plaintiff finds himself as a consequence of the negligence; and (3) goes to the same loss.

I preface my comments on this submission by observing that there is a danger in elevating a particular description of a factual situation in a judgment in a particular case to a general principle.

The first condition suggested by Mr Johnson is taken from what Mustill LJ said when distinguishing *Pagnan* from the case before him. I cannot accept that thereby he was laying down a condition for the application of the principle of *British Westinghouse*, nor has he been understood by any text-book writer as so doing. I would add that in *Pagnan* the subsequent purchase only flowed inexorably from the original transaction in the sense that it formed part of a continuous dealing with the situation in which the buyers found themselves; the buyers chose to sell as one of the options open to them.

The second condition is again a statement of the factual situation described in *British Westinghouse* and *Pagnan*. It serves to illustrate what, in my judgment, is a more general principle. As is stated in *McGregor op cit* at para 336, the basic rule is that the benefit to the plaintiff, if it is to be taken into account in mitigation of damage, must arise out of the act of mitigation itself. Similarly in *Benjamin op cit* 16-046:

> The benefit to the plaintiff must arise *out of* his attempts to mitigate his potential loss resulting from the breach: if it arises from his actions which were independent of his mitigating steps, it should not lead to a reduction in his damages.

See also *Chitty op cit* para 26-055. In *Hussey* v *Eels* the obtaining of planning permission and the sale with that permission meant that the benefit arose from actions independent of mitigation. As one commentator (AJ Oakley (1980) 49 CLJ at p396) has observed, the key to the conclusion reached by this court in that case appears to have been the fact that the plaintiffs had had to wait a considerable time and engage in a considerable effort in order to obtain the planning permission which was a prerequisite of any resale. Mr Oakley suggested that the result reached in *Hussey* v *Eels* was likely to prove to be the exception rather than the rule and that it was probable that a profitable resale by a party claiming damages would normally have to be taken into account when those damages were being assessed. I agree that *Hussey* v *Eels* was an exceptional case, turning on its special facts. Similarly in *Hodge* v *Clifford Cowling & Co* [1990] 2 EGLR 89 (in which through the defendant solicitors' negligence the plaintiff tenants lost the right to apply to the court for a new tenancy, but subsequently purchased other larger premises from which they traded) this court held that, like the planning permission in *Hussey* v *Eels*, that purchase was a fortuitous opportunity the advantages of which the plaintiffs were entitled to keep. Bingham LJ at p93 accepted that if it were shown to be a fact that the plaintiffs acted in a manner which, although independent of the plaintiffs' negligence, had the effect of mitigating their loss, it might be proper to take account of that fact.

As for the third condition, I accept that the action must go to the same loss in the sense that (to use the words of Mr Johnson's own premise) it must affect the amount of the diminution in value or remedy the defect.

The common sense of the situation in the present case is that once the plaintiffs were aware that they had purchased a structurally defective property of less value than the price they had paid as a result of the defendants negligence, they sensibly and promptly took steps to eliminate their loss by procuring the remedying by the freeholder of the defect. That seems to me plainly an act of mitigation resulting in a benefit to the plaintiffs which eliminated their loss. I repeat what the judge said, that they have a rectified property worth the equivalent of what they had paid for it without any extra cost to them. The significant point is that this occurred as a result of the pressure applied to the freeholder by the plaintiffs. To take an example suggested by Mr Brunner, if the plaintiffs had sued both the freeholder under the Defective Premises Act 1972 and the defendants in negligence in the same action they could not expect to recover damages in full from the freeholder as well as damages in full from the defendants. Once the property had been put in repair at no cost to the plaintiffs, in my judgment, they cannot be allowed to obtain double recovery by an award of damages against the defendants. To adapt the words of Salmon LJ in *Pagnan*, to allow the plaintiffs' claim would be contrary to justice, common sense and the *British Westinghouse* principle.

For my part, therefore, I would allow this appeal on Mr Brunner's second point.

Agreeing with Hirst LJ that the appeal should be dismissed **Pill LJ** said: On the first point in the appeal, I agree with Hirst LJ and Peter Gibson LJ. Mr Taylor adopted the residual method of valuation because he considered that potential purchasers seeking a house to live in would not have been prepared to purchase a property with the structural defect which was present. The residual valuation was based upon the calculation a property developer would make with the object of obtaining a return on an investment in the property. The judge was entitled to accept Mr Taylor's evidence as to the appropriate method of valuation and as to the figure which, subject to adjustment, was reached.

I find the second point more difficult. If the plaintiffs succeed, there is double recovery in the sense that they retain damages based on the failure of the chartered surveyor to detect a structural defect in 1985 which was rectified in 1990 without cost to them. It is common ground that while the approach to the question whether damages are payable in a particular case must be principled, there are dangers in elevating a decision of fact into a statement of principle. There are dangers in attempting statements of principle in this branch of the law and cases will often be determined upon questions of fact. In the present case, for example, had the plaintiffs found it necessary to sell the property, with defect, for the best price they could obtain in 1988 there is no doubt that they could have obtained damages upon the *Philips* v *Ward* basis. Had there been such a dramatic improvement in the property market that they could have sold the property at a substantial profit even with the defect, they would not ordinarily have had to set off that profit against the *Philips* v *Ward* loss. The windfall would have resulted from the state of the market. On the other hand, had the tortfeasor himself rectified the defect before completion in 1985, it would have been difficult to sustain a *Philips* v *Ward* claim. Whether that claim can be sustained depends on the actual sequence of events and the facts found. In the event, the tortfeasor seeks to take advantage of action taken under a contractual obligation in a contract to which he was not a party. Had the sale to the plaintiff been of the freehold, no such contractual obligation would have arisen and the sequence of events must have been different.

In *Admiralty Commissioners* v *SS Chekiang (The Chekiang)* [1926] AC 637 Lord Sumner stated at p643 that:

> The measure of damages ought never to be governed by mere rules of practice, nor can such rules override the principles of law on this subject.

The defendants rely upon the statement of principle by Viscount Haldane LC in *British Westinghouse Electric Co Ltd* v *Underground Electric Railways of London Ltd* [1912] AC 673 at p689:

> when in the course of his business [the plaintiff] he has taken action arising out of the transaction, which action has diminished his loss, the effect in actual diminution of the loss he has suffered may be taken into account even though there was no duty on him to act.

At the end of his speech Lord Haldane stated that the increase in the power of the replacement turbines in that case

> formed part of a continuous dealing with the situation in which they [the plaintiffs] found themselves, and was not an independent or disconnected transaction.

That there may be situations in which what may be regarded as double recovery can occur is clear. In *Bradburn* v *Great Western Railway* (1874) LR 10 Exch 1, an injured plaintiff who was unable to work and sustained damage, did not have to deduct a payment to him under a policy of insurance against accidents. Pigott B stated at p3:

> The plaintiff is entitled to recover the damages caused to him by the negligence of the defendant and there is no reason or justice in setting off what

> the plaintiff had entitled himself to under a contract with third persons, by which he has bargained for the payment of a sum of money in the event of an accident happening to him. He does not receive the sum of money because of the accident but because he has made a contract providing for the contingency; an accident must occur to entitle him to it, but it is not the accident, but his contract, which is the cause of him receiving it.

Bradburn was applied in the House of Lords, by a majority, in *Parry* v *Cleaver* [1970] AC 1 where it was held that a police pension should be ignored in assessing the financial loss of a police constable injured in a road accident.

The facts in *Bradburn* and in *Parry* are of course very different from those in the present case but at common law the same principles should govern questions of mitigation and remoteness whatever the context. The cases do illustrate that there may be exceptions to the broad general principle which governs the assessment of damages namely that "A successful plaintiff is entitled to have awarded to him such a sum as will, as far as possible, make good to him the financial loss which he has suffered and will probably suffer as a result of the wrong done to him for which the defendant is responsible": Lord Reid in *British Transport Commission* v *Gourley* [1956] AC 185 at p212. Lord Reid added that the general principle was subject to one qualification namely that: "A loss which the plaintiff has suffered, or will suffer, or a compensatory gain which has come or will come to him following on the accident, may be of a kind which the law regards as too remote to be taken into account".

Gourley was also a personal injury case and the issue was whether the tax position should be taken into account in the assessment of damages. At p214 Lord Reid stated:

> I do not think that it is possible to formulate any principle by which it can be determined what is and what is not too remote. Mayne on Damages, 11th ed, p151, refers to "matter completely collateral," and for a general description of what is too remote I cannot find better words, but I do not think that every case can be solved merely by applying those words to it.

Earl Jowitt stated, at p199:

> In all such cases the real issue seems to be whether the facts relied upon as affecting the measure of damages are too remote to be taken into consideration.

Lord Pearson dissented in *Parry*, believing that the tax position should have been taken into account. In his review of the authorities upon the proper method of assessing damages, Lord Pearson cited at p50 the judgment of Andrews LCJ in *Redpath* v *Belfast & County Down Railway* [1947] NI 167 where a voluntary fund was established to relieve the distress of passengers injured on the railway. Andrew LCJ stated at p172:

> The important consideration, to my mind, common to all these cases, is that the circumstance relied upon in mitigation of damages arose independently of the cause of action, and was not naturally attributable to it. Whilst admittedly a sequence it was not a consequence. It arose really as the result of a *novus actus interveniens*, and was not the outcome of the relations between the plaintiff and the defendants which gave rise to the cause of action. The defendants' wrongful act may in each case have been a *causa sine qua non*, but in no true sense was it the *causa causans* of the circumstances relied upon in mitigation of damages. In the present case the *causa causans* of the Fund was not the accident, but the bounty or charitable motives of the subscribers.

The defendants submit that the plaintiffs having required the freeholder to meet his contractual obligation to them to rectify the defect, a reasonable action in the circumstances, they can, in the words of Lord Haldane in *British Westinghouse*, take into account the "diminution of the loss" resulting from that action. In *British Westinghouse*, Lord Haldane distinguished *Bradburn*, which he described as a case from a quite different class, on the ground that: "The reason of the decision was that it was not the accident, but a contract wholly independent of the relation between the plaintiff and the defendant, which gave the plaintiff his advantage". However, in my view, Lord Haldane had in mind the *Bradburn* class of case when categorising the plaintiffs' conduct in *British Westinghouse*, to which the principle he stated applied, as forming "part of a continuous dealing with the situation with which they found themselves, and was not an independent or disconnected transaction". That expression distinguished the approach in *British Westinghouse* from that in *Bradburn*. The expression has been repeated in later cases including *Pagnan* v *Corbisa Industrial Agropacuaria* [1970] 1 WLR 1306 (*per* Salmon LJ at p1315) and *Hussey* v *Eels* [1990] 2 QB 227 (*per* Mustill LJ at p241).

The facts in *Hussey* have been set out by Hirst LJ and Peter Gibson LJ, and Hirst LJ has set out the relevant parts of the judgments in both cases. In holding that the negligence which caused the damage did not also cause the profit, Mustill LJ in *Hussey* referred to the lapse of time between purchase and sale and to the fact that the plaintiffs unlocked the development value of the land. That was done for their own benefit and not as part of a continuous transaction of which the purchase following misrepresentation was the inception. In distinguishing *British Westinghouse,* Mustill LJ said (at p236D) that "there was no question of [that] case being concerned with a chain of disconnected transactions".

In my judgment, the present case, on its facts, is on the *Hussey* side of the line. Hirst LJ has described the intervening events. Years after the defendants' negligence, the freeholders performed their obligation to the plaintiffs under a contract which the plaintiffs had negotiated with them. That had the effect of rectifying the damage resulting from the defendants' negligence. The benefit came by reason of the performance of a contractual obligation by a third party. The plaintiffs had to undertake protracted negotiations with that third party and other third parties, the other tenants in the building. Before that obligation was performed by the freeholder, there was a considerable lapse of time in the course of which the plaintiffs, because of the structural defect, were unable to sell the property when they wished to do so in 1988. In my judgment, the facts relied upon as affecting the measure of damages are too remote to be taken into consideration and, on the facts, the judge was entitled to find for the plaintiffs as he did. On the sequence of events as it occurred, and not as it might have occurred, I do not regard the decision reached by the judge as contrary either to principle or to common sense.

I would dismiss the appeal.

Appeal dismissed with costs.

Omega Trust Co Ltd and another *v* Wright Son & Pepper and another

COURT OF APPEAL

December 17 1996

LEGGATT, HENRY and HUTCHISON LJJ

Estates Gazette May 3 1997

[1997] 18 EG 120

Negligence — Valuation report — Whether disclaimer effective against unknown lender — Whether duty of care negatived — Whether disclaimer satisfied reasonableness test under Unfair Contract Terms Act 1977

In July 1991 N Ltd sought a loan from the first plaintiff, who requested the second plaintiff bank to participate in providing part of the advance. At the first plaintiff's request, the defendant valuers provided a retyped copy of a valuation report they had prepared in January of three short leasehold interests owned by N Ltd and to be provided as security. The report made clear that it was for the first

A plaintiff's purpose "only", and it contained a disclaimer that the report was for private and confidential use of the clients for whom the report was undertaken, that it should not be reproduced or relied on by third parties for any use whatsoever without express authority. The defendant valuers were not informed that the second plaintiffs would rely on the report. N Ltd defaulted on the loan. The plaintiffs claimed that the valuation was negligent. On the hearing of an application by the defendant valuers under RSC Ord 14A, Morland J held that they owed a duty of care to the second plaintiff, that that duty was negatived by the disclaimer, but the disclaimer did not satisfy the test of reasonableness under the Unfair Contract Terms Act 1977. The B valuers appealed.

***Held*: The appeal was allowed. There was an express and effective disclaimer, and no duty of care could be implied on behalf of the defendant valuers to the second plaintiff, as an unknown lender. The disclaimer notice was effective as against the second plaintiff; they were not the valuers' client, however close the relationship was between the two plaintiffs. The valuers were entitled to know who their clients were and to whom the duty was owed. They were entitled to refuse to assume liability to any known lender to whom they had not agreed, and entitled to increase the fee. The disclaimer was not unfair, it was not C unreasonable, it was fatal to the second plaintiff's claim; there was no public policy objection to the disclaimer under the terms of the Unfair Contract Terms Act; the conclusion the judge reached was outside the margin of appreciation afforded to him.**

The following cases are referred to in this report.

Candler v *Crane, Christmas & Co* [1951] 2 KB 164; [1951] 1 All ER 426, CA
Caparo Industries plc v *Dickman* [1990] 2 AC 605; [1990] 2 WLR 358; [1990] 1 All ER 568, HL
Galoo Ltd v *Bright Grahame Murray (a firm)* [1994] 1 WLR 1360; [1995] 1 All ER 16, CA
D *Goodwill* v *British Pregnancy Advisory Service* [1966] 1 WLR 1397; [1996] 2 All ER 161, CA
Hedley Byrne & Co Ltd v *Heller & Partners Ltd* [1964] AC 465; [1963] 3 WLR 101; [1963] 2 All ER 575; [1963] 1 Lloyd's Rep 485, HL
Henderson v *Merrett Syndicates Ltd* [1995] 2 AC 145; [1994] 3 WLR 761, HL
Smith v *Eric S Bush (a firm)* [1990] 1 AC 831; [1989] 2 WLR 790; [1989] 2 All ER 514; (1989) 87 LGR 685; [1989] 1 EGLR 169; [1989] 17 EG 68 & 18 EG 99, HL
White v *Jones* [1995] 2 AC 207; [1995] 2 WLR 187; [1995] 1 All ER 691, HL

This was an appeal by the second defendants, Barker & Co, from a decision of Morland J, who had given judgment for the plaintiffs, E Omega Trust Co Ltd and Banque Finindus, on a summons issued by the second defendants under RSC Ord 14A in proceedings by the plaintiffs against the first defendants, Wright Son & Pepper, and the second defendants for damages for negligence.

Rupert Jackson QC and Alison Millar (instructed by Cameron Markby Hewitt) appeared for the appellant second defendants; Roger Smith (instructed by Hugh Cartwright & Amin) represented the second plaintiff; Michael Gadd (instructed by Ince & Co) represented the first defendant; the first plaintiffs, Wright Son & Pepper, did not appear and was not represented.

F Giving first judgment at the invitation of Leggatt LJ, **Henry LJ** said: Nantell plc owned three short leasehold interests in small supermarkets in central London. Mr Sharif controlled Nantell and he became the guarantor of certain loans made to Nantell secured on those three properties. Mr Barker FRICS, was a partner in the second defendants, Barker & Co, who were professional valuers. In connection with the indebtedness of the company to the bank, he was instructed by the company's clearing bank to value these short leasehold interests on the three properties on an open market bricks and mortar basis. He had previously valued certain residential properties for Mr Sharif. He was instructed by the bank (this was in January 1991) that the valuations were to be sent to the bank and that the invoice was to be sent to Mr Sharif.

On January 15 1991 he performed the task. He valued the properties G together at £945,000 and he sent a valuation, expressed to be for Nantell plc, containing a disclaimer, the terms of which must be examined later, as those terms are attacked under the provisions of the Unfair Contract Terms Act 1977.

In June 1991 Mr Sharif applied on Nantell's behalf for a loan from the first plaintiffs, Omega Trust Co Ltd ("Omega"). He intended to consolidate his existing borrowings from the National Westminster and Barclays with this new loan which was to replace them. The second plaintiffs, a bank called Finindus, were closely linked with Omega. They were linked through a Mr Dubois who was a director of both and the chairman of one. In the past (though of course the valuers knew H none of this) they had participated together in making loans, those loans being called syndicated loans in the sense of having two or more financiers to lend the money to the borrower.

In July of that year Omega asked the Finindus bank to participate in the loan. The bank agreed subject to obtaining an independent valuation (this is what they told Omega), valuing the three properties at not less than £750,000. The proposed division of the loan was that 40%, or £150,000 should be lent by Omega, and the remaining £200,000 by the bank. That loan was agreed subject to valuation. On July 30 Omega rang Barker & Co to ask to be provided with the valuation. They did not say anything about Banque Finindus. Barker & J Co knew the purpose and the terms of the loan, but they had no actual knowledge of Banque Finindus nor of any lender other than the first plaintiff, Omega, nor did they know of any connection between Omega and Finindus.

The valuation that had originally been prepared for the clearing bank was retyped; a paragraph relating solely to requirements of the bank was omitted; the heading was changed; it was now made clear that the valuation was "for Omega" and the disclaimer stood in exactly the same terms as before, namely it said:

> This report shall be for private and confidential use of the clients for whom the report is undertaken and should not be reproduced in whole or in part or relied K upon by third parties for any use whatsoever without the express written authority of the surveyors.

As part of the transaction whereby Omega were made the client, they were invoiced for some £30 plus VAT, and the loans were completed on October 28. It was pleaded on the plaintiffs' behalf that at that date the security was insufficient, indeed it was pleaded that the securities were valueless, and shortly thereafter the company went into liquidation without repaying any of the loan; the securities proved valueless and the lenders sought to recover their loss from two defendants (and we are only concerned here with the valuers) alleging L negligence against them.

The valuers applied under RSC Ord 14A, on the question as to whether they owed a duty of care to the undisclosed lender, Finindus. The judge at first instance, Morland J, after a contested hearing, agreed that it was a proper case for Ord 14A, and the parties went on to argue that point as to whether a duty was owed by the second defendants, the valuers to the bank.

There were three issues raised before the judge. The first issue was: absent an effective disclaimer, did the valuers owe a duty of care to the bank in respect of the three valuation reports sent to Omega? The judge found that they did owe such a duty. The second issue related to M the disclaimer: would the disclaimer, if effective, negative such a duty of care? The judge found that it would. The third issue was: does the disclaimer notice satisfy the requirement of reasonableness under the Unfair Contract Terms Act 1977? The judge found that it did not. He was not satisfied that the disclaimer was fair and reasonable and therefore the disclaimer notice was not effective.

In relation to the first issue, the duty of care, we are of opinion that as, in our judgment, there was a good and effective disclaimer in this case, it is hypothetical as to whether a *Hedley Byrne* duty would have been owed to the unknown lender in the absence of such a disclaimer, because, as the headnote in *Hedley Byrne* puts it, "we find there to have been an express and effective disclaimer. No duty can be

A implied", but we consider it to be useful to draw attention to a point as to why there would not in any event have been a duty made by Mr Rupert Jackson QC on behalf of the valuers in this case.

The judge found that there was such a duty in the following passage. He first made the point that the second defendants were professional valuers, valuing to bankers:

> . . . whom they knew would rely upon them for the express purpose of which they knew in deciding whether to grant loans with an upper limit of £750,000 to their clients, Nantell Plc, controlled by Mr Sharif, who was also in effect their client.

B He then continued in this way:

> Viewed objectively the second defendants ought to have foreseen the reasonable *possibility* that the grant of the loans might not be made solely by the first plaintiffs but might have been syndicated by them or made in association with another bank whom the second defendants ought to have foreseen would rely upon the valuations for the express and very same purpose for which they were provided to the first plaintiffs. In my judgment, it is immaterial that the second defendants did not know the name of the second plaintiffs or even the actual existence of the second plaintiffs. What is material is that they should have foreseen the existence of some such bank whom they ought to have known would be likely to have relied upon the valuations for the purpose for which they were provided.

C (*Emphasis supplied.*)

Mr Jackson submitted that as a matter of law, the bare possibility that the first plaintiffs' loan might not be made solely by the first plaintiffs was not enough. He submitted that under the *Hedley Byrne* test there had to be a probability that the grant of the loans might not be made solely by the first plaintiffs. He addressed us with a forcible and sustained submission on that point, and it may be useful for any future occasion when the point may be considered, simply to refer to the authorities he cited: *Candler* v *Crane, Christmas & Co* [1951] 2 KB 164; *Hedley Byrne & Co Ltd* v *Heller & Partners Ltd* [1964] AC
D 465; *Smith* v *Eric S Bush (a firm)* [1990] 1 AC 831*; *Caparo Industries plc* v *Dickman* [1990] 2 AC 605; *Galoo Ltd* v *Bright Grahame Murray (a firm)* [1994] 1 WLR 1360; *Henderson* v *Merrett Syndicates Ltd* [1995] 2 AC 145; *White* v *Jones* [1995] 2 AC 207; *Goodwill* v *British Pregnancy Advisory Service* [1996] 1 WLR 1397.

Those were the authorities that he relied on. That point is for decision another day. On the findings we have made it does not become a live issue in this case.

Turning to the second issue, the effect of the disclaimer, it is submitted by Mr Roger Smith for the bank, is that there was no breach of the disclaimer in this case. That submission turns on a proper
E construction of the disclaimer. He submits that, properly construed, "clients" must be taken to mean not only Omega, but any closely related party to Omega, and would only be excluded in relation to a total stranger. He makes the point that there was a close relationship between Omega and the bank, as I have already summarised, and therefore they came within the ambit of clients. But the bank was a separate legal entity; the borrower was entirely ignorant of it. As a matter of construction, the only clients in relation to the valuation were the clients named as the clients (who the valuation was for), that is to say, Omega. On the facts of this case, a lender unknown to the valuer is not the valuer's client, however close his relationship with the
F known lender client (Omega) might have been. Accordingly, in relation to the second issue, the disclaimer notice, it effectively would negative any duty of care that there might be.

The next question relates to whether the disclaimer notice satisfied the requirements of reasonableness under the Unfair Contract Terms Act 1977. That was an Act imposing limits on the extent to which liability for, *inter alia*, breach of contract could be avoided by contractual terms. The context set by the Act in section 1(3)(a) is in relation to things done by a person in the course of a business. We are here concerned with negligent liability under section 2(2) which reads:

> In the case of other loss or damage, a person cannot so exclude or restrict his liability for negligence except in so far as the term or notice satisfies the requirement of reasonableness. G

The reasonableness test is dealt with in section 11(3) which reads as follows:

> In relation to a notice (not being a notice having contractual effect) the requirement of reasonableness under this Act is that it should be fair and reasonable to allow reliance on it, having regard to all the circumstances obtaining when the liability arose or (but for the notice) would have arisen.

That then is the statutory background.

The next matter is to distinguish the commercial setting between H
commercial bodies in this case from the ordinary domestic householder purchasing his home such as was considered in *Smith* v *Bush* to which I have already referred. There Lord Griffiths in his speech at p859 said:

> But I expressly reserve my position in respect of valuations of quite different types of property for mortgage purposes, such as industrial property, large blocks of flats or very expensive houses. In such cases it may well be that the general expectation of the behaviour of the purchaser is quite different. With very large sums of money at stake prudence would seem to demand that the purchaser obtain his own structural survey to guide him in his purchase and, in such circumstances with very much larger sums of money at stake, it may be J
> reasonable for the surveyors valuing on behalf of those who are providing the finance either to exclude or limit their liability to the purchaser.

The point made on behalf of the second plaintiff to us was that Lord Griffiths was there dealing with a lender/purchaser situation and so that passage does not apply by way of analogy in this case. In my judgment it does. The point that Lord Griffiths was making was that in a transaction involving a lot of money in a commercial context, both parties are well able to look after themselves, and it is not necessary to have any statutory protection for them. It is that point which, in my judgment, distinguishes this commercial case from those purchases of domestic houses with which other cases have dealt. K

Lord Griffiths in his speech dealt with various matters for and against that the court would need to take into consideration. The judge considered them. He pointed out, rightly, that the burden of proof to sustain the exemption clause was on the valuer. Next he found that the parties were of equal bargaining power. That conclusion is not challenged. In a commercial context, it would seem to me to point to upholding the exclusion clause rather than not.

The next point raised by Lord Griffiths and considered by the court was whether it would have been reasonably practicable for the bank to have obtained a valuation for themselves. Obviously it would have been. It would have been the easiest thing in the world either to go and L
get their own independent advice, or to do what the disclaimer would, in the ordinary course of business, prompt them to do, that is to say, to get in touch with the valuer and ask the valuer as to whether they too could rely on the valuation given. Accordingly, the judge found that it was reasonably practicable and that, in my judgment, would also be another pointer in relation to the valuer's assertion in submissions in support of this clause.

Next, here there was no hurry or cost imperative that would point against that conclusion. The fourth point is that this valuation was a straightforward, easily duplicated task. There could be no question of the bank being unable to get a valuation done, and done in the time. M

The next and important point to be considered is the practical consequences of the matter one way or the other. Viewed at the moment of breach under section 13(3) when the liability, but for the notice, would have arisen, at that time the valuers would have discovered for the first time that there was a second lender, that that lender had taken responsibility for 60% of the loan without either getting in touch with them or anyone notifying them or seeking their permission, or making any payment to them. But for that lender appearing on the scene, the valuer's liability if negligent would have been to Omega alone, but that liability would have been for the same sum of money whether there was one lender or two.

---

*Editor's note: Also reported at [1989] 1 EGLR 169

At this moment it is necessary to look at the purpose of the disclaimer. The first and obvious purpose of the disclaimer, as obtained by construction of the document, is to limit the assumption of responsibility to Omega and to no one else. It was clearly entered into to assure clarity, to assure transparency and to assure certainty. The valuer was entitled to do all that could be done to prevent himself having to fight a difficult law suit as to whether he owed a duty to an unknown lender. If his disclaimer had been complied with by either Omega or by the bank, that would have been the position. If his document had been complied with and consent from him had been sought, he could, had he wished, have declined to assume the additional responsibility. It can, in certain circumstances, be more onerous in fact to face potential claims from two plaintiffs who may be separately represented, and in relation to whom there may be two measures of damage appropriate, than to face claims from one, though that point pales into insignificance beside the point that what he had been deprived of was a position where everything would have been clear and certain and he had, through no fault of his own, a position where he could only rely on the terms of the disclaimer after fighting a difficult law suit in relation to it.

Against this the bank submits that this is simply an uncovenanted benefit to the valuer, and it would be unreasonable to let him rely on it. Unreasonable because, had in fact they asked permission of him, he probably would have granted permission. As to whether he would have granted permission or not, we will never know because he was never asked. Certain it was that no fee was paid to him and that he would have been entitled to a fee had permission been sought of him.

It seems to me that this professional valuer, valuing expensive properties in a commercial context, was entitled to know who his client was and to whom his duty was owed. He was entitled, it seems to me, to refuse to assume liability to any unknown lender, indeed, I would go further and say that he is entitled to refuse to assume liability to any known lender to whom he had not agreed. He was entitled to increase the fee (or would have been) as a term of permitting the second lender to rely on the valuation because, as I have said, potentially it can be more expensive to be sued by two lenders rather than one. And if the second lender was not prepared to pay what is asked, it seems to me that the valuer would have been entitled to refuse to assume that liability to the bank, and the 1977 Act would not have required the contract to be rewritten. Here it may be said by way of comment that Omega themselves when they were introduced recognised the valuer's right to know who his client was. They acted properly, sought his permission and became clients. They received his report and the disclaimer; they never objected to the terms of the disclaimer; they never asked for permission to disclose it to their co-lenders; they were prepared to approbate the contract in the sense of acting and relying on the disclaimer, but they seem to have reprobated it by disclosing it to their co-lenders without having obtained the permission of the valuers first. If they had raised the question with the valuers, the bank could have been accepted as clients and risk could have been assumed to them, but only if the valuers were prepared to assume responsibility, and not otherwise.

It seems to me *prima facie* that Omega were only entitled to put the valuation before the bank for reliance if they had sought permission from the borrowers, and the fact they did not would, on the face of it, seem to be a *prima facie* breach of contract on their part unless the clause was unlawful under the Act.

On the basis of all those considerations, I am satisfied: first that the disclaimer is not unfair; second, that the disclaimer is not unreasonable; third, that the disclaimer is fatal to the bank's claim; fourth, that there is no public policy objection to the disclaimer under the terms of the Unfair Contract Terms Act or otherwise; and, fifth, (though this is a matter of balancing conflicting interests and the appreciation of that balance is first and foremost to the trial judge) that the conclusion that the judge here reached was outside the margin of appreciation that is afforded to him, and in the circumstances, in my judgment, wrong. Therefore, for those reasons, I would allow this appeal.

HUTCHISON LJ agreed and did not add anything.

Also agreeing, LEGGATT LJ said: In my judgment, the appellants, by the words which each of their valuation reports contain, effectively disclaimed any assumption of a duty of care in relation to any person other than Omega. I have no doubt that, having regard to all the circumstances specified by Henry LJ, it is fair and reasonable to allow reliance on the disclaimer. The appeal will therefore be allowed. We will declare that Barker & Co owed no duty of care to the bank in respect of the valuation reports, and unless cause be shown to the contrary, the bank's claim against Barker & Co will be dismissed with costs.

Appeal allowed.

Routestone Ltd *v* Minories Finance Ltd and another

Same *v* Bird and others

CHANCERY DIVISION

May 16 1996

JACOB J

Estates Gazette May 24 1997

[1997] 21 EG 148

Negligence — Exercise of mortgagee's power of sale — Whether estate agent instructed by mortgagee and receivers carried out negligent marketing — Whether opinion of expert on ultimate question of negligence admissible

The plaintiff company, a subsidiary of the G group of companies, was the owner of three properties mortgaged to the defendant bank (the first defendant in the first of the two actions). By reason of the indebtedness of the plaintiff to the bank, the first and second defendants in the second action were appointed receivers and in respect of two of the properties exercised a direct power of sale. In respect of the third property the bank exercised the power of sale conferred by section 101 of the Law of Property Act 1925. The bank and the receivers used the services of KFR, a firm of estate agents, a defendant to both actions. The plaintiff claimed: (1) in the first action, that the bank ought not to have exercised the power of sale because the plaintiff and G were about to redeem ("the imminent redemption issue"); and (2) in both actions, that the advice about marketing given by KFR was negligent and that they, the bank and the receivers were liable to the plaintiff ("the negligent marketing issue").

***Held*: The actions were dismissed. (1) The claim, that the bank and the receivers knew that redemption was imminent and ought not to have made the sales concerned, was rejected on the facts. In any event, as a matter of law, the security documents gave the receivers an unrestricted right to sell at any time, and the fact that the plaintiff claimed that they would shortly be able to redeem could not give the plaintiff a right in law to restrict those powers. (2) The evidence did not remotely establish even an error by KFR, still less an error of the kind amounting to negligence. Having regard to the nature of the interests being marketed, there was no obligation on KFR to market more widely than they did; there was also no obligation on KFR to recommend that properties should be subject to a programme of "working" by dealing with rent reviews and refurbishment. The fact that another estate agent would have marketed the property differently did not indicate negligence. An**

A **expert's opinion on the ultimate question as to whether the defendant had been negligent was admissible in evidence. However the opinion on its own (unless uncontested) would be empty rhetoric and what really mattered in most cases was the reasons given for the opinion. It was a good thing for an expert to be given the legal test and the facts as alleged by the plaintiff.**

The following cases are referred to in this report.

Bown v *Gould & Swayne* unreported January 24 1996
Commercial & General Acceptance Ltd v *Nixon* (1981) 152 CLR 491
Cuckmere Brick Co Ltd v *Mutual Finance Ltd* [1971] Ch 949; [1971] 2 WLR 1207; [1971] 2 All ER 633; (1971) 22 P&CR 624; [1971] RVR 126, CA
B *Downsview Nominees Ltd* v *First City Corporation Ltd* [1993] AC 295; [1993] 2 WLR 86; [1993] 3 All ER 626
Duchess Theatre Co v *Lord* [1993] NPC 163; The Times December 9 1993, CA
Liddell v *Middleton* The Times July 17 1995, CA
Midland Bank Trust Co Ltd v *Hett, Stubbs & Kemp* [1979] Ch 384; [1978] 3 WLR 167; [1978] 3 All ER 571
Mölnlycke AB v *Procter & Gamble (No 5)* [1994] RPC 49, CA
Tomlin v *Luce* (1889) 43 ChD 191
Whitehouse v *Jordan* [1981] 1 WLR 246; [1981] 1 All ER 267, HL

This was the hearing of two actions by the plaintiff, Routestone Ltd, the first against Minories Finance Ltd, and second against Colin C Graham Bird, Andrew Holman and the firm Knight Frank & Rutley.

Anthony Trace and Richard Morgan (instructed by Holman, Fenwick & Willan) appeared for the plaintiff; Michael Crystal QC, Richard Adkins and Roxanne Ismail (instructed by Freshfields) represented Minories Finance Ltd, Colin Graham Bird and Andrew Mark Holman; Robert Moxon Brown QC and Clare Brown (instructed by Kennedys) represented Knight Frank & Rutley.

Giving judgment, **Jacob J** said: The plaintiff, Routestone, is a company within the Gomba Group of companies. In 1985 this group was ultimately controlled by a Mr Shamji. It still is, though during the D events which I have to consider receivers had been appointed pursuant to powers in mortgages over the properties owned by Routestone: other companies in the group were in receivership at the same time The Routestone properties included the freehold reversions in the three estates with which I am concerned, namely North End Parade ("NEP"), 128-132 Grosvenor Road ("GR") and the St George's Estate ("SGE"). The receiverships came about because Gomba was heavily indebted (in a sum of about £21m) to its bankers then called Johnson Matthey Bankers Ltd and now called Minories Finance Ltd ("the bank").

The bank commenced appointing receivers over the Gomba E companies and their assets on October 25 1985. Appointments continued during early November 1985. The bank's right to do so was disputed. Gomba said that the bank had agreed not to appoint receivers provided certain conditions were fulfilled. There were conditions as to payment of certain sums by a certain date which was agreed to be extended by 14 days provided negotiations with Lonrho (who Mr Shamji was claiming would provide a rescue package) were "proceeding in a proper and expeditious manner". The sums were not paid within the initial period and receivers were appointed during the period of potential extension. The dispute was whether the condition for an extension had been fulfilled, for if so the receivers should not F have been appointed. The action commenced with a claim for interlocutory relief on about November 8 1985. A speedy trial was ordered and happened. Judgment was given by Hoffmann J (as he then was) on January 16 1986. The Gomba Group's claim that the condition had been fulfilled was roundly rejected. For present purposes it suffices to say that Hoffmann J found Mr Shamji an unreliable witness and that he had at various points made unfulfilled promises to the bank about repayment or negotiations with third parties which were supposed to lead to repayment.

I have two actions before me, 1989 No 3764 and 1989 No 3765. 3764 is about the receiverships over NEP and SGE, 3765 about the receivership over GR. In each case (and indeed in all other cases) the receivers were Messrs Holman and Bird of Price Waterhouse (the "receivers"). It is common ground that at all material times the position G in relation to SGE and GR was the mortgages concerned conferred a direct power of sale on the receivers. It was they who effected the sales of these properties. The mortgage over NEP did not itself contain any power to appoint a receiver or for the mortgagee (ie the bank) to sell. But the power to appoint a receiver was conferred by section 101(1)(iii) of the Law of Property Act 1925. And the bank itself, as mortgagee, had the power to sell conferred by section 101(1)(i) of that Act. In practice the bank and the receivers used the services of the well known firm Knight Frank & Rutley ("KFR") in relation to the sales of all three properties in ways which I will have to explore in detail later.

In the first action the defendants are the bank and KFR. It is H claimed against the bank alone that it knew that Routestone and the other Gomba companies intended to and were prepared to do their utmost to cause the redemption of all the receiverships and that it was highly probable that they would be discharged. It is suggested that accordingly the bank ought to have stayed its hand on the sale of the relevant property the subject of the mortgage. This property was SGE, the other two properties having been sold at an earlier stage. I will call this claim the "imminent redemption" issue.

The claims in the first action also relate to the way in which SGE and NEP were in fact sold. It is said that the advice given by KFR and accepted by the bank via the receivers in relation to the sales was J negligent and that they and the bank are both liable for such negligence. In the second action the claim is that KFR were negligent in relation to the advice given in relation to the marketing of GR. The claim is made against the receivers and KFR. I call the claims involving alleged negligence the "negligent marketing" issue and it will be necessary to examine what was done in relation to each of the three properties separately. I should perhaps record that at one point there was a third issue involving a suggestion that the powers of sale had come to an end before SGE was sold. The point was dropped after some discussion, not surprisingly in view of the decisions of Mummery J and the Court of Appeal in *Duchess Theatre Co* v *Lord* K The Times, December 9 1993 (litigation also involving Routestone, the bank and the receivers). It was partly as a result of waiting for the result of this that these cases have taken so long to come to trial, having been commenced in 1989.

Concessions of law

It is, of course, settled law that in exercising a power of sale a mortgagee or receiver owes a duty to the mortgagor to take reasonable care to obtain a proper price: *Downsview Nominees Ltd* v *First City Corporation Ltd* [1993] AC 295 at p315 *per* Lord Templeman (PC). On the pleadings two unsettled points of law emerge. First, is it a defence for a mortgagee or receiver to say "I instructed a competent L estate agent"? Second, does any such agent owe a direct duty to the mortgagor, either as a result of an equitable duty or as a duty of care for the purpose of the tort of negligence? The second point raises a subpoint, "Is any duty of care of the agent wider than the duty of care of the mortgagee".

There is no settled authority on the first point. It was conceded at first instance in *Cuckmere Brick Co Ltd* v *Mutual Finance Ltd* [1971] Ch 949 that the defence did not run. The Court of Appeal would not consider it because the evidence might have turned out differently if the point had been alive. The point has a long history (going back at M least to *Tomlin* v *Luce* (1889) 43 ChD 191) and has been the subject of considerable debate in Commonwealth decisions, particularly *Commercial & General Acceptance Ltd* v *Nixon* (1981) 152 CLR 491 in the High Court of Australia. The second point may be related to the first, for if the mortgagee is not liable for the agent's negligence and the agent himself owes no duty of care then the mortgagor would be without any remedy against anyone when his property is sold at an undervalue through an agent's negligence.

It became clear that full argument on even the first point would take a considerable time. And in relation to the second, Mr Robert Moxon Browne QC (for KFR) and Mr Anthony Trace (for Routestone) were preparing to lock horns on whether the alleged duty of care existed and if so how wide it may be. Interesting though complex legal argument may be for some lawyers, it is not the function of the court to hear it at great expense to the parties if it can be avoided. I therefore inquired of the parties whether or not for the purposes of this case only, sensible concessions could be made. It is to the credit of all that they were. I must record the concessions.

(1) Mr Michael Crystal QC for the bank and the receivers accepted that if KFR were negligent in the performance of their duty to the bank or the receivers, the bank and the receivers would be answerable for that negligence to Routestone. The concession was made solely for the trial before me, the point being reserved for argument in the Court of Appeal if necessary. That would be possible because, (unlike *Cuckmere Brick*) in this case the point remained alive until all the evidence was in.

(2) Mr Trace accepted that if KFR were not in breach of any duty to the bank or the receivers, then they would not be liable to Routestone. In other words they were under no wider duty than the duty they owed to those who instructed them.

Imminent redemption?

Routestone's only witness on this issue was a Mr Robinson, a senior partner in the firm of Holman, Fenwick & Willan ("HFW"). Mr Shamji, who could have given more direct evidence, or provided corroboration in respect of some of the matters to which Mr Robinson testified did not give evidence although he had given a witness statement and was present throughout most of the hearing. A large part of Mr Robinson's proposed evidence (as contained in his witness statement) was that he thought that the bank, the receivers and their solicitors (Freshfields) had behaved unfairly. I considered that some of the allegations could not be material to this case and ruled out parts of the statement. In the end it was suggested by Mr Crystal that Mr Robinson should give his evidence in chief orally and, Mr Trace not demurring, this was the course that was followed.

I have to say that I found Mr Robinson's evidence did not remotely support the factual allegations made in the pleadings which I have summarised above. Although I acquit him of any intention deliberately to mislead the court, I do think that his whole thinking about the case — from very early on — was coloured by an over-zealous passion for his clients' cause. Such an attitude manifested itself from the very outset of his involvement with Gomba group. Mr Robinson (and HFW) were, I think party to a number of instances where they were not as full and frank as they ought to have been or wrongly exaggerated the real position. For instance in relation to a potential saviour called Mr Abili, Mr Robinson telexed Freshfields on February 26 that his clients were in a position to discharge their liability. At that point he had nothing whatever in writing and when a document did emerge (on March 3) it was far from being a cast-iron offer of the necessary funds and Mr Robinson did not disclose it to the bank or receivers. Nor was it disclosed to Hoffmann J in the further proceedings then pending before him. Likewise, in September, when some negotiations with another potential saviour were taking place, Mr Robinson allowed the saviour's solicitors to continue to consider the proposed deal as though it still included the SGE when he knew it had been sold. He thought his client would not want this information revealed, even though the opposite party was plainly being misled and may well have been incurring expenditure in relation to the proposed acquisition of this property.

Mr Robinson's and HFW's attitude throughout seems to have resulted from a misplaced enthusiasm for the client, which I think affected the objectivity which a solicitor should provide. In particular they were not willing to provide full details of various attempts at putting packages for redemption together. Instead assurances that a package would be available shortly were given in woolly terms — terms which the receivers, the bank and Freshfields were fully entitled to regard and did regard with considerable suspicion. I do not go so far as to hold that Mr Robinson's evidence was coloured further by the fact that HFW have taken the remarkable step of giving the Gomba group £2m worth of credit in respect of legal fees secured in part by a charge over certain claims made in a separate account, though the giving of that amount of credit may well be another manifestation of an over-zealous involvement on behalf of the client.

To a large extent Mr Robinson originally sought to traverse ground already covered by Hoffmann J in the action to which I have already referred and in later rulings on various applications made by the Gomba group. In the end Mr Robinson was prepared to accept the statements of fact made by Hoffmann J as correct and so I can, with gratitude, adopt them for the early part of the history of attempts to prevent sales by the bank/the receivers by promises of some sort of rescue. I have already summarised the position held by Hoffmann J on January 16 1986. On March 24 1986 he rejected motions relating to claims for redemption and the disclosure of information by the receivers. Hoffmann J said, in passages finally accepted by Mr Robinson under cross-examination:

> Part of the background is that the relationship between Mr Shamji's solicitors, Messrs Holman, Fenwick & Willan ("HFW") on the one hand and the receivers and their solicitors Messrs Freshfields on the other has not been easy. Few disposals of assets have not provoked threats of legal proceedings for breach of duty in selling at an undervalue. HFW have been free with accusations of bad faith against the receivers and the bank. Letters of complaint dealing with every aspect of the receivership have been written to the receivers' professional body. The first two and a half months of the receivership were taken up with heavy litigation, fought to a finish, in which the validity of the receivers' appointment was challenged. I am not suggesting that the receivers are entitled to penalise Mr Shamji for being difficult. But I do not think it is unreasonable for them to be wary about the disclosure of even apparently innocuous information about their activities in case it should precipitate legal proceedings in which they would have to disclose more confidential information in order to defend themselves.
>
> Shortly before the trial of the earlier proceedings to which I have referred, Mr Shamji brought a motion for the disclosure of information rather similar to this one. The basis of the motion was that a third party, which on that occasion he did identify, was ready and willing to provide funds to pay the group's debts to the bank in full. The receivers provided a redemption statement and offered to provide the third with information. On this basis, the plaintiffs consented to their motion being dismissed with costs. The receivers did enter into negotiations with the third party but it appeared that he was not willing to pay the indebtedness in full and no agreement was reached.
>
> Further requests for information began immediately after the failure of the action to remove the receivers when HFW asked for another redemption statement. Freshfields wrote on 28th January 1986 enclosing such a statement but saying:
>
> "When you are in a position to demonstrate that the necessary funds are available for settlement and to identify a specific date for payment, we will arrange for new figures to be calculated to the agreed date."
>
> By a letter of the same date, HFW asked for "full details . . . of the disposals made by the Receivers and the terms upon which those disposals have been made, or upon which the Receivers have contracted to make such disposals."

I can omit an irrelevant paragraph. Going on with Hoffmann J's judgment:

> On 20th February Freshfields wrote giving some and refusing some of the information which had been requested but adding:
>
> "If we were convinced that repayment was imminent, the Receivers would of course provide more details."
>
> On 26th February HFW sent a telex saying that the companies were now in a position to discharge their liabilities and asking for an urgent meeting to arrange for agreements bringing the receiverships to an end. They asked for Mr Shamji's accountants to be given access to all files to verify the balance due. Freshfields replied on the same day saying a minimum of £11.4 million was needed to discharge the indebtedness and adding —
>
> "As we have emphasised before our clients are not prepared to enter into detailed discussions about the sums needed to discharge the total amounts owing to the bank until you can provide us with firm evidence satisfactory to our clients that your client has sufficient monies held to his order to satisfy the debt in full . . . Our clients are not prepared to give access to the files to your accountant."
>
> This telex was met by an accusation of bad faith and a claim of legal right. On 28th February 1986 HFW telexed saying that —

A “Mr Shamji is a director of the companies in receivership and is entitled to information regarding their affairs and full access to their records including all contracts and correspondence.”

This assertion, as I have already said, was quite wrong. The telex did however add the information that Mr Shamji was not in a position to redeem from his own resources but that —

“The payment will be made by a third party who will have assigned to him the benefit of all sales not complete. A substantial deposit is immediately available to be paid on exchange of contracts.”

On 3rd March 1986 HFW followed this up with a telex saying that they held £500,000 on a joint account to be available as a deposit until completion and that the remaining balance up to £12 million was “available”. They also asked
B for copies of all contracts to be assigned to the third party at completion to be made available to a Mr G D Silver who was acting as his solicitor.

On the same date Freshfields sent a telex asking for more information about the third party. Who was he and who was his banker? There were also questions about the proposed form of the transaction. Was the third party going to buy the assets (including the uncompleted contracts) from the receiver? Or was be going to pay off the debts and take over the bank’s security? Or was he going to take over the plaintiff companies and then redeem? The telex ended:

“It is not our intention to frustrate a genuine offer by Mr Shamji to settle in full all outstanding liabilities owed to the bank. As we have repeatedly told you, we are prepared to have a meeting but only if your client can show to our clients’ satisfaction that the potential purchaser has the funds available to
C discharge the full indebtedness. In view of your telex we would also need to know the identity of the third party to which you refer.”

On the following day, 4th March, there was a further exchange of telexes. Freshfields asked for information about the joint account and added

“we anticipate that it is unlikely that our clients will be prepared to release [the requested contracts] at least until the points referred to above and in our yesterday’s telex have been met and it is apparent that bona fide and realistic negotiations are proceeding”

HFW said that they confirmed that it was the intention to discharge the full indebtedness “by means of a sale of assets to a third party”. No further details about the third party, his means or the nature of the transaction were supplied.

I have set out the correspondence leading up to the 4th March in some detail
D because that was the date of issue of the writ and notice of motion claiming information which are now before me. The first question is whether the plaintiffs had at that date an arguable case for saying that they were entitled to more information than they had been given. All the relevant facts are in the correspondence between the solicitors. In my judgment the plaintiffs had not demonstrated any need to know more facts than they had already been given. I am willing, as I have said, to assume that a company mortgagor is entitled for the purposes of exercising its right to redeem to be given particulars of the amount still due and the assets which would be returned on redemption. These would include particulars of the uncompleted contracts made by the receivers. But I do not think that for this purpose it is sufficient for the mortgagor merely to say that he wants to redeem or that he has some mysterious third party who
E will put up £12 million. The history of this case, both before and after the appointment of the receivers, is a chronicle of unfulfilled assurances by Mr Shamji that someone was just about to provide the money to pay his debts to the bank. In my judgment the receivers were under no obligation to provide any information until they had firmer evidence that there was a realistic prospect of redemption.

This case is of course concerned with the period beyond that covered by Hoffmann J’s judgment of March. He gave a further judgment on May 23, refusing an interlocutory order for redemption, when it remained the position that the terms of some rescue package were too vague. And there was a similar judgment on July 2. The fact
F is that right up until the sale of the last of the three properties, SGE, on September 25 1986 the same pattern continued. At no point did any firm, credible, evidence of a realistic prospect of redemption emerge. And still less did Mr Robinson or HFW provide any reasonable information upon which the receivers or the bank might reasonably have held their hand. The nearest point was in June/July when an unnamed client of Ashurst Morris Crisp emerged as a possible saviour. But he quickly withdrew, giving to Mr Bird as his reason:

Dealing with Mr Shamji and his advisers was “like hitting your head against a brick wall”.

I do not think it necessary to go further into Mr Robinson’s evidence — it simply is of no value. But I should record that Mr Harper of the bank gave unchallenged evidence of a string of G
unfulfilled promises by Mr Shamji during 1984 and 1985. And Mr Homan (one of the receivers) gave unchallenged evidence about promises of “white knights” during the first half of 1986. The attitude taken in the first receiver’s report of February 28 1986 was that:

. . . the possibility of some sort of a deal with Larco, Vado or others lingers in the background and cannot be totally dismissed, but it now seems more likely that Mr Shamji’s current approaches are designed to create a smoke screen in relation to the bank’s pursuit of its judgment debt against him

I think that was the only rational view to take. And nothing changed until well after the sale of SGE in September. H

Accordingly I reject on the facts the claim based on a claim that the bank and the receivers knew that redemption was round the corner and ought not to have made the sales concerned. The whole plea amounts to no more than a claim that receivers and the bank should have paid heed to a series of vague Micawberesque promises.

Further, however the claim is simply bad in law. As Hoffmann J said in his judgment of March 24:

The security documents give the receivers an unrestricted right to sell at any time. Until actual redemption or at least until a valid tender of the redemption price, these powers continue to exist. The fact that the plaintiffs claim that they will shortly be able to redeem cannot give them a right in law to restrict the J
powers granted to the receivers.

It cannot be wrong to exercise a power of sale which exists. Alleging that a decision to exercise a power is negligent in itself is tantamount to saying that the mortgagee or receiver is a trustee of the power of sale which he is admittedly not. No duty of care is owed by the mortgagee or receiver in relation to the actual decision to sell. Mr Crystal made it clear that in practice if a credible real offer of redemption had been made at any time then the sales might have been delayed. That would be a sensible thing to do but there would be no legal obligation so to do.

I therefore reject the claim based on alleged imminent redemption. I K
do not think it ever had any prospect of succeeding and ought not to have been advanced.

Negligent marketing?

The relevant legal standard was not in dispute. Routestone had to show that the advice given by KFR was more than erroneous — it had to amount to an error of the sort that would not have been made by a reasonably competent estate agent professing himself to have the standard and type of skill that KFR held themselves out as having, and acting with ordinary care, *per* Lord Fraser in *Whitehouse* v *Jordan* [1981] 1 WLR 246 at p263. L

In the end I heard from three witnesses. There was Mr Christopher [FRICS] Kilduff of KFR who had actually given the advice, Mr Derek Ames [FRICS] of a firm called York Montague (and formerly of Druce & Co) and a Mr Graham Vivian [FRICS] of Chestertons. The latter two witnesses were called as independent experts, Mr Ames for Routestone and Mr Vivian for KFR. After Mr Vivian had given his evidence, Mr Crystal sensibly decided that further expert evidence on behalf of the bank and receivers would serve no useful purpose and so he did not call his experts.

Before I embark on a detailed consideration of what happened in relation to the three properties I propose to state my conclusion. I do M
not think the evidence remotely establishes even an error by KFR, still less an error of the kind amounting to negligence. I say this because I simply have no evidence that the advice given by KFR was wrong. All that I have is some evidence from Mr Ames that he would have advised differently and some from Mr Vivian that he would have done some things differently but not others. As Mr Vivian (whom I found to be a fair, objective and reliable witness) said:

Major firms such as Chestertons, Knight Frank & Rutley, do not always do things in the same way. A lot of investments are sold by letter and never advertised. So each specialist has his own way of doing things, and as long as it gets the best price the aim is achieved.

A And I would add this: some of Mr Ames' proposed way of going about sales (given necessarily with hindsight) would have involved major delays and would have involved the bank and receivers in either playing the property market themselves or asking KFR to do so for them (inevitably involving the expense of a lot of work). That seems to me to be inconsistent with the right of sale. It was for purchasers to take a long term view — all that could be expected of KFR in seeking the best price under instructions to find a buyer was to point out to purchasers (who indeed were sophisticated property owners) that the estates concerned had potential. The advice Mr Ames would have given would have been more appropriate to a property company.

B I must also consider an important question of the law of evidence. Is an expert's opinion on the "ultimate question" admissible? In this case the ultimate question is the *Whitehouse* question. Now it is clear that an expert's opinion as to what the legal test is must be inadmissible. As Oliver J observed in *Midland Bank Trust Co Ltd* v *Hett, Stubbs & Kemp* [1979] Ch 384 at p402:

> The extent of any legal duty in any given situation must, I think, be a question of law for the court

It is by no means self-evident that an expert's opinion on whether or not that test is satisfied is also inadmissible. Section 3 of the Civil
C Evidence Act 1972 provides:

> (1) Subject to any rules of court made in pursuance of Part I of the Civil Evidence Act 1968 or this Act, where a person is called as a witness in any civil proceedings, his opinion on any relevant matter on which he is qualified to give expert evidence shall be admissible in evidence.
>
> . . .
>
> (3) In this section "relevant matter" includes an issue in the proceedings in question.

I think this provision makes the opinion admissible. It by no means follows that the court must follow it. On its own (unless uncontested) it would be "a mere bit of empty rhetoric" Wigmore, *Evidence*
D (Chadbourn rev) para 1920. What really matters in most cases is the reasons given for the opinion. As a practical matter a well constructed expert's report containing opinion evidence sets out the opinion and the reasons for it. If the reasons stand up the opinion does, if not, not. A rule of evidence which excludes this opinion evidence serves no practical purpose. What happens if the evidence is regarded as inadmissible is that experts' reports simply try to creep up to the opinion without openly giving it. They insinuate rather than explicate. And there is this further consideration, of considerable significance in this case. Suppose the expert is *not* prepared to say that the conduct complained of fell short of the *Whitehouse* test? It would be bizarre for
E the case to go to court at all. I think it is a good thing for an expert to have given to him the legal test and the facts as alleged by the plaintiff. If his own expert is not prepared to say that the defendant fell short of the relevant standard, then the plaintiff knows where he is. And the expert himself will be faced directly with what he is doing — saying another fell below an acceptable standard and knowing that he himself will be challenged on that opinion. If the expert merely says he would have done things differently the case may yet end up in court when it should not, as this one has, and I suspect others do.

The point is of significance here precisely because Mr Ames was not prepared to say (despite being pressed a number of times in his
F evidence in chief both by me and Mr Trace) that KFR fell below the standards of any competent valuer or estate agent. He thought other marketing strategies should have been followed in the sense that he would have done something different. He thought there were matters of criticism but no more. Oliver J described that sort of evidence in *Midland Bank*:

> Evidence which really amounts to no more than an expression of opinion by a particular practitioner of what he thinks that he would have done had he been placed, hypothetically and without the benefit of hindsight, in the position of the defendants, is of little assistance to the court.

Because Mr Ames would not say that what KFR advised fell below the standard of any reasonably competent valuer Mr Trace was driven to argue that opinion evidence on the ultimate question was G
inadmissible. I have already said that I think it is (as opposed to evidence of what the legal test is) but I should deal with the cases suggested to show otherwise. Only one of them refers to section 3 of the 1972 Act. And as it seems to me they are not dealing with the precise point. Thus, in *Midland*, Oliver J was dealing with evidence as to the legal test (inadmissible) and evidence as to what a particular expert would have done (admissible but of "little assistance"). He was not dealing with an opinion as to whether particular conduct satisfied the legal test. That is surely an "issue in the proceedings in question" within the meaning of the statute. Again, in *Liddell* v *Middleton*
unreported, CA (Stuart-Smith, Gibson and Hutchison LJJ) July 7 1995, H
a traffic accident case, the Court of Appeal made it clear that there was over-use of expert evidence in road traffic cases, when the question is essentially whether or not the defendant was negligent. In that sort of case, where the court has its own experience to go by, the expert can only help when considering things like inferences to be drawn from tyre marks. You do not need an expert to tell you whether or not someone should have sounded their horn or driven slower. And in *Bown* v *Gould & Swayne* unreported, CA (Leggatt, Simon Brown and Millett LJJ) January 24 1996, expert evidence (in a claim for professional negligence) as to what a solicitor should do was excluded.
Such evidence would be evidence as to the legal test to be applied and J
was plainly inadmissible. All the matters of evidence sought to be admitted "proves, on analysis, to involve either a question of law or a question of fact" *per* Stuart-Smith LJ. *Bown* does not assist either side, though both sought to rely upon it.

By way of contrast, and in support of the conclusion that an expert opinion on whether a legal test has been complied with is admissible is what was said by Sir Donald Nicholls V-C giving the judgment of the Court of Appeal in *Mölnlycke AB* v *Procter & Gamble No 5* [1994] RPC 49 at p113

> In applying the statutory criterion [ie as to whether an alleged inventive step
> was obvious] and making these findings [ie as to obviousness] the court will K
> almost invariably require the assistance of expert evidence. The primary evidence will be that of properly qualified expert witnesses who will say whether or not in their opinions the relevant step would have been obvious to a skilled man having regard to the state of the art.

Further, the current edition of *Cross and Tapper on Evidence* 1995 at p552 is clearly of the view that opinion evidence relating to the ultimate question is admissible by virtue of the 1972 Act. It is noteworthy that that Act was passed following the 17th report of the Law Reform Committee 1970 Cmnd 4489 which specifically recommended that an expert opinion on the ultimate question should
be made admissible. Mr Trace submits that *Cross* is irrelevant because L
it was published before *Liddell* and *Bown* but as I have indicated I do not think those decisions are on the point. As to the Law Reform report, Mr Trace is driven to say it is irrelevant even though it is clearly the basis upon which the 1972 Act was passed.

So I think the fact that Mr Ames was prepared to criticise but was not prepared to say that KFR fell short of the standard of any competent estate agent is itself fatal to Routestone's case. But that is far from an end of the matter. I next observe that Mr Ames did not give any evidence of a kind recognised by Oliver J as admissible:

> Clearly, if there is some practice in a particular profession, some accepted M
> standard of conduct which is laid down by a professional institute or sanctioned by common usage, evidence of that can and ought to be received.

So all I have is that Mr Ames would, with hindsight, have advised differently. That is far from proving that KFR were wrong or negligent. This is again a clear further reason for rejecting the claim. Mr Trace says that having heard what KFR did and what Mr Ames would have done I can use my common sense. I suppose there may be cases where that would be enough to establish a case of negligence, but they would have to be so plain that they would probably never come to court. Here I am dealing with a fairly specialised field — the sale of investment properties of a rather unusual kind (very unusual in the case of SGE and GR). I do not think my "common sense" provides any light as to what would amount to going about things the best way still less whether a particular way fell below an acceptable standard.

Finally I must turn to Mr Ames' criticisms as what was done in relation to each of the three properties and Mr Ames' criticisms of this.

St George's estate

I begin with this although it was the last to be sold. The property consisted of a large number of freeholds in Pimlico. At the time of the receivership the property was producing an income of about £30,000 pa in all. It was unusual, partly in being a large estate in London (such estates come on the market rarely) but more so because it was subject to a 115-year lease to interests controlled by the Berger family. So all that was for sale was the reversion after 115 years. This reversion was not of great interest for itself. In particular ownership of the freehold would not give any development control over the land. To make the property attractive financially would need the co-operation of the headleaseholder. But it seems to be common ground the Berger interests were renowned for being unco-operative and Mr Vivian had personal experience of this. Any potential buyer would have known (or learned very quickly) that for him to make any money on the property he would need Berger co-operation which would be unlikely to be forthcoming.

That the co-operation of the headlessee would be important for a good price was apparent from an early stage. KFR reported on November 18 1985:

> . . . it is to be appreciated that a break-up of the Estate will require the co-operation of the headlessee and indeed any other parties holding an interest in the property.

Mr Trace fastened on this as an example of wrong advice — technically it was not necessary to have any co-operation. But I am quite satisfied both that what was intended to be conveyed was that "a break-up for a good price will require co-operation" and that that is how the letter was understood. Mr Kilduff made it clear he had explained the position to the receivers over the telephone and in any event the position was fairly self-evident, both at that time and thereafter.

What was done to sell the estate was as follows. First over a period of some months, it was necessary simply to put together the details of all the properties in the estate and what rents they were producing. I heard unchallenged evidence from the receivers that this task was extensive. If and to the extent there is criticism of what was done, I reject it. Once the information was accurate enough (and there were still minor discrepancies) KFR prepared sales particulars (750 copies) and circulated it widely to their client list. That 1ist would have included all or nearly all the sorts of potential buyer of this very specialised investment. Probably KFR also arranged for circulation by other agents through a co-operative scheme, but they certainly circulated 250 to 300. It is firmly established that substantially all potential purchasers were circulated or would have come to know of the offer. It should be remembered that the receivership of the Gomba Group was itself widely known. For that reason even before sales particulars were circulated inquiries were made of the receivers which were duly passed on to KFR.

The result of all this was that KFR received eight offers which they reported to the receivers on July 8 1986. Apart from one, the offers were in two groups. One group was around £6-750,000 and the other at around £300,000. It is noteworthy that Mr Berger, who by putting the freehold with his long leasehold would have been able to put together a package of real interest to his subtenants or others, only offered £325,000. But there was one offer on a different scale. It was from a company called Jaymont and was in the sum of £3.5m. Mr Kilduff, having checked that the offer was from an apparently genuine source with proper finance available advised by letter of July 8:

> The offer for the Estate represents a years purchase figure of 119.41 and I recommend that it is accepted. In my view this is an extremely good offer from an able purchaser and reflects the break-up potential for sales to special purchasers coupled with the ultimate reversionary value on expiry of the head-lease. This obviously results in a considerably higher level of offer than those from the traditional ground rent purchasers who view it as a fixed interest investment.
>
> I would also add that in the event that this purchaser does not proceed, then the next highest offer is still at a very attractive level.

Once the offer had been accepted, Jaymont resiled from the price. They indicated that they would be willing to proceed at £800,000. After some haggling £850,000 was agreed, although the ultimate price was £831,672. This allowed for some adjustments covering some parts of the estate which were not certainly determined. Contracts were exchanged on September 25.

Mr Ames's principal criticism was that the estate should not simply have been sold as a whole but that active attempts should have been made to exploit special situations first, breaking up at least some of the estate. He would have advised the receivers to try to contact all the subtenants and to sell to them — indicating that if they did not buy, then the estate would be sold. He accepted that without the co-operation of Mr Berger this would have been of little value to them. For instance service of a schedule of dilapidations on him by a subtenant who had bought the freehold would have resulted in his serving a similar notice on the subtenant. Mr Ames none the less thought that some subtenants would have liked the idea of owning their freeholds and have accordingly been prepared to pay more than simply their investment value. The exercise suggested would have been difficult and expensive. Without a list of these subtenants identifying them all would have been a considerable exercise. Then there would have had to be negotiations with each. The number involved would have been about 600. Various estimates as to how long such an exercise would have taken were made. They varied from a year to five years. And it is very likely that in the end there would have been a significant number of cases where the tenants would not buy. I am far from satisfied that the Ames strategy is the only one a competent estate agent would have advised in 1986. It should be remembered that the strategy is one that could equally have been followed by a buyer hoping to make a profit from the estate and the price in the market would reflect this potential. Indeed it probably did, because the value of the estate on a simple income producing basis was about 10 years' purchase, namely £300,000.

Mr Ames further says the decision to sell "seems hasty". But that was a decision the receivers were entitled to make. And in any event I reject the suggestion that the decision was hasty. Mr Trace wanted me to find otherwise on the basis of a file note of an assistant to the receivers. The note was not put to anyone who gave evidence — most importantly to Mr Kilduff whose words are supposed to be recorded in the note. By the time concerned, the estate had been widely marketed. It is self evident that a failure to close a sale reasonably soon after then would be likely to lead to a fall in price. You do not normally get a higher price later for something which has proved to be a drag on the market. So as a matter of prudence there would naturally be some urgency to close a deal after the advertising.

I should also consider the price achieved. For if it was obviously too low that might provide some evidence that the marketing was conducted negligently. But the price was the best of the offers made after wide marketing — this is obviously of great significance for I am not dealing with a valuation but a real marketing exercise. And there is alternative evidence from which I can infer the price was reasonable, even good. Routestone bought the estate in 1981 for £500,000. By September 1986 prices of this sort of investment property had not changed much. And about 30% of the estate had been sold through a passive sales policy. Looking at these figures suggests that a good price was obtained. Of course in retrospect a lot more could have been obtained a year or so later. For a boom in prices occurred, commencing in late 1986. But who was to know that? Mr Ames suggested he had knowledge that a boom was on by earlier in 1986, but I do not find that this was common knowledge.

Another way of testing whether there was a sale at an obvious undervalue is by looking at the transaction through the eyes of the purchaser, Jaymont. Their agent, Mr Champness, an experienced estate agent and a partner in Gerald Eve & Co said that he had valued SGE at between £850,000 and £1,300,000 when the particulars were received from KFR. The valuation was hedged in the sense that Mr Christopher Champness [FRICS] recognised that there were no near comparables. His client, a Mr Ross, instructed him, notwithstanding the valuation, to offer £3.5m. Mr Champness said he was "stunned". In his witness statement he gave hearsay evidence as to his client's thinking. I refused to admit this, but even without that I think it fair to infer that the offer was intended to be what in the trade is sometimes called a "blow-out" — an offer intended to cause the seller to discard other potential purchasers but not in itself genuine. Mr Kilduff was not surprised when it was withdrawn later. In the event the blow-out worked, but Jaymont still ended up as the highest bidder.

One of the factors which may have been taken into account by Jaymont was that the 1981 price had been £500,000. During the trial Mr Trace twice applied to amend the pleadings to allege that Jaymont only found this out through the negligence of one of the defendants. I refused to allow the amendment. It was not simply that it was outrageously late, but that it would have been very unfair. The defendants would surely have wanted to go into this issue further — for instance by obtaining from Jaymont directly the very evidence which I excluded as hearsay. In any event it is not shown that the £3.5m offer was ever intended to be serious, even though it was got up as such for Mr Kilduff.

I should also deal with some lesser criticisms levelled by Mr Ames. He said the KFR brochure should have been grander — glossy paper and an aerial view of the estate would have been what he would have done. Mr Vivian would have done the same, but specifically indicated that other agents might well have taken a different course. This evidence does not establish negligence. Mr Ames also thought that the brochure should have made much more of the fact that a "great London estate" was for sale and that there should have been international circulation. That may well be what he would have done, but again that does not establish negligence. It is not as though the buyer would be likely to be anything but a hard-nosed businessman buying as an investment. He would surely take his own advice and so, if he did not realise it was the position at the outset, would quickly find out that the estate was subject to the 115-year headlease of the Berger interest. Actually Mr Kilduff told me that he had a number of calls from people showing initial interest who dropped it as soon as the nature of that interest was made clear. I would add that Mr Ames, although he recognised that interest in the body of his report, to my mind did not give it the prominence which it deserved.

Neither error nor negligence is made out in relation to the sale of the SGE.

128-132 Grosvenor Road

This was part of the SGE. So there was exactly the same 115-year Berger lease millstone round the property. The rental income was £2,389.30. Before the receivers were appointed a Mr Haffenden of a company called Albany were already in negotiations for sale to a company called Hanover. Albany was, in effect, an in-house estate agent for the Gomba Group. So to the extent that the receivers or KFR picked up negotiations or advice already put in train by Gomba, Routestone can hardly complain of what was done.

Hanover were the subtenants at no 128. It seems plain that they had entered into some sort of arrangement with the Berger interests. In 1984 they offered £20,000 and received a counteroffer of £500,000 for no 128. On December 5 1985 Hanover offered the receivers £20,000 for 128-132 inclusive. The receivers had discussed with KFR the SGE generally and the result of that discussion is recorded in a letter of December 6 from the receivers:

> You [ie KFR] would give consideration, as a matter of urgency, to any special situations, for example development properties, where the freehold of individual properties might be sold before seeking to market the remainder.

Hanover's position was special. Mr Kilduff co-operated with Mr Haffenden who suggested asking £1.3m. This was on any view wildly optimistic for an interest producing only just over £2,000 pa. Mr Kilduff wrote to Hanover's agents saying:

> From copy correspondence with which I have been provided I note the Receiver has been advised to give serious consideration to an offer in the region of £1.3m [plus fees].

Mr Kilduff was there passing on the Haffenden view. His own view is also to some extent recorded inasmuch as he also added that:

> I do not consider that the offer submitted of £20,000 fairly reflects the benefit to your client of owning the freehold interest.

It is pleaded that the Haffenden view of £1.3m is the true value. Mr Haffenden was not called and I am bound to say that the idea that anyone would pay £1.3m for a property producing only £2,000 pa for the next 115 years is nothing short of fantastic.

Following this letter, KFR played no part in the further negotiations. In February Hanover offered £75,000. KFR were asked about this, advised that the offer represented 31.8 years' purchase and said:

> In view of the amount of rent involved I consider this to be an attractive offer for this interest and I recommend that it is accepted subject to the purchasers [paying fees]

There was a little more haggling during which Hanover refused to pay fees and the offer was accepted without them, KFR having advised, by letter of March 4, that the offer without fees was nevertheless attractive and should be accepted.

Routestone's allegation of negligence here is KFR should have advised that the property ought to have been marketed, ie widely advertised. This was put in two alternatives first that this ought to have been done in respect of all the properties and second (and more weight was put on this) in relation to those properties (four out of the five but only just over half in apportioned rental value) in which Hanover were not subtenants and had no obvious special interest. It is further said that in giving their advice KFR should have had regard to comparables, particularly nos 135 and 137, it being said that information about the price of these was available to Mr Kilduff. Mr Ames supported these contentions in the sense that he said that he would have taken these steps. He did not say that the steps concerned are those which any reasonable estate agent in the circumstances would have taken.

Now there is a certain inconsistency about the contentions concerning no 128 and those concerning the SGE as whole. In relation to the latter the contention was that special situations should be actively sought and exploited before going to the market as a whole. Yet in relation to no 128 the complaint is that the special situation was adopted without going to the market as a whole. I think the complaint about no 128 is hopeless. Hanover were pushed as far as they would go and that was that.

As to the remaining parts of Grosvenor Road, the idea that there should be an extensive marketing programme for properties generating an income of just over £1,000 pa seems to me to be absurd. There were in fact special difficulties about the properties for any would-be developer over and above the Berger problem. It seems Westminster Council had rights which would make things even more difficult. It should be remembered that the Berger interest was itself a special interest. It clearly knew about the property but was not interested.

Mr Vivian would not support Mr Ames' view about marketing. And although he would have liked comparables, it is not, in my judgment, established that Mr Kilduff knew there were any or that if they had been available any different advice would have been given.

Further I have evidence that the purchaser, Hanover, thought the price fair. In unchallenged evidence, Mr Farr of Vigers (Hanover's agents) said:

> £75,000 was a fair reflection of the value of the freehold to Hanover as a special purchaser. The open market value would have been lower.

A Mr Trace said this evidence was inadmissible as being expert evidence for which no leave was given. I do not agree. This was plainly part of Mr Farr's thinking at the time. And in any event Mr Trace had a full opportunity of cross-examining this evidence but chose not to do so or to indicate that he had any objection to this self-evidently crucial sentence when it was let in subject to objections about admissibility.

Mr Ames said the proper value was £250,000. But he provided no real reasons for this view ("it is a purely reasonable supposition") and I do not think that is good enough.

The case on GR is not made out.

B

North End Parade

This consisted of a parade of shops north of the West Cromwell Road. It was valued by KFR for Price Waterhouse (ie before they were appointed receivers) on February 5 1985 at £640,000. Mr Kilduff was not challenged on this valuation. Mr Kilduff explained how when KFR were instructed to sell (on December 5 1985) the valuation was reconsidered and that by then KFR would take "a slightly stronger view on price". Offers were being received from the market already and the receivers passed these to KFR. KFR advised the receivers that NEP should be marketed widely seeking offers of about £1m and that
C the receivers should accept £825,000 if necessary. KFR prepared sale particulars, and circulated them to their client base in the same way as they did with SGE. I am satisfied that all likely purchasers knew of the property.

The highest offer that came in was from Merivale Moore, specialists in this sort of property, who had similar property nearby. KFR recommended acceptance and the sale went through at £800,000. D

It is alleged that KFR gave the wrong advice. What they should have advised is a programme of "working" the property by dealing with all rent reviews (some were overdue), refurbishing and so on, thereby increasing the rental income which in turn would have increased the capital value. This would have meant not selling the property for some time. Yet the receivers were entitled to sell. I do not think this is a legitimate criticism in those circumstances. The bank was not obliged to "work" before exercising its right of sale.

Moreover it is quite clear that the "working up" would have E involved a lot of expenditure and skill. A Ms Joyce (who was praised as very skilful) was the one who, for Merivale, did that working up. She gave unchallenged evidence that the price paid was in line with other acquisitions being made by Merivale at the time. Plainly the price paid took into account that the property had not been well managed (Ms Joyce's view) and that there would have to be working up.

Mr Ames accepted that the price obtained, "£800,000, was disappointing but not totally out of court". His criticism was of the timing. Mr Vivian would have valued the property at the time at £900,000 but accepted that the market was the "acid test". F

Again I do not think a case of error by KFR is made out, still less negligence.

In the result the actions fail.

REAL PROPERTY AND CONVEYANCING

A

R *v* Suffolk County Council, *ex parte* Steed and another

COURT OF APPEAL

July 11 1996

BUTLER-SLOSS, PILL and SCHIEMANN LJJ

Estates Gazette March 8 1997

B

[1997] 10 EG 146

Commons — Village green — Whether council entitled to refuse registration where right claimed based on 20 years' use after January 2 1970

During the 19th century, and in pursuance of powers in a local Act, two areas of land which became known as the "People's Park" were acquired by the local borough council. The land was to be held for the freemen in replacement for the redemption of rights of shackage. It was grazed by the freemen until 1944 when they agreed to vest their
C rights in trustees. Thereafter the trustees let the grazing. There was also some evidence of recreational use by the public from time to time. An application to register the park under the Commons Registration Act 1965 was withdrawn in 1971. In 1987 an area of 15.8 acres of the uncultivated land was acquired by the Secretary of State for Health for the purposes of constructing a hospital. In 1992 the applicants in the present case applied to the respondent county council for the registration of the land as a village green under the 1965 Act on the basis that the park had been used as of right for a period of 20 years by local inhabitants. Following an objection by the Secretary of State, the application was rejected on the grounds that the usage was not
D referable to a "locality" within the meaning of section 22 and because the usage relied on was not the result of any change of circumstances since 1970. The applicants appealed against the decision of Carnwath J ([1995] 2 EGLR 233), who had dismissed their application for judicial review.

***Held*: The appeal was dismissed. Where registration of a new town or village green is sought based on 20-years use, and use post-January 1970, the evidential safeguards present in the authorities dealing with the establishment of a customary right should be imported. The application failed on the evidence of use as of right. The use did not demonstrate belief in a legal right to**
E use and, even if it did, there was no sufficient definition of the right asserted. The failure to register a green under section 1(2)(*a*) of the 1965 Act does not have the effect of extinguishing the customary rights over the land. An unregistered green would not be deemed to be a green, as registered land would be, but it may still be proved by evidence to be a green. Amendment of the register is possible under section 13 when a green is established as one by 20-years use after January 2 1970, even if it could have been, but was not, registered as a green by customary rights under section 1(2)(*a*). *Quaere* whether the effect of registration creates rights to indulge in sports and pastimes.

The following cases are referred to in this report. F

Bolton Metropolitan District Council v *Secretary of State for the Environment* [1995] 1 WLR 1176; (1995) 71 P&CR 309; [1995] 3 PLR 37; [1995] JPL 1043, HL
Brocklebank v *Thompson* [1903] 2 Ch 344
Corpus Christi College, Oxford (President and Scholars) v *Gloucestershire County Council* [1983] QB 360; [1982] 3 WLR 849; [1982] 3 All ER 995; (1982) 81 LGR 29; 46 P&CR 9, CA
Forbes v *Ecclesiastical Commissioners* (1872) LR 15 Eq 51
Hammerton v *Honey* (1876) 24 WR 603
Hue v *Whiteley* [1929] 1 Ch 440
Jones v *Bates* [1938] 2 All ER 237
Mercer v *Denne* [1905] 2 Ch 538 G
Mills v *Silver* [1991] Ch 271; [1991] 2 WLR 324; [1991] 1 All ER 449; (1990) 61 P&CR 366, CA
New Windsor Corporation v *Mellor* [1975] Ch 380; [1975] 3 WLR 25; [1975] 3 All ER 44; (1975) 73 LGR 337; 30 P&CR 429, CA
O'Keefe v *Secretary of State for the Environment* [1996] JPL 42
R v *Joliffe* (1823) 2 B&C 54
Turnworth Down, Dorset, In re [1978] Ch 251; [1977] 3 WLR 370; [1977] 2 All ER 105; (1976) 33 P&CR 192
Tyson v *Smith* (1838) 9 A&E 406

This was an appeal by David Donald Steed and Colin Malcom Steed from the decision of Carnwath J ([1995] 2 EGLR 233) who had dismissed their application for judicial review of the decision of the H respondents, Suffolk County Council, declining to register land as a village green under the Commons Registration Act 1965.

Gerard Ryan QC and Martin Porter (instructed by Barker Gotelee, of Ipswich) appeared for the appellants; Nigel Thomas (instructed by the solicitor to Suffolk County Council) represented the respondents; Vivian Chapman (instructed by Mills & Reeve, of Cambridge) represented the Secretary of State.

Giving judgment at the invitation of Butler-Sloss LJ, **PILL LJ** said: On February 27 1992 Mr David Steed and Mr Colin Steed (the appellants) applied to Suffolk County Council (the respondents) for J registration of the People's Park, Sudbury, Suffolk ("the land"), as a town green under the Commons Registration Act 1965 by way of amendment of the register under section 13. The land is about 15.8 acres in extent and is within the urban fabric of the town of Sudbury as depicted on the current ordnance survey plan and to the north-east of the town centre. The land had been purchased by the Borough of Sudbury in the 1870s and renamed People's Park. Its use was later regulated by a scheme approved by the High Court and administered by trustees. There was grazing and recreational use. In 1987 the land, together with other land, was acquired from the trustees by the Secretary of State for Health with the object of the construction on it K by the regional health authority of a hospital.

The Secretary of State objected in writing to the application, which was refused by the respondents on September 3 1992. The planning committee resolved that the application "be rejected on the grounds that it failed to establish that all the statutory criteria for amendment of the register of town and village Greens were met".

This is an appeal against the dismissal by Carnwath J of the appellants' application to quash that decision of the respondents. The court heard submissions on behalf of the Secretary of State (who was heard in opposition to the motion under Ord 53 r 9(1)) and the regional health authority as well as from the appellants and the respondents. A substantial sum of money has been spent on site for preparatory work for the hospital.

The application

The application to the respondents was made on form 30 annexed to the Commons Registration (New Land) Regulations 1969 (SI 1969 No 1843). Those regulations were made by virtue of powers conferred in sections 13 and 19 of the Act. Form 30 is headed "Application for the registration of land which became a town or village green after 2 January 1970". The question "On what date did the land become a town or village green?" is answered "1st August 1990 on completion of 20 year period from 31st July 1970". The question "How did the land become a town or village green?" is answered as follows:

> By the use thereof by the local inhabitants ie the inhabitants of Sudbury. Lawful sports and pastimes including amongst other things football, dog walking, mushroom picking, fetes, children's games, kite flying, cricket, neighbourhood access, rounders.

In answer to a later question inquiring if there are any other facts relating to the application which ought to be brought to the attention of the registration authority, it is stated:

> The inhabitants have used People's Park for lawful sports and pastimes "as of right" for the 20 year period 1970-1990. Same identification of usage pre-1970. Commencement date from time immemorial.

The application was supported by 39 forms, six of which were signed by more than one person, completed by local residents and giving particulars of use of the land. A detailed statement in support was submitted by Mr SJ Blackwell, resident and former borough surveyor. The detailed written objection of the Secretary of State was supported by statements from four trustees of the Sudbury Common Lands Charity, who sold the land to the Secretary of State. The appellants and the Open Spaces Society commented in writing and in detail upon the objections.

When the planning committee considered the application, all the above material was available to them together with a report prepared by the county secretary and solicitor, who had obtained counsel's opinion. The recommendation in the report was for rejection of the application.

The appellants claim that the respondents were misdirected in law in the solicitor's report. Upon the evidence and a proper legal direction, the respondents ought to have registered the land as a town green, it is submitted. Further, procedural impropriety is alleged in a failure of the respondents to give the consideration to the application which the Act and regulations required. The application should be remitted to the respondents for reconsideration.

Town or village green

The Act followed in time the report in 1958 of a Royal Commission on Common Land. Further legislation, which has not since been enacted, was plainly contemplated. The Act has been subject to considerable judicial criticism for its uncertainty, for example in *President & Scholars Corpus Christi College, Oxford* v *Gloucestershire County Council* [1983] QB 360.

The essential characteristic of a town or village green as stated in *Halsbury's Laws*, vol 6, para 525, is that:

> by immemorial custom the inhabitants of the town, village, or parish should have acquired the right of playing lawful games thereon and enjoying it for purposes of recreation. The nature of the enjoyment is a matter for proof in each case, and the custom must be limited to the inhabitants of the parish, manor or other district for which it is claimed.

The classic statement of the nature of a customary right is that of Sir George Jessel MR in *Hammerton* v *Honey* (1876) 24 WR 603:

> There are certain rights which may be claimed by custom as distinct from prescription. What is called a right of recreation and amusement, of air and exercise, or the playing of all manner of lawful games and pastimes, is a right in the nature of an easement which may well be claimed by custom.
>
> Now, what is a custom? A custom, as I understand it, is local common law. It is common law because it is not statute law; it is local law because it is the law of a particular place as distinguished from the general common law. Now, what is the meaning of local common law? Local common law, like general common law, is the law of the country as it existed before the time of legal memory, which is generally considered the time of Richard I. Therefore, when people allege a custom they allege that which they call a custom as having been the law of the place before the time of legal memory . . .
>
> After all, whatever definition you may give to custom and prescription, we all know that they are legal fictions invented by judges for the purpose of giving a legal foundation or origin to long usage. I do not suppose anybody imagines that in the case of prescription there has been a lost grant, or of a custom that there has been from time immemorial a common law of the particular place, and the meaning and substance of the old law is merely this — that when you find long-continued usage which can have a legal origin, then, with the view of preserving to the people claiming them the acquired possession of the rights of property or rights of easement which they have so long enjoyed, you shall attribute these rights, if possible, to a legal origin so as to support them. . . .
>
> I agree that, the nature of custom being, as it is, local law, you can only get rid of it in the same way as other local laws are got rid of, namely, by Act of Parliament. I am not aware of any other means. Of course, it is opposed to the true notion to suppose there is such other means, but that is an additional reason why the judges have been very careful, not only in restricting the custom to certain classes of cases, but also in seeing that the custom is properly and strictly proved. If there is such a custom in the inhabitants of a vill you never can get rid of it, as far as I know, by any means short of an Act of Parliament.

Sir George Jessel also stated that when it is said that "custom must be certain, reasonable, and continuous — that all relates to the evidence of a custom". Tindal CJ had stated in *Tyson* v *Smith* (1838) 9 A&E 406 at p421 that:

> It is an acknowledged principle that, to give validity to a custom —which has been well described to be an usage, which obtains the force of law, and is, in truth, the binding law, within a particular district or at a particular place, of the persons and things which it concerns— it must be certain, reasonable in itself, commencing from time immemorial, and continued without interruption.

Custom, stated Stirling LJ in *Mercer* v *Denne* [1905] 2 Ch 538 at p582 is "truly described as a local law".

The 1965 Act became law against the background that such customary rights existed. The central difficulty about construing the Act is in considering whether it was intended merely or primarily to provide a mechanism for registering rights or whether it has the potential to destroy existing rights and permit the creation of new rights.

In section 22 of the Act, the interpretation section, it is stated that:

> "town or village green" means land which has been allotted by or under any Act for the exercise or recreation of the inhabitants of any locality or on which the inhabitants of any locality have a customary right to indulge in lawful sports and pastimes or on which the inhabitants of any locality have indulged in such sports or pastimes as right for not less than twenty years.

(Lord Denning in the *Corpus Christi College* case referred to the three categories within the definition as classes a, b and c and I will use the same terms.)

As the judge pointed out, the statutory definition follows the recommendations of the Royal Commission in adopting a three-part definition but class c is different from that recommended by the commission which was to include:

> in a rural parish any unenclosed open space which is wholly or mainly surrounded by houses or their curtilages and which has been continuously and openly used by the inhabitants for all or any such purposes (lawful sports and pastimes) during a period of at least 20 years without protest or permission from the owner of the fee simple or the Lord of the Manor

(para 403 of report). It may be noted that it has been held that a regular usage of 20 years unexplained and uncontradicted may be sufficient to justify a finding of the existence of an immemorial custom: (*Brocklebank* v *Thompson* [1903] 2 Ch 344, citing *R* v *Joliffe* (1823) 2 B&C 54.) As will have appeared from the

A form of the application, form 30, the appellants however sought to rely upon the third part of the definition, class c, use for lawful sports and pastimes by the inhabitants as of right for the 20 years after January 2 1970, the 20 years being said to commence from July 31 1970.

Statutory framework

It is necessary to set out further extracts from the Act and Regulations. Section 1, in so far as is material provides:

> (1) There shall be registered, in accordance with the provisions of this Act and subject to the exceptions mentioned therein, —
>
> (*a*) land in England or Wales which is common land or a town or village
> B green;
>
> . . .
>
> (2) After the end of such period, not being less than three years from the commencement of this Act, as the Minister may by order determine — [in the event July 31 1970]
>
> (*a*) no land capable of being registered under this Act shall be deemed to be common land or a town or village green unless it is so registered; and
>
> (*b*) no rights of common shall be exercisable over any such land unless they are registered either under this Act or under the Land Registration Acts 1925 and 1936.
>
> (3) Where any land is registered under this Act but no person is registered as the owner thereof under this Act or under the Land Registration Acts 1925
> C and 1936, it shall —
>
> (*a*) if it is town or village green, be vested in accordance with the following provisions of this Act . . .

(Section 8 provides that in certain circumstances the Commons Commissioner shall, in the case of a town or village green, direct the registration authority to register as the owner of the land a local authority.)

Section 3 provides:

> (1) For the purposes of registering such land as is mentioned in section 1(1) of this Act and rights of common over and ownership of such land every
> D registration authority shall maintain —
>
> (*a*) a register of common land; and
>
> (*b*) a register of town or village greens;
>
> and regulations under this Act may require or authorise a registration authority to note on those registers such other information as may be prescribed.
>
> (2) Any register maintained under this Act shall be open to inspection by the public at all reasonable times.

Section 10 provides:

> The registration under this Act of any land as common land or as a town or village green, or of any rights of common over any such land, shall be conclusive evidence of the matters registered, as at the date of registration,
> E except where the registration is provisional only.

Section 13 provides:

> Regulations under this Act shall provide for the amendment of the registers maintained under this Act where —
>
> (*a*) any land registered under this Act ceases to be common land or a town or village green; or
>
> (*b*) any land becomes common land or a town or village green; or
>
> (*c*) any rights registered under this Act are apportioned, extinguished or released, or are varied or transferred in such circumstances as may be prescribed.

F

Regulation 3 under the heading "Land becoming common land or a town or village green", provides, in so far as is material:

> (1) Where, after 2nd January 1970, any land becomes common land or a town or village green, application may be made subject to and in accordance with the provisions of these Regulations for the inclusion of that land in the appropriate register and for the registration of rights of common thereover and of persons claiming to be owners thereof.

Regulation 3 and subsequent regulations state how an application for registration is to be made and what action is required of the registration authority.

Regulation 6(1) provides:

> As soon as possible after the date by which statements in objection to an G
> application have been required to be submitted, the registration authority shall proceed to the further consideration of the application, and the consideration of statements (if any) in objection thereto, in accordance with the following provisions of this regulation.

The rest of the regulation provides for the consideration by the authority of written and signed statements of objection and the right of applicants to an opportunity to deal with the contents of objections. Regulation 8 specifies the action to be taken by a registration authority which has . . . "disposed of an application . . ."

There are detailed notes to form 30 prescribed by the regulation. Note 5, under the heading "How land can become a town or village H
green" provides:

> Land can become a town or village green after 2nd January 1970 in one of the following ways: —
>
> (1) By or under an Act of Parliament otherwise than as substituted land (as to substituted land, see category (4) below).
>
> (2) By customary right established by judicial decision.
>
> (3) By the actual use of the land by the local inhabitants for lawful sports and pastimes as of right for not less than 20 years.
>
> (4) By substitution or exchange for other land which has ceased to be a town or village green under — (enactments) . . .

Consideration J

The evidence submitted to support the application to the respondents consisted of questionnaires completed by residents of Sudbury between May and December 1991. The printed document is headed, "Questionnaire for collecting evidence to create a new green". The uses mentioned are summarised in the written submissions of counsel to this court as "The playing of football, cricket and rounders, walking with and without dogs, picnics, flying kites, mushroom picking, jogging, children's playing and cycling". The "walking of dogs" features prominently on the forms and "fetes" and "neighbourhood access" are also mentioned in the application. Almost
all the forms include negative answers to questions whether the K
owner's permission to use the land had been requested and whether the writer had ever been prevented from using the land. In their statements, the former trustees say that permissions to use were given to the local football club and for activities such as fairs and galas.

In his report, the county solicitor identified the issues as follows:

> In considering whether the criteria in the third limb of section 22 are met, there are a number of issues between the applicant and the objector, namely: —
>
> (a) Is there a specific locality the inhabitants of which have indulged in lawful sports and pastimes?
>
> (b) Has the use been for lawful sports and pastimes?
>
> (c) Has the user been as of right? L
>
> (d) Has there been 20 years user?

The four issues are then considered, with considerable references to counsel's opinion. The solicitor concludes:

> Counsel has advised that the application does not succeed on the 1st and 4th of the statutory criteria. In addition, the evidence that it meets the 2nd criteria is scant while in respect of the 3rd it is inconclusive either way.

Counsel had advised that most of the evidence came from residents of the area immediately surrounding the land and that did not in law amount to a "locality". He also advised that because the uses relied on
are claimed to have existed before 1970 the land cannot have M
"become" a town green after 1970 within the meaning of the Act.

The minute of the meeting of September 3 1992 set out the background and continued:

> The applicant claimed that People's Park had become a village green by virtue of 20 years use by local inhabitants. The County Secretary and Solicitor, however, had accepted counsel's advice that the applicant had failed to satisfy the four criteria needed to establish a new village green. The local member urged the committee to hold a public inquiry into the application. The committee considered the arguments that had been put forward by both supporters of and objectors to the application.

The resolution already noted is then set out.

A *Judgment and submissions*

The judge expressed his agreement with the respondents' decision. He stated:

> First, the evidence presented by the applicants in the present case did not in terms give any indication of a right arising since 1970. There was evidence of recreational use since 1970. But there was no indication that those using the land did so by virtue of a right claimed to have arisen since 1970. On the contrary the use is alleged to have continued virtually unchanged since as far back as the 1920's or before. There is no hint of anything having happened since 1970 which might be thought to give rise to a right which did not previously exist.
>
> B Secondly, the evidence presented in the forms does not link the use to a right claimed by reference to any particular locality. The use appears to have been mainly by those living in the nearby streets but the evidence does not show that they were asserting a right defined by reference to that, or any other, area.

Dealing with evidence of earlier use, Carnwath J added:

> Such a claim, if well founded, could have been registered in 1970. In the absence of registration it can no longer be relied on. It is clearly not something by virtue of which the land can be said to have "become" a village green since 1970.

C When dealing earlier in his judgment with the word "locality", the judge stated:

> Whatever its precise limits, it should connote something more than a place or geographical area — rather, a distinct and identifiable community, such as might reasonably lay claim to a town or village green as of right. In the present case the "locality" on which the application for judicial review and the supporting affidavit rely is Sudbury itself; I agree that this is the only realistic basis on which to proceed.

For the appellants, Mr Gerard Ryan QC submits that what constitutes a "locality" for present purposes is a question of fact in
D every case subject only to the requirement that the extent of the locality is ascertainable. A pragmatic approach is required. In proving the right enjoyed, it is not necessary to show its exercise by inhabitants of every part of the locality.

The fact that, on the evidence, the land was a town green before 1970 does not prevent registration as a third category town green by virtue of use for 20 years from 1970, it is submitted. The effect of section 1(2)(*a*) is that the status of land which qualified for registration but was not registered is permanently disregarded. Section 13(*b*) does not require reconsideration of the earlier qualification; it is made irrelevant. Completion of the new qualifying period is an event
E independent of the original qualification and the right to have the register amended is not defeated by the presence of the word "becomes". For the purposes of registration, the previous qualification should be ignored. In cases where there was a failure to register customary rights by 1970, parliament did not intend so drastic a sanction as to prevent registration for all time when pre-1970 rights continue to be exercised after that date.

Mr Ryan submits that there was a sufficient assertion by the informants of use "as of right". As well as the denial of permission granted, there was the signed assertion that the land was the "People's Park".

F The procedural challenge is on the basis that the respondents, as registration authority, gave no sufficient attention to the facts. They were not entitled to devolve their duty to consider the evidence to counsel. "Consideration" of the application under regulation 6 requires an "element of inquisition". The respondents must analyse the evidence and, if necessary, make inquiries of their own. It was impermissible to relate propositions of law to unconsidered facts.

Mr Ryan submits that the application should be remitted to the respondents for reconsideration first because they acted on an erroneous view of the law and, second, and in any event, because there was no "consideration" of the application within the meaning of the regulations.

On the effect of the word "becomes" in section 13(*b*) of the Act Mr G
Vivian Chapman, for the Secretary of State and the Health Authority, as on other points, seeks to uphold the finding of the judge. Because the evidence showed no difference in the use before 1970 from that after 1970, the land could not "become" a town green registerable under the Act during the period after 1970. There must be some change in the facts on the ground, it is submitted. Mr Chapman relies, though the judge did not, upon the decision of this court in *President & Scholars Corpus Christi College, Oxford* v *Gloucestershire County Council* [1983] QB 360. It is submitted that the word "ceases" in section 13(*a*) was construed in that case as postulating "a beginning and ceasing to be" (Oliver LJ at p378B), and "becomes" requires a H
factual change on the ground.

Mr Thomas, for the respondents, submits that the evidence demonstrates a sufficient consideration of the application. The material submitted by the appellants was available to members along with the solicitor's report. It was necessary for counsel to assess the evidence in order to give legal advice.

Locality and "as of right"

The application must be treated for what it purports to be, that is, an application for the registration for a "new" class c town green for the
town of Sudbury by virtue of the third part of the definition in section J
22 of the Act. It is new not only as being within class c, but in relying on post-January 1970 use. However, I approach the issue on the basis that it is no trivial matter for a landowner to have land, whether in public or private ownership, registered as a town green and that the evidential safeguards present in the authorities already cited dealing with the establishment of a customary right (class b) should be imported into a class c case. Use, as of right, and as inhabitants of Sudbury, for sports and pastimes must be "properly and strictly proved". Mr Ryan rightly agreed that the requirements of reasonableness, certainty and uninterrupted use are imported into class
c cases. Considering usage, Sir George Jessel MR stated in K
Hammerton:

> Again what must be the usage proved? It must not only be consistent with the custom alleged, but, if I may use the expression, not too wide. For instance, if you allege a custom for certain persons to dance on a green, and you prove in support of that allegation, not only that some people danced, but that everybody else in the world who chose danced and played cricket, you have got beyond your custom.

That the locality must be certain was confirmed by Lord Denning
MR in *New Windsor Corporation* v *Mellor* [1975] Ch 380, at p387B. L

In my judgment, the application inevitably failed on the evidence submitted. I would accept that the application does not necessarily fail because the users of the land are in the main those residents of Sudbury nearest to it, nor does it fail for lack of proof of continuous use. Where the evidence fails, and fails by a long way in my judgment, is in establishing use of the land for sports and pastimes as of right as inhabitants of Sudbury. The expression "as of right", in this context, requires a belief in the users that they are exercising a right enjoyed by the inhabitants of Sudbury to use the land for sports and pastimes. The expression "as of right" has the same meaning in this Act as it has in the highways Acts, as to which see *Hue* v *Whiteley* [1929] 1 Ch 440, at
p445 *per* Tomlin J, *Jones* v *Bates* [1938] 2 All ER 237 and *O'Keefe* v M
Secretary of State for the Environment [1996] JPL 42, citing *Mills* v *Silver* [1991] Ch 271. There must, in my judgment, be an honest belief in a legal right to use, the legal right in this case being the right as an inhabitant of Sudbury, and not merely a member of the public, to indulge in sports and pastimes on the land. I do not propose to dwell on how much of the use described in the answers to questionnaires can be described as "sports and pastimes" because, on the view I take of the expression "the inhabitants . . of any locality as of right", the evidence falls far short of establishing the requisite use. The use does not demonstrate belief in a legal right to use and, even if it did, there is no sufficient definition of the right asserted. It follows that I agree with

the judge on the second of his reasons and his comments upon "locality" and would dismiss the appeal on that ground.

"Becomes"

Consideration of the first of the judge's grounds is however appropriate, especially having regard to the most helpful submissions addressed to the court. Had the land become a town green within the meaning of that term in section 13(*b*) of the Act? Construing the word "becomes" in section 13(*b*) does involve some assessment of the general effect of the Act, upon which differing views have been expressed. I agree with the judge and with Mr Ryan, that the *Corpus Christi College* case does not assist in the present context. The decision was peculiar to the two-limb definition of "common land" in the context of registration of common rights. On the question whether "becomes" in section 13(*b*) requires some change in the facts on the ground, the case does not assist.

Having regard to the nature of customary rights, as explained in *Hammerton*, I do not consider that a failure to register a town green under section 1(2)(*a*) of the Act has the effect of extinguishing the customary rights over the land. The unregistered green shall not be "deemed" to be a green, as registered land would be, but it may still be proved by evidence to be a green. That view is supported by note 5 to form 30 (above) which contemplates the existence of customary rights after 1970.

In *Mellor* Lord Denning MR, at p387E, cited *Hammerton* as authority for the proposition that a right, once acquired by custom, can only be extinguished by Act of Parliament and added that, "no statute can take away that right by a side-wind. It can only be done by express words: see *Forbes* v *Ecclesiastical Commissioners* (1872) LR 15 Eq 51".

In re Turnworth Down, Dorset [1978] Ch 251 Oliver J stated at p260G:

> It is not said, it is true, what is the consequence of non-registration of a town or village green, except to the extent that section 1(2)(*a*) provides that, after the end of the specified period:
>
> "... no land capable of being registered under this Act shall be deemed to be common land or a town or village green unless it is so registered".
>
> On the face of it it seems that the local inhabitants are merely left to bear with such fortitude as they can command the fact that the land ... is not deemed to be a town or village green.

I agree with Mr George Laurence QC, in his opinion to the Countryside Commission as published by the Open Spaces Society (*Getting Greens Registered* 1995), that Oliver J was indicating in that passage that, "the inhabitants will need very little fortitude indeed because, despite the fact that their town or village green is not on the register, their rights to enjoy it remain undisturbed". Oliver J also stated (at p260E) that the primary purpose of the Act is "not to derogate from the [common] rights but to provide for their registration and indeed to protect them".

The 1969 Regulations make provision for the registration of "new" greens, that is, greens which did not exist on January 2 1970. Given that customary rights continue notwithstanding non-registration, and given the apparent purpose of the Act to encourage the registration of town greens, I cannot conclude that parliament intended in the case of continued use after 1970 to prevent the registration of a new green merely because a customary green had not been registered in time. The Act envisages the creation of class c greens. The word "registered" appears in both paras (*a*) and (*c*) of section 13. In the context I read the expression "land becomes a town or village green" in section 13(*b*) as "land becomes registrable as a town or village green". The way in which form 30 requires the application to be made supports that view: "Application is hereby made for the registration as a town or village green of the land described below, which *became so registrable* after 2nd January 1970" (emphasis supplied). Amendment of the register is possible under the section when a green is established as a class c green after January 2 1970, even if it could have been but was not registered as a class b green under section 1(1)(*a*).

Creation of rights

Mr Chapman submits that registration of a class c green does not confer on or confirm in the inhabitants of the locality the right to indulge in lawful sports or pastimes on the land. Registration deals not with rights but with registration. The owner can remove the town cricket team from a registered town green.

In *Mellor* application was made, within the time-limit imposed by section 1 of the Act, to register land as a village green by virtue of a customary right and also by virtue of 20 years user as of right. The Court of Appeal held that the Chief Commons Commissioner was entitled to find a customary right and it was not necessary to consider the 20-year user. However Lord Denning MR stated at p391H:

> But the difficulty about this 20-year user is that the Act does not tell us what rights, if any, ensue to the inhabitants by virtue of a 20-year user. It enables the land to be registered as a town or village green, but that mere fact of registration confers no rights. And at common law 20-year user gives no rights.

Under the heading *The effect of registration*, Lord Denning stated at p392E-G:

> ... there is nothing [in the Act] to tell us what the effect of registration is. It confers no rights in itself. All is left in the air. The explanation is that Parliament intended to pass another statute dealing with these and other questions on common land and town or village greens. This Act twice refers to matters which "Parliament may hereafter determine": see section 1(3)(*b*) and 15(3). I hope that another statue will not be long delayed. But, if there should be delay, I would be tempted to infer from this Act of 1965 that Parliament intended that all land registered as a "town or village green" should be available for sports and pastimes for the inhabitants ...

Browne LJ stated, at p395G:

> I also agree that as the Act stands, and without further legislation, such user (20 year user as of right) confers no rights on the public.

On my earlier findings, the point does not arise for decision in this case but I would respectfully question whether it is correct that registration of a "new" green under section 13 of the Act and the regulations does not have the effect of confirming in the relevant inhabitants a right to indulge in lawful sports and pastimes on the land. The point was not argued in *Mellor* which in any event concerned pre-1970 and not post-1970 use. Lord Denning recognised that the class c green in section 22 was "a new thing which is not known to the common law". That proposition cannot, with respect, be challenged but it leaves open the question whether rights may have been created by statute.

Town or village greens undoubtedly have a legal existence. They were expressly exempted from inclosure under section 15 of the Inclosure Act 1845. Section 12 of the Inclosure 1857 provided, "summary means of preventing nuisances in town greens and village greens". Section 29 of the Commons Act 1876 provided that, "an encroachment on or inclosure of a town or village green, also any erection thereon or disturbance or interference with or occupation of the soil thereof which is made otherwise than with a view of the better enjoyment of such town or village green or recreation ground, shall be deemed to be a nuisance". Section 15 of the Commons Act 1889 brings town and village greens within the regulatory provisions contained in Part I of that Act. Once it is established that land is a town or village green, there are protectable rights to use it.

While I accept that, by its title, the Act is said to be "an Act to provide for the registration ... of town or village greens" and not an Act to amend the law relating to public rights, section 22 does provide that "in this Act unless the context otherwise requires", class c land is included within the definition "town or village green". There is no express limitation upon the purposes for which the land is included within the definition. There is no doubt that the inhabitants have rights over class a and class b greens and the effect of land being a class c green should be considered in that context. I find it difficult to conclude other than that parliament intended, in section 22, to open the way to the creation of new rights. The right is "to indulge in lawful

A sports and pastimes" while avoiding the need to prove an immemorial custom or legal origin which would establish a class a or class b green. The limitation of the period of user to 20 years does not offend against principle having regard to *Brocklebank* (*supra*). The land is a town green by virtue of statute and, once it is a town green, should bear the same incidents as any other town green. It is the relevant use which, under the statute, creates the right but once it is registered under section 13, section 10 provides that registration shall be conclusive of the matters registered. Classes b and c should be considered together and just as the evidential requirements of class b are imported into class c, the substantive effect of the required proof is, in my view, the B same in both classes.

The analogy is not exact but I see class c as a way of establishing rights just as section 1(1) of the Rights of Way Act 1932 (now section 31 of the Highways Act 1980) provided a means of proving the existence of a highway. Omitting words of limitation and exception, section 1(1) provided that, "where a way . . . upon or over any land has been actually enjoyed by the public as of right and without interruption for a full period of 20 years, such way shall be deemed to have been dedicated as a highway". In *Jones* v *Bates*, Scott LJ stated that the effect of the section was to give "a new statutory effect to mere proof of actual user as of right and without interruption". An actual C dedication need not be proved. I would construe the class c definition as having the same effect in making proof of the appropriate user sufficient to create a right.

The test for considering whether the rights are established is a stringent one, as stated earlier in this judgment. That provides protection for those with an interest in the land. The registration procedure, as laid down in the regulations, provides further protection by imposing duties upon a registration authority to send, publish and affix notices (regulation 5(4)) and conferring the right to object. The regulations, together with the prescribed form 30 and its notes, appear to me to be drafted on the basis that what is to be registered, if the application is successful, is a right.

D *Respondents' procedure*

Had a different view of the facts by the respondents been a real possibility, I would not have quashed their decision on the ground of procedural irregularity. Regulations require them to consider the application and the material supplied by the applicants and the objectors was available to members. In his report, the respondents' solicitor correctly identified the issues. The regulations do not, in my view, impose on the respondents in present circumstances a duty to make their own investigations into the use made of the land.

It is right to acknowledge that the solicitor's report did lay stress on counsel's view of the evidence, which the solicitor adopted, but the E members were not precluded from making their own assessment of the evidence and, according to the council minute, they considered the arguments on both sides. The issues upon construction of the Act are such that legal advice was not only helpful but essential. There was a "consideration" sufficient to satisfy the requirements of regulation 6. Nothing I say derogates from the duty of council members to assess the submitted evidence when performing the duty under the regulations.

Costs

The judge ordered the appellants to pay the costs of the respondents F and one half of the costs of the Secretary of State. According to counsel's note, the Secretary of State was deprived of half his costs on the ground that the judge took into account "that part of the problem is due to the failure of successive Secretaries of State to implement legislation".

I do not regard that as a relevant consideration. The Secretary of State for Health cannot be responsible for parliament's failure, if it rightly be regarded as failure, to pass legislation on the management of town greens.

This court should, in my view, exercise the discretion afresh. The Secretary of State appeared as landowner and developer of land over which rights were claimed by the appellants. His interest was quite different from that of the respondents as registration authority and whose duties were to be performed, it is agreed, in an impartial manner G under the Act and regulations.

The mere fact that the party claiming costs is the developer will not of itself normally justify a second set of costs (*Bolton Metropolitan District Council* v *Secretary of State for the Environment* [1995] 1 WLR 1176, *per* Lord Lloyd at p1178H). In this case however, in addition to claiming relief against the registration authority, the appellants sought a declaration that the land was a town or village green. The Secretary of State appeared at the hearing to protect his interest as landowner against a claim to rights over that land. In my judgment, this is one of those cases where the Secretary of State, as well as the respondents, should be awarded costs. H

I would dismiss the appeal and allow the Secretary of State's appeal on costs.

Butler-Sloss and **Schiemann LJJ** agreed and did not add anything.

Appeal dismissed.

J

Louis and another *v* Sadiq

COURT OF APPEAL

November 12 1996

EVANS, HENRY and ALDOUS LJJ

Estates Gazette April 19 1997

[1997] 16 EG 126 K

Party wall — London Building Acts (Amendment) Act 1939 — Works commenced and damage caused prior to party wall award — Whether building owner liable in nuisance to adjoining owner for damages continuing after Act complied with

In August 1988, and without complying with the procedures under the London Building Acts (Amendment) Act 1939, the appellant commenced works consisting of the demolition and reconstruction of his two-storey end-of-terrace house. These works interfered with the party wall and caused damage to the structure of the respondents' adjoining house and cracks to its front wall. Following the grant of an L injunction in October 1988 restraining further works, the appellant served the appropriate notice under the 1939 Act in December 1988. The appointed surveyors' final party wall award was made in October 1991. The appellant then repaired the damage to the respondents' house at his expense, and completed the reconstruction of his property by October 1992. The respondents had accepted an offer to purchase their house in August 1988, but because of the works it remained unsold and they were unable to start building on a plot of land they had acquired in July 1988. In a claim in nuisance for their losses, the county court judge awarded the respondents damages for, *inter alia*, M mortgage interest payable from the date when the house would have been sold until trial, and the extra building costs in respect of the plot of land attributable to price increases over the period. The appellant appealed contending that once the provisions of the London Building Acts (Amendment) Act 1939 had been invoked, any liability at common law had been excluded or reduced.

***Held*: The appeal was dismissed. The appellant committed a nuisance because of the work which he did without obtaining statutory authority under the London Building Acts (Amendment) Act 1939, and that liability continued until the works were completed in 1992 because the nuisance continued for that period in fact, rather than because the statutory procedures took that**

A **time to run their course. The decided authorities establish that the appellant would not have been liable in nuisance if he had given notice, or obtained consent, in accordance with the Act and then done what was agreed or was approved by the surveyors. The appellant's liability at common law was not excluded or reduced by the provisions of the Act which he eventually invoked after the nuisance had arisen. The adjoining owner's common law rights are supplanted when the statute is invoked, which can have the effect of safeguarding the building owner from common law liabilities when he complies with the statutory procedures, just as he may incur liabilities under the statute which did not exist at** B **common law. But if he commits an actionable nuisance without giving notice and without obtaining consent, he cannot rely upon a statutory defence under procedures with which *ex hypothesi* he has failed to comply. If he then does give notice he will then in due course acquire statutory authority for whatever works are approved or agreed, but this does not relieve him from liability for the continuing nuisance which he has unlawfully committed, until such time as and to the extent that such authority is obtained. The works which created the nuisance to the respondents' house were not subsequently authorised; the appellant was liable for the losses awarded, the loss of the sale of the house being caused by the** C **unauthorised works and were not too remote. The respondents had not failed to mitigate.**

The following cases are referred to in this report.

Adams v *Marylebone Borough Council* [1907] 2 KB 822
Major v *Park Lane Co* (1866) LR 2 Eq 453
Selby v *Whitbread & Co* [1917] 1 KB 736
Standard Bank of British South America v *Stokes* (1878) 9 ChD 68
Thompson-Schwab v *Costaki* [1956] 1 WLR 335; [1956] 1 All ER 652, CA
Upjohn v *Seymour Estates Ltd* (1938) 54 TLR 465

This was an appeal by the defendant, Mohammed Sadiq, against a decision of Judge Platt, who in Shoreditch County Court had awarded D damages to the plaintiffs, Paul and Paula Louis, in their claim for damages for nuisance.

Ann McAllister (instructed by Dowse & Co) appeared for the appellant; David Gerrey (instructed by Laurence Brass Conway) represented the respondents.

Giving judgment, EVANS LJ said: This appeal raises issues concerning the rights and duties of neighbours whose properties are divided by a party wall. Being in London, these are regulated by statute, as they have been since the 18th century. The statute currently in force is the London Building Acts (Amendment) Act 1939.

E

Facts

The appellant Mohammed Sadiq is the owner of 50 Jennifer Road, London N16. This is a two-storey end-of-terrace house. He has lived there since 1985. His neighbours at no 52 are the respondents, Mr and Mrs Louis, who bought their property in 1979.

Sometime in the early summer of 1988 there was an informal conversation between the appellant and Mrs Louis. He told her that he wanted to build a small further extension to the rear of his property where there was already an extension known as the rear addition. She said that she had no objection, and she heard no more about it.

Then suddenly, as she put it, in August 1988 the appellant began F what proved to be works of demolition and reconstruction involving the whole of his house. First, without warning, the garden wall was demolished. Then the rear wall of the rear addition and the rear wall of the house itself were demolished, the new extension was constructed and the rear wall partly rebuilt. Then the flank wall, furthest away from no 52, was demolished and partly rebuilt. Finally, the front wall was demolished also. This interfered with the party wall, as the demolition at the rear had done, and damage was caused to the structure of no 52, including cracks in its front wall.

At this stage, as the judge put it, "The plaintiffs who had up to this point suffered more or less in silence decided to take legal advice". A writ was issued and the plaintiffs sought and obtained an interlocutory injunction. The injunction G was dated October 14 1988. This restrained the appellant from carrying out any works of demolition or reconstruction at no 50, save as to the flank wall, until such time as he had complied with the provisions of the 1939 Act. He had not given the notice required by the Act nor attempted to comply with its provisions although he was advised by the borough surveyor that he should do so, before the work began. He was also ordered to provide support for the front elevation of no 52 at the junction with the party wall and make good any damage caused.

Work stopped, except for the shoring-up operation, and the statutory notice was given in December 1988. The Act provides for the H appointment of two surveyors and in case of dispute they appoint an umpire. The umpire's award was made on July 30 1991 and the surveyors made their final award on October 22 1991. Meanwhile, no 50 remained open to the elements with no front wall and a partly built flank and rear walls which somehow managed to support the roof. The injunction was lifted in December 1991 and work began in February 1992. The appellant repaired the damage caused to no 52 at his own expense, something in excess of £10,000, and the reconstruction of his own property was complete by October 1992.

The respondents claimed damages for losses which they suffered by reason of this extraordinary catalogue of events. These were J substantial. They put no 52 on the market for sale in June 1988 and accepted an offer of £143,000 in August. The purchaser required a mortgage advance of £95,000. Valuers who reported on August 13 1988 saw that "the adjoining property no 50 is being improved and extended at the present time . . . Solicitors should obtain full details . . . and should confirm that all such [party wall] notices were properly served and agreement obtained for the approved plans". Subsequently, the purchaser withdrew.

The judge found:

> I am satisfied that all the other matters raised in their report could have been resolved within a few weeks. K

and

> The property market in the middle of 1988 was extremely buoyant. Prices were rising and continued to do so well into 1989. . . . If the defendant had properly complied with the Act I am satisfied that the plaintiffs would have still been able to find a buyer and to complete the sale of their property by the end of October 1988.

So the property remained unsold. It is doubtful whether it was saleable at any price until the repairs to it and the rebuilding at no 50 were complete in 1992. By May 1994 due to the fall in the market its value was estimated as £100,000 and in February 1995 when the L plaintiffs served a "Provisional Schedule of Special Damage" it had fallen still further to £78,000.

Their failure to sell no 52 affected the plaintiffs in other ways. They planned to move to Guadeloupe and on July 22 1988 they purchased a building plot there for 250,000ff, about £25,000. They had arranged a mortgage of their London house, no 52, borrowing £65,000 from Barclays Bank in April 1988. Part of this sum was used to repay earlier loans for smaller amounts. Their building costs in Guadeloupe were estimated in May 1988 as 680,428ff, but by February 1995 these had risen to 985,500ff, an increase of 305,072ff over 81 months. Three-fifths of the total costs was in fact incurred during this period, leaving M two-fifths outstanding at the date of trial.

Trial and judgment

The respondents abandoned their claim for the fall in value of no 52 as this was due to the general operation of market forces. However, they maintained a claim for special damages in respect of mortgage interest which they paid or were liable to pay to Barclays Bank from the date when no 52 would have been sold, up to the date of trial. They also claimed special damages for the extra costs of building in Guadeloupe, and general damages for nuisance etc.

These claims succeeded, for reasons which are clearly set out in the careful and comprehensive judgment of Judge Platt, which was

A followed by a supplementary ruling as to the amount of interest on May 16 1995. He awarded the following sums:

(a) Interest paid from November 1 1988 to March 31 1993 £22,762.07

(b) Extra building costs, calculated as $^2/_5$ths of the price increase during the same period: 79,500ff.

(c) General damages: £12,500 (plus £1,500 for a separate item).

The period under (a) and (b) ran to March 1993 because that allowed the respondents six months to sell no 52 after the work was complete.

B *Appeal*

This appeal is against the awards of special damages under (a) and (b) above. Miss Ann McAllister on behalf of the appellant submits that there was no basis for the award of such damages and that in any event the respondents' losses were not caused by him, or were too remote, or excessive in amount.

Liability

This makes it necessary to quote from the judgment the basis on which liability was found to exist. Having held that the defendant's conduct between July and October 1988 amounted to a nuisance, the
C judge continued:

2. Apart from the manner in which the work was done did the actions of the defendant amount to a nuisance?

On this issue specific complaint is made of a number of matters.

1. That the defendants interfered with the rear end of the party wall by demolishing the rear addition wall to no 50 without complying with the provisions of section 47 of the Act.

2. That the defendant interfered with the foundations of the party wall by digging foundations for the new extension and the rebuilt rear addition wall of no 50 without complying with the provisions of section 50 of the Act.

3. That the digging of foundations for the new extension which encroached on to the adjoining property caused damage to the foundations to the rear
D addition to no 52 necessitating underpinning.

4. That the demolition and rebuilding of the rear addition wall to no 50 crossed over the boundary line at first floor level into the rear addition wall of no 52 and must have involved physical damage to the rear addition wall of no 52...

5. That the defendant interfered with the front of the party wall by demolishing the front wall of no 50 without complying with the provisions of section 47 of the Act.

6. That the demolition of the front wall in particular by cutting into the front end of the party wall caused physical damage to the front wall of no 52 by separating it to a significant extent from its junction with the party wall.

E In relation to each of these factual issues, the judge found that the appellant's activities had encroached upon the respondents' property and had caused physical damage to it. This can constitute a private nuisance in law for which damages are recoverable: see *Clerk and Lindsell on Torts* (17th ed) para, 24-03, approved by Lord Evershed MR in *Thompson-Schwab* v *Costaki* [1956] 1 WLR 335, which the judge quoted in full:

A private nuisance may be and usually is caused by a person doing on his own land something which he is lawfully entitled to do. His conduct only becomes a nuisance when the consequences of his acts are not confined to his own land but extend to the land of his neighbour by (1) causing an encroachment on his
F neighbours land, when it closely resembles trespass, (2) causing physical damage to his neighbours land or building or works or vegetation upon it, or (3) unduly interfering with his neighbour in the comfortable and convenient enjoyment of his land.

He then held that the appellant was liable under each of the six heads to which he had referred:

Complaints 1, 2 and 5 are effectively admitted in the sense that the defendant simply cannot deny that he did carry out work affecting the party wall and its foundations without serving notice under either section 50(2) or section 47 at least two months before commencing work. The nub of each complaint is the failure to comply with the statutory procedures laid down by the act before commencing work. I have come to the conclusion that this is a nuisance under both heads (2) and (3) of the passage from *Clerk and Lindsell* cited above. G

The mischief which results is that the plaintiffs as adjoining owners are left in the position where work has been done to part of their own property without their having any precise information as to the nature of the work, without their having any information as to whether and if so how much damage has been caused to the party wall and without their having any knowledge that the work has been done either on instruction or advice from a properly qualified surveyor.

and

In my judgment complaints 3 and 4 are true. They both fall under the headings both of encroachment and physical damage and therefore constitute a H nuisance.

He made a similar finding regarding complaint, and he concluded:

It follows in my judgment that each of the six acts complained of created nuisances for which the defendant is *prima facie* liable to the plaintiffs.

As regards the mortgage interest claim he made the following specific finding:

It is in my judgment interference with the convenient enjoyment of land to act in such a way as to make the neighbours property unsuitable security for a J mortgage advance certainly at a time when it is known to be on the market for sale. No 52 was known to the defendant to be on the market for sale from June 1988. It was overwhelmingly likely that any purchaser of no 52 would require a mortgage to assist in the purchase. It is equally clear from the mortgage survey and in my judgment equally foreseeable that the failure to comply with the procedures of the Act would constitute a fundamental obstacle to the property being judged suitable security for a mortgage advance.

It is important to note, as the judge did, that there was no claim for the cost of repairs. This was because the respondents did not incur such costs. The repairs were done by the appellant himself and at his own expense. "In each case . . . the nuisance has so far as possible been K abated and the damage repaired so that there is no continuing problem". The issue, therefore, concerned the claim for special damages under these heads, both involving financial loss.

Liability

Miss McAllister's submission is that the appellant is under no liability at common law, because the statutory procedures and remedies for breaches of them depend entirely on the statute itself: common law rights, duties and remedies are excluded by the Act. The respondents, represented by Mr David Gerrey at the trial, expressly abandoned any claim for damages for breach of statutory duty. There was no finding that the appellant ". . . exerted his statutory privileges L so as to inflict injury on the plaintiff by negligence, improper obstructiveness, avoidable nuisance or unreasonable delay" *per* McCardie J in *Selby* v *Whitbread & Co* [1917] 1 KB 736 at p753. Without such a finding, she submits, the respondents cannot succeed in this part of their claim. The Act required the appellant to give prior notice, which he did not, but "failure to comply with the Act does not *per se* give rise to an action in nuisance". Moreover, the appellant did give notice in December 1988, and the subsequent delay was caused by the length of time which the statutory procedures took. He should not be liable for that delay, when he was complying with the Act rather than otherwise. M

This analysis of the basis on which liability exists, in my judgment, is wrong, although it might be said to be supported by the judge's statement "The nub of each complaint is the failure to comply with the statutory procedures laid down by the Act before commencing work", which was reflected in Mr Gerrey's submission to us that the appellant was liable for failing to give the notice which the statute required. I would prefer to say that the appellant committed a nuisance because of the work which he did without obtaining statutory authority, and that liability continued until the works were completed in 1992 because the nuisance continued for that period in fact, rather than because the statutory procedures took that time to run their course.

If this view is correct, then it is unnecessary to refer to the statutory provisions in any detail, but in deference to Miss McAllister's submissions I will summarise them as follows.

Section 46 specifies the "Right of owners of adjoining lands where junction line built on". These rights are given to "the building owner" and they include demolition as well as strengthening etc of party structures. However, by section 47, "Before exercising any right conferred on him by section 46 . . . a building owner shall serve on the adjoining owner . . . a 'party structure notice' (section 47(1)) and the notice must be served at least two months (sometimes one month) "before the date stated therein as that on which the work is to be begun" (section 47(2)). The adjoining owner may then serve a counternotice (section 48) and where differences arise between them they may appoint an agreed surveyor or each appoint their own surveyor (section 55(a)). The parties' surveyors then appoint a third surveyor (section 55(a)(ii)) and there is a short and, as the present case demonstrates, optimistic time table for the eventual publication of an award, which may "determine the right to execute and the time and manner of executing any work and generally any other matter arising out of or incidental to the difference" (section 55(k)). There is provision for an appeal to the county court or to the High Court (section 55(n)(ii)).

These procedures, however, are all subject to section 47(4):

> (4) Nothing in this section shall prevent a building owner from exercising with the consent in writing of the adjoining owner . . . any right conferred on him by section 46 . . .

So the statutory scheme is clear. The building owner has certain express rights, but these can only be exercised (1) with the adjoining owner's written consent, or (2) in accordance with a valid award by the surveyor or surveyors appointed under section 55.

The effect of these provisions and their predecessors has been considered in a number of authorities. No notice is required in the case of the "mere removal of a building from an adjoining building without disturbing the party structure": *Major* v *Park Lane Co* (1866) LR 2 Eq 453. In *Standard Bank of British South America* v *Stokes* (1878) 9 ChD 68 the building owner undertook works with the *bona fide* intention of supporting the party wall, and at common law the adjoining owner had no right of action (see p73). Under the Act, however, the defendant "had no right to proceed with his work until the direction of the surveyor had been obtained", and he was liable to be restrained from doing so (p77).

Sir George Jessel MR said (at p76):

> I think it the duty of the Court to read these Acts in a reasonable manner, with a view of carrying out the manifest intention of the Legislature that these party structures should not be interfered with by building owners without due notice to the person or persons other than the building owner interested in the party structure . . . .

In *Selby* v *Whitbread & Co* (cited above) McCardie J considered the relationship between the statutory scheme and the parties' rights at common law. He said (at p752):

> An examination of the code at once shows that common law rights are dealt with in a revolutionary manner. The two sets of rights, namely, the rights at common law and the rights under the Act of 1894 (which followed the Act of 1855), are quite inconsistent with one another. The plaintiffs' common law rights are subject to the defendants' statutory rights. A new set of respective obligations has been introduced. The common law was seen to be insufficient for the adjustment of modern complex conditions. Hence I think that the Act of 1894 is not an addition to but in substitution for the common law with respect to matters which fall within the Act. It is a governing and exhaustive code, and the common law is by implication repealed. (p752).

In the particular case, he held ". . . the plaintiffs cannot succeed upon their claim at common law inasmuch as the defendants' party wall notice had been duly given under the provisions of the Act . . ." (p753). He then referred to actions which might be brought outside the Act if the building owner abused his statutory rights (quoted above).

These authorities establish, in my judgment, that the appellant would not have been liable in nuisance if he had given notice, or obtained consent, in accordance with the Act and then done no more than was agreed or was approved by the surveyors. But then, no damage would have been caused to the respondents' house, save in the party wall itself, and in that respect no liability would have arisen. The issue raised in the present case is whether the appellant's liability at common law is either excluded or reduced by the provisions of the Act which he invoked, eventually, after the nuisance had arisen.

I would have no hesitation in rejecting this submission even without reference to authority, because, in my judgment, there is nothing in the Act which can be said to have this effect. The adjoining owner's common law rights are supplanted when the statute is invoked, which can have the effect of safeguarding the building owner from common law liabilities when he complies with the statutory procedures, just as he may incur liabilities under the statute which did not exist at common law (the *Standard Bank* decision). But if he commits an actionable nuisance without giving notice and without obtaining consent, he cannot rely upon a statutory defence under procedures with which *ex hypothesi* he has failed to comply. If he does then give notice he will in due course acquire statutory authority for whatever works are approved or agreed, but, in my judgment, this does not relieve him from liability for the continuing nuisance which he has unlawfully committed, until such time as and to the extent that such authority is obtained.

This conclusion is supported by the judgment in *Adams* v *Marylebone Borough Council* [1907] 2 KB 822. The decision of the Court of Appeal was that surveyors appointed to resolve differences between the parties had no jurisdiction to award compensation to the adjoining owner for inconvenience and loss of trade caused by works which were authorised by the statute: ". . . in my opinion, the mere exercise of such rights cannot afford any ground for compensation under the Act" (*per* Vaughan Williams LJ at p835; compare p841 *per* Fletcher Moulton LJ ". . . the adjoining owner has no right to compensation for any inconvenience caused by operations which the Act has given the building owner a right to carry out, so long as the latter acts within the provisions of the Act, for, so long as he does so, his action is lawful.").

The case was one where unauthorised works were begun before any statutory notice was served. The adjoining owner, who was a restaurant owner, claimed an injunction (see [1905] 1 KB 661) and in due course was awarded damages for loss caused by the operations of the building owners prior to the giving by them of the party wall notice (p824). It was not clear to the court what further claim was then made before the surveyors, but it was assumed that the adjoining owner had suffered further loss of trade after the notice was given and the statutory procedures began, by reason of work which was already done (see p828). The court held that no such claim could be made, and Vaughan Williams LJ in the leading judgment expressed the view that no claim would lie "before any tribunal whatever" (p829). That was because the work was "in its nature rightful" (*ibid*).

In the present case, therefore, Miss McAllister submits that no claim lies for damage caused after the notice was given, in December 1988. If *Adams* v *Marylebone* was authority for that proposition then of course it would be binding on us, even if that result seemed not in accordance with principle, for the reasons stated above. But, in my judgment, there were two significant features of that case which are not present here. First, there was an agreement between the parties when the notice was given "that the work already done should thenceforward be treated as if it had been done under a notice duly given under the Act" (p828). No such agreement is alleged or found to have been made in the present case. Therefore, the basis for the court's decision that what was done was lawful under the Act does not exist here. Second, a related point is that there is no finding in the present case that the statutory approval when it was given covered all the works which created the nuisance in 1988. It would be surprising if that finding was even suggested, because the surveyors could hardly have agreed to approve works which caused such widespread damage

A to the adjoining house. For example, the appellant failed to provide shoring when the front wall of his house was removed. So it cannot be said, in my judgment, that the works which created the nuisance were subsequently authorised, whether by agreement or by surveyors under the statutory procedures.

Without such justification, the appellant has no defence, in my judgment, to the claim for damages for the nuisance which he created in 1988 and which continued, on the judges's findings, until the works were complete.

One authority, on which we did not hear argument, appears to be
B directly in point. In *Upjohn* v *Seymour Estates Ltd* (1938) 54 TLR 465 the owner of a house, 92 Camden Road, began to demolish his property with the result that his half of the party wall collapsed. The plaintiff lived next door at no 94 and as a result his property was damaged and he suffered business losses. ("His stock suffered severely, as did his business . . ." (p466).) The defendants gave notice under the London Building Act 1930 and surveyors were appointed, but before their award was published the demolition took place without further notice to the plaintiff. Goddard J held that no trespass had taken place but the defendants were in breach of duty by withdrawing support from the party wall, and they were liable for "all such damage
C to the structure as can be proved to have resulted from the withdrawal, and also for the damage to the stock and disturbance of the business which was caused by the collapse . . ." (p467). In short, damages were at large and they were directed to be assessed by an official referee. It was not suggested that they should be reduced by reference to any award, if there was one, made by the surveyors after the disturbance took place.

Causation

The judge found that "if the defendant had properly complied with the Act . . . the plaintiffs would have still been able to find a buyer and
D to complete the sale of their property by the end of October 1988". There was evidence that the mortgage surveyors, who saw the extent of the demolition and rebuilding that the appellant was carrying out, were concerned to know that proper notice had been given and the statutory procedures complied with. The judge was entitled to hold that the sale actually negotiated with a prospective purchaser, Miss Knight, or with some other purchaser would have been concluded, even if the works were in progress, provided they were carried out lawfully under the Act, *a fortiori*, if a notice had been served and as a result the works themselves had not begun. The appellant submits that the loss of a sale was not shown to have been caused by his failure to serve the statutory
E notice, but that is not what the judge found. "If the defendant had properly complied with the Act" means "if he had not carried out unauthorised and therefore unlawful work", and undoubtedly it was those works which were the effective, if not the only cause of the respondents' failure to sell their house in 1988.

Remoteness

It is sufficient, in my judgment, that the appellant could reasonably foresee that the respondents might be unable to sell their house because of the (unlawful) works which he carried out, at least until statutory
F authority was obtained and the works so authorised were complete, and that they might suffer financial loss in the form of a continued liability to pay mortgage interest on a loan or loans which would otherwise be repaid. Moreover, the respondents might intend to build a new property, in this country or overseas, and increased building costs might be the consequence of delay. None of these was bound to happen, and it might also have been the case that the respondents would have obtained a fresh mortgage advance on their new property, for an equal or greater amount, and therefore would not have suffered any loss of mortgage interest, by reason of the delay. Both possibilities were reasonably foreseeable and the respondents did in fact incur this type of loss by reason of their inability to sell the house for the period in question.

Length of period/failure to mitigate G

The period covered by the judge's award was from November 1 1988 until March 31 1993, made up as follows:

December 1988	Notice given.
October 1991	Surveyors' final award.
October 1992	Rebuilding work complete. This was the work "necessary to repair the front wall of number 52 and tie it in to the rebuilt party wall".
March 1993	Six months for sale.

The bulk of this period was taken up with obtaining the surveyor's final award. The judge examined the reasons for delay, one of which was that the respondents' surveyor held certain views as to the content of an agreed award with which the appellant's surveyor disagreed. H
This led to the appointment of a third surveyor. A second reason was "the obvious inexperience bordering at times on incompetence of the Defendant's surveyor". He held that the appellant could not rely on delay caused by his own appointee, and that the respondents' surveyor's conduct had not been so unreasonable as to be unforeseeable. "The truth is that he was throughout as anxious to resolve matters as speedily as was compatible with his professional duty to protect his clients' interest."

I can see no ground on which these findings can be challenged, and in the light of them the appellant's submissions that the period was excessive and that the respondents failed to mitigate their loss must, in J
my judgment, fail.

I therefore would dismiss this appeal.

Henry and **Aldous LJJ** agreed and did not add anything.

Appeal dismissed.

K

Observatory Hill Ltd *v* Camtel Investments SA

CHANCERY DIVISION

July 15 1996

Judge Levy QC, sitting as a judge of the High Court

Estates Gazette May 3 1997

[1997] 18 EG 126 L

Party Walls — London Building Acts (Amendment) Act 1939 — Surveyor's party wall award — Damage to adjoining owner's property to be made good at cost of building owner — Whether adjoining owner entitled to register caution against title of building owner

By a party wall award dated December 23 1992, and made under the London Building Acts (Amendment) Act 1939, the plaintiff building owner was permitted to carry out work on land which is partly the property of the plaintiff and partly the property of the defendant M
adjoining owner. The award imposed on the plaintiff an obligation at its expense to make good damage to the defendant's building, and in the event the plaintiff fails to do so, the defendant is entitled to carrying out such works at the plaintiff's expense and upon the plaintiff's land. The defendant's premises were damaged by the plaintiff's excavations and the defendant registered a caution against the plaintiff's title at HM Land Registry. The plaintiff applied to have the caution removed on the ground that the defendant did not have a cautionable interest.

***Held*: The application was allowed and the caution ordered to be removed on a cross-undertaking by the plaintiff as to damages. There is no right in an "adjoining owner" who has the benefit of a**

party wall award under the London Building Acts (Amendment) Act 1939 to become a person or body who can enter a caution on the register as a person having an interest in land belonging to the building owner. Although in certain circumstances the defendant adjoining owner might in due course become entitled to enter the plaintiff's land, such right is not sufficient to entitle the defendant to have such an interest as can be protected by a caution.

The following cases are referred to in this report.

Clayhope Properties Ltd v *Evans* [1986] 1 WLR 1223; [1986] 2 All ER 795; [1986] 2 EGLR 34; (1986) 279 EG 855, CA
Linton International Ltd v *Noble* (1991) 63 P&CR 452

This was a hearing of a notice of motion in an application by originating summons by the plaintiff, Observatory Hill Ltd, against the defendant, Camtel Investments SA, for the removal of a caution.

Wayne Clark (instructed by Simmons & Simmons) appeared for the plaintiff; Alan Steyner (instructed by Cumberland Ellis) represented the defendant.

Giving judgment, **Judge Levy** said: Observatory Hill Ltd ("the plaintiff") is a company which owns property in London W8, which includes Observatory 1, 2, 3, 5, 5A, 6 and 12 of Observatory Gardens. The ultimate owner of the plaintiff company is a company which is, I think, incorporated in Luxembourg.

The defendant company, Camtel Investments SA ("the defendant"), is a company incorporated in Panama, which owns property adjoining the plaintiff's premises in such a way that the two properties share a party wall.

The difficult London Building Acts (Amendment) Act 1939, has been invoked by the plaintiff and there has been an award made under the provisions thus invoked. It is common ground between counsel appearing for the plaintiff and the defendant that the substantial renovations which have been made by the plaintiff on the plaintiff's premises have caused more than minimum damage on the land owned by the defendant.

The defendant has placed on Her Majesty's Land Registry a caution in respect of the interest which it says it has on the plaintiff's land and the plaintiff has commenced proceedings by way of originating summons ordering that the caution entered by and on behalf of the defendant on or about June 19 in the proprietorship register of its title be vacated and the defendant to take all such steps and effect all such things as necessary to ensure that the caution is vacated.

Soon after the originating summons, the plaintiff issued and served a notice of motion and affidavit seeking to remove the caution summarily. The motion first came before Harman J on July 2. On that day, he gave liberty to the defendant to file evidence in answer to the evidence of the plaintiff by 5:30 last Friday and ordered that the motion should come before the court today.

The defendant has so filed an affidavit and the plaintiff has filed an extra affidavit beyond that which was before Harman J on July 2, which I have read without objection from the defendant's counsel.

The situation is abundantly clear: the defendant's premises has been damaged because of the work done by the plaintiff in excavation. The award made by the very experienced arbitrator makes it quite clear that work will have to be done not only on the defendant's land but also on the plaintiff's land to remedy the damage caused by the plaintiff. In such circumstances the question is raised as to whether there is a cautionable interest in the defendant.

In his latest skeleton argument prepared over the luncheon adjournment, Mr Alan Steyner, who has put the matter persuasively on behalf of the defendant, says this: the defendant has an interest in the plaintiff's land sufficient to lodge a caution under section 54(1) of the Land Registrations Act 1925, by virtue of the party wall award made on December 23 1992, under the provisions of the London Building Acts (Amendment) Act 1939. The award: (a) permits the plaintiff to carry out work on land which is partly the property of the plaintiff and party the property of the defendant; (b) imposes on the plaintiff an obligation to make good at his own expense all structural and decorative damage that the defendant's building occasioned by the works; and (c) in the event that the plaintiff fails to do so, the defendant is entitled to carry out such works at the plaintiff's expense and upon the plaintiff's property.

For present purposes Mr Wayne Clark, who appears for the plaintiff, accepts the fact of the award and the consequences of it as spelled out by Mr Steyner. The issue between the parties is as to whether that amounts to a cautionable interest.

In that respect I have been referred to passages both in the current edition of *Emmet on Title* and to sections of *Ruoff and Roper on Registered Conveyancing*, and to several decided cases. It is common between counsel that the facts on which I am asked to pronounce do not fall between any of the decided cases.

I am, to some extent, encouraged by the stance taken by Noggs J in giving his decision in *Linton International Ltd* v *Noble* (1991) 63 P&CR 452, where he looked closely at the rights of a party to have a caution on the register before coming to the conclusion on the facts of that case that there were no such rights.

So far as *Emmet* is concerned, there is this passage at 10.008:

> Caution against dealings: A caution against dealings has a similar effect.

That is referring to the paragraph above, which reads:

> A caution provides a means whereby a person interested may be warned before registration of title is effected or before any disposition is registered which might adversely effect him. There are two types of caution: (i) against first registration, and (ii) against dealings with registered land or a registered charge.

And then the passage deals first of all against first registration. Then the passage on 10.008 continues:

> It is substantially a hostile action, and the procedure is designed accordingly. A person interested under an unregistered instrument, or interested as a judgment creditor or otherwise in any land or charge registered in the name of another, may lodge a caution to the effect that no such dealing with such land or charges to be registered until notice has been served on him.

A question arises as to what is an unregistered instrument; in that regard, I have been referred to section 3 of the Land Charges Act, which does not give an enormous amount of help and it is clearly the position at the moment that the defendant is neither a judgment creditor nor anything approaching one. The best claim that the defendant has is a potential interest to enter on to the plaintiff's land if and so far as the award of the arbitrator is not in due course observed and put into effect by the plaintiff.

The passage in *Emmet* continues by giving examples of many cases, and then at the foot of page reads:

> The interest of a contributory and creditor in land of a company in liquidation is *not* a minor interest which can be protected by caution . . . Nor is a provision for a sum additional to the purchase price to be payable if development occurred.
>
> Similarly, it is not thought that a *Mareva* injunction, which does not strictly involve any interest in land, can properly be protected by a caution . . . However a receivership order obtained by a tenant against his landlord could be so protected.

And there is a reference to *Clayhope Properties Ltd* v *Evans* [1986] 1 WLR 1223*, one of the cases to which I have been referred.

Having regard to the foregoing, in my judgment, there is no right in an adjoining owner who has the benefit of an award under the 1939 Act to become a person or a body who can enter a caution on the register as a person having an interest in land. Such a person or body does not come within any of the categories which are now shown to exist as a person having an interest in land; I think it would be unwise to enlarge that category.

---

*Editor's note: Also reported at [1986] 2 EGLR 34

In the circumstances it seems to me that the submissions made by Mr Steyner cannot be right. I recognize that in certain circumstances the defendant might in due course become entitled to enter on to the plaintiff's land but such a right is not sufficient as now to entitle the defendant to have such an interest in the plaintiff's land as can be protected by a caution on the register. The fact that there is a party wall there, in some part of which the defendant may have some interest, does not seem to me to be a sufficient interest in land any more than does the fact that the implication of the 1939 procedures have been invoked.

In the circumstances it seems appropriate to me to give the relief sought by Mr Clark in his notice of motion, that is to remove the caution on the register, on the cross-undertaking proffered by the plaintiff. I accept as suitable the words in the fifth exhibit to Mr Gordon Hunter's affidavit:

> Upon the plaintiff by its counsel undertaking to abide by any Order this court may make as to damages in case this court should be of the opinion that the defendant shall have sustained any, by reason of this Order, which the plaintiff ought to pay

I should make it clear that although the defendant does have ample other remedies in its hands if the caution is removed, as I shall order, I recognize that the existence of a caution on the register of title of the plaintiff's land improves the bargaining position of the defendant in his dispute with the plaintiff, but that is no reason for a remedy for which there is no statutory, legal or equitable right.

Orders accordingly.

Cheltenham & Gloucester plc *v* Booker and another

COURT OF APPEAL

November 14 1996

MILLETT and POTTER LJJ

Estates Gazette May 10 1997

[1997] 19 EG 155

Mortgages — Power of sale — Whether jurisdiction at common law to postpone execution of warrant for possession until completion of sale — Whether jurisdiction properly exercised

By a legal charge dated June 22 1989 the defendants charged their property to the plaintiffs to secure a loan of £38,000. Because the defendants fell into arrears, an order was made by the county court for possession of the property to the plaintiffs on October 1 1990, suspended on terms as to the payment of current instalments and arrears. After a number of appointments were made, and vacated, for the execution of the warrant for possession, by an order dated April 30 1996 the county court judge ordered that the warrant for possession should lie in the court office conditionally upon the plaintiffs having the conduct of the sale with vacant possession to be provided upon completion. The plaintiff mortgagee appealed.

***Held*: Appeal allowed. Where the court has no jurisdiction under the Administration of Justice Acts 1970 and 1973, as were the circumstances before the judge, there is a limited jurisdiction at common law to defer giving of possession for a short time to enable the property to be sold. The court must be satisfied that possession will not be required by the mortgagee pending completion, the presence of the mortgagor will enhance, or at least not depress, the sale, the mortgagor will co-operate in the sale, and he will give up possession on completion. However because these conditions are seldom likely to be satisfied, the jurisdiction should be sparingly exercised, and only in rare cases. There were disadvantages to a mortgagee trying to sell property with the mortgagor still in occupation because of the risk that possession would not be given up on completion. However the jurisdiction should not have been exercised in this case.**

The following cases are referred to in this report.

Birmingham Citizens' Permanent Building Society v *Caunt* [1962] Ch 833; [1962] 2 WLR 323; [1962] 1 All ER 163

Royal Trust Co of Canada v *Markham* [1975] 1 WLR 1416; [1975] 3 All ER 433; (1975) 30 P&CR 317, CA

This is an appeal by the plaintiffs, Cheltenham & Gloucester plc, from a decision of Judge Mitchell, in proceedings against the defendants, Robert Charles Booker and Susan Jane Booker, for possession.

Elizabeth Ovey (instructed by Church Adams Tatham) appeared for the appellant; the respondents, Robert and Susan Booker, did not appear and were not represented.

Giving judgment, **MILLETT LJ** said: This is an appeal by the plaintiffs, Cheltenham & Gloucester Building Society, from an order of Judge Mitchell in Stoke-on-Trent County Court on April 30 1996. The judge gave a full reasoned judgment for the order he had made on that date when he refused leave to appeal on May 15 1996. Leave to appeal was subsequently given by the single lord justice.

The question which arises on the appeal is a question of general importance. It is whether the court has jurisdiction, and if so in what circumstances should it exercise such jurisdiction, to give conduct of a sale to a mortgagee while at the same time postponing the execution of a warrant for possession until completion of the sale.

The plaintiffs claim as mortgagee under a legal charge dated June 22 1989, granted to secure a loan of £38,000. As at July 1990 the required monthly payment was £587 and the defendants' arrears amounted to £5,212.94. A possession order was made on September 13 1990 for delivery of possession on October 11 1990. The order was suspended on terms that the current instalments be maintained and the arrears, then £4,899.94, should be paid by additional monthly instalments of £500. Mr Booker appeared in person at the hearing. The terms of the suspension were not complied with. Appointments for execution of a warrant for possession were subsequently obtained for no fewer than 14 dates between May 3 1991 and April 30 1996.

The earlier appointments were vacated by agreement with the plaintiff. Later appointments were vacated as a result of orders of the court made on six different occasions between February 3 1994 and February 22 1996 following applications by the defendants for suspension of execution of the order. Of those six applications for suspension, two were eventually dismissed on the defendants' failure to appear and one was adjourned generally with liberty to restore, but was never restored.

Following an order made on March 28 1996 dismissing the defendants' application for further suspension, an appointment for execution of the warrant was made for April 30 1996. On that day, following a further application by the defendants for suspension of the order, the judge made the order which is now appealed against.

The material terms of his order are:

> The warrant for possession do lie in the court office conditionally upon the plaintiffs having conduct of the sale of the property with vacant possession to be provided by the defendants upon completion, and the defendants fully cooperating with the plaintiffs in the marketing of the property.

The defendants' application for suspension of execution of the warrant was made on the grounds that they were still seeking to sell the property and were considering reducing the asking price. They offered to pay the plaintiffs £600 per month pending sale, and relied upon the fact that they had three children who were in occupation of the property, the eldest of whom was shortly to be sitting GCSE examinations. In rejecting the plaintiffs' subsequent application for leave to appeal against the judge's order that the warrant for execution should lie in the office, the judge explained that his order had been made on the footing that:

1. The defendants accepted that they were no longer in a position to maintain the mortgage payments, let alone pay off the arrears.

2. It was not right to exercise the powers under section 36 of the Administration of Justice Act 1970 or section 8 of the Administration of Justice Act 1973.

(I interpose to say that it is plain that in the circumstances the judge had no jurisdiction to exercise the statutory powers referred to.)

3. The purpose for which the plaintiffs sought possession was to realise the security at the best possible price and that this would best be done by early sale while the property was still occupied by the defendants;

4. Accordingly he provided for what he described as a temporary deferment of execution of the warrant under what he considered to be the long established practice of the county courts, supported by the notes to section 21 of the County Courts Act 1984 in the *County Court Practice* under the heading "Execution".

At April 1 1996 the defendants' monthly instalments were £562.57 together with £14.24 for property insurance; the arrears totalled £11,860.09 and the last payment of £460 had been paid on January 16 1996. The evidence before the judge was that the property might fetch £65,000 and that the total indebtedness was then £58,395. At September 4 1996 the arrears amounted to £14,178.94 and the total indebtedness, including interest to the end of September 1996 was £63,291.42. It is therefore apparent that at the present time, after allowing for the costs of sale, the likelihood is that there is little, if any, equity in the property.

The appeal is said to raise the following questions of principle.

1. Where the power to suspend execution given by section 36(2) of the Administration of Justice Act 1970 cannot be exercised because it does not appear to the court that the requirements of section 36(1) of that Act, read in conjunction with section 8 of the Administration of Justice Act 1973, are satisfied, or where the court decides in the exercise of its discretion that the power should not be exercised, does the county court nevertheless have a residual inherent jurisdiction to suspend execution of a warrant for possession?

2. If it has such a jurisdiction, can it be exercised so as to suspend the execution for an indeterminate period?

3. Is it in principle appropriate to suspend execution pending sale when the conduct of the sale is entrusted to the mortgagee rather than to the borrower?

I am not entirely satisfied that the second question arises since the judge did not intend to suspend execution of a warrant for an indeterminate period, but rather for a short period pending completion of a sale in the near future by the plaintiffs.

I will deal first with the question of jurisdiction. The Administration of Justice Act 1970 was enacted because it was considered by parliament that the jurisdiction of the courts to deny mortgagees possession was too limited. It was extremely limited at common law but nevertheless it did exist. In *Birmingham Citizens' Permanent Building Society* v *Caunt* [1962] Ch 883 at p912 Russell J stated the position as follows:

> Accordingly, in my judgment, where (as here) the legal mortgagee under an instalment mortgage under which by reason of default the whole money has become payable, is entitled to possession, the court has no jurisdiction to decline the order or to adjourn the hearing whether on terms of keeping up payments or paying arrears, if the mortgagee cannot be persuaded to agree to this course. To this the *sole exception* is that the application may be adjourned for a short time to afford to the mortgagor a chance of paying off the mortgage in full or otherwise satisfying him; but this should not be done if there is no reasonable prospect of this occurring. When I say the sole exception, I do not, of course, intend to exclude adjournments which in the ordinary course of procedure may be desirable in circumstances such as temporary inability of a party to attend, and so forth.

In *Royal Trust Co of Canada* v *Markham* [1975] 1 WLR 1416, Sir John Pennycuick V-C, when delivering the leading judgment of this court, cited that passage from the judgment of Russell J, with approval and continued at p1420C:

> A characteristic instance in which that sole exception is applicable is where the mortgagor has entered or is about to enter into a contract for the sale of the property at a price which will enable the mortgage to be paid off in full.

So, as the law stood before 1970 the mortgagee had an immediate right to possession, subject only to the single exception mentioned by Russell J, and it was not in the power of the court to refuse it.

In a case where the court has no jurisdiction under the 1970 and 1973 Acts to defer the order for possession, the position is as stated by Russell J and Sir John Pennycuick. There is a residual jurisdiction in the court, but a strictly limited one, to postpone the giving of possession to the mortgagee for a short period in order to enable the property to be sold by the mortgagor.

If so, it appears to me in principle difficult to deny the existence, at least in theory, of a similar jurisdiction to defer the giving of possession for a short time in order to enable the property to be sold by the mortgagee. If the court is satisfied: (a) that possession will not be required by the mortgagee pending completion of the sale but only by the purchasers on completion; (b) that the presence of the mortgagor pending completion will enhance, or at least not depress, the sale price; (c) that the mortgagor will so co-operate in the sale by showing prospective purchasers round the property and so forth; and (d) that he will give possession to the purchaser on completion, it seems to me that there is no reason in principle why the court should accede to a mortgagee's insistence that immediate possession prior to the sale should be given to him.

However, while the jurisdiction exists, experience shows that these conditions are seldom likely to be satisfied. Accordingly, in my judgment, the jurisdiction should be sparingly exercised, and then exercised only with great caution. If the conditions which I have mentioned exist, the court is likely to entrust the conduct of the sale to the mortgagor. There is an inherent illogicality in entrusting conduct of the sale to the mortgagee and yet leaving the mortgagor in possession pending completion unless the mortgagee has agreed to this course. The obtaining of possession with a view to giving it to the purchaser is part of the necessary arrangements for sale. In my opinion, the party having conduct of the sale ought normally to have the right to decide when it is desirable for him to obtain possession from those in occupation in order to enable the sale to be effectively carried through.

As the plaintiffs observe, in what I would wish to describe as a most impressive skeleton argument, if the contractual obligation to give vacant possession, which the mortgagee will wish to assume in order to obtain the best price reasonably obtainable, is separated from the ability to give immediate vacant possession, the mortgagee is put at risk of being in breach of contract through circumstances beyond his control. Moreover, if the conduct of the sale is given to the mortgagee, any prospective purchaser will become aware prior to exchange of contracts that the property is being sold by a mortgagee who has not yet obtained vacant possession. The risk that the borrower will not vacate the property on completion will become apparent and the purchaser may be deterred from proceeding. He will also be aware that the sale is a forced sale, the advantage of achieving a better price through continued owner occupation is unlikely to be realised.

Accordingly, while I would not wish to hold that the court has no jurisdiction in an appropriate case to make an order such as the judge made in the present case, it is hard to see the advantages of such an order and easy to see the disadvantages. I find it difficult to envisage circumstances in which such a course would be appropriate unless the mortgagee consented. It ought to be a rarity and taken only if the necessary conditions are satisfied.

In the present case the judge made the order on April 30 1996 on the very day fixed for execution of the warrant when the bailiff was waiting to take possession. It was the ninth appointment for execution of the warrant. On the first, second and third occasions the warrant had been withdrawn by the plaintiffs; on the fourth and fifth it had been

A suspended on terms with which the defendants had not complied; and on the sixth and eighth it had been suspended during an adjournment of an application by the defendants upon the adjourned hearing of which they had not attended.

The history of the case shows repeated attempts by the defendants to avoid execution with no real prospect of doing so, and no object other than to postpone the evil day. The plaintiffs point to the defendants' repeated failures to meet the monthly repayments they expressed themselves as able to make; their failure to attend adjourned hearings after they had achieved their objective of obtaining a further suspension of the warrant; and the extreme lateness of the final
B application which was made, at least in part, on the basis of a proposal for monthly payments which Mr Booker acknowledged in court he could not make.

The judge said that he made the order because the purpose for which the plaintiffs required possession was to realise the security at the best possible price. He said that the means by which that might be done would be to effect an early sale of the property at the highest price which might be obtainable. But the evidence demonstrated that the defendants were unlikely in the extreme to co-operate with the plaintiffs. They had failed to sell the property for more than two years or even to produce evidence of serious offers. It was self-evidently
C unlikely that the defendants would assist the plaintiffs in achieving an early sale.

The judge assumed that in the event of non co-operation by the defendants the plaintiffs could readily return to court and apply for immediate execution. But he overlooked a number of factors. In the first place, non co-operation by the defendants might not be readily susceptible of proof; the plaintiffs would not easily be able to monitor what the defendants were saying to prospective purchasers; the plaintiffs would not know until the last moment whether the defendants would co-operate in giving vacant possession on completion; and, finally, even if an application was made at very short
D notice and was successful, immediate execution would not follow.

The plaintiffs were unable to obtain access to the property for the purpose of valuing it until the end of October. They have since received two valuations and are now at last in a position to put the property on the market. At the time of the present hearing, the defendants have not referred any potential purchaser to the plaintiffs.

I agree with the submission made on behalf of the plaintiffs that the judge's order, contrary to his expressed intention, was calculated to and did delay the sale of the property. If continued it might depress the price and create a deficiency in the net proceeds of sale.

No doubt the judge would say, and there is some truth in this, that if
E the plaintiffs had returned to him after the expiry of, say, three months instead of pressing ahead with the present appeal, he would have lifted his order and enabled possession to be obtained earlier than will have been achieved by a successful appeal. But such a course would only confirm the order which the judge made and encourage the lower courts to make similar orders more commonly in future.

In my judgment, on the facts of the present case, the judge was plainly wrong to have made the order he did. I would allow the appeal and direct that the warrant for execution should issue without delay.

Potter LJ agreed and did not add anything.

F *Appeal allowed.*

Re Nichols' Application G

LANDS TRIBUNAL

February 3 1997

PH Clarke FRICS

Estates Gazette May 17 1997

[1997] 20 EG 150

H *Restrictive covenants — Application under Law of Property Act 1925, section 84(1)(a), (aa), (b) and (c) — Whether restriction obsolete — Whether objectors enjoyed practical benefits of substantial value or advantage — Whether tribunal's discretion can be exercised in absence of planning permission — Failure to provide agreed statement of facts*

The application land was part of the garden of a large detached house. A conveyance dated July 24 1953 of, *inter alia*, the application land contained a covenant prohibiting any building or erection on the land. The applicants applied to the tribunal to discharge the covenant so as to permit the erection of a detached two-storey dwelling-house.
J At the hearing they were given leave to amend their application to modify the covenant to permit the erection of a bungalow. Their application was under section 84(1)(*a*), (*aa*), (*b*) and (*c*) of the Law of Property Act 1925; the objectors were owners of several adjoining properties. The applicants had not applied for planning permission for the erection of the bungalow. At the hearing the member was informed that the experts had been instructed not to agree the agreed statement of facts.

Decision: **The application was allowed. There had been material changes in the character of the property and the neighbourhood such that the restriction was now obsolete for the**
K **purposes of section 84(1)(*a*) of the 1925 Act. For the purposes of section 84 (1) (*aa*), the proposed use of the land was a reasonable user impeded by the restriction, and the restriction did not secure practical benefits of substantial value or advantage to the objectors. There was no injury to the objectors' properties for the purposes of section 84(1)(*c*). The application therefore succeeded on paras (*a*), (*aa*), and (*c*) and no compensation was awarded to the objectors. It did not succeed under para (*b*). The expert witnesses were under an obligation to agree material facts, and the failure to agree was a matter to be taken into account in the award of costs. The objectors were ordered to pay half the applicants' costs.** L

The following cases are referred to in this report.

Bowden's Application, Re (1983) 47 P&CR 455; [1983] JPL 819, LT
Bradley Clare Estate's Application, Re (1988) 55 P&CR 126, LT
Chapman's Application, Re (1981) 42 P&CR 114; [1981] JPL 520, LT
Gilbert v *Spoor* [1983] Ch 27; [1982] 3 WLR 183; [1982] 2 All ER 576; (1982) 44 P&CR 239, CA
Dransfield's Application, Re (1975) 31 P&CR 192; [1975] JPL 539, LT
Gaffney's Application, Re (1974) 35 P&CR 440; [1975] JPL 159, LT
Henderson's Conveyance, Re [1940] Ch 835
Ling's Application, Re (1955) 7 P&CR 233; [1956] JPL 380, LT
Truman, Hanbury, Buxton & Co Ltd's Application, Re [1956] 1 QB 261; [1955] 3 WLR 704; [1955] 3 All ER 559; (1955) 7 P&CR 348; [1955] EGD 169; 166 EG 564, CA M

Suzanne Ornsby (instructed by Horwood Patterson, of Leicester) appeared for the applicants, Frank Edward and Sarah Jane Nichol; Nicholas George (instructed by Harvey Ingram Owston, of Leicester) appeared for the objectors.

Giving his decision, **Mr PH Clarke** said: This is an application under section 84 of the Law of Property Act 1925 by Dr FE and Mrs SJ Nichol ("the applicants") for an order modifying a restrictive covenant affecting freehold land at the rear of 27 Elms Road, Stoneygate, Leicester ("the application land"), to permit the erection of

A a detached bungalow and garage. The objectors are the owners of houses in Sackville Gardens.

Suzanne Ornsby, of counsel, appeared for the applicants and called Simon Spencer FRICS, senior partner of Spencer Birch, chartered surveyors, of Nottingham. Nicholas George, of counsel, appeared for the objectors and called John MS Whitehead CBE DL, William M Simpson FRICS, a partner in Snow & Astill, chartered surveyors, of Leicester, and G Wyatt, A D Crofton, G Jewell and P Day, objectors.

Facts

When I inquired at the start of the hearing as to a statement of
B agreed facts I was told that the expert witness for the objectors had been instructed not to agree such a statement. I did not pursue the matter but, if this is true, I would say that it is improper for an expert witness to be given or to accept instructions not to agree facts. This is a matter which the parties may wish to refer to in their representations on costs. The absence of a statement of agreed facts increases the length of a hearing and the time taken to prepare the decision. Every hearing should be accompanied by a comprehensive statement of agreed facts and an agreed bundle of documents. It is the responsibility of the solicitors to the parties to ensure that these documents are prepared before the hearing.

C From the evidence I find the following facts:

1. The application land is in the Stoneygate district of Leicester, about two miles south of the city centre. Stoneygate was the principal residential suburb of Leicester in the 19th century and comprised mainly large houses set in extensive grounds. Development from about 1880 to 1940 resulted in four distinct areas: large houses set in their grounds, medium-sized houses in large gardens, inter-war suburban villas and tightly built-up streets of mainly terraced houses. Later post-war development included the conversion of large houses to student accommodation for the two universities in Leicester, the construction of purpose-built halls of residence, the provision of nursing homes, (purpose-built or converted houses), the building of modern high
D density residential estates and infill or backland development.

2. Elms Road is in Stoneygate and runs in a north to south direction from Knighton Road to Knighton Church Road. It is a wide road for most of its length, with a mix of housing, nursing homes and student halls of residence.

3. The application land is situated on the west side of Elms Road at the rear of no 27, a large detached house and garden, which is now empty but was part of Clare Hall, a hall of residence of the University of Leicester. The application land was formerly part of the garden of no 27. The application land is an irregular shaped plot with an area of
E about 1,725 sq yds. It is bounded on the east by the gardens of 27 to 31 Elms Road, on the south and west by the gardens of 22 to 28 Sackville Close (including the objectors' houses at nos 24, 26 and 28) and to the north by a part made-up access road to Elms Road and the site of a new house under construction on land at the rear of 25 Elms Road (now Harley Grange Nursing Home). Access to the application land is along the access road referred to above, which joins Elms Road between nos 25 and 27. The western boundary of the application land and the rear gardens of the objectors' houses is part-post and wire fence and part brick wall. There are trees, bushes and shrubs along this boundary. The site is mainly rough grass with several mounds of earth.
F There are two derelict builders' huts on the land. It has an unsightly appearance. It was formerly a garden and orchard, but the fruit trees have now been grubbed up.

4. By a conveyance of the application land dated July 24 1953 between Isalene Whitehead (vendor) (I) and Charles Viccars Goddard (purchaser) (2), the purchaser covenanted to observe and perform the restrictions and stipulations in the second schedule to the conveyance. The benefit of this covenant was enforceable by the vendor and her personal representatives and successors in title, assignees of the benefit and the owners and occupiers of any part of the vendor's adjacent or adjoining property, which was then a T-shaped parcel of land now comprising 25 Elms Road, the land at the rear of no 25, the access road to Elms Road and 16-28 Sackville Gardens. The purchaser of the application land was the owner of 27 Elms Road, which immediately G
adjoins the land to the east, and which then included the property which is now 29 Elms Road.

5. Para 1 of the second schedule to the above conveyance prohibits any building or erection on the application land except the existing shed and triangular building and a green-house. This is the restriction which the applicants seek to modify ("the restriction").

6. In about 1957, 25 Elms Road (including the rear land) was sold to the University of Leicester.

7. In or about 1957 the land at the rear of the application land and at the rear of 25 Elms Road, known as the Ratcliffe House Estate and later developed by the building of 16 to 28 Sackville Gardens, was H
sold for development.

8. In or about 1957 the University of Leicester purchased 27 Elms Road (including the application land).

9. In or about 1990 the University of Leicester sold 25 and 27 Elms Road (including the application land and the land at the rear of no 25).

10. In or about 1992 the applicants acquired the land at the rear of 25 Elms Road.

11. By a transfer dated April 27 1994 the applicants acquired the application land.

12. On February 7 1995 the applicants applied to this tribunal for an order discharging the restriction to allow the erection of a good quality J
detached two-storey dwelling-house containing four or five bedrooms. The application was amended at the hearing to the modification of the restriction to permit the erection of a detached bungalow and garage ("the proposed bungalow"), subject to conditions regulating height, siting, overlooking and landscaping. The proposed bungalow comprises: lobby and hall, living room, dining room, study, kitchen/breakfast room, family room, laundry room, four bedrooms, two bathrooms and cloakroom. The application was made under section 84(1)(*a*), (*aa*) and (*c*) of the Law of Property Act 1925 (as amended) ("the 1925 Act"), and amended at the hearing to include section 84(1)(*b*). K

13. Objections have been made by the following residents of Sackville Gardens: Mr and Mrs G Wyatt (20), Mr and Mrs A Crofton (24), Dr GJ Jewell (27) and Mr P Day (28). All objections are admitted.

14. The following planning permissions and refusals are relevant:

December 13 1957	outline planning permission granted for the construction of 41 houses in Sackville Gardens.
March 1 1988	conditional planning permission granted for the development of the application land and land at the rear of 25 Elms Road by the erection of two detached bungalows. L
June 1 1989	conditional planning permission granted for the formation of an access road (between 25 and 27 Elms Road) to service the above bungalows.
October 5 1989	planning permission refused for the construction of two bungalows with garages on the application land and land at the rear of 25 Elms Road.
November 13 1990	conditional planning permission granted for the erection of a bungalow and garage at 33 Elms Road. M
February 26 1991	conditional planning permission granted for the erection of a detached house and garage on the application land and land at rear of 25 Elms Road.
February 25 1992	conditional planning permission grunted for the erection of two two-storey detached houses on the application land and land at the rear of 25 Elms Road.
March 3 1995	conditional planning permission granted for the construction of a detached house with

	double garage on land at the rear of 25 Elms Road.
July 24 1995	conditional planning permission granted for the construction of a new bungalow with detached double garage on 33 Elms Road (renewal of planning permission).

15. Planning permission has not been granted (or applied for) for the proposed bungalow on the application land.

16. In a letter to the applicants' solicitors dated January 27 1995 from the Director of Environment and Development of Leicester City Council he indicated that he "would have no objection in principle, to the building of a 2-storey house" on the application land:

> This is subject to normal development control criteria (relating to design, garden size, overlooking etc) being met. It may also depend on how the other half of the site is developed; . . . approval may not be given for more than two dwellings on the whole site. This may necessitate an agreement not to proceed with the extant planning permission (91/1779) for two houses on the northern part of the site if the southern part is developed.
>
> The foregoing advice is in any event without prejudice to my final recommendation or any decision of the Planning Committee in respect of any formal applications for development of the site.

17. The application land is within a primarily residential area as defined in the Knighton-Stoneygate local plan. Policy H6 in the written statement to the City of Leicester Local Plan is relevant and is as follows:

> Within the Primarily Residential Areas defined on the Proposals Map, planning permission will normally be granted for development for residential purposes (Class C3 uses) except where:
> (a) A plot is too small to accommodate a dwelling satisfactorily;
> (b) It would be likely to prove detrimental to the amenities of occupiers of nearby properties by reason of substantial loss of privacy or light;
> (c) The proposed access to the site or car parking arrangements are unsatisfactory;
> (d) The proposed design of the development is unsatisfactory;
> (e) It would result in the loss of garages, parking space, social, community or recreational facilities for which there is a need in the area;
> (f) It would result in the loss of trees or land of high amenity or ecological value;
> (g) Development would be likely to prejudice the possible comprehensive development of a larger area;
> (h) Infill development or extensions would prevent proper maintenance of a building or the adjoining properties;
> (i) It would be unacceptably close to an existing or proposed employment area falling within Classes B2-B8 of the Town and Country Planning (Use Classes) Order 1987, or any other use which would be detrimental to residential amenities.
>
> There are a number of small plots of land within the existing housing areas which could be suitable for residential development, and which, in total, could make a significant contribution to housing needs in the area. The above conditions are designed to ensure, however, that development does not take place in unsuitable locations.

18. The application land is within the Stoneygate Conservation Area.

19. The agreed values of the objectors' houses in Sackville Gardens are:

20 (Wyatt)	£150,000
24 (Crofton)	£120,000
26 (Jewell)	£155,000
28 (Day)	£127,500

Decision

I inspected the application land and the surrounding area on a dry day at the beginning of October. The trees were in almost full leaf I also viewed the application land from the objectors' houses and gardens.

This application is made under section 84(1)(*a*), (*aa*), (*b*) and (*c*) of the 1925 Act. The originating application omitted para (*b*) but leave was requested to add it and I give leave for this amendment. I now consider each of the four grounds.

Section 84(1)(a)

The question here is whether by reason of material changes in the character of the property or the neighbourhood the restriction ought to be deemed obsolete? In this application this question can be divided into three parts:

(i) Have there been material changes in the character of the application land since the restriction was imposed in July 1953?

(ii) Have there been material changes in the neighbourhood since that date?

(iii) If there have been such changes, ought the restriction to be deemed obsolete?

I look first at changes in the character of the application land. Mr Spencer, for the applicants, put in evidence two Ordnance Survey maps showing the land and the surrounding area in 1953 and at the present time respectively. When the restriction was imposed in 1953 the land formed part of a larger parcel of land enjoyed with a detached house, 25 Elms Road. The application land was garden or orchard. It was sold to the owner of 27 Elms Road and became part of that property. Today the application land is separate from no 27 and access to it is along the new connecting road to Elms Road. The orchard has been grubbed up and the application land has the unsightly appearance of a building plot waiting for development to begin. I am satisfied that there have been material changes in the character of the application land since July 1953.

The second question relates to material changes in the neighbourhood. What is the neighbourhood? Miss Ornsby, for the applicants, submitted that it is either the area immediately surrounding the application land or a larger area bounded by a circle with a radius of 650 ft from the land and by Carisbrooke and Ratcliffe Roads. She said that, whatever area is taken, however, material changes have occurred within it. Mr George, for the objectors, took a narrow view of the neighbourhood and said that it related solely to the area in close proximity to the application land and the objectors' houses. The extent of the neighbourhood is a question of fact. I find on the evidence and particularly my inspection, that the neighbourhood is the wider area bounded by Ratcliffe Road, Carisbrooke Road, Knighton Church Road and the properties on the eastern side of Elms Road.

Mr Spencer listed the changes which have occurred in this neighbourhood since 1953. They are:

(i) the conversion of large houses to student accommodation;

(ii) the construction of purpose-built halls of residence;

(iii) the construction of residential estates on former open land, including Sackville Gardens, Carisbrooke Gardens, and along Carisbrooke Avenue and at Parklands, Ratcliffe Road;

(iv) infill and backland development on large plots, eg at nos 25, 28, 29, 30 and 33 Elms Road.

Mr Simpson, for the objectors, did not analyse the changes in the neighbourhood, but said that he has lived in the area for the past 16 years and is of the opinion that it has retained its residential character, albeit with greater density of housing.

Miss Ornsby invited me to accept the evidence of Mr Spencer and find that the character of the neighbourhood had materially changed since 1953. Mr George conceded that there is now greater density of residential use in the neighbourhood, but submitted that there has not been any change of use from residential and no change in the character of that residential use.

I find that the character of the neighbourhood has materially changed since July 1953. I agree with Miss Ornsby that this conclusion holds good whether the neighbourhood is defined narrowly or widely. If one looks at the area immediately surrounding the application land then the following material changes have occurred:

(i) 25 Elms Road, the house for the benefit of which the restriction was imposed, became a hall of residence and was then demolished and replaced by Harley Grange Nursing Home;

(ii) the land at the rear of no 25 has been severed and a new house is nearing completion on the land, connected to Elms Road by a new access road which also serves the application land;

A (iii) 27 Elms Road ceased to be a house and became part of Clare Hall, a hall of residence of the University of Leicester, and is now vacant;

(iv) 29 Elms Road was built in the grounds of no 27;

(v) Sackville Gardens, including the objectors' houses, has been built on the former open land (the Ratcliffe House Estate), which was originally part of 25 Elms Road, at the rear of the application land.

Looking at the neighbourhood in a wider perspective, I accept Mr Spencer's evidence that four broad changes have occurred. Stoneygate is still an attractive residential area and has retained a predominantly residential use, but its character has changed, particularly close to the
B application land, by the growth of halls of residence and by greater density of development. I should perhaps make the point that the construction of Sackville Gardens, including the objectors' houses, has had a particular impact on the character of the area immediately surrounding the application land.

I find that there have been material changes in the character of the application land and the neighbourhood since the restriction was imposed in July 1953. This leads to the third and crucial question: ought therefore the restriction to be deemed obsolete?

The meaning of "obsolete" was considered by Romer LJ in *Re Truman, Hanbury, Buxton & Co Ltd's Application* [1956] 1 QB 261.
C He said at pp271-272 that " the meaning of the term 'obsolete' may very well vary according to the subject-matter to which it is applied". When applied to restrictive covenants "it may said that the covenants have become obsolete, because their original purpose can no longer be served and, in my opinion, it is in that sense that the word 'obsolete' is used in section 84(1) (*a*)". Later he said (at p272): "If serious injury would result to the opponents and others if the covenant was discharged . . . I cannot see how . . . the covenant can be described as obsolete, because the object of the covenant is still capable of fulfilment, and the covenant still affords a real protection to those who are entitled to enforce it".

D The test is therefore whether the restriction can still serve its original purpose? This begs the further question: what was the original purpose? I heard evidence on this point from Mr JMS Whitehead.

Mr Whitehead said that his father purchased 25 Elms Road in 1927, then 27 Elms Road in 1936 and the Ratcliffe House Estate in 1939. He later sold no 27 but kept half the garden for a tennis court, orchard, stables and stables yard. Mr Whitehead's father died in 1946. Mr Whitehead acted on behalf of his mother when she sold part of the garden of no 27 (the application land) to the owner of that house in July 1953 and imposed the restriction (which is the subject of this application) on the land. The purpose was to protect the Ratcliffe
E House Estate, which was then owned by trustees. The application land was sold at a reduced price to reflect the restriction. The purpose of the restriction was to preserve the ambience of the locality so that the building plots in Sackville Gardens would be large and with desirable views.

Mr Whitehead said that his father commenced work on the development of the Ratcliffe House Estate in 1939 by instructing an architect, demolishing Ratcliffe House and building a road on the land (now Sackville Gardens). When the land was sold for development in or about 1957, the intention of the restriction was achieved because the price obtained was a substantial sum.

F Mr Whitehead concluded that the reasons for the imposition of the restriction on the application land were:

(i) to protect 25 Elms Road;

(ii) to maintain the price on the sale of the rear land, the Ratcliffe House Estate, for development;

(iii) to protect the purchasers of the houses built on the Ratcliffe House Estate (Sackville Gardens) from overdevelopment.

I am not wholly persuaded by Mr Whitehead's evidence as to the three-fold purpose of the restriction. I can accept reason (i). I think this was the main, perhaps the only, reason for the restriction, and allied to a lower price as garden land. Reason (ii) may be correct. With regard to reason (iii), I am doubtful whether the restriction was imposed to protect the future purchasers of houses to be built on the Ratcliffe House Estate. In my view, the restriction was imposed to reflect the G
low price on the sale of the application land as garden land, to protect the amenity of no 25 (then owned by Mrs Whitehead, the vendor) and, possibly, to increase the price on the subsequent sale of the Ratcliffe House Estate. I do not think that Mrs Whitehead, or her solicitors, had the later owners of houses in Sackville Gardens in mind as beneficiaries of the restriction (in the way they now oppose the modification), although I accept that they have the benefit in law as subsequent purchasers of part of the Ratcliffe House Estate.

Can these purposes still be served by the restriction? Miss Ornsby submitted that the restriction is now obsolete with regard to the protection of the enjoyment and amenity of 25 Elms Road. I H
agree. The house which was no 25 when the restriction was imposed has been demolished and replaced with Harley Grange Nursing Home. The land at the rear is the site of a new house. The original purpose of the restriction cannot be served in this respect. With regard to the second purpose, the maintenance of the price on the sale of the Ratcliffe House Estate, I agree with Miss Ornsby that this purpose has also been fulfilled and the restriction is obsolete in this respect.

If I am right therefore, that the original purpose of the restriction was two-fold only, then it follows that the restriction is obsolete. I will, however, consider the third reason, because it will be seen that, even if I accept it as one of the original purposes of the restriction, it cannot J
now be fulfilled and the restriction is still obsolete.

Miss Ornsby submitted that, assuming that one of the purposes of the restriction was to protect the owners of houses in Sackville Gardens from overdevelopment, then the restriction is obsolete for two reasons. First, the houses in Sackville Gardens were in fact erected at a higher density than envisaged in 1953 and the building of one bungalow on the application land would not constitute overdevelopment in that it would be at a lower density than Sackville Gardens. Second, that the proposed modification of the restriction would not constitute overdevelopment and, in so far as it now prohibits development which does not constitute overdevelopment, it is no K
longer fulfilling any purpose. Miss Ornsby referred to *Re Ling's Application* (1955) 7 P&CR 233 which she said established the principle that a restriction should be modified if the restrictions necessary to preserve the character of the neighbourhood when imposed were far more stringent than required at the time when modification is sought. She said that, in order to maintain the character of the area, it is no longer necessary to have a blanket rejection of all development on the application land. The proposed bungalow would not adversely affect the amenity of adjoining land. The restriction is unnecessarily severe in relation to permitted surrounding development. Thus, it is now obsolete and should be modified. She also referred to L
Re Bradley Clare Estate's Application (1988) 55 P&CR 126.

Mr George submitted that the original purpose of the restriction is not exhausted because it can still properly be served by preventing development at the rear of the objectors' houses. He referred to *Re Henderson's Conveyance* [1940] Ch 835 and particularly part of the judgment of Farwell J at p845, where he said:

> This is purely a question between two private owners, one of whom desires to erect a house on a particular portion of his property, and the other of whom desires to keep that which he has long enjoyed and that for which presumably he paid when he acquired his house and grounds, namely, the benefit of having at the bottom of his garden an open space, a space which, so long as the M
> covenant remains, will be always an open space, and one which, being an open space, will not have upon it a dwelling or dwellings from which his garden can be overlooked.

I accept Miss Ornsby's submissions on this point. Even if the protection of future owners of unbuilt houses in Sackville Gardens was an original purpose of the restriction, then I am still satisfied that it is obsolete. That protective purpose cannot be served because the proposed bungalow will not injure those with the benefit, by over development or, as I shall explain later in this decision, by obstructing a view. In *Truman* Romer LJ found that the covenant in that case was still capable of fulfilment because it still afforded "a real protection to those who are entitled to enforce it" (p272). It follows therefore that, if

A the restriction does not afford protection, by not stopping overdevelopment because the proposed bungalow would not constitute such development, or by not stopping the obstruction of a view, because the bungalow would not obstruct the view, then it is not serving its original purpose and is obsolete.

I distinguish the decision in *Re Henderson* on three grounds. First, in *Henderson* there was no evidence on any changes in the property or neighbourhood. Second, the applicant required the discharge or modification of the covenant to enable him to site his proposed house on a particular part of his land (burdened with the covenant), although his other land was free from any restriction. As Farwell J said, the
B question was whether this personal benefit is sufficient "to justify him in asking that the benefit under the covenant which the appellant has long enjoyed shall be taken away from him" (p846). Third, the proposed house in *Henderson* would have caused overlooking of the house and garden with the benefit of the covenant; that is not the position with the proposed bungalow in this application.

I am satisfied that, by reason of material changes in the property and neighbourhood, the restriction can no longer fulfil its original purpose and is obsolete. The application succeeds under section 84(1)(*a*).

Section 84(1)(aa)

The issues under this ground can be expressed in the form of three
C questions. There is no issue between the parties as to whether the restriction is contrary to the public interest.

The first question is whether the proposed user of the application land for the erection of a bungalow and garage is a reasonable user of the land which is impeded by the restriction? The parties are agreed that the proposed bungalow is a reasonable user of the land and clearly it is impeded by the restriction. I should, however, add that I do not accept Miss Ornsby's submission that it is a corollary of a finding of reasonable user that planning permission must necessarily be assumed to be granted for that use.

I proceed to the second question. This is whether the restriction, in
D impeding the proposed user of the application land, secures practical benefits to the objectors which are of substantial value or advantage to them?

Miss Ornsby submitted that the evidence clearly shows that the view from the rear of the objectors' houses will not be adversely affected by the proposed bungalow. The height will be only 18 ft, there are mature trees along the rear boundary of the objectors' gardens and the application land and landscaping is proposed on that land to further screen the proposed building. Miss Ornsby referred to *Re Dransfield's Application* (1975) 31 P&CR 192 (view from the objectors' house not a substantial practical benefit: ". . . in the summer months the new
E house would scarcely be visible at all . . ." (p192)) and *Re Gaffney's Application* (1974) 35 P&CR 440 (a case with similar facts). In both cases the application succeeded. She said that Mr Simpson conceded that the modification of the restriction would have only a marginal effect on the value of the objectors' houses, up to a maximum of 10 %. Clearly therefore the benefit derived from the absence of the bungalow cannot be substantial.

Mr George said that it is established law that a view can be a practical benefit: *Gilbert* v *Spoor* [1983] Ch 27. It is a question of fact and degree for the tribunal as to whether a view exists and the extent to which the proposed bungalow will interfere with it. Mr George
F submitted that a further benefit is that the restriction was in part taken to keep the application land free from development. This, in itself, is a substantial practical benefit to the objectors. The objectors gave evidence, to the effect that they regarded the view from the rear of their properties and the existence of an orchard at the rear of their gardens as benefits attaching to their houses. They understood when they bought their houses that the application land could not be developed due to the restriction. The benefits claimed to be secured to the objectors by the restriction are: (1) the view from the rear rooms and gardens over the application land; (2) the maintenance of open land at the rear of their gardens. It was not claimed that the restriction prevented overlooking from the proposed bungalow or that this property would produce noise and disturbance. I now deal with each of the alleged benefits.

I accept that a view can be a practical benefit. I viewed the G
application land from each of the objectors' houses and gardens and my decision on this issue must rest largely on this inspection. It is a matter of impression.

I am satisfied that Mr Wyatt at 20 Sackville Gardens obtains no benefits from the restriction due to the position of his house immediately to the rear of 25 Elms Road, a short distance to the north of the application land. The view from Mr Wyatt's house and garden has been greatly affected by the house now nearing completion on the land at the rear of 25 Elms Road but this is unconnected with this application. I am satisfied that the proposed bungalow will not be seen, or at most not seen to any extent, from Mr Wyatt's property. H

The houses of the other objectors share a common boundary with the application land (the rear boundary of the gardens and the western boundary of the land). From my inspection I am satisfied that the proposed bungalow will hardly be seen, if at all in the summer, from the houses or the gardens, except at the very rear of the gardens of 27 and 28 Sackville Gardens. In my opinion, the objectors have greatly exaggerated the effect of the bungalow on the rear view from their houses and gardens. Although I note that the original application envisaged the erection of a two-storey house on the application land and this would clearly be far more intrusive.

The restriction can only secure a substantial practical benefit to the J
objectors if it prevents a substantial deterioration in the view they enjoy over the application land. I do not think that this test is satisfied. I am not persuaded that the proposed bungalow will materially affect the rear view from the objectors' properties. Accordingly, I find the restriction does not secure a substantial practical benefit in terms of view.

The second benefit claimed by the objectors is that the keeping of the application land free from development by the restriction is a substantial practical benefit in itself. Miss Ornsby counted this argument on two grounds. First, that the purpose of the restriction was not to prevent development in itself but, on the evidence of Mr Whitehead, it was to K
prevent *overdevelopment* and this would not arise by the erection of one bungalow. Second, if it is right that the restriction itself can constitute a substantial practical benefit, then it follows that no application can ever succeed under section 84(1)(*aa*). Mr George's submissions on this point represent a circular argument.

I agree with Miss Ornsby, at least on her second ground of rebuttal. I cannot see how the mere existence of the restriction can itself be a substantial benefit, otherwise, as Miss Ornbsby says, every application under section 84(1)(*aa*) is bound to fail. The benefit to be considered under this ground must be the effect of the restriction on the objector and his land. Does the consequence of the restriction to the objector L
provide a benefit to him? Applying that test to this application, I am satisfied that the consequence of the restriction, that is to say the prevention of the erection of the proposed bungalow, is not a substantial practical benefit to the objectors. The view is not safeguarded because it would not be impaired by the bungalow. Overlooking would not be prevented because this cannot arise out of the construction of the bungalow and it is not suggested that the occupation and use of the bungalow would produce unreasonable noise or disturbance, which would be prevented by the restriction.

I am satisfied that the restriction, in impeding the user of the application land for the erection of the proposed bungalow, does not M
secure practical benefits which are of substantial value or advantage to the objectors.

I can now proceed to my third question, which is whether money will be an adequate compensation for the loss or disadvantage (if any) which the objectors will suffer from the modification of the restriction and, if so, what amounts should be awarded? The objectors, through the evidence of Mr Simpson, have claimed compensation of between 5 and 10% of the agreed values of their houses.

Miss Ornsby accepted that loss of a view can be compensated by a monetary award, but submitted that, in this case, the damage is too small to be quantified. The effect of the bungalow would be negligible or minimal and no compensation should be payable. She said that Mr Simpson's figures are flawed. The true comparison of values is between the value of the house with the existing view and with the view it will have after the construction of the bungalow. Mr Spencer expressed the opinion that the bungalow would not cause any diminution in value of the objectors' houses.

I agree that money will be adequate compensation for loss of a view and that therefore the requirement of section 84(1A) as to compensation is satisfied in this respect. Compensation is, however, not necessarily payable in every case and I am of the opinion that it is not appropriate in this application. I have found that the restriction is obsolete and I will find later that the objectors will not be injured by its modification. It must follow therefore that the restriction does not add to the value of their houses because it is obsolete and its modification would not reduce the value because this will not cause injury. I am satisfied that Mr Simpson's figures are wrong and that no diminution in value will arise solely as the result of the modification of the restriction. If the application had succeeded only under section 84(1)(*aa*) then modest amounts of compensation might be appropriate for any insubstantial loss or damage due to the modification of the restriction. The applicants' success under section 84(1)(*a*) and (*c*), however, necessarily rules out any compensation payments as being inconsistent with success on those grounds.

The application succeeds under section 84(1)(*aa*).

Section 84(1)(b)

I give leave for this ground to be added to the original application but it must fail on the facts. Seven houses in Sackville Gardens have the benefit of the restriction. Objections have been made by four owners; objections were not made by the owners of nos 16, 18 and 22. Miss Ornsby invited me to find that these owners have agreed, by implication, to the modification of the restriction. Even if I accept this argument the application must fail on this ground because, it seems to me, that all those with the benefit of the restriction must consent to its modification. This is clearly not the position here and the application fails under section 84(1)(*b*).

Section 84(1)(c)

To succeed on this ground the applicants must show that the proposed modification of the restriction will not injure the objectors.

Miss Ornsby submitted that the objectors will not be injured by the proposed modification due to the existing screening of the application land from the objectors' houses and the existing view of 27 Elms Road, which is larger and more intrusive than the proposed bungalow will be. She referred to *Re Chapman's Application* (1981) 42 P&CR 114 and *Re Bowden's Application* (1983) 47 P&CR 455. The bungalow will not materially affect the character of the surrounding area. It cannot set a precedent for further development.

Mr George said that injury is not simply a question of compensation. It must be determined by reference to the purpose for which the restriction was imposed, which, in accordance with the evidence of Mr Whitehead, was to prevent overdevelopment to the detriment of the houses in Sackville Gardens. He said that the reference to 27 Elms Road is a false point because this property was in existence when the restriction was imposed.

As I said earlier in this decision, I am satisfied that the proposed bungalow will hardly be seen, if at all in the summer, from the objectors' houses and gardens and it follows therefore that they will not be injured by the modification of the restriction to permit the erection of the bungalow. Furthermore, I have found that the mere prevention of development on the application land is not, in itself, a benefit and it follows that the modification to allow that development will not injure the objectors The application succeeds under section 84(1)(*c*).

Discretion

I have found that the requirements of section 84(1)(*a*), (*aa*) and (*c*) of the 1925 Act are satisfied and I therefore have jurisdiction to modify the restriction. I must now consider whether, in the exercise of that jurisdiction and of my discretion, I should do so. I am particularly required by section 84(1B), when exercising my discretion, to take into account the planning position, the period at which and the context in which the restriction was imposed and any other material circumstances.

This is an unusual case because planning permission has not been granted for the proposed bungalow. The explanation I was given for this absence of planning permission was that Dr Nichol did not wish to upset the objectors further by applying for such permission before he had obtained a modification of the restriction. Although I think that this application should have been accompanied by planning permission for the proposed bungalow, I do not think that lack of permission is necessarily fatal due to the late amendment of the application to provide for the modification of the restriction to permit the construction of a particular building governed by conditions. Further, I am satisfied on the evidence, that planning permission is likely to be granted for the construction of a bungalow within the scope of the modification now sought.

I have already considered the period and context of the imposition of the restriction and concluded that the purpose of the restriction has been fulfilled and is now obsolete.

I must however deal with certain allegations made by the objectors against Dr Nichol which arose from time to time during the hearing. It is not for me to make a judgment on their accuracy, except in so far as it might influence my discretion. I can understand the concern felt by the objectors about the changes taking place at the rear of their houses, but I would make the following comments.

First, parliament has given an owner of land burdened with a restrictive covenant the right to apply to the Lands Tribunal to have that covenant discharged or modified. It is for the tribunal to decide whether the applicant has satisfied one or more of the requirements of section 84(1) of the 1925 Act. I do not think that any criticism can be attached to the applicants for making this application. Second, much of the development at the rear of the objectors' houses is not connected to this application and has arisen after various grants of planning permission for backland development by the local authority. Third, the removal of the trees from the application land by Yellowhammer Ltd was not, as far as I am aware (and it was certainly not proved), on the instructions of Dr Nichol. This is not a case where I should exercise my discretion against an applicant because he has proceeded with development in flagrant breach of a restrictive covenant. It appears that Dr Nichol stopped all activity on the land when he became aware of the restriction and the strength of the objections.

Having considered all these matters I am satisfied that I should exercise my discretion to order the modification of the restriction. If I refuse the application the land will remain an unsightly vacant plot of land. It is unlikely that it will ever revert to the garden and orchard, occupied with the adjoining house, which the objectors remember from past years.

Order

During the hearing the application was amended to provide for the modification of the restriction to permit the erection of a bungalow and garage, subject to conditions which were agreed by Mr George. One of these related to landscaping and in the following terms: "the western boundary of the said land to be landscaped with a mixture of deciduous and evergreen trees". This imposes a positive obligation on the owner. In my view, I do not have the power to impose such an obligation on the land. Section 84(1C) of the 1925 Act provides that:

> It is hereby declared that the power conferred by this section to modify a restriction includes power to add such further provisions restricting the user of or the building on the land affected as appear to the Lands Tribunal to be reasonable in view of the relaxation of the existing provisions, and as may be accepted by the applicant; and the Lands Tribunal may accordingly refuse to modify a restriction without some such addition.

This subsection gives the tribunal power to add provisions "*restricting* the user of or the building on the land", but this does not

A extend to the addition of positive obligations. Accordingly, I propose to vary the proposed landscaping condition (condition (v)) to provide for the preservation of the existing trees and bushes along the western and south western boundaries of the application land, and this decision can, at this stage, be only an interim decision.

I propose that para 1 of the second schedule to the conveyance of the application land dated July 24 1953 shall be modified to read as follows:

> 1. That no building or erection shall at any time hereafter be built or erected on the property hereby conveyed or any part thereof save and except one B detached bungalow and detached garage subject to the following conditions
>
> (i) no part of the bungalow shall exceed a height of 18 feet (5.5 metres) at its highest point
>
> (ii) the gross floor area of the bungalow shall not exceed 2,900 square feet (176 square metres)
>
> (iii) no point on the western elevation of the bungalow shall be less than 45 feet (13.7 metres) from the western boundary of the property
>
> (iv) any windows in the western elevation of the bungalow shall not be above ground floor level
>
> (v) no existing trees, shrubs, bushes or hedges along the western and south western boundaries of the property shall be topped lopped uprooted felled wilfully damaged or removed
>
> C (vi) no point on the western elevation of the garage shall be less than 20 feet (6.10 metres) from the western boundary of the property
>
> (vii) no part of the garage shall exceed a height of 15 feet (4.6 metres) at its highest point

This can only be an interim decision in the absence of acceptance by the applicants of varied condition (v) above. The applicants are invited to inform the registrar within four weeks of this interim decision that they accept this condition. The original positive obligation regarding landscaping was, however, agreed by the objectors, and, in view of the proposed change of wording, I should give them the opportunity to make representations on revised condition (v) before I issue my final D decision.

Accordingly, this decision is an interim decision, dependent only on the form of condition (v) of the proposed modification to the restriction. The parties are invited to agree or make submissions on this matter as referred to above and also now to make representations as to costs based on this interim decision. A letter accompanies this decision as to the procedure for the making of submissions in writing. I will, in due course, issue a final decision incorporating orders as to the form of the modification of the restriction and as to costs. Rights of appeal under section 3(4) of the Lands Tribunal Act 1949 and Rules of the Supreme Court, Ord 61 will not accrue until the date of the final and E completed decision.

Addendum

The applicants accept condition (v) referred to above but suggest the addition of a proviso to allow topping, lopping, etc, where this is required to deal with disease or dangerous condition. The objectors have not made submissions on this matter. I accept the applicants' proposed amendment to condition (v). Accordingly, I order that para 1 of the second schedule to the conveyance of the application land dated July 24 1953 shall be modified to read as follows:

> That no building or erection shall at any time hereafter be built or erected on F the property hereby conveyed or any part thereof save and except one detached bungalow and detached garage subject to the following conditions:
>
> (i) no part of the bungalow shall exceed a height of 18 feet (5.5 metres) at its highest point
>
> (ii) the gross floor area of the bungalow shall not exceed 2,900 square feet (269.41 square metres)
>
> (iii) no point on the western elevation of the bungalow shall be less than 45 feet (13.7 metres) from the western boundary of the property
>
> (iv) any windows in the western elevation of the bungalow shall not be above ground floor level
>
> (v) no existing trees shrubs bushes or hedges along the western and south western boundaries of the property shall be topped lopped uprooted felled wilfully damaged or removed unless such topping lopping uprooting felling or removal is required on the written advice of an aboriculturist where the tree G shrub bush or hedge has become diseased or dangerous
>
> (vi) no point on the western elevation of the garage shall be less than 20 feet (6.10 metres) from the western boundary of the property
>
> (vii) no part of the garage shall exceed a height of 15 feet (4.6 metres) at its highest point

The metric conversion figure of 176 m^2 in condition (ii) as set out earlier in this decision, given by the applicants and agreed by the objectors, is incorrect and should be 269.41 m^2. The correct figure is included in revised condition (ii) set out above.

I have received written submissions on costs. The applicants ask for their costs on an indemnity basis or, at least, on the High Court scale. H The objectors seek their costs on the High Court scale.

I have considered these submissions. The current practice of the Lands Tribunal as to costs in section 84 applications is set out in Preston and Newsom, *Restrictive Covenants* (8th ed) para 16-11. The relevant parts of this para are as follows:

> The practice in respect of cases which go to a substantive hearing appears to have settled down along the following lines: . . .
>
> 2. An applicant who wins and does not have to pay any compensation may well recover some or all of his costs from the objectors or there may be no order as to costs. The more closely the case is analogous to hostile litigation, the greater is the chance that costs will follow the event. J
>
> 3. . . . If a sealed offer exceeds the sum awarded this may influence the decision on costs . . .

A similar statement appears in Scamell *Land Covenants* at p485. The former practice of this tribunal, not normally to award costs against an objector unless the objection was frivolous or vexatious (Practice Note of May 27 1954), referred to by the parties, was superseded in the late 1970s by the practice of treating each case on its merits and the 1954 note was revoked by Practice Direction of March 24 1982. The current Practice Direction on costs (8/1997), which took effect on January 1 1997, does not refer specifically to costs in section K 84 applications but confirms the general principle that "in general orders for costs are made by the tribunal in the exercise of its discretion in accordance with the principles applied in the High Court and County Court".

In my view, this is a case where the applicants should recover part of their costs. I do not think that they should recover all their costs nor should they recover them on an indemnity basis. The main reasons why they should not recover all their costs are: late amendment of the application at the hearing, from one to discharge the restriction to allow the erection of a two-storey house (which I would have refused) to one seeking a conditional modification to permit the erection of a L bungalow and to the lack of planning permission for the proposed bungalow. By a letter dated September 24 1996 marked "without prejudice save as to costs" the applicants offered "to settle the matter" with a total payment to the objectors of £5,000, each party paying their own costs. This offer was not accepted. I have taken it into account, but it does not persuade me to give the applicants the whole of their costs.

I have also taken into account the submissions of the parties regarding the agreement of facts. I refer to this at the beginning of my decision. From the submissions and evidence I cannot find that the failure to agree facts was solely the fault of one party. My conclusion M is that the objectors and their advisers were mainly to blame for the non-agreement of facts. I cannot accept the statement by the objectors' solicitors in their submissions on costs that "the expert witness of the Objectors was not under any obligation to agree facts with the Applicants expert witness". Both witnesses were under an obligation to agree material facts. These were largely uncontentious but they were numerous (19 paragraphs and five pages in this decision). I draw attention to Practice Direction 5/1997 (Expert Evidence) para 4:

> Disclosure and exchange of experts reports will usually be only a first step. Thereafter, the Tribunal will normally require experts of like discipline to meet and reach agreement as to facts, to agree any relevant plans, photographs etc,

A and to define those facts upon which they are unable to agree and the issues remaining unresolved between them. It has been known occasionally for retained experts to be instructed not to agree; such instructions are to be deprecated. The Tribunal will regard failure to cooperate in reaching agreement as to the facts and issues as incompatible with the expert's duty to the Tribunal.

A similar obligation was contained in the previous Practice Direction on Expert Evidence (November 8 1988).

In the exercise of my discretion as to costs, and taking into account B the matters set out above, I have decided that the objectors should pay half the applicants' costs. Accordingly, I order that the objectors shall be jointly and severally liable to pay half the applicants' costs, such costs, if not agreed, to be taxed by the Registrar of the Lands Tribunal on the High Court standard basis.

A

Chattey and another *v* Farndale Holdings Inc and others

COURT OF APPEAL

October 11 1996

KENNEDY, MORRITT and POTTER LJJ

Estates Gazette February 8 1997

B [1997] 06 EG 152

Vendor and purchaser — Purchasers making deposits in respect of proposed flats in refurbishment scheme — Vendor developer insolvent — Whether purchasers entitled to lien on property in priority to successor in title to vendor to protect deposits

On April 1 and May 28 1987, respectively, the two appellant purchasers signed contracts with PWA to take subunderleases of flats to be created in an existing building. The premiums were £381,500 and £329,500. On May 27 1987 PWA applied to the Land Registry to register the title to the underlease it had acquired. The contracts, which C were conditional on the grant of planning permission, made provision for a deposit of 20% which was paid in two tranches, and provided that the purchasers would not register any caution, notice or other entry on the vendor's title. The contracts became unconditional in August 1988 with the grant of planning permission. PWA became insolvent and in due course the chargees of the superior titles sold these to the respondent company. The plaintiffs appealed the decision of Blackburne J, who had dismissed their claims for the return of their deposits, contending that they had liens on the property to protect the deposits, alternatively there was a constructive trust.

***Held*: The appeal of the first appellant was allowed, and that of D the second appellant dismissed. The circumstances in which a purchaser's lien will arise are not limited to those in which the contract is or has been specifically enforceable, but include those in which there is or has been a right to call for the legal estate whether presently, in the future or conditionally so as to give rise to an equitable interest or estate. The provisions of the contracts against protecting the interests of the purchasers did not modify those interests. When PWA applied to register the title to its underlease on May 27 1987, the Land Registry was bound to note on the title the rights of those purchasers whose contracts were signed before that application; the consequence was that the first, E but not the second, appellant had priority over the rights of the respondent. The respondent was not subject to any obligation by way of a constructive trust to give effect to the second appellant's contract.**

The following cases are referred to in this report.

Ashburn Anstalt v *Arnold (No 2)* [1989] Ch 1; [1988] 2 WLR 706; [1988] 2 All ER 147; (1988) 55 P&CR 137; [1988] 1 EGLR 64; [1988] 23 EG 128, CA

Hewett v *Court* (1983) 149 CLR 639

Levy v *Stogdon* (1898) 1 Ch 478

London & South Western Railway Co v *Gomm* (1882) 20 ChD 562

Lyus v *Prowsa Developments Ltd* [1982] 1 WLR 1044; [1982] 2 All ER 953; (1981) 44 P&CR 213 F

Middleton v *Magnay* (1864) 2 H&M 233

Rose v *Watson* (1864) 10 HLCas 672

Sainsbury (J) plc v *O'Connor (Inspector of Taxes)* [1991] 1 WLR 963, CA

Whitbread & Co Ltd v *Watt* [1902] 1 Ch 835, CA

This was an appeal by Mr Chattey and Mr Strebel against the decision of Blackburne J, who had dismissed their claims against the respondent, Farndale Holdings Inc, for the return of deposits paid under contracts for the grant of subunderleases.

John Macdonald QC and Nicholas le Poidevin (instructed by Davis Arnold Cooper) appeared for the appellants; Kim Lewison QC and G John Wardell (instructed by Jay Benning & Peltz) represented the respondent.

Giving judgment at the invitation of Kennedy LJ, **MORRITT LJ** said: Each of the plaintiffs, Mr Chattey and Mr Strebel, contracted to buy from the then owner of the former West London Air Terminal one of a number of flats being constructed in it. Each of them paid to the vendor deposits totalling 20% of the purchase price. The construction work necessary for the creation of the flats was paid for out of the deposits paid by the plaintiffs and purchasers of other flats and money borrowed from banks on the security of the development. These H sources proved to be insufficient and the vendor became insolvent. The development was then sold at the instance of the banks holding the security over the development and the plaintiffs and others claimed to recover their deposits from the subsequent owners of the development, ultimately the first defendant, Farndale Holdings Inc, in priority to the banks and their successors in title by means of the purchasers' liens to which they claimed to be entitled or a constructive trust they claimed to bind the owner of the development from time to time. The plaintiffs' claims were dismissed by Blackburne J on May 21 1996 and this is an appeal by the plaintiffs from his decision.

Though the nature of the action may be simply described the facts, J though not in dispute, and the issues the judge was called on to decide are a good deal more complicated. In this court the areas of dispute have been narrowed and four of the parties previously concerned have settled their differences, leaving only the plaintiffs and Farndale Holdings Inc to contest those which remain. In those circumstances the material facts may be more shortly stated.

On February 12 1987 Manhattan Development plc was registered in HM Land Registry as the proprietor of the headlease of West London Air Terminal, being a lease for 140 years from May 1986, and two small freehold plots which formed part of the terminal, subject, *inter alia*, to a sublease of part to Sainsbury plc. By an underlease dated K March 25 1987 ("the underlease"), in consideration of a substantial premium, Manhattan Development plc let those parts of the terminal not occupied by Sainsbury's and known as the Pointwest Apartments to Pointwest Apartments plc ("PWA") from March 1987 to September 2113 at a rent equal to the occupation rents for the flats to be constructed and the cost of insurance. There were to be some 398 flats, together with a car park, health centre and commercial areas in that part of the old terminal building.

The flats were marketed for sale by PWA on the basis of a standard form contract to grant a subunderlease of a flat and an associated

A parking space in consideration of a premium and an annual rent. The contracts were in terms conventional for such a transaction. However it is necessary to refer to three provisions. First, clause 3 stipulated for a deposit of 10% on the signing of the contract and a further 10% on March 25 1988 to be paid to the vendors' solicitors as agents for the vendors. Second, it was provided by clause 11.01 that the agreement was conditional upon the vendor obtaining a satisfactory planning consent, as defined, prior to September 29 1988. Third, clause 21 provided that the purchaser would not "at any time register or cause or permit to be registered any caution notice or other entry on the Vendor's title or any title superior thereto".

B On April 1 1987 Mr Chattey contracted with PWA for the grant of an subunderlease of flat 941W and the associated parking space for a total premium of £381,500. In accordance with the terms of the contract he paid a 10% deposit on signing the contract and on March 25 1988. This was before the solicitors for PWA applied for registration of the title to the underlease on May 27 1987. That application disclosed, as required by form 3F as prescribed by the Land Registry Rules, incumbrances arising from agreements for underleases entered into by PWA as set out in an attached schedule. In fact there was no such schedule, but there is no doubt which agreements were referred to nor that they included that with Mr C Chattey.

The agreement between Mr Strebel and PWA was made after the application for registration of the title to the underlease. On May 28 1987 Mr Strebel contracted to buy a subunderlease of flat 373E and its associated parking space for the total sum of £329,500. He duly paid the two deposits required by the contract amounting to £65,900.

By an agreement made on September 14 1987 Manufacturers Hanover Trust Co agreed to provide a facility of £15m to PWA as working capital to assist in the acquisition and development of the former terminal. The facility was secured, as required by the agreement, by a debenture ("the debenture") granted by PWA D conferring fixed and floating charges on all its property including the benefit of contracts with purchasers of flats and the deposits paid or payable thereunder. It was further secured by a guarantee supported by a fixed charge over the headlease granted by Manhattan Developments plc and the guarantee of a major shareholder in PWA, Land & Property Trust Ltd. The charge and debenture were lodged for registration in HM Land Registry on October 27 1987 and June 8 1988 respectively.

On August 11 1988 planning permission for the construction of the flats complying with the contractual definition of a satisfactory planning consent was granted and notice to the effect that the contract E was unconditional was sent later that month to Mr Chattey and Mr Strebel.

In view of the submission of the plaintiffs relating to a constructive trust it is necessary to trace briefly the subsequent history of the development. In June 1989 Manufacturers Hanover Trust Co called for repayment of the facility from PWA and each of the guarantors. The debt was paid in full by one of the guarantors, Land & Property Trust, and the benefit of the charge over the headlease and the debenture granted by PWA were transferred accordingly to Land & Property Trust in August 1989. PWA was compulsorily wound up on October 25 1989.

F The development and its finance was restructured by Land & Property Trust in October 1989. Land & Property Trust enforced its rights under the charge and debenture by transferring the headlease for £1m to Pointwest Properties Ltd ("PWP") and the underlease for £35m to Pointwest Developments Ltd ("PWD"). The respective recipients of the headlease and the underlease charged them to secure a syndicated loan facility of £95m to PWD. In November 1989 there was a confirmatory deed entered into between Land & Property Trust and PWD to the effect that the transfer of the underlease included the benefit and burden of the flat purchasers' contracts. But the development so restructured was no more successful and on December 7 1990 receivers were appointed of, *inter alia*, the property and undertaking of PWD and PWP. By then the development was 75% G complete. On January 6 1993 the headlease and the underlease were transferred by the chargees thereof to the defendant, Farndale Holdings Inc.

The writ in this action was issued on November 3 1993 claiming against Farndale the return of the deposits paid by the plaintiffs to the original vendor PWA. On October 13 1995 a further writ was issued by a further 239 flat purchasers claiming similar relief. Of the many issues originally raised and decided by Blackburne J only four now survive. They are:

1. whether in view of the fact that the contracts were (a) conditional until the grant of satisfactory planning consent on August 11 1988, and H (b) for the grant of leases not previously in existence the plaintiffs could ever have been entitled to a purchasers' lien to secure the return of the deposits paid exercisable in priority to the debenture and those, such as Farndale, claiming under it; and if so

2. whether such lien was modified or excluded by the provisions of clause 21 of the contracts entered into by the plaintiffs; and if not

3. whether r40 Land Registration Rules or section 70(2) Land Registration Act 1925 requires there to be noted on the title to the underlease (which will confer priority over Farndale) not only those flat purchase contracts, such as that of Mr Chattey, as were concluded before the application for registration of the title to the underlease was J submitted but also such of them as, like Mr Strebel's, were not concluded until after the application was made; and if not

4. whether Mr Strebel, as well as Mr Chattey, is entitled to the return of his deposit because Farndale, as the successor in title of PWD to the underlease, is bound to give effect to the constructive trust found by Blackburne J to be binding on PWD.

Blackburne J decided the first issue in favour of the plaintiffs and second, third and fourth issues in favour of Farndale. Each party contends that the judge was wrong in respect of such of those issues as he decided against them.

The first issue necessitates consideration of the basic principles on which the purchaser's lien is founded. K Farndale submits that it is essential that the equitable interest in the property contracted to be purchased should have passed to the purchaser. The rationale of that requirement, it is suggested, is that in that event if the contract goes off otherwise than through the default of the purchaser he should be entitled to retain the property until moneys paid by him in respect of the contractual price for it have been repaid. Farndale submits that if the contract is conditional it is not specifically enforceable until the condition has been satisfied so that until then the beneficial or equitable interest in it cannot have passed to the purchaser, and, in consequence, there cannot be a purchaser's lien. Starting from the L same proposition Farndale also submits that there can be no lien if the contract was for the grant of property not previously existing, such as a new lease, for there can have been no subsisting beneficial interest capable of being vested in the purchaser even if the contract was unconditional and therefore susceptible of specific performance.

Farndale supports each of these submissions by reference to the decision of the House of Lords in *Rose* v *Watson* (1864) 10 HLCas 672. In that case the purchaser had paid deposits under an unconditional contract for the sale of land induced by the misrepresentations of the vendor. On discovering the falsity of the representations the purchaser rescinded the contract and successfully M resisted a claim for specific performance made against him. In subsequent proceedings against the vendor's successors in title to the land contracted to be sold the purchasers claimed a lien over that land to secure the repayment to them of the deposits paid under the contract. The purchasers' claim was upheld. Lord Westbury, at p678 said:

> . . . I think that your Lordships will agree with me that the case is determinable by principles which are very simple and very clear, and which have long been established in the Courts.
>
> When the owner of an estate contracts with a purchaser for the immediate sale of it, the ownership of the estate is, in equity, transferred by that contract. Where the contract undoubtedly is an executory contract, in this sense, namely, that the ownership of the estate is transferred, subject to the payment of the

> purchase-money, every portion of the purchase-money paid in pursuance of that contract is a part performance and execution of the contract, and, to the extent of the purchase-money so paid, does, in equity, finally transfer to the purchaser the ownership of a corresponding portion of the estate.
>
> My Lords, that being so, we have only to inquire under the terms of the present contract whether the sums of money paid by the Respondent were, or were not, paid in pursuance of that contract. About that, my Lords, there is no controversy whatever. They were *bona fide* payments made by the Respondent, in conformity with the contract which required such payments to be made in part of the purchase-money; and they were accepted by the vendor as portions of that purchase-money. In conformity, therefore, with every principle, the purchaser paying the money acquired an interest in the estate by force of the contract and of that part performance of the contract, namely, the payment of that portion of the purchase-money.
>
> Then, my Lords, if that contract fails, and the failure is not to be attributed to any misconduct or default on the part of the purchaser, the obvious question arises, Is the purchaser to be deprived of the interest in the estate which he has acquired by that *bona fide* payment? And yet, my Lords, that he ought to be so deprived is the whole controversy of the Appellants at your Bar.

At p683 Lord Cranworth said:

> There can be no doubt, I apprehend, that when a purchaser has paid his purchase-money, though he has got no conveyance, the vendor becomes a trustee for him of the legal estate, and he is, in equity, considered as the owner of the estate. When, instead of paying the whole of his purchase-money, he pays a part of it, it would seem to follow, as a necessary corollary, that, to the extent to which he has paid his purchase-money, to that extent the vendor is a trustee for him; in other words, that he acquires a lien, exactly in the same way as if upon the payment of part of the purchase-money the vendor had executed a mortgage to him of the estate to that extent.
>
> It seems to me that this is founded upon such solid and substantial justice, that if it is true that there is no decision affirming that principle, I rejoice that now, in your Lordships' House, we are able to lay down a rule that may conclusively guide such questions for the future.

In addition, in relation to the argument based on the conditionality of the contract, Farndale supports its argument with the decision of this court in *J Sainsbury plc* v *O'Connor (Inspector of Taxes)* [1991] 1 WLR 963. In that case the court was concerned with the question whether the taxpayer remained the beneficial owner of certain shares notwithstanding the existence of an unexercised option to purchase them given to another. It decided that it did. At p979 Nourse LJ said:

> Then take the previous example [company A enters into an unconditional contract to sell shares in company B to company C], but suppose that the contract is subject to a condition precedent. Until the condition is satisfied the equitable interest in the shares will not pass to company C. It will remain in company A. What ground is there for thinking that the beneficial ownership of the shares will not also remain in company A? In order to answer that question we must look to *Wood Preservation Ltd* v *Prior* [1969] 1 WLR 1077. That is a difficult decision. Goff J at first instance did not distinguish between the beneficial ownership of the shares and the equitable interest in them. In my view he was right not to make that distinction. However, he thought that, because the purchaser could obtain specific performance of the contract by waiving the condition precedent at any time, "the beneficial interest had sufficiently passed to the purchaser." I respectfully think that that was an error on the part of the judge. Unless and until the condition was either waived or satisfied there could be no right to specific performance and no passing of the equitable interest.

Farndale does not contend that it is necessary that the contract should be specifically enforceable at the time the benefit of the lien is claimed nor when the money for which it is claimed to be security was paid, rather that unless and until the contract is specifically enforceable by the purchaser the lien cannot arise. Thus Farndale accepts, subject to its other submissions, that a lien arose when the contract became unconditional on August 11 1988, but by then it would necessarily be subject to the debenture. They also accept that if the contract is once specifically enforceable so that the purchaser's lien arises the lien is not lost if the right to specific performance is subsequently lost by, for example, delay.

The plaintiffs submit that a lien arises in favour of one who has a contractual right, whether present, future or conditional to call for the transfer to him of the legal estate in the land in respect of all sums paid by him on the faith of that contract and as part of the purchase price. They rely on the well known statement of Sir George Jessel MR in *London & South Western Railway Co* v *Gomm* (1882) 20 ChD 562 at p581 that:

> The right to call for a conveyance of the land is an equitable interest or equitable estate. In the ordinary case of a contract for purchase there is no doubt about this, and an option for repurchase is not different in its nature. A person exercising the option has to do two things, he has to give notice of his intention to purchase, and to pay the purchase money; but as far as the man who is liable to convey is concerned, his estate or interest is taken away from him without his consent, and the right to take it away being vested in another, the covenant giving the option must give that other an interest in the land.

The plaintiffs submit that the statements of principle in *Rose* v *Watson* (*supra*) were directed to the facts of that case and were not exhaustive. So far as what has been described as the conceptual objection is concerned the plaintiffs adopt the decision of Blackburne J that:

> the reasoning in *Rose* v *Watson* was directed to the case, there being considered, in which the subject matter of the sale was an existing estate which, if the contract is completed, would pass to the purchaser but if the contract went off would remain in the vendor. Whereas, if the contract is to create and convey a derivative interest, the lien attaches to whatever interest the vendor has, if any, out of which the derivative interest is to be created so that the vendor's interest becomes encumbered by that obligation in favour of the purchaser when the contract is made and the deposit paid. This means that where the contract is to grant a lease of a freehold plot the lien attaches to the vendor's interest in that plot. That seems to me to be entirely in conformity with the approach of Lords Westbury and Cranworth in the passages that I have quoted.

I propose to deal with the conditionality argument first. It seems to me that the argument for Farndale would give rise to anomalies. First, the lien would not be directly related to the purchaser's payments sought to be secured and would be denied in those circumstances where such protection was most required. Thus on the facts of this case no lien would have arisen when the deposits were paid on the exchange of contracts and on March 25 1988, notwithstanding that in accordance with the statement of Sir George Jessel MR the plaintiffs then had an equitable interest in the property by virtue of their respective conditional rights to call for the legal estate in their respective flats. Second, if, as Farndale contends, the lien arises at the time of and in consequence of the accrual of the right to specific performance of the contract it is hard to understand why the lien does not cease if the right to specific performance is subsequently lost. That the lien is not lost is conceded by Farndale and in any event established by the decision of Stirling J in *Levy* v *Stogdon* (1898) 1 Ch 478. In that case the claim for specific performance of the contract by assignees of the purchaser was dismissed on the grounds of delay, but the claim for the lien was upheld. It is true that a similar anomaly would arise on the plaintiffs' rival submission, but that does not detract from the anomalous nature of the case for Farndale. Third, it is hard to see why the substantive right of the purchaser to a lien should depend on the availability to him of a remedy, particularly one which if successfully invoked would negate the need to rely on the right at all.

In my view, both principle and authority support the argument for the plaintiffs. The statement of Sir George Jessel shows that the purchaser has an equitable interest or estate in the land if he has a right to call for the legal estate, albeit future and conditional, which the vendor has no right to refuse. In this case the vendor was contractually bound to use his best endeavours to obtain a satisfactory planning consent on the grant of which the contract became unconditional. The equitable interest or estate of the purchaser was one which entitled him to seek specific relief in the form of injunctions so as to protect that right notwithstanding that a claim for specific performance might have been premature.

Rose v *Watson* (*supra*) clearly establishes that the purchaser's beneficial or equitable ownership of the property arising in consequence of a contract to purchase land which is from the outset specifically enforceable is sufficient for the creation of the purchaser's lien. In my view, it does not establish that such an interest is not only sufficient but necessary for the establishment of a lien in favour of the purchaser. The necessity for such an interest was not in issue and no consideration was given to the lesser interest or right to which Sir George Jessel referred.

Further, the decision of this court in *Whitbread & Co Ltd* v *Watt* [1902] 1 Ch 835 establishes that the decision in *Rose* v *Watson* was not an exhaustive statement of principle and, in my view, indicates the tacit acceptance of the Court of Appeal of the submissions for the plaintiffs made in this case. In that case the purchaser contracted to purchase a plot on a building estate belonging to the vendor. The contract provided that "the purchase is to be completed as soon as 300 houses shall have been erected on the said estate".Thus the contract was one under which completion was conditional. It also contained a provision entitling the purchaser to rescind the contract by giving notice to that effect "if 300 houses shall not be erected on the said estate within two years from the date of this agreement". The purchaser rescinded the contract and claimed a lien to secure repayment to him of the deposit paid by him on signing the contract. The vendor contended that there was no lien as the contract had not gone off due to any default of his. At first instance Farwell J said [1901] 1 Ch 911, at p915:

> The lien is created by the contract under which the money is paid as part of the purchase-money, and on the faith that the contract will be carried out, and not by the default of the vendor. The default gives rise to the necessity for enforcing the lien, but the lien arises from the contract. I see no reason why a condition that, if 300 houses are not built, the purchaser may rescind should be held to differ in any way from the ordinary condition in a contract that, if the purchaser makes or insists upon any requisition or objection to the title which the vendor is unable or unwilling to comply with, the vendor may rescind.

In the Court of Appeal the judgment of Farwell J was specifically approved. All three lord justices accepted that it was not necessary for the contract to go off because of the default of the vendor, thus going further than the statements in *Rose* v *Watson*. In addition Vaughan Williams and Stirling LJJ recognised a far wider principle than that expressed in *Rose* v *Watson*. Thus, at p838 Vaughan Williams LJ said:

> The lien which a purchaser has for his deposit is not the result of any express contract; it is a right which may be said to have been invented for the purpose of doing justice. It is a fiction of a kind which is sometimes resorted to at law as well as in equity. For instance, when an action is brought for money had and received to the use of the plaintiff, it is not true that the money has been so received, but that is the way in which the law states the case in order to do justice. When Lord Westbury in *Rose* v *Watson* speaks of "a transfer to the purchaser of the ownership of a part of the estate corresponding to the purchase-money paid," and Lord Cranworth speaks of the purchaser being exactly in the position of a mortgagee of the estate to the extent of the purchase-money which he has paid, those expressions are merely verbal vehicles to carry the right which justice demands that the purchaser should have. Having read the report of *Rose* v *Watson*, I must say that, speaking for myself, I agree with Mr Brinton to this extent, that that decision does not expressly carry the purchaser's lien beyond a case in which the contract has gone off through the default of the vendor.

Similar points were made by Stirling LJ in the Court of Appeal at p840 where he observed that:

> It is, I think, quite true, as Mr Brinton has contended, that the question of the existence of the purchaser's lien for his deposit arises in the present case in circumstances which differ from those of all previous reported cases. The contract has here been brought to an end, not by any act or default of the vendor, but by reason of the purchaser's exercising a power of rescinding it which is reserved to him by the contract itself. This does not seem to have occurred in any previous case. Nevertheless, in the judgments in the two leading cases on the subject, *Wythes* v *Lee* and *Rose* v *Watson*, the rule is stated in terms which cover the present case. And, if we look at that which is really the foundation of the doctrine, namely, the desire to do justice as between vendor and purchaser, it appears to me that that reason applies no less forcibly in the present case than in the ordinary case in which the rescission of the contract takes place by reason of some default on the part of the vendor. In a case in which the vendor had rescinded under a power reserved to him, it would, I think, be absolute injustice if the purchaser were not allowed to have a lien for the purchase-money which he had paid, and which was the security on his part for the performance by him of the contract. I think also the justice of the case requires that the purchaser should have a lien when the contract reserves to him a power to rescind.

In my judgment, the circumstances in which a purchaser's lien will arise are not limited to those in which the contract is or has been specifically enforceable but include those in which there is or has been a right to call for the legal estate whether presently, in the future or conditionally so as to give rise to the equitable interest or estate to which Sir George Jessel referred. I accept the submissions for the plaintiffs in this respect. This conclusion is in line with that of Blackburne J in dealing with the issue of conditionality. I have not previously referred to the judge's conclusion for he did not deal with the point expressly in connection with the submission for Farndale that the existence of the lien depended on the specific enforceability of the contract. At p76, having referred earlier to *London & South Western Railway Co* v *Gomm* and *Whitbread & Co Ltd* v *Watt* he said:

> In my view, the plaintiffs became owners in equity of the premises (of which they were contracting to take subunderleases) as soon as their contracts were entered into and, subject to the effect of clause 21, became entitled to liens on those properties on payment of their initial deposits, and it matters not that, until August 1988, their contracts remained purely conditional.

Provided that ownership in equity is understood to refer to the equitable interest or estate to which Sir George Jessel referred in *London & South Western Railway Co* v *Gomm* I agree with the judge. My conclusion is supported by the decision of the majority in *Hewett* v *Court* (1983) 149 CLR 639. I have not referred to it earlier because the decision was reached by a bare majority and the relevant issue appears to have been conceded in the opposite sense to the majority decision.

I can deal with what was described as the conceptual objection more shortly. The objection, which arises from *Rose* v *Watson*, is to the effect that because the leasehold interest contracted to be sold had no previous separate existence, for it was to be granted on completion of the contract, the purchaser never could have been the beneficial owner of such a term even when the contract became unconditional and therefore specifically enforceable. Once it is established that the existence of the lien is not restricted to cases where the purchaser has been entitled to specific performance the concept on which the objection is based disappears too. What is important is that the purchaser shall have had the right, whether present, future or conditional, to call for the legal estate. It would be absurd if the lien should be denied merely because that legal estate did not exist but another out of which the vendor would grant it did.

In *Middleton* v *Magnay* (1864) 2 H&M 233 the vendor agreed to grant a 21-year lease over certain land. In fact he only had an eight-year lease in half the land. When the contract went off on account of the inability of the vendor to grant the lease contracted for the purchaser claimed a lien to secure repayment to him of certain expenditure on improvements undertaken by him pursuant to a term in the contract. The claim for a lien was upheld by Page-Wood V-C in respect of that interest which the vendor did have. It is true, as counsel for Farndale submitted, that in law the contract took effect as one to assign such interest as the vendor did have. I entirely agree with the conclusion of Blackburne J, at p49, that:

> that consequence cannot, as it seems to me, affect the vendor's contractual obligation. That obligation was to grant a 21 year term. The agreement went off because the vendor could not fulfil that obligation. In my view, whilst it is true that the conceptual difficulty which has been raised in argument before me does not appear to have been considered in that case, the decision is nevertheless an example of a lien over an interest which was not itself the subject matter of the contract but out of which a derivative interest was agreed to be created. cannot think that the decision would have been any different if the vendor's interest had been a freehold in a moiety only of the premises or a 22 year term in such moiety.

For all these reasons I would answer the first issue in the affirmative. Accordingly, the second issue on the basis of which the judge dismissed the action arises. That issue depends on the effect of clause 21 of the contract which I have already quoted. The judge's conclusion, at p44, was that:

> a lien arises by implication of law for the purpose of doing justice between vendor and purchaser. The essence of the matter is that the bargain between PWA and Mr Chattey was that the rights he acquired in the underlease should not be protectable, and should not therefore be capable of being asserted, against subsequent disponees because protection against such persons could only be achieved by the entry of a caution notice or other entry on PWA's title, which is what Mr Chattey was prohibited by clause 21 from causing or permitting. Whilst, therefore, it would be perverse, in my view, that the law should imply a lien to protect Mr Chattey's rights in the underlease against the very persons against whom the contract has impliedly stipulated that he should have no protection, I can see no reason why, for what it is worth, the law should not continue to imply a lien to protect his deposit against PWA.

The judge had earlier defined subsequent disponees as "persons acquiring for value an interest in the underlease under a registrable disposition and those claiming under such persons". The plaintiffs submit that Blackburne J was wrong. They contend that he confused the substantive right to a lien with the statutory machinery for its protection by registration. They criticise the judge for accepting that one reason for the inclusion of clause 21 must have been the common intention of the parties that the vendor should be able to offer an unincumbered title to a secured lender when the other reason, to ensure that the vendor would not have to warn off cautions registered by all other purchasers in respect of the underlease as a whole when completing the contract to one of them for the grant of a subunderlease of a specific flat and parking space, was correct and sufficient.

Farndale supports the decision of the judge for substantially the reasons he gave. They maintain that the substance of the right and the machinery for preserving its priority are so closely allied that by agreeing not to cause or permit registration the purchasers must be taken to be agreeing to the modification of the lien rather than to its postponement where registration is necessary to the preservation of its priority. On the assumption that the third issue is decided in their favour they take the case of Mr Strebel and point out that because of the date of his contract it could not be protected by notice on the title pursuant to the request of PWA. They conclude that Mr Strebel must, therefore, have accepted that his lien should not have priority over a subsequent disponee for value under a registered disposition and should be modified accordingly.

I have no hesitation in preferring the submissions for the plaintiffs. It is not disputed that the purchasers' lien arises by operation of law from the contract unless it is modified or excluded by express agreement of the parties or by necessary implication from the contractual arrangements the parties have entered into. The lien so arising is an unqualified equitable right. In common with all other equitable rights it is capable of being subordinated to the rights of a subsequent purchaser for value of a legal estate without notice of the equitable right. But that inherent weakness is no reason to modify the nature of the right itself. That is particularly so in this case where the contract precludes the registration of any caution, notice or other entry and the effect of clause 21 contended for by Farndale would involve a modification of the contract from which the purchasers' lien arises.

Obviously there is no express agreement that the lien itself should be modified and I can see no necessary implication from the terms of the contract. The information pack supplied to the purchasers stated that the title of PWA to the underlease "is presently the subject of registration at" the Land Registry. In that event it would have been the duty of PWA as the applicant under Land Registration Rules r 19 and the prescribed form to disclose, as incumbrances, at least all contracts for subunderleases concluded at the time of the application, and as conceded by Farndale, for the registrar to note them on the register. Such notice would have protected such incumbrances against subsequent disponees. Thus clause 21 would not necessarily result in all purchase contracts being subordinated to subsequent disponees.

But, in any event, I see no reason to give to clause 21 any greater effect or force than that which is entailed by its express terms. If the result of complying with clause 21 is that the rights of the purchaser are subordinated to those of the subsequent disponee so be it; but that is no reason for going further and treating it as a sufficient reason for modifying the right itself. In that event it is unnecessary to do so. If, contrary to the assumption, compliance with clause 21 does not have that effect in all circumstances then that is good reason for not attributing a common intention that the right itself should be modified. Clause 21 assumes an unqualified right so that it would be inconsistent with the terms of the contract that the purchasers' right should be modified as contended for by Farndale.

Further, I agree with the criticisms levelled by the plaintiffs at the reasons suggested by the judge for justifying his conclusion. In a development scheme such as this I do not think that it can be predicated that it must have been the intention of the parties that the title available to a secured lender to the vendor should be unincumbered by any purchase contracts capable of binding a lender. In many cases, as in fact happened in this, the lender will wish to have a charge over the purchasers' contracts. He could not achieve this if the contracts themselves were modified rather than the rights thereunder postponed. Accordingly I agree with the plaintiffs that the first reason given by the judge as accounting for the inclusion of clause 21 does not in fact do so. I also agree that the second reason accounts for the presence of clause 21 but does not lead to the conclusion that the purchasers' rights, rather than their priority, were intended to be modified. In my view, the judge was wrong on this issue which I would answer in the affirmative.

I turn then to the third issue. It was not disputed that the judge was right to have concluded that if he was wrong on the second issue then the rights of purchasers whose contracts preceded the application of PWA for the registration of its title to the underlease were incumbrances which the registrar was entitled and bound to note on the register when he completed the registration of the title of PWA. The only dispute on this issue was whether that right and duty extended to rights arising under contracts concluded after the application to register the title to the underlease. Reliance was placed on Land Registration Rules r 40 and, not unprompted by the court, section 70(2) Land Registration Act 1925. I will consider r 40 first.

The Land Registration Rules, which include the prescribed forms (r 74), have the same force as if enacted in the Land Registration Act 1925; section 144(2). R 19 to 42 cover the registration of the underlease for which PWA applied on May 27 1987. The scheme provided for requires the applicant to use one of the prescribed forms, in this case form 3A, and to send with it "all such original deeds and documents relating to the title as the applicant has in his possession": see rr 19 and 20. Form 3A requires the applicant to disclose any agreement for a sublease entered into by him. In the case of a first application, which that of PWA was, r 24 prescribes that on delivery of the application it should be allotted "a title number" and dated as of the day on which it is deemed by that rule to have been delivered. Rules 25 to 29 deal with examination of the title "shown by the documents accompanying the application". Rules 31 to 35 require advertisements in certain cases "before any registration is completed" inviting objections to the registration. Any objections thus elicited must have been withdrawn or otherwise disposed of before the title is registered. Rules 40 and 42 provide:

> 40. Incumbrances, conditions, and other burdens (including fee farm grants, or other grants reserving rents or services) to which the land may be subject, shall be entered in the register and may be entered either directly, or by reference to the instruments by which they are created or by setting out extracts therefrom . . .
>
> 42. When —
>
> (*a*) all objections (if any) have been disposed of,
>
> (*b*) the time fixed by the advertisements and notices (if any) has expired,
>
> (*c*) it has been determined with what title the registration is to be made, and the requirements of sections 14 and 16 of the Act have been complied with,

> the registration shall be completed as of the day on which, and of the priority in which, the application was delivered, and the documents of title, other than such as have under the Act or rules to be retained in the Registry, shall be delivered to the applicant.

Blackburne J concluded that there must be a cut-off point which, in accordance with r 42, he found to be the date of the application for registration and that that cut-off point applied for the purposes of r 40 as well as for r 42. At p87 he said:

> Where LRR 40 applies, the Registrar's obligation is to note on the register of title to the land, in respect of which the application for first registration is made, incumbrances to which the land may be subject as at but not subsequent to the date of the application. Protection of interests subsequent to that date is a matter for the person entitled to the interest: he can seek to protect himself by lodging a caution against dealings or in any of the other ways which may be available to him (depending on his interest) under the Act.

This conclusion accorded with the practice of the Land Registry but had the consequence that those purchasers, like Mr Strebel, whose contracts were concluded after the date of the application for registration made by PWA would not be protected by registration of the incumbrance constituted by the rights under their contracts. Before Blackburne J Farndale had submitted that r 40 did not apply to the application by PWA for first registration of what is known as a dispositionary lease, ie one by which a new registrable term of years is created. That contention was abandoned before us.

The conclusion as to the cut-off point for the purpose of r 40 is disputed by the plaintiffs, not because it is wrong in itself but because it does not go far enough. They submit that the cut-off date is the date of registration so that as the registrar is well aware of the contracts entered into after the application for registration they must be noted on the register in the same way as those entered into before the application. It was suggested that the conclusion adopted by the judge would deny to the registrar a "snapshot" of the state of the title being registered because of the inevitable gap between the application and the actual registration. It was pointed out that there was nothing to restrict the objections elicited by the advertisements to those arising from acts and events occurring before the application for registration. It was also suggested, not unprompted by the bench, that unless the plaintiffs were right the action of registration would alter the pre-existing priorities which it was not to be presumed that the draftsman intended.

I do not accept these submissions. I will take the question of priorities first. The point arises from the fact that neither the underlease nor the debenture (nor for that matter all the subsequent dealings with the underlease) was registered in the appropriate register in the Land Registry before the writ was issued. Accordingly all dealings took effect in equity and the rights of PWA, the purchasers and MHT ranked in accordance with the order of their creation. If the judge was right as to the cut-off point then until registration of the title to the underlease all purchase contracts had priority to the debenture but after registration only those concluded before the application for registration of the underlease would continue to enjoy such priority. This is true so far as it goes. But as counsel for Farndale pointed out it does not go far enough. If the land were unregistered the grant of the underlease to PWA and the debenture granted by PWA to MHT would in each case have transferred a legal estate. Thus unless the purchase contracts were registered as class C(iv) land charges the debenture would have obtained priority over all of them. None of them was then registered anywhere and the purchasers were precluded from doing so. Even if the object of the land registration system is to reproduce the equivalent priorities to those which would have prevailed in unregistered conveyancing, which is not always achieved anyway, it is still not necessary to construe r 40 in the way the plaintiffs contend.

The fact is that the only point of time referred to in the rules is the date of the application. This governs what documents the applicant must submit (r 20), the date when a title number is allotted (r 24) and by the express terms of r 42 the day as of which the title is to be registered. It is true that the provisions for advertisement inviting objections are silent as to any restriction on the period in which the objection must have arisen. I accept the submission of counsel for Farndale that the objection must be to the registration of the title as of the date of the application for registration; otherwise there will be no correspondence between the proposed registration and the objection. This conclusion is also supported by the mundane provision that the title number is allotted at the date of the application to register the title. Thereafter there is a title number against which to lodge a caution. It is consistent with that scheme that incumbrances previously arising should be disclosed and noted on the register when the title is subsequently registered but that those arising thereafter should be protected by the registration or lodgment of the appropriate instrument. It would cause unnecessary duplication and potential confusion if those arising later should also be noted on the title when registered. Accordingly, in my view, the judge was right as to the scope of r 40.

This leaves the alternative contention that section 70(2) of the Land Registration Act 1925 imposes a duty on the registrar at the time of first registration to note on the register "any easement, right, privilege, or benefit created by an instrument and appearing on the title [which] adversely affects the land". Section 70(1) prescribes what are overriding interests subject to which all registered dispositions take effect. The contractual rights of a purchaser not in actual occupation of the land are not overriding interests. I have considerable doubt whether the rights of a purchaser under a contract for the purchase of registered land are encompassed by the subsection at all; but I do not find it necessary to decide that point for I accept the argument for Farndale that it does not advance matters even if they are. This is because the provisions of the rules would apply on the first registration to which the subsection refers. Thus, section 70(2) would call for notation on the register in accordance with rr 40 and 42 which, as I have already held, import the cut-off date of the application for registration.

In my judgment, Blackburne J was right in the conclusion to which he came that whereas the registrar is duty bound to note on the title to the underlease when registered the rights of those purchasers whose contracts were signed before the application for registration he is not entitled to include a similar note in respect of those contracts signed after that date. The consequence is that the rights of Mr Chattey have priority over those of Farndale, but the rights of Mr Strebel do not. Accordingly I answer issue 3 in the negative.

In the result therefore issue 4 is only of relevance to Mr Strebel. The argument before Blackburne J was described by him in the following terms:

> It had as its starting point the submission that PWD took its assignment of the underlease on 16 October 1989 upon trust to give effect to the flat purchase contracts. The nature of the trust was one entitling the flat purchasers to enforce against PWD all of the vendor's obligations under the flat purchase contracts, ie to proceed with and complete the conversion of the Pointwest building and grant to the purchasers their respective subunderleases, alternatively, in the event of the flat purchase contracts going off otherwise than through the purchasers' default, to repay the deposits with interest. The next stage in the argument was that BBL had actual notice of that trust when, by the PWD debenture, PWD charged the underlease in favour of BBL to secure the monies due under the loan agreement. That knowledge was said to derive from clause 15(A)(12) of the loan agreement. The third stage in the argument was that, as a result of stages one and two, BBL thereupon became subject to the same constructive trust as that which bound PWD. The final stage was that, assuming that the conclusion in the third stage of the argument was correct, Farndale, on taking an assignment of the underlease on 6 January 1993 (which was expressed to be with the benefit of and, to the extent that BBL's interest was so subject, subject to the flat purchase contracts: see paragraph 15 of the background facts) likewise constituted itself a constructive trustee for the flat purchasers to give effect to the vendor's obligations under the flat purchase contracts. The result, so it was submitted, is that, now that the plaintiffs have lawfully terminated their flat purchase contracts, BBL and Farndale are separately obliged to repay to the plaintiffs their deposits with interest.

In the case of PWD the judge found that:

> by taking an assignment of the underlease from LPT, entering into the loan agreement and charging the underlease as security for its obligations under the

A loan agreement, PWD was undertaking to carry out PWA's obligations under the purchase sale contracts. It was doing so because PWA was insolvent and indeed facing a winding-up petition on which, only nine days later, a winding-up order was made.

For what the point is, therefore, worth, I am of the view that PWD's conscience was affected in such a way as to give rise to an obligation to the flat purchasers to complete the development and fulfil the vendor's obligations under the flat purchase contracts.

But in the case of BBL the judge decided:

> The only matter relied upon as subjecting BBL to a constructive trust in favour of the flat purchasers to give effect to their flat purchase contracts was B BBL's knowledge, said to derive from clause 15(A)(12) of the loan agreement, that PWD held the underlease upon trust to give effect to the flat purchase contracts. I am of the view, however, that BBL's knowledge of PWD's trusteeship, if such it was, provides no basis for the suggestion that BBL's conscience was affected so as to subject it to an obligation to give effect to the flat purchase contracts. It was no part of the purpose of the loan agreement that BBL should undertake any obligations to the flat purchasers. Its purpose was the provision of finance by BBL and the other banks to enable PWD to acquire and build out the development and complete the flat purchase contracts.

The judge considered that that was enough to dispose of the constructive trust claim. But he noted an alternative and more limited C way of putting the claim:

> was that, as BBL only ever had an equitable charge over the underlease (because the charge, like the underlease over which it was granted, was not registered and therefore operated only in equity) the underlease was subject, in BBL's hands, to all of the equities affecting it created prior in time to the charge in favour of BBL. Those equities included the constructive trust binding on PWD to give effect to the rights against the underlease of the purchasers under the many flat purchase contracts (including those of the plaintiffs). The result, so [counsel for the plaintiffs] submitted, is that, when BBL exercised its power of sale by selling the underlease to Farndale, Farndale took the underlease subject to those prior equities. Now that the plaintiffs have D lawfully terminated their flat purchase contracts the underlease now owned by Farndale is burdened by the plaintiffs' liens to secure the repayment to them of their deposits with interest. Since BBL has parted with its interest in the underlease to Farndale this alternative claim is maintained only against Farndale.

The judge disposed of the alternative argument on the grounds that:

> In truth the claim is to no more than a minor interest in the underlease. Like any minor interest it will not prevail against Farndale's entitlement, if and when registered, as proprietor of the underlease except and to the extent that the plaintiffs and the other flat purchasers succeed in having their interests noted on the register of title to the underlease. In short, the alternative way of E putting the claim adds nothing.

Counsel for Mr Strebel challenges this conclusion. He submits, and I quote from his written argument, that:

> the judge overlooked the fact that Farndale was not entitled to use its entitlement to be registered in order to defeat the plaintiffs' priority. That was the decision in *Lyus* v *Prowsa*, itself a land registration case.

In *Lyus* v *Prowsa Developments Ltd* [1982] 1 WLR 1044 the plaintiffs had contracted to buy from a developer a plot of registered land with the house to be built on it, the developer having already charged the estate as a whole to a bank to secure the development F finance. The developer became insolvent and the bank sold the estate as mortgagee to the first defendant "subject to and with the benefit of" the plaintiff's contract. Three months later the first defendant resold the estate to the second defendant subject to the plaintiffs' contract so far if at all as it might be enforceable as against the first defendant. The transfer to the second defendant, which was duly registered, did not refer to the plaintiffs' contract. It was common ground that if the provision in the contract for the sale by the bank to the first defendant was adequate to impose a constructive trust on the first defendant then the effect of the provision in the contract for sale by the first defendant to the second defendant was to impose a similar trust on him. Dillon J, as he then was, concluded that both of them had that effect.

At p1053 he said: G

> Bearing in mind that there is no basis on which it could be suggested that the bank could be under any obligation to the plaintiffs to complete the house on Plot 29 for them, and bearing in mind the first defendant's solicitors' letter to Messrs, Strutt and Parker, to which I have referred, I conclude that clause 11 was not inserted in the agreement of October 18, 1979, solely for the protection of the bank, like clause 7 of that agreement which sets out other matters subject to which the property was sold, and I conclude that it was a stipulation of the bargain between the bank and the first defendant that the first defendant would give effect in relation to Plot 29 to the contract which had been made between the vendor company and the plaintiffs.
>
> If that is correct, it would follow, in my judgment . . . unless the Land H Registration Act 1925 requires a different conclusion, that the first defendant, having accepted the land under the agreement of October 18, 1979, and the consequent transfer, holds Plot 29 on a constructive trust in favour of the plaintiffs to give effect to the plaintiffs contract. That trust is also imposed on the second defendants by virtue of condition (b) of their agreement with the first defendant.

In my view, the crucial distinction between that case and this is the finding that the trust was also imposed on the second defendants by the terms of their contract. There is no finding in this case that there is imposed on Farndale a constructive trust similar to that which the judge found to be imposed on PWD.

In dealing with the effect of the Land Registration Act 1925, at J p1054, Dillon J said:

> It seems to me that the fraud on the part of the defendants in the present case lies not just in relying on the legal rights conferred by an Act of Parliament, but in the first defendant reneging on a positive stipulation in favour of the plaintiffs in the bargain under which the first defendant acquired the land. That makes, as it seems to me, all the difference. It has long since been held for instance, in *Rochefoucauld* v *Boustead* [1897] 1 Ch 196, that the provisions of the Statute of Frauds 1677 (29 Car 2 c3), now incorporated in certain sections of the Law of Property Act 1925, cannot be used as an instrument of fraud, and that it is fraud for a person to whom land is agreed to be conveyed as trustee for another to deny the trust and relying on the terms of K the statute to claim the land for himself. *Rochefoucauld* v *Boustead* was one of the authorities on which the judgment in *Bannister* v *Bannister* [1948] 2 All ER 133 was founded.
>
> It seems to me that the same considerations are applicable in relation to the Land Registration Act 1925. If, for instance, the agreement of October 18 1979, between the bank and the first defendant had expressly stated that the first defendant would hold Plot 29 upon trust to give effect for the benefit of the plaintiffs to the plaintiffs' agreement with the vendor company, it would be difficult to say that that express trust was over-reached and rendered nugatory by the Land Registration Act 1925. The Land Registration Act 1925 does not, therefore, effect the conclusion which I would otherwise reached in reliance on *Bannister* v *Bannister* and the judgment of Lord Denning MR in *Binions* v L *Evans* [1972] Ch 359 had Plot 29 been unregistered land.

Though Dillon J does not in that passage refer expressly to the necessity for a constructive trust to have been imposed on the second defendant he must have intended to apply the same principles to him for otherwise that defendant would not have been "a person to whom land is agreed to be conveyed as trustee for another" and it could not have been a fraud for the second defendant "to rely on the terms of the statute to claim the land for himself".

It is true, as counsel for Mr Strebel contended that the decision in *Lyus* v *Prowsa* was approved by the Court of Appeal in *Ashburn Anstalt* v *Arnold (No 2)* [1989] Ch 1*. But that approval was limited as M the following passages from the judgment of the court indicate. At p24 G Fox LJ said of the decision:

> This again seems to us to be a case where a constructive trust could justifiably be imposed. The bank were selling as mortgagees under a charge prior in date to the contract. They were therefore not bound by the contract and on any view could give a title which was free from it. There was, therefore, no point in making the conveyance subject to the contract unless the parties intended the purchaser to give effect to the contract. Further, on the sale by the bank a letter had been written to the bank's agents, Messrs Strutt & Parker, by

---

*Editor's Note: Also reported at [1988] EGLR 64; [1988] 23 EG 128

> A the first defendant's solicitors, giving an assurance that their client would take reasonable steps to make sure the interests of contractual purchasers were dealt with quickly and to their satisfaction. How far any constructive trust so arising was on the facts of that case enforceable by the plaintiffs against owners for the time being of the land we do not need to consider.

Later, at p25H, he added that:

> The court will not impose a constructive trust unless it is satisfied that the conscience of the estate owner is affected. The mere fact that that land is expressed to be conveyed "subject to" a contract does not necessarily imply that the grantee is to be under an obligation, not otherwise existing, to give effect to the provisions of the contract. The fact that the conveyance is
> B expressed to be subject to the contract may often, for the reasons indicated by Dillon J, be at least as consistent with an intention merely to protect the grantor against claims by the grantee as an intention to impose an obligation on the grantee. The words "subject to" will, of course, impose notice. But notice is not enough to impose on somebody an obligation to give effect to a contract into which he did not enter.

In my view, these observations apply as much to the position of the second defendant in *Lyus* v *Prowsa* as to Farndale in this case. The fact that they had notice of the prior contracts or obligations was insufficient to impose an obligation on them and unless there was such an obligation it could not be a fraud on their part to rely on the
C provisions of the Land Registration Act 1925 as conferring on them an unincumbered title.

In my judgment, Blackburne J was right for the reasons he gave in concluding that Farndale was not subject to any obligation by way of constructive trust to give effect to the Mr Strebel's contract. Accordingly I would answer the fourth issue in the negative.

Before Blackburne J there were two issues as to the property over which the lien existed. The first arose out of what was called the conceptual objection to which I referred earlier. The consequence of dismissing that objection was that the lien existed over the interest to which the vendor was entitled out of which the derivative interest was
D to be created. There was no appeal in that respect. The second issue was as to the geographical extent of the land in which the interest subsisted which was subject to the lien. Blackburne J concluded that the lien was confined to the area of land which was the subject-matter of the contract. At p51 of his judgment he said:

> The reason why, in my judgment, the lien is confined to the vendor's interest in the area of land which is the subject matter of the contract and does not extend to any greater area is because the payment of the deposit is regarded as a part payment for an interest in that land and for no other with the result that, by force of that payment, the purchaser acquires an interest in the land in question. Where, therefore, the contract goes off, the interest does not revert to
> E the vendor but is retained as security by the purchaser. The security therefore is co-extensive with the acquisition of an interest in the land by force of the payment. The interest so acquired is in the land which is the subject matter of the contract and not in any other. There is, therefore, no principled basis upon which, if the contract goes off otherwise than for the purchaser's default, the lien should be held to attach to any other land of the vendor.

The plaintiffs appealed from that finding contending that the lien extended to all the land comprised in the same interest of the vendor. During the course of the hearing of the appeal the plaintiffs abandoned that contention. Accordingly, it is now common ground that the lien to which Mr Chattey is entitled is exercisable over the property
F comprised in the contract of sale to him for the interest therein conferred by the underlease. We were informed that the development had advanced sufficiently far when work ceased in December 1990 to enable that property to be identified physically without any difficulty. Accordingly there should be no difficulty in making declarations giving effect to the rights of the parties in accordance with the judgments of this court. The question of how to give effect to a purchases lien in cases in which the relevant building or part does not exist does not arise.

I would allow the appeal of Mr Chattey, but not of Mr Strebel, and make a declaration to the effect that Mr Chattey is entitled to a lien over the property comprised in the contract made between him and PWA on April 1 1987 for all the interest therein of Farndale created by the underlease to secure repayment to him of the deposits paid by him G
pursuant to that contract.

KENNEDY and POTTER LJJ agreed and did not add anything.

Appeal allowed.

Lloyds Bank plc *v* Carrick and another H

COURT OF APPEAL

February 28 1996

BELDAM and MORRITT LJJ and SIR RALPH GIBSON

Estates Gazette March 15 1997

[1997] 11 EG 155 J

Contract for sale of land — Unregistered and void against chargee — Whether chargee bound by interest of purchaser under bare trust, constructive trust or proprietary estoppel

In February 1982 the first defendant acquired by assignment a lease for 99 years of a maisonette; at all material times the title was unregistered. Following the death of his brother, the first defendant allowed the second defendant, his sister-in-law, to move into the maisonette following the payment to him of £19,000, the first defendant saying that the maisonette would be hers. In 1983 the second defendant carried out improvement works to the maisonette. On K
November 25 1986 the first defendant executed a charge over the property in favour of the plaintiff bank to secure a loan. The first defendant failed to repay the loan, and the bank brought proceedings for an order for possession. The recorder found that there was an uncompleted contract for the sale of the maisonette to the second defendant; the contract had not been protected by registration. He dismissed the claim and made a declaration that the first defendant held the lease on trust for the second defendant, so that her interest was not subject to the legal charge. On the bank's appeal, the parties agreed that the bank was not bound by the unregistered, and uncompleted, contract, but that the bank would be bound by any interest the second defendant might have which was separate and distinct from the L
unregistered contract. The second defendant contended she had such an interest under a bare trust, a constructive trust and/or by virtue of proprietary estoppel.

***Held*: The appeal was allowed. (1) The second defendant had no interest under a bare trust for the trust had no existence except as the equitable consequence of the contract, and the contract was avoided by section 4(6) of the Land Charges Act 1925. (2) There is no room for the implication for any further trust, such as a constructive trust, where a bare trust already existed by virtue of the contract. (3) There is no room for the application of the** M
principles of proprietary estoppel when, at the time of the relevant expenditure, there was already a bare trust arising in consequence of an enforceable contract to the same effect as the interest sought pursuant to the proprietary estoppel. It would not be right to confer on the second defendant by means of proprietary estoppel binding on the bank that which the 1925 Act prevented her from obtaining by the contract the Act has declared void.

The following cases are referred to in this report.

Austin v *Keele* (1987) 61 ALJR 605
Bridges v *Mees* [1957] Ch 475; [1957] 3 WLR 215; [1957] 2 All ER 577
Inwards v *Baker* [1965] 2 QB 29

A *Ives (ER) Investment Ltd* v *High* [1967] 2 QB 379; [1967] 2 WLR 789; [1967] 1 All ER 504, CA

Lloyds Bank plc v *Rosset* [1991] 1 AC 107; [1990] 2 WLR 867; [1990] 1 All ER 1111; (1990) 60 P&CR 311; 22 HLR 349, HL

Shiloh Spinners Ltd v *Harding* [1973] AC 691; [1973] 2 WLR 28; [1973] 1 All ER 90; (1973) 25 P&CR 48, HL

Taylors Fashions Ltd v *Liverpool Victoria Trustees Co Ltd* [1982] QB 133; [1981] 2 WLR 576; [1981] 1 All ER 897; [1979] 2 EGLR 54; (1979) 251 EG 159

Western Fish Products Ltd v *Penwith District Council* [1981] 2 All ER 204; (1978) 77 LGR 185; 38 P&CR 7; [1978] JPL 623, CA

Willmott v *Barber* (1880) 15 ChD 96; 49 LJ Ch 792; 43 LT 95; 28 WR 911

B This was an appeal by the plaintiff, Lloyds Bank plc, against the decision of Mr Recorder Holmes in Cambridge County Court, who on July 5 1994 dismissed their claim for possession of premises against the first defendant, Michael Carrick, and the second defendant, Mrs Margaret Carrick, his sister-in-law.

Josephine Hayes (instructed by Taylor Vintners, of Cambridge) appeared for the appellant; Neil Vickery (instructed by Quirke & Co, of Croydon) represented the respondents.

Giving judgment at the invitation of Beldam LJ, **MORRITT LJ** said: This is an appeal of Lloyds Bank plc, to which I shall refer as the bank, from the order of Mr Recorder Holmes made in Cambridge County
C Court on July 5 1994. By that order he dismissed the bank's claim as mortgagee for possession of leasehold property known as 7 Derby Way, Newmarket, Suffolk, and made a declaration that the mortgagor, the first defendant Mr Carrick, held the lease of that property in trust for the second defendant, Mrs Carrick, so that her interests in and rights over that property were not subject to the bank's charge dated November 25 1986.

The facts are simple and may be shortly stated: 7 Derby Way, Newmarket, is a maisonette the title to which is unregistered. By a lease dated August 4 1971 it was demised for a term of 99 years from March 25 1971 in consideration of a premium and a relatively nominal
D rent. There were no restrictions on the lessee's ability to assign the term, which, on February 2 1982, was assigned to the first defendant Mr Carrick, who then lived in the maisonette.

Mr Jeffrey Carrick was the brother of Mr Carrick and the husband of the second defendant. Mr Jeffrey Carrick and Mrs Carrick, together with their two small boys, lived in Edinburgh Road, Newmarket. Mr Jeffrey Carrick died in March 1982. After his death, but before November 1982, there were discussions between Mr Carrick and Mrs Carrick as to where she should live. Those discussions and their aftermath were described by the recorder in the following terms:

> E the first defendant says that he told Mrs Carrick that, effectively, "If you like, you can come and live in this property [ie 7 Derby Way, Newmarket]". She was concerned whether she could afford to do so and whether she could sell her house and what would happen. Suffice it to say that Mr Michael Carrick, the first defendant, said to his sister-in-law: "Put the property on the market and what you get from the net proceeds of sale you can pay to this property, which will become yours.

Effectively, that is what happened. Mrs Carrick, the second defendant, put her property on the market. She realised a figure, which after deductions of repayment of mortgage (no doubt estate agent's commission and legal expenses and the usual outgoings that one incurs on a sale), amounted to around £19,000. As a result of completing, the
F £19,000 was paid over to the first defendant. The first defendant, being a building contractor, had a number of other properties at the material time and, in fact, went to live at another property, taking his then wife and child with him, leaving Mrs Carrick to come and live in the property. Mrs Carrick and her two children moved into the maisonette in about November 1982.

In the course of his judgment the recorder referred to the existence of a charge over the maisonette in favour of the bank existing at that time, but that seems to be an error for the original document is dated November 10 1983. At all events Mrs Carrick became aware of it and raised the matter with Mr Carrick on a number of occasions when he told her not to worry as he would sort it out. In the event it was redeemed on November 11 1986 some two weeks before the execution G
of the charge with which this appeal is concerned. At some time before it was redeemed works were carried out in the maisonette consisting of a kitchen extension, central heating and damp proofing. They were effected by Mr Carrick at a cost of about £5,000, but paid for by Mrs Carrick's father.

The legal charge on which the bank relies is dated November 25 1986 and was made between Mr Carrick and the bank to secure all moneys due on any account by the former to the latter. Mr Carrick as beneficial owner charged the lease of the maisonette as security for those moneys. The charge was preceded by a questionnaire signed by Mr Carrick to the effect that, to the best of his knowledge, there were H
no persons other than the mortgagor who would then or thereafter occupy the maisonette.

On January 16 1991 and again on August 9 1991 the bank demanded from Mr Carrick the substantial sums then due from him to the bank. They were not paid and the bank commenced these proceedings in Cambridge County Court by a summons issued on February 27 1992 seeking against Mr Carrick a judgment for £89,010.95 and interest thereon accruing at the rate of £53.18 per day and an order for possession of the maisonette.

By his answer dated March 8 1992 Mr Carrick admitted the money claim but, in respect of the possession claim, he alleged that he was J
and had been since a time before the execution of the legal charge relied on by the bank a bare trustee of the maisonette for his sister-in-law. Accordingly, Mrs Carrick was joined as the second defendant and ordered to serve a defence. By her defence and counterclaim served on April 27 1992 Mrs Carrick claimed that Mr Carrick had been a bare trustee for her of the long leasehold interest in the maisonette. The material paragraphs are paras 5, 6, 9, 10 and 11 which are in the following terms:

> 5. Pursuant to a suggestion made by the First Defendant to the Second Defendant, the Second Defendant in or about October 1982 sold her former matrimonial home and took up occupation at 7 Derby Way, Newmarket which K
> she has occupied at all material times since that date.
>
> 6. On the direction of the Second Defendant the Solicitors who acted on the sale of the said matrimonial home paid the resulting proceeds of around £19,000 to the order of the First Defendant. The said sum was for the acquisition of 7 Derby Way, Newmarket by the Second Defendant
>
> ...
>
> 9. After the said acquisition of 7 Derby Way Newmarket the second defendant was responsible for and paid all outgoings relating to 7 Derby Way.
>
> 10. The second defendant in or about 1983 had an extension built to 7 Derby Way at a cost of about £5,000 which the second defendant paid.
>
> 11. It was the common intention of both the first and second defendants L
> that 7 Derby Way, Newmarket should become the property of the second defendant free of all incumbrances.

The bank replied to the effect that the interest of Mrs Carrick was registrable as a land charge of class C(iv), namely an estate contract, and was void against the bank for want of registration. The bank sought particulars of the defence so as to require Mrs Carrick to spell out the origin of the interest she claimed. The particulars served by Mrs Carrick on September 17 1993 contended:

> 1. that no contract of sale was concluded between the First and Second Defendant in respect of the First Defendant's interest in 7 Derby Way; M
>
> 2. that upon the Second Defendant entering into occupation of 7 Derby Way and paying to the First Defendant the sum of about £19,000 as pleaded in paragraph 6 of the Defence and Counterclaim, the First Defendant held his interest in 7 Derby Way upon bare trust for the Second Defendant;
>
> 3. that further and alternatively, the Second Defendant paid the said sum and entered into occupation of 7 Derby Way and acted as pleaded in paragraphs 9 and 10 of the Defence and Counterclaim, with the intention, which was a common intention of the First and Second Defendants, that 7 Derby Way was to be the Second Defendant's permanent home and that she was to be absolutely entitled to the beneficial interest in it;

Yet further particulars were sought from Mrs Carrick and given by her on December 20 1993. These included the following passage:

It was agreed that the Second Defendant would buy 7 Derby Way for whatever she received upon the sale of 93 Edinburgh Road (in the event about £19,000). The First Defendant told the Second Defendant not to worry about the matter further and that he would sort things out. The Second Defendant thereafter left it up to the First Defendant to sort out the legal formalities of the sale. The Second Defendant moved in to 7 Derby Way on 19th November 1982. She paid the First Defendant the proceeds of sale from 93 Edinburgh Road on about that date. Since then the Second Defendant has lived at 7 Derby Way and paid all the outgoings on it. The agreement between the Defendants is not recorded in writing.

The action was heard by the recorder on July 4 1994. He heard evidence from two bank officials and from Mr Carrick and Mrs Carrick. He gave judgment on July 5 1994. The recorder set out the history of the case, the state of the pleadings and the documents of title relating to the maisonette. He expressed the view, at p12 of the transcript of his judgment, that the case turned on the law so far as it relates to registrable interests and later, at p13 of the transcript, whether there was a contract between Mr Carrick and Mrs Carrick. In respect of these issues his conclusions set out on p15 of the transcript were:

In my view, there was a contract that upon payment of the £19,000 by the second defendant there arose, in my view, a bare trust. In other words, there was nothing left vested in Mr Carrick (the first defendant) other than the legal title. Mr Carrick had no right to the property. He simply held the legal title upon trust. Mrs Carrick could have called for the legal title to be conveyed to her by the first defendant as bare trustee. If, as I do, I follow that through the interest behind the bare trust is not registrable.

On the basis that there was no registrable interest the recorder then went on to consider whether the bank had notice of the interest of Mrs Carrick so that the legal charge took effect subject to it. He concluded on all the evidence that the bank was not entitled to rely simply on the answer to the questionnaire for one or more of its officers knew that Mrs Carrick was resident in the maisonette and had constructive notice of her beneficial interest for any inquiry of her would have revealed it. The recorder declined to deal with a submission based on proprietary estoppel as, in his view, it had not been pleaded and did not arise.

The bank appeals with the leave of Neill LJ and Cazalet J granted on March 31 1995. It contends that the recorder should have found that the only interest of Mrs Carrick in the maisonette was an estate contract within Land Charges Act 1972 and accordingly was void for want of registration as against the bank as the purchaser of a legal estate for valuable consideration.

By a respondent's notice served on May 1 1995 Mrs Carrick contends that the decision of the recorder should be upheld on the additional ground that her expenditure on the improvements to the maisonette prior to the legal charge was incurred in the belief encouraged by Mr Carrick that she was the owner of the property and thus gave rise to an equity in her favour. She submits that the bank had notice of it and that such equity should be satisfied by the affirmation of the recorder's order. It was accepted by counsel on her behalf that she could not go outside the facts pleaded in her defence of which I have quoted the relevant paragraphs. The other points referred to in the respondent's notice were abandoned at the hearing.

It is convenient to consider first the relevant statutory provisions. Land Charges Act 1972 replaced Land Charges Act 1925. So far as relevant it provides:

The register of land charges

2. — (1) If a charge on or obligation affecting land falls into one of the classes described in this section, it may be registered in the register of land charges as a land charge of that class.

. . .

(4) A Class C land charge is any of the following [not being a local land charge], namely —

. . .

(iv) an estate contract;

and for this purpose —

(iv) an estate contract is a contract by an estate owner or by a person entitled at the date of the contract to have a legal estate conveyed to him to convey or create a legal estate, including a contract conferring either expressly or by statutory implication a valid option to purchase, a right of pre-emption or any other like right.

. . .

4. — (6) An estate contract and a land charge of Class D created or entered into on or after 1st January 1926 shall be void as against a purchaser for money or money's worth [or, in the case of an Inland Revenue charge, a purchaser within the meaning of the Capital Transfer Tax Act 1984] of a legal estate in the land charged with it, unless the land charge is registered in the appropriate register before the completion of the purchase.

Section 17 incorporates the definitions contained in the Law of Property Act 1925 of legal estate and purchaser, which, it is common ground, include a charge by way of legal mortgage and a mortgagee such as the bank.

So far as relevant section 199 of the Law of Property Act 1925 provides:

Restrictions on constructive notice

199. — (1) A purchaser shall not be prejudicially affected by notice of —

(i) any instrument or matter capable of registration under the provisions of the Land Charges Act 1925, or any enactment which it replaces, which is void or not enforceable as against him under that Act or enactment, by reason of the non-registration thereof;

In the light of these provisions and, no doubt, the consideration by each party of the written argument of the other the issues between them were narrowed considerably. First, it is accepted by Mrs Carrick that if her only interest in the maisonette was derived from the contract, which she accepts is void as against the bank as an unregistered estate contract, then the appeal succeeds. Second, Mrs Carrick accepts that the original contract between her and Mr Carrick, as found by the recorder, was a valid open contract for the purchase of the maisonette; that it became enforceable by her when she partly performed it by entering into possession and paying the whole of the purchase price, but that it remained executory, that is to say uncompleted, at the time of the legal charge to the bank granted in November 1986. Third, the bank accepts that if Mrs Carrick had an interest in the maisonette not arising but separate and distinct from the unregistered contract it was and is binding on the bank for, as found by the recorder, the bank had notice of it.

Thus the issue argued on this appeal was whether Mrs Carrick had an interest in the maisonette separate and distinct from that which arose under the unregistered estate contract which was capable of binding the bank as successor in title to Mr Carrick. For Mrs Carrick it was submitted that she did. It was contended that she was entitled to such an interest under a bare trust, a constructive trust and by virtue of a proprietory estoppel.

I shall consider each of these points in due course. But before doing so it is necessary to consider the position of Mr Carrick and Mrs Carrick before the charge to the bank was executed. At the time it was made the contract was valid but, as provided by section 40 Law of Property Act 1925, unenforceable for want of a memorandum in writing or part performance. It became enforceable when in or about November 1982 Mrs Carrick paid the purchase price to Mr Carrick and went into possession. One consequence of the contract becoming enforceable was that it was specifically enforceable at the suit of Mrs Carrick. Accordingly, Mr Carrick became a trustee of the maisonette for Mrs Carrick. Normally such trusteeship is of a peculiar kind because the vendor himself has a beneficial interest in the property as explained in *Megarry & Wade on The Law of Real Property* 5th ed, p602. But in this case as Mrs Carrick had paid the whole of the purchase price at the time the contract became enforceable Mr Carrick as the vendor had no beneficial interest. Thus he may properly be described as a bare trustee *cf Bridges* v *Mees* [1957] Ch 475, at p485. It follows that at all times after November 1982 Mrs Carrick was the absolute beneficial owner of the maisonette and Mr Carrick was a trustee of it without any beneficial interest in it.

The argument for Mrs Carrick relied on the relative position at law and in equity as I have described it to found the argument that such an absolute equitable interest was not itself registrable but bound the bank as they had constructive notice of it. Counsel for Mrs Carrick accepted that such interest came or started from the contract but, he contended, it matured into an interest separate and distinct from the contract as soon as the purchase price was paid in full.

For my part, I am unable to accept this analysis. The payment of £19,000 by Mrs Carrick to Mr Carrick did not as such and without more give her any interest in the maisonette. Nor, prior to the conclusion of the contract, were the circumstances such that Mrs Carrick could assert that her brother-in-law held the maisonette on any trust for her benefit. The source and origin of the trust was the contract; the payment of the price by Mrs Carrick served only to make it a bare trust by removing any beneficial interest of Mr Carrick. Section 4(6) Land Charges Act 1972 avoids that contract as against the bank. The result, in my judgment, must be that Mrs Carrick is unable to establish the bare trust as against the bank for it has no existence except as the equitable consequence of the contract. Accordingly, I reject the contention founded on the bare trust.

The second contention for Mrs Carrick was to the effect that she was entitled to the whole beneficial interest in the maisonette arising under a constructive trust and that that interest was not registrable so that the bank having had constructive notice of it took subject to it. For this proposition her counsel relied on the speech of Lord Bridge of Harwich in *Lloyds Bank plc* v *Rosset* [1991] 1 AC 107. That case was concerned with the question of what must be established to entitle a wife to an equitable interest in registered land the title to which is registered in the sole name of her husband. At p132E Lord Bridge of Harwich said:

> The first and fundamental question which must always be resolved is whether, independently of any inference to be drawn from the conduct of the parties in the course of sharing the house as their home and managing their joint affairs, there has at any time prior to acquisition, or exceptionally at some later date, been any agreement, arrangement or understanding reached between them that the property is to be shared beneficially. The finding of an agreement or arrangement to share in this sense can only, I think, be based on evidence of express discussions between the partners, however imperfectly remembered and however imprecise their terms may have been. Once a finding to this effect is made it will only be necessary for the partner asserting a claim to a beneficial interest against the partner entitled to the legal estate to show that he or she has acted to his or her detriment or significantly altered his or her position in reliance on the agreement in order to give rise to a constructive trust or a proprietary estoppel.

Counsel recognised that in this case the contract between Mr Carrick and Mrs Carrick was entered into after Mr Carrick had taken an assignment of the lease into his own name. But he submitted that the same principle applied and for that purpose relied on the statement of Lord Oliver of Aylmerton in giving the opinion of the Privy Council in *Austin* v *Keele* (1987) 61 ALJR 605, at p609, where he said

> Although Lord Diplock [*Gissing* v *Gissing* [1971] AC 886 at p905] referred to the formation of a common intention "at the time of acquisition", the Court of Appeal expressed the view, with which their Lordships agree, that although it may be more difficult to prove the requisite intention in relation to property already held beneficially by the trustee, there is no reason in principle why the doctrine should be limited to an intention formed at the time of the first acquisition of the property — an opinion echoed by Mustill LJ in his judgment in *Grant* v *Edwards* [1986] Ch 638 at 651. In essence the doctrine is an application of proprietary estoppel and there is no reason in principle why it should be confined to the single event of acquisition of the property by the owner of the legal estate.

Counsel for Mrs Carrick submitted that if there had been no contract then on the proper application of these principles there would have been a constructive trust in favour of Mrs Carrick. From this he argued that Mrs Carrick should not be in any worse position just because there was a contract.

In this case there was a trust of the maisonette for the benefit of Mrs Carrick precisely because there had been an agreement between her and Mr Carrick which, for her part, she had substantially if not wholly performed. As between her and Mr Carrick such trust subsisted at all times after November 1982. I agree with counsel for the bank that there is no room in those circumstances for the implication or imposition of any further trust of the maisonette for the benefit of Mrs Carrick In *Lloyds Bank plc* v *Rosset* there was no contract which conferred any interest in the house on the wife. As with all such statements of principle the speech of Lord Bridge of Harwich must be read by reference to the facts of the case. So read there is nothing in it to suggest that where there is a specifically enforceable contract the court is entitled to superimpose a further constructive trust on the vendor in favour of the purchaser over that which already exists in consequence of the contractual relationship.

It is true that on this footing the ultimate position of Mrs Carrick with the benefit of a specifically enforceable contract may be worse than it would have been if there had been no contract. But that is because she failed to do that which parliament has ordained must be done if her interest is to prevail over that of the bank, namely register the estate contract. Her failure in that respect cannot, in my view, justify the implication or imposition of a trust after the execution of the charge when the dealings between Mr Carrick and Mrs Carrick before such execution did not. For these reasons I would reject the second point on which Mrs Carrick relied.

The third contention was that Mrs Carrick is entitled to the benefit of a proprietary estoppel. Counsel on her behalf submitted by reference to the principles set out in *Snell's Equity* 29th ed pp 574 to 576, that such an estoppel arose in her favour by virtue of the facts pleaded in her defence. He submitted that Mrs Carrick had paid the purchase price and carried out the improvements to the maisonette in the belief common to both her and Mr Carrick and to that extent encouraged by him that she either did or would own it. Reliance was placed on the decisions of this court in *Inwards* v *Baker* [1965] 2 QB 29 and *Ives (ER) Investment Ltd* v *High* [1967] 2 QB 379 as establishing that such an estoppel gives rise to an interest in land capable of binding a successor in title with notice.

This was disputed by counsel for the bank. She submitted that such principles could not be applied to cases in which there was no belief or expectation of having or acquiring an interest in someone else's land. In this context she relied on *Western Fish Products Ltd* v *Penwith District Council* [1981] 2 All ER 204. Further she submitted that the facts did not warrant such an estoppel as they did not cover all the elements referred to as "probanda" in *Willmott* v *Barber* (1880) 15 ChD 96 and were otherwise insufficient. In addition she submitted that such an estoppel cannot give rise to an interest in land capable of binding successors in title with notice.

I would observe at the outset that it is a matter of some doubt whether the principles of proprietary estoppel differ from those of that species of constructive trust which was referred to by Lord Bridge of Harwich in *Lloyds Bank* v *Rosset*. In the passage from his speech which I have already quoted he treated the two labels as interchangeable. To the like effect is the passage in the advice of the Privy Council in *Austin* v *Keele* given by Lord Oliver of Aylmerton which I have also quoted. However that may be the case under this head was put somewhat differently and should be considered on its own merits.

With regard to the second submission of counsel for the bank I think that it is now clear that to constitute the requisite estoppel it is not necessary to establish all of the five elements or "probanda" referred to by Fry J in *Willmott* v *Barber* (1880) 15 ChD 96 at pp 105-106. In his judgment in *Taylors Fashions Ltd* v *Liverpool Victoria Trustees Co Ltd* [1982] QB 133* Oliver J traced through the subsequent cases in which this question had been considered. I do not propose to repeat the process but would respectfully agree with the conclusion of Oliver J that proof of all those elements or "probanda" is not necessary to found an estoppel. For my part, I agree with the proposition stated by Oliver J at p151-152 that:

> . . . the more recent cases indicate, in my judgment, that the application of the *Ramsden* v *Dyson*, LR 1 HL 129 principle — whether you call it proprietary estoppel, estoppel by acquiescence or estoppel by encouragement is really

---

*Editor's note: Also reported at [1979] 2 EGLR 54

> A immaterial — requires a very much broader approach which is directed rather at ascertaining whether, in particular individual circumstances, it would be unconscionable for a party to be permitted to deny that which, knowingly or unknowingly, he has allowed or encouraged another to assume to his detriment than to inquiring whether the circumstances can be fitted within the confines of some preconceived formula serving as a universal yardstick for every form of unconscionable behaviour.
>
> So regarded, knowledge of the true position by the party alleged to be estopped, becomes merely one of the relevant factors it may even be a determining factor in certain cases — in the overall inquiry.

In *Western Fish Products Ltd* v *Penwith District Council* [1981] 2
B All ER 204 the plaintiff carried out works on its own land in reliance on statements made by an officer of the local planning authority that permission would be granted for its development. Planning permission was in due course refused and enforcement notices were served. The plaintiff then instituted proceedings for a declaration that it was entitled to the permissions the officer had represented that it would obtain in reliance on which the paintiff had carried out the works on its own land. The claim failed on a number of grounds. In relation to the claim to a proprietory estoppel Megaw LJ giving the judgment of the court said at p218h

> We know of no case, and none has been cited to us, in which the principle
> C set out in *Ramsden* v *Dyson* and *Crabb* v *Arun District Council* has been applied otherwise than to rights and interests created in and over land. It may extend to other forms of property: see per Lord Denning MR in *Moorgate Mercantile Co Ltd* v *Twichings* [1975] 3 All ER 314 at 323-324, [1976] QB 225 at 242. In our judgment there is no good reason for extending the principle further. As Harman LJ pointed out in *Campbell Discount Co Ltd* v *Bridge* [1961] 2 All ER 97 at 103, [1961] 1 QB 445 at 459, the system of equity has become a very precise one. The creation of new rights and remedies is a matter for Parliament, not the judges.
>
> In his reply counsel for the plaintiffs seemed to recognise that the reported cases did put limits to the application of the so-called concept of proprietary estoppel. He submitted that the plaintiffs' case was within that concept because
> D what the defendant council, by their officers, had represented had, to their knowledge, caused the plaintiffs to spend money on or in connection with their own land which they would not otherwise have spent. On their own case they have spent money in order to take advantage of existing rights over their own land which the defendant council by their officers had confirmed they possessed. There was no question of their acquiring any rights in relation to any other person's land, which is what proprietary estoppel is concerned with.

In my judgment, the claim of Mrs Carrick fails on a number of grounds. First, as in the case of the constructive trust, I do not see how there is any room for the application of the principles of proprietary estoppel when at the time of the relevant expenditure there was already
E a bare trust arising in consequence of an enforceable contract to the same effect as the interest sought pursuant to the proprietary estoppel. As the evidence showed Mrs Carrick knew of the need for a conveyance and was content that it should be deferred. Thus at the time that she paid the price and committed herself to the expenditure on the subsequent improvements she believed, rightly, that she was spending the money in respect of her own property albeit under an uncompleted contract. In this respect I see no relevant distinction between this case and that of *Western Fish Products Ltd* v *Penwith District Council.*

Second, this is not a case in which the expectations of Mrs Carrick
F have been defeated by Mr Carrick seeking to resile from the position he had encouraged her to expect. As far as he is concerned he has always accepted that she had contracted to buy the maisonette and had paid the price in full. As against him the contract is still binding and enforceable, although as he is unable to redeem the mortgage he is in breach of contract for having charged the maisonette and in breach of trust for failing to account to Mrs Carrick for the money raised on the security of the maisonette. Mrs Carrick's expectations have been defeated because the contract was not registered at any time before the charge was granted and parliament has decreed that in those circumstances the contract is void against the bank.

Third, it was common ground that the right arising from a proprietary estoppel cannot exceed that which the party sought to be estopped encouraged the other to believe that she had or would G
acquire. The party sought to be estopped is Mr Carrick. In so far as he encouraged Mrs Carrick to believe that she was or would become the beneficial owner of the maisonette there is no further right to be obtained for she was, and, subject to the charge, still is. But counsel for Mrs Carrick submits that Mr Carrick went further and encouraged her in the belief that she was or would become the legal owner of the maisonette. Apart from the facts that this was never alleged in the defence of Mrs Carrick nor explored in evidence at the trial I do not think that it could avail Mrs Carrick. Section 4(6) of the Land Charges Act 1972 invalidates, as against the bank, any unregistered contract by the estate owner for the conveyance of the legal estate. It cannot be H
unconscionable for the bank to rely on the non-registration of the contract. I do not see how it could be right to confer on Mrs Carrick indirectly and by means of a proprietary estoppel binding on the bank that which parliament prevented her from obtaining directly by the contract it has declared to be void. To avoid any future misunderstanding I would emphasise that there was and is a valid and enforceable contract as against the vendor. Accordingly this case is quite unlike those which may become more prevalent where there is no contract at all not because there was no agreement but because the agreement was not in writing as now required by section 2 of the Law of Property (Miscellaneous Provisions) Act 1989. J

In my judgment, the claim based on proprietary estoppel fails. In the circumstances it is unnecessary to consider further the submission of counsel for the bank to the effect that a proprietary estoppel cannot give rise to an interest in land capable of binding successors in title. This interesting argument will have to await another day, though it is hard to see how in this court it can surmount the hurdle constituted by the decision of this court in *Ives (ER) Investment Ltd* v *High* [1967] 2 QB 379.

For all these reasons I consider that the recorder was wrong to have held that Mrs Carrick had any interest valid against the bank sufficient to constitute a defence to the claim against her for possession of the maisonette. I would allow this appeal. K

This result seems to me to be inevitable in the light of the provisions of the Land Charges Act 1972 and of the Law of Property Act 1925 I quoted earlier. However it should be noted that the result would have been different if the title to the maisonette had been registered. In such a case the interest of Mrs Carrick, who was in possession of the maisonette and of whom no inquiry had been made, would have been an overriding interest under section 70(1)(*g*) of the Land Registration Act 1925. As such it would have been binding on the bank.

As the authors of *Megarry & Wade on The Law of Real Property* 5th ed point out at pp186-187 the same position would have been achieved under the Law of Property Act 1922 for what is now section L
14 of the Law of Property Act 1925 was then in a part which also contained the legislation which subsequently became the Land Charges Act 1925.

In my view, it is beyond doubt that section 14 Law of Property Act 1925 does not achieve for unregistered land that which section 70(1)(*g*) of the Land Registration Act 1925 achieves for registered land but whether that was originally intended or is a quirk of the process of breaking up the Law of Property Act 1922 into, among others, the Law of Property Act 1925 and Land Charges Act 1925 is unclear. What is certain is that it must be for others to consider and for parliament to decide whether this distinction between registered and unregistered M
land should continue, particularly as the system for the registration of incumbrances in the case of unregistered land is by no means complete as shown by *Inwards* v *Baker* [1965] 2 QB 29; *Ives (ER) Investment Ltd* v *High* [1967] 2 QB 379 and *Shiloh Spinners Ltd* v *Harding* [1973] AC 691.

Sir Ralph Gibson agreed and did not add anything.

Also agreeing, **Beldam LJ** said: For the reasons given by Morritt LJ, I agree that this appeal should be allowed.

The order which we shall make is that the appellant bank have possession of the property; the order of the recorder be set aside and

A the counterclaim of the respondent and declarations made are dismissed. In so far as the recorder made an order for costs in favour of the respondent that order is set aside; the appellants may have their costs before the recorder in the county court, such order for costs not to be enforced without the leave of the court. So far as the costs of the appeal are concerned the appellants are entitled to their costs of and occasioned by the appeal, although we would hope very much that B
before any such order was enforced against Mrs Carrick consideration would be given to it at the highest level in the bank; and we make an order for possession on May 28 1996. Leave to appeal is refused.

Appeal allowed.

LANDS TRIBUNAL
COMPENSATION

Barrett *v* Mole Valley District Council

May 8 1996

JUDGE RICH QC, sitting as a member of the Lands Tribunal

Estates Gazette February 22 1997

[1997] 08 EG 144

Compensation for deemed revocation of planning permission — Reference to Lands Tribunal compromised — Whether claimant entitled to costs

The claimant had made a reference to the tribunal for the determination of the compensation payable under sections 107 and 108 of the Town and Country Planning Act 1990 following the deemed revocation of planning permission to use the claimant's land for 28 days in each year for clay pigeon shooting. The claimant's valuations ranged between £100,000 and £221,000; the compensating authority contended the compensation should be nil. The parties compromised the reference by the claimant accepting the sum of £30,000 as compensation, with the costs to be determined by the tribunal. In relation to the question of costs the compensating authority contended that the usual order that the authority should pay the costs should not apply because: (1) the claimant had failed to provide details of valuation calculations until his expert's proof was filed some three years after the claim, contrary to section 4 of the Land Compensation Act 1961; (2) there had been a three-year delay in making discovery of accounts; (3) there had been protracted delays relating to a sound test; (4) there had been an order providing for a preliminary point of law or a split trial, which was not proceeded with; and (5) the claimant made a late change in the nature of his case.

Decision: **The claimant was entitled to his costs of the reference and of the hearing of the application for costs. Section 4 of the Land Compensation Act 1961 had no direct application. In any event although the details were eventually provided in 1993, no offer of compensation was made by the compensating authority until 1996. The claimant was not entitled to any increase in costs attributable to the delay in providing details between 1990 and 1993. In respect of the other matters, the existing orders as to costs should not be disturbed. There was no reason why the usual order for costs should not be made, namely that the claimant is entitled to his costs. It was the responsibility of the compensating authority to offer compensation at the proper time.**

No cases are referred to in this report.

Robin Purchas QC (instructed by Gersten & Nixon) represented the claimant; Keith Lindblom QC (instructed by the solicitor to Mole Valley District Council) represented the compensating authority.

Giving his decision, **JUDGE RICH QC** said: This is a claim for compensation under sections 107 and 108 of the Town and Country Planning Act 1990 for the loss or damage directly attributable to the deemed revocation of the planning permission to use the claimant's land for 28 days in each year for the purpose of clay pigeon shooting.

The parties have sensibly compromised the proceedings by, on the compensating authority's part offering and on the claimant's part accepting, the sum of £30,000 as compensation under the statute. That figure compares with a valuation on the part of the compensating authority of nothing and on the part of the claimant variously ranging from well over £100,000 up to £221,000 according to the basis upon which the valuation is made. Had such an award been made after hearing unless there were special circumstances for an order in a different form, the costs would have followed the event notwithstanding the fact that the claimant has recovered only a small proportion of his total claim, and the claimant would have had his costs of the claim.

Mr Keith Lindblom QC has sought to put before me special reasons why that usual form of order should not be followed in this case. Of the five reasons that he has adduced only one has seemed to me to be a proper basis upon which I should vary the usual form of order. That reason is that he complains that the claimant, having made his claim in the sum of £221,454 on December 20 1989 and then given particulars of the heads under which he arrived at that figure on October 24 1990, failed to provide calculations as to how that series of figures amounting to £221,000 odd had been calculated. The figures which were thus given were given immediately before the reference to the Lands Tribunal in December 1990 but particulars of the calculation were provided only on January 18 1993 when the claimant's valuer's report was filed. Mr Lindblom draws my attention to section 4 of the Land Compensation Act 1961 which provides that where the Lands Tribunal is satisfied that the claimant has failed to deliver to the acquiring authority in time to enable them to make a proper offer, notice in writing of the amount claimed by him containing the particulars mentioned in subsection 2, the Lands Tribunal shall, unless for special reason it thinks proper not to do so, order the claimant to bear his own costs and to pay the costs of the acquiring authority so far as they were incurred after whatever was the time when in the opinion of the Lands Tribunal particulars should have been delivered. Subsection 2 requires the particulars to consist of not only the amount of compensation claimed distinguishing the amounts under separate heads, but also showing how the amount claimed under each head is calculated.

The terms of section 4 do not, as I think, apply directly to this case for two reasons. First, I am doubtful that it applies in terms to the assessment of compensation under the particular provisions to which I have drawn attention, but even if it does the situation with which I am dealing is not one of inability of the compensating authority to make an offer by reason of the withholding of this information. The information was provided, as I have indicated on January 18 1993, but no offer was forthcoming until 1996. Nevertheless, it does seem to me to be proper conduct in a claimant particularly when specifically asked, and indeed repeatedly requested, to provide information which would satisfy the requirements of section 4, to provide such information. In so far as the claimant has increased his own costs by his failure to do so, and that must be the case in so far as he has spent time replying to letters explaining why he was not doing what was asked and promising but failing then to provide the information required, he, the claimant, should not be entitled to recover such additional costs. In the end this failure led to the compensating authority seeking to strike out the reference, but on February 1 1993 such application to strike out was, as I understand it, treated as an application for directions in the case, and in the course of making directions the registrar provided that the costs of the application were to be costs in the reference. That order of the registrar in regard to the costs of that application on February 1 1993 should not be interfered with. Subject however to that, the costs which I otherwise award to the claimant should be limited to exclude any respect in which those costs have been increased by reason of the failure of the claimant to provide the calculations requested between October 24 1990 and January 18 1993. I do not, for reasons which I have sought to indicate as to the direct applicability of section 4, regard this as a case where I should make an order in respect of that period or that failure in favour of the compensating authority.

The other matters which Mr Lindblom has raised are first, a complaint that accounts were sought over a long period of time and only in August of 1995 was discovery finally completed after the claimant ultimately told the compensating authority in July 1995 that there were no such accounts as the compensating authority had been vigorously pressing for over a period of some three years before that. I accept that the claimant's conduct in this respect is open to criticism, nevertheless I do not think that it requires any special order as to costs to protect the compensating authority. The costs which can be claimed on behalf of the claimant are limited to those that are properly and reasonably incurred. In so far as he has incurred costs unreasonably or improperly in failing to disclose as early as he could have done the extent of the material that he had available he will not be able to recover them. In so far as the compensating authority may have incurred costs which they might have avoided, had discovery been given earlier, the compensating authority made their own election in failing to make an application for discovery until May 1995, and then accepting on the order that was made no order as to costs. It is not now an appropriate moment, it appears to me, to seek costs in respect of that application.

Criticism has also been made by Mr Lindblom that the claimant, or those advising him, protracted negotiations for agreeing terms before carrying out a noise test which it had been ordered on February 1 1993 should be agreed and carried out. Those negotiations had for the most part preceded the making of that order. If the compensating authority wished to claim that the claimant had behaved unreasonably and wished therefore to be protected as to costs, that was the moment at which to draw the attention of the tribunal to that claim and to seek an appropriate order as to costs. I do not think that the undoubted changes of mind, I think on both sides, as to the terms of testing, is a reason now, for doing otherwise than making the usual order in favour of the claimant.

It is also right that there was a period from before that order of February 1 1993 and after it, when it was contemplated on both sides that there should be either a preliminary point of law or a split trial in order to separate the question whether or not the claimant was debarred from receiving compensation by the provisions of r 4 of section 5 of the 1961 Act, which provides that where the value of the land is increased by use thereof in a manner which could be restrained by any court or is contrary to law or is detrimental to the health of the occupants of the premises or to the public health, the amount of the increase shall not be taken into account. Whether or not it might have been convenient if it had been necessary to try out this claim to have had a split trial of some kind is not now, in my judgment, to the point.

The order of February 1 1993 provided that an application for a hearing on a preliminary point of law or a split trial was to be made after the joint statement of agreed facts with regards to the amount of noise being measured on the tests to which I have made reference already, had been filed. On that application the costs as I have indicated were decided to be costs in the reference. Once again I would not think it appropriate now to interfere with the effect of that order in regards to the costs up to February 1. In regards to the period thereafter it is right to say that the claimant who had previously thought that it would be convenient to have a split trial, after attempting to agree the terms of the split trial came to the conclusion that it would not result in a saving of costs or greater convenience in the trial. Since what was provided for in the order was an application at a particular time, it was open to the compensating authority, if they thought differently, so to apply, and if the application had been resisted by the claimant but had none the less been successful, the compensating authority could then have sought the costs of making such application. The compensating authority did not take those steps and in those circumstances in so far as costs were properly and reasonably incurred in a discussion of whether or not there should be a split trial those should be costs in the reference and the claimant should be entitled to receive them, in so far as reasonably and properly incurred.

Finally, the complaint is made that the claimant very late in the day proposed to change his case in respect of the extent to which clay pigeon shooting could be carried on on the site without contravention of r 4 of section 5 of the 1961 Act. The agreed test indicated how, with the particular precautions that were agreed for the purposes of the test, the adjoining residential properties were adversely affected by carrying out the shooting that was agreed to be tested on that occasion. That no doubt would have made a good foundation on which to draw conclusions as to whether or not shooting could profitably have been carried on on the site without causing nuisance to the neighbourhood. Whether or not the particular shooting that was tested on that particular occasion in the conditions that were then agreed did or did not cause a nuisance would not be determinative of the true question which would have fallen for decision in the course of the reference had I had to proceed to hear it. No doubt the compensating authority would have been able to object to the introduction of evidence by the claimant's expert not included in his report which could not be reasonably considered without an adjournment or without new testing. Had such evidence been introduced at the hearing such objection would have had to be considered and if upheld its admissibility on terms would have then had to be considered and it may be that if additional costs were incurred the compensating authority would have been entitled to an order in their favour in respect of them. All that is speculative. It does not, in my judgment, go to the question whether or not the usual order as to costs, should be varied in the particular circumstances of the acceptance of an agreed sum in compensation in the manner to which I have already referred. In those circumstances I do not regard any of the other matters as matters which would affect the way in which I would on this occasion exercise my discretion with regards to costs and the order that I would make as to costs is the order which I have already indicated.

After further submission as to the costs of the current hearing

I take the view that it is the responsibility of the compensating authority to offer proper compensation at the proper time. They decided that they would offer compensation which proved acceptable only upon terms that the costs should be determined by this tribunal. They were determined by this tribunal in favour of the claimant. The attempt after the initial offer to negotiate in regards to what those costs should be, placed a totally unfair strain upon the claimant's ability to deal with the matter fairly, and I see no reason why the costs of the argument today should not be included in the costs of the reference. It is ordered that such costs if not agreed shall be taxed by the registrar of the Lands Tribunal on the High Court Scale.

A

Crown House Developments Ltd *v* Chester City Council

July 13 1996

MR JC HILL FRICS

Estates Gazette March 1 1997

B

[1997] 09 EG 155

Compensation for compulsory purchase — Valuation — Effect of "no-scheme world" — Ransom strip — Effect of alternative means of access

In July 1990 the acquiring authority made a general vesting declaration in respect of the reference land, an area of 190m² forming part of a narrow and unmade cul-de-sac known as Leech Lane, stating that the land was required in order that 1.42 acres of land ("the red land") belonging to the acquiring authority could be developed. The red land had been acquired in June 1987. Prior to its purchase, the red land and an C adjoining area of land ("the yellow land"), which extended through to a public highway, were in the same ownership. The red and yellow lands were allocated in the local plan for housing purposes, the red land having a development value of £345,000 assuming access was available. The cost of constructing an access to the red land over the reference land was £3,500, whereas the costs of constructing an access over the yellow land was £13,500, and over two other alternative routes, £30,000 and £14,800 respectively. The claimants sought compensation of £176,000 on the basis of ransom value. The acquiring authority's valuer spoke to £10,000 on the basis that in the "no-scheme world" the red and yellow lands would have been developed as one entity, and there was access to a public highway D through the yellow land.

Decision: **The value of the reference land was £158,125. The tribunal was not convinced that the *Pointe Gourde* principle can be deployed to disregard the actual extent and the boundaries of the red land bought by the acquiring authority. In the "no-scheme world" a purchaser would have bid for an area of land similar, if not identical to, the red land for development purposes. There had been a consistent planning requirement that access to the red land for development should come through the reference land, and strong opposition to an access between the public highway and the yellow land. Development would have been most likely in the area E of the red land. Although a developer of the red land would probably accept that access through the reference land was the most suitable, he would have investigated other means. The alternative accesses advanced by the acquiring authority all had disadvantages and practicable problems; 30% of the development value of the red land was therefore appropriate to attribute to an access.**

The following cases are referred to in this report.

Batchelor v *Kent County Council* (1989) 59 P&CR 357; [1990] 1 EGLR 32; [1990] 14 EG 129; RVR 181; [1990] JPL 571, CA; [1992] 1 EGLR 217; F [1992] 03 EG 127; [1992] 04 EG 138; (1991) 31 RVR 239, LT

Challinor v *Stone Rural District Council* (1972) 27 P&CR 244, LT

Margate Corporation v *Devotwill Investments Ltd* [1970] 3 All ER 864; (1970) 22 P&CR 328; 69 LGR 271; [1971] EGD 313; 218 EG 559, HL

Pointe Gourde Quarrying & Transport Co Ltd v *Sub-Intendent of Crown Lands* [1947] AC 565, PC

Stokes v *Cambridge Corporation* (1961) 13 P&CR 77; [1961] EGD 207; 180 EG 839, LT

Wards Construction (Medway) Ltd v *Barclays Bank plc* (1994) 68 P&CR 391, CA

Nicholas Nardecchia (instructed by Harrison Clark) appeared for the claimants; Nicholas Burton (instructed by the solicitor to Chester City Council) represented the acquiring authority.

Giving his decision, MR JC HILL said: This is a reference to G determine the compensation payable to the claimants following the acquisition by Chester City Council of a small area of land in Malpas, Cheshire.

Nicholas Burton of counsel appeared for the compensating authority and called Mr PJ Hall ASVS, Mr DF Evans Dip TP MRTPI and Mr DJ Roberts. Nicholas Nardecchia of counsel appeared for the claimant company and called Mr Michael Guise BSc FRICS IRRV and Mr Graham Eves BSc C Eng MICE MIHT.

The subject land has an area of about 190m² and at the acquisition date formed part of a narrow and unmade cul-de-sac known as Leech Lane. Entry to Leech Lane was from the east side of Chester Road H close to the junction with High Street, Malpas, and the lane extended eastwards for some 200m terminating at a property known as The Leech. The land is coloured blue on the plan attached at appendix 1 [not reproduced here].

By the end of the hearing the following facts had been agreed:

1. The freehold interest in an area of land immediately to the north of the subject land and having an area of about 1.42 acres was bought by the acquiring authority on June 16 1987. This land ("the red land") is edged in red on appendix 1. Prior to the purchase the latter land and a further area of land ("the yellow land") edged in yellow on appendix 1, situated immediately to the north of the red land and extending J through to Chester Road, were both in the same ownership.

2. On December 29 1989 in respect of the subject land the acquiring authority made the "Council of the City of Chester, land off Mercer Close, Malpas Compulsory Purchase Order 1989" the stated reasons for the order being:

(a) The land is required in order that the 1.42 acres of adjacent land currently owned (and currently lying unused) by the city council can be developed for residential purposes;

(b) it is expedient to acquire the order land in order to allow the early development of the city council's adjoining land.

The order was subsequently confirmed by the Secretary of State for the Environment. K

3. The freehold interest in the property known as The Leech at the eastern end of Leech Lane was bought by the acquiring authority on January 31 1990. The owner of The Leech had a right of way over Leech Lane.

4. On July 19 1990 the acquiring authority made a general vesting declaration in respect of the order land: on August 22 1990 the land vested in the acquiring authority and entry was taken. The latter date is the valuation date.

5. The claimants have at all material times claimed to be the freehold owners of the order land and are the registered owners with absolute title of that part of the order land situated between Mercer L Close (part of the housing development immediately to the south) and Leech Lane. As regards that part of the order land which comprises part of Leech Lane, HM Land Registry refused to register the claimants' title. No other party or person has claimed to be the owner of the order land or any part thereof.

6. The planning history of the red land is as follows:

(i) On July 18 1978 outline planning clearance was given by the acquiring authority for residential development subject to a condition that all vehicular access to the site should be taken from Mercer Close only. The permission was not implemented and therefore expired.

(ii) On June 4 1980 outline planning permission was granted by the M acquiring authority to Lansdowne Rodway Estates Ltd for residential development. This permission was never implemented and therefore expired. It contained no condition relating to access except for a condition requiring the subsequent approval by the local planning authority before development commences of the construction and drainage of the means of access thereto (including any estate roads).

(iii) On February 13 1986 outline planning clearance was given by the acquiring authority for residential development and formation of access across Leech Lane subject to a condition that access to the site should be from Mercer Close and the extended roadway to serve the proposed development should marry into line and level with the existing short section of road at Mercer Close. On September 27 1989 planning clearance was given by the acquiring authority for erection of category 2 sheltered housing for the elderly subject to a condition that the proposed development should be carried out in accordance with a revised plan deposited on August 22 1989. This development was subsequently built and completed in October 1992; it comprises 34 flats for elderly persons with warden's accommodation and the sole highway access is from Mercer Close.

7. In October 1985 the acquiring authority adopted the Chester Rural Area Local Plan as the statutory local plan for the area. Policy H4 thereof allocated the red land and the yellow land for housing development providing:

(a) A range of dwelling sizes up to a maximum size of 93m² (1,000 sq ft) gross;

(b) A reasonable mix of two and three bedroom and wherever practical one-bedroom dwellings.

In addition inset map no 30 showed the red land, the yellow land and The Leech situated within the settlement boundary of Malpas and showed the red land and the yellow land subject to policy H4.

8. The existing use value of the red land at the date of entry was £3,550 (ie £2,500 per acre).

9. The red land had a development value irrespective of the acquiring authority's proposal for sheltered housing development and at the date of entry the development value was £345,000 assuming that access was available.

10. The cost of constructing an access and access road from Chester Road to the red land was:

(i) £13,500 through the yellow land.

(ii) £30,000 through land owned by Cheshire County Council to the west.

(iii) £14,800 through land owned by British Telecom and partly occupied by a telephone exchange.

11. The cost of constructing an access from Mercer Close over the subject land was £3,500.

Case for the claimants

Mr Nardecchia said that the subject land was a valuable ransom strip which unlocked the development value of the red land. The claimants' valuer would contend that a developer would have paid the owner of the subject land one-third of the development value of the red land in accordance with the principle adopted in *Stokes* v *Cambridge Corporation* (1961) 13 P&CR 77*. It was not in issue that at the valuation date the red land was ripe for development. It was allocated for housing development on the adopted local plan of 1985; permission had been granted for its development both for the council's scheme and for private schemes. All that was required to enable development to proceed was the provision of satisfactory access to the red land. This was not a case where the council's scheme involved the development of a piece of land which would not otherwise have been developed and it had been agreed that the red land had development value irrespective of the council's sheltered housing scheme.

In the planning permission which the council granted to themselves in 1986 there was an express condition that access to the development should be only from Mercer Close. If the red land was to have an access from Mercer Close then that access could only be achieved through the subject land. He accepted that the condition formed part of the scheme and that the decision in *Pointe Gourde Quarrying & Transport Co Ltd* v *Sub-Intendent of Crown Lands* [1947] AC 565 required a disregard of an increase in the value of the subject land which was entirely due to the underlying scheme. This decision did not, in his view, require the evidence of what actually happened to be ignored. If it could be seen that the events would have been the same without the scheme then that was evidence to which the tribunal could have regard. That the only acceptable and available highway access to the red land was from Mercer Close was emphasised by the fact that from documents disclosed by the council and from information within the claimant's knowledge, apart from brief references in council internal memoranda in 1977, no one ever contemplated any other access to the red land other than from Mercer Close until after the compulsory purchase order had been made and confirmed and the claimants had formulated their claim and in particular no one had contemplated an access from Chester Road. It was the council's case however that in the "no-scheme world" the red land would have been developed jointly with the yellow land with an access from Chester Road. As a consequence the council argue that the subject land is not a ransom strip in the true sense because another highway access would have been used. He (Mr Nardecchia) would contend that in the real world a developer would have done exactly what the council did — ie would have sought to acquire the red land and the subject land. The price for the red land would have been reduced because it was land-locked and the price for the subject land would have reflected its geographical position and value as a ransom strip.

Mr Guise said that the claim had initially been based on a development land value of £200,000 per acre. At the time the claim was prepared in 1990 there was no evidence of sales of comparable building land at or around the August 1990 valuation date. The figure of £200,000 had been arrived at by adjusting the price of £75,000 paid in June in 1987 by the council for the red land to allow for the three-year time differential. Subsequently, however, documents disclosed by the council on discovery had revealed that the council had agreed to purchase The Leech in the latter half of 1989 for £215,000 before provision for access. As a consequence, agreement was reached by the parties during the hearing that the development value of the red land was £345,000 per acre.

Not only was the subject land the key to the development of the red land but the existing access to The Leech via Leech Lane was both inadequate and unacceptable in planning terms in the event of the development of the site of The Leech and the acquisition of the subject land would allow the development potential of the latter to be released. The additional value accruing to the subject site did not arise from the underlying scheme but from the geographical location of the site and in his view the *Pointe Gourde* principle did not therefore apply.

As far as the alleged defect to the title to part of the subject land was concerned he considered that this could be dealt with by title indemnity insurance and he had made the appropriate provision in his valuation.

Having regard to the decision in *Stokes* v *Cambridge* he considered that in the present case the appropriate proportion of the development value of the red land to be attributable to the subject land was one-third.

With regard to the three alternative access routes put forward by the acquiring authority it was his opinion that the possibility of achieving any access from Chester Road would have been considered by potential developers as being too uncertain and difficult and that the natural choice would have been the Mercer Close access over the subject land, which was the shortest and cheapest route and only involved dealing with one landowner.

The fire station access would require negotiations with three separate owners. Its actual availability was uncertain and the presence of a public car park on part of the fire station land would need to be taken into consideration. Access over the British Telecom land would also involve dealing with three separate owners and there would probably be highway objections in connection with junction spacings. Access over the yellow land involved at least two owners and the effect of the presence of the adjoining haulage depot on house prices would need to be considered. In addition the yellow land would lose at least one-dwelling plot which would be reflected in the acquisition cost.

Potential developers would not have compulsory powers and would be aware of the concern expressed by the highway authority regarding access from Chester Road and would be aware of the planning history with the imposition of conditions limiting access to Mercer Close.

In the case of each of the three alternative access points put forward by Mr Hall he considered the latter's costs to be seriously flawed as no

*Editor's note: Also reported at (1961) 180 EG 839

A provision was made for land acquisition costs. The *Stokes* v *Cambridge* principle was well known and it was inconceivable that the three major owners concerned would have been prepared to sell land without securing a share in the development value released by such a sale. His final valuation is attached at appendix 2.

Mr Eves said that access from Chester Road had consistently been resisted by the planning authority. In 1978 outline planning permission was granted for residential development on a site substantially the same as the combined red and yellow lands with a condition that access was to be from Mercer Cose. In 1980 permission was granted for the development of a site including but slightly larger than the red B land; the plan accompanying the application showed the access to be from Mercer Close so that a condition requiring access from the latter would have been otiose. The 1986 planning consent granted to Chester City Council in respect of the red land also included a condition that access should be from Mercer Close. Subsequently in 1990 permission was given for development for residential purposes of a site approximating to the yellow land. The highways department recommended that permission be refused for the latter development on the ground that the residential access way would be served from a road which would be primarily an access into a haulage depot and that this would undoubtedly lead to undue conflict between residential vehicles C and large haulage vehicles and would not be in the best interests of highway safety. In his report to the planning committee however the planning officer identified that it was considered unreasonable to withhold consent "particularly when the access position is the only reasonable point to gain access to the land".

Given this background it seemed to him (Mr Eves) absolutely clear that a planning permission for residential development in this area would not have been granted with an access from Chester Road while an access to Mercer Close was achievable.

Although he did not accept that access from Chester Road was feasible, he had considered the design criteria which would have D applied if such access had been possible, the relevant requirements being set out in a Department of Transport Advice Note TA20/84, Planning Policy Guidance Note 13 and the Chester Design Guide.

Applying the criteria to the fire station access, even with some relaxation of the recommended site line distance, additional land would be required both to the south west (ie part of the frontage of the adjoining dwelling-house) and to the north east (ie parts of both the British Telecom land and the yellow land. In addition he considered that there would be a substantial planning objection to a residential development having an access through a public car park as such a mix would conflict with the general philosophy enshrined in *Design* E *Bulletin* 32, which in such cases sought to segregate "through" traffic seeking general access to the residential development from that traffic seeking access solely to the car park and the various vehicle manoeuvrings which inevitably take place within a car park.

Access through the yellow land raised a number of separate issues. If in the "no-scheme world" it was to be assumed that the yellow and red lands were a single entity, then there would have been an alternative access via Mercer Close and it was therefore reasonable to assume that in view of the previous history there would have been objections by the highway authority to an access from Chester Road. F Any proposal to use the existing access shared with large haulage vehicles for the substantially increased traffic which would result from the development of the combined site, would have produced substantially stronger objections from the highways authority that those voiced in respect of the 1990 application or the development of the yellow land alone. Even if the access had been located some 80m to the south west of the existing access, there would still have been objections from the highway authority and to achieve a visibility splay to the south-west part of the British Telecom frontage would have been required. If it was to be assumed that the "scheme" and "no-scheme" worlds were the same, ie that the red land was to be developed separately from the yellow land, access through the latter would because of the number of units involved, have required a 5.5m-wide road. To link any development on the yellow land with the G access land would result in the loss of at least one dwelling on the yellow land.

An access through the British Telecom land would require the provision of site lines through both the fire station and the yellow land. In addition, a new junction here would change the status of the existing British Telecom entrance from a "private access" to a full road junction. This would then be comparable with any access to serve the fire station and the proposed car park. The maximum junction spacing which could be achieved would only be some 40m, which, in his view, was totally unsatisfactory and would generate a strong objection.

H

Case for the acquiring authority

Mr Hall was of the view that since the land immediately to the north of the order land and extending through to Chester Road was within the notation H4 on the Malpas village inset no 30 from the written statement to the Chester rural area local plan (which indicated that the site was allocated to housing), the order land had a potential value prior to the initiation of the council's residential scheme as one of several potential means of providing access to the latter scheme.

Taking the red and yellow lands as a whole, the most likely access would have been via the existing frontage to Chester Road. He considered that the cost of provision of such an access would have J been relatively minimal and that only the additional costs in providing adequate site lines needed to be considered. To provide site lines the acquisition of a strip of land from British Telecom might be necessary. In this context British Telecom had recently agreed to sell part of their frontage for identical purposes at £760 for approximately 60 sq yds.

An alternative route for access would have been over the land to the west of the red land owned by the county council and earmarked for a fire station and a public car park. An access across this land to the red land was in fact obtained during the development of the red land on the basis of a temporary licence. To provide site lines for this access it was likely that land would need to be acquired from British Telecom to the K north and from the owner of the adjoining house to the south. An access over the land owned by British Telecom would also have been feasible, although land for site lines would need to be acquired from Chester County Council and from the owner of the yellow land.

He considered that the condition in the consent granted to the acquiring authority requiring access from Mercer Close was an integral part of the scheme underlying the acquisition and was therefore to be disregarded for valuation purposes having regard to the *Pointe Gourde* principle.

In reality the most likely access route to the red and yellow lands would have been from Chester Road. The Mercer Close access would L have had a nominal value only, say £100. To allow a premium to represent any "hope value" and to allow an appropriate margin of variance on the construction and land acquisition costs for the alternative access routes, he had however put the maximum value for the order land at £10,000.

Although he did not accept Mr Guise's valuation approach, which was restricted to the assumption that the no-scheme world and the scheme world were identical he had nevertheless produced valuations of the subject land for each of the alternative means of access. His valuations are reproduced at appendix 3.

Mr Evans confirmed the planning history in respect of the red and M yellow lands. His conclusion was that if the red land had not been developed for housing by Chester City Council, there would have been no planning objection to taking access to the red land from Chester Road. This would have applied regardless of whether the red land had been developed independently or in conjunction with the yellow land to the north.

Mr Roberts said he was satisfied that the alignment of Chester Road was such that given control of sufficient land there would be several possible locations along the Chester Road frontage where satisfactory means of access could be provided to serve development of the red land together with any other adjacent land. From a highways viewpoint he considered that there could be advantages from serving a residential

A development from Chester Road rather than from Mercer Close. This would provide the opportunity to avoid the existing traffic difficulties faced in negotiating the very busy High Street and to avoid the rather lengthy vehicular route through the existing residential areas of Leech Road, Well Street and Springfield Road. Any resident wishing to gain vehicular access from Mercer Close, or indeed other parts of Malpas itself, would have to use these streets for each journey if there was no access to Chester Road. He would not however wish to see a road linking Mercer Close with Chester Road as this could in fact become a "mini bypass" of Malpas to avoid the busy high street commercial area with a resulting danger to residents in the residential road which would
B form the route.

The following cases were referred to during the hearing:

Stokes v *Cambridge Corporation* (1961) 13 P&CR 77; 180 EG 389
Challinor v *Stone Rural District Council* (1972) 27 P&CR 244
Batchelor v *Kent County Council* (1989) 59 P&CR 357; [1990] 1 EGLR 32; [1990] 14 EG 129, CA; [1992] 1 EGLR 217; [1992] 03 EG 127, (1991) 31 RVR 239, LT
Pointe Gourde Quarrying & Transport Co Ltd v *Sub-Intendent of Crown Lands* [1947] AC 565
Margate Corporation v *Devotwill Investments Ltd* [1970] 3 All ER 864; (1970) 218 EG 559

C Subsequent to the hearing I visited the order land and the other parcels of land referred to in evidence.

Decision

Although the parties were generally in agreement that the scheme underlying the acquisition included the acquisition and development of the red land for sheltered housing together with the formation of an access from Mercer Close, there was a major difference of opinion as to the situation which existed in the no-scheme world at the valuation date.

The claimants maintained that there was effectively no difference between the scheme world and the no-scheme world and that the value
D of the order land had to be considered in the context of the red land only, ie a land-locked site with access only over the order land. The acquiring authority contended that in the no-scheme world the affect of disregarding the underlying scheme was that the red land did not have to be considered in isolation and that both the red and the yellow lands had to be regarded as one entity and that as the combined site had a frontage to Chester Road the order land was, at best, only to be considered as a potential secondary entrance. It was not however in issue that at the valuation date both the red and the yellow lands were ripe for residential development and that the order land, independently of the underlying scheme, had at least some value in excess of existing
E use value (agreed at £100) as a potential access to the red or to the red and yellow lands combined. Mr Hall put the value of the order land at £10,000 to allow for any potential "hope value"; Mr Guise valued the land at £176,000.

For the acquiring authority it was argued that the application of the principle in *Pointe Gourde*, expressed by the Privy Council in the following terms:

> . . . compensation for the compulsory acquisition of land cannot include an increase in value which is entirely due to the scheme underlying the acquisition

F resulted in the whole of the progression of the council's scheme for the development of the red land from the original vague proposals, which included land in addition to the red land, through to the scheme as it was eventually built having to be disregarded, including the various planning permissions and the conditions requiring access to be taken off Mercer Close only.

I am not wholly convinced however that the *Pointe Gourde* principle can be applied so comprehensively in the present case. In the second decision by this tribunal in *Batchelor* v *Kent County Council*, the member, Dr Hoyes, said at p246:

> Although the statutory provisions and authorities caused the parties to notionally reside in a world conditioned by assumption and a measure of unreality they do not, in my judgment, require the abandonment of the actual facts as subsisting at the valuation date. To cast aside the actual facts and G
> substitute a different matrix founded upon speculation and supposition is contrary to *West Midlands* and a principle stated by Lord Morris in *Margate* when he said in [1970] RVR 770 at p775:
>
> "The decisions to be made by the Lands Tribunal were decisions in the field of fact. Some of the witnesses appear to have felt that they were asked to apply their minds to situations having an unreality only to be expected in fairyland. I do not think that their approach should have been so hesitant. Their evidence had to be directed to the realities of the conditions that existed and to the actual facts of the period so that the Lands Tribunal should be assisted to reach decision on the basis of the assumptions that had to be made".

Since the latter decision this case has again been before the Court of H
Appeal, this time as *Wards Construction (Medway) Ltd* v *Barclays Bank plc* (1994) 68 P&CR 391 and the appeal and cross-appeals against Dr Hoyes' decision were dismissed.

The acquiring authority bought the red land by private treaty in the open market and there does not appear to have been any suggestion that if terms had not been agreed the council would have sought compulsory powers for the purchase of the red land. It was a fact that the planning authority consistently required the access from any development on the red land to be from Mercer Close and again there was absolutely no suggestion that the same condition would not have been inserted if an application to develop the red land had emanated J
from a private developer rather than from a public authority. Permission was in fact granted on June 4 1980 to a private developer in respect of the red land, albeit with a slightly different northern boundary. No Mercer Close condition was imposed but since the accompanying plans showed that the access would be from Mercer Close, such a condition would have been unnecessary. Nevertheless the recommendation of the highways department in respect of the application was that the access "should marry into line and level with the existing short section of road at Mercer Close". Potential purchasers considering the purchase of the red land prior to the acquisition of the land by the council would have been aware either K
from discussions with the council's officials or from inspection of the Register of Planning Decisions of the existence of the access condition and it seems to me that this is a reality which cannot be erased by adopting a restricted interpretation of the *Pointe Gourde* principle.

It appears that prior to the valuation date there was no physical division between the red and the yellow lands. Both were within that part of Malpas allocated for residential development in the written statement in the Chester rural area local plan. On the Malpas map referred to as inset 30 in the local plan, the land allocated to residential development appeared under the notation H4 which effectively covered the whole of the red and the yellow lands. On the map the H4
area is divided horizontally, the division being similar to, but not identical with, the eventual division between the red and yellow lands. L
No explanation for the division of the H4 area was forthcoming during the hearing although since the local plan was adopted on October 7 1985 it may be that the division was intended to reflect the possibility of separate developments. What is clear from the evidence, however, is that, certainly from 1985 onwards, permission would have been readily forthcoming for development for residential purposes of either the red land, or some variation of the red land, or the red and yellow lands combined. Again, I am not convinced that the *Pointe Gourde* principle can be deployed to argue that although the council bought the red land, M
the valuation effect of the actual extent and the boundaries of the red land should be ignored. I have however reached the conclusion from the evidence that even if in the no-scheme world the actual purchase by the council of the red land is to be ignored, a purchaser coming fresh on the scene in 1990 would have bid for part only of the red and yellow lands and that that part would have been similar to, even if not actually identical with, the red land.

Prospective purchasers in the no-scheme world would know that the whole of the red and yellow lands were ripe for residential development. Inspection would show the existence of a partly formed access from Mercer Close at the southern end and they would be aware or become aware that although the yellow and red lands were in one ownership, there was a section of Leech Lane which was in the claimants' ownership. As I have said earlier discussions with the planning department or inspection of the register would have revealed that from as early as 1978 there had been a consistent requirement that access in the event of development of the southern part of the combined site (ie the red land or an area similar to the red land) access should only be from Mercer Close. Inquiries of the highway department would most likely have indicated that there was opposition to access being taken from Chester Road not only in respect of the red land but in respect of the whole of the red and yellow lands despite the yellow land having a frontage to Chester Road. When commenting on the application by the council to develop the southern part of the site in 1985 the highways department said:

> As the piece of land immediately to the north west (ie the yellow land) "of the application site" (ie the red land) is also designated for housing and the highway authority would not wish to see access to it taken directly from the B5069 Chester Road then provision should be made in the road design in connection with this application for eventual extension into that area.

I am in no doubt from the evidence that interested parties would have concluded that there would be strong opposition from the planning authority to any access from Chester Road and that planning applications based on a Chester Road access were likely to have been met by refusals. Mr Roberts said, in cross-examination, that he was not aware that the council had a policy of opposing access from Chester Road. To the extent that there may have been no formal policy he may well be correct, but we are concerned here with market perception which could only have been that the council did not wish to see access from Chester Road.

I attach no significance to the fact that in June 1990 permission was granted for the development of the yellow land with access from Chester Road. By granting themselves permission for the development of the red land with access from Mercer Close and making no provision in the layout of the development for future access for the yellow land, the council effectively prevented the development of the latter event though it was allocated in the town plan for residential purposes and despite the previous opposition to access from Chester Road unless access was in fact forthcoming from Chester Road. This was acknowledged in the report by Mr Evans to the planning committee which recorded that the highway department had recommended refusal (on the ground that the proposed access was to be used by resident vehicles and large haulage vehicles) and in which he recommended that:

> It is considered unreasonable to withhold consent particularly when the access position is the only reasonable point to gain access to the land.

Since in the no-scheme world both the red and yellow lands had potential for residential development the probability is that any market appraisal would have initially involved both sites. When the council first became interested in buying a site for residential development an approach was made to the owner of the red and yellow lands (Mr Corbett) in respect of the whole of these lands. Subsequently, Mr Corbett refused to deal with the whole on the ground that for security reasons he did not want residential development immediately adjoining the land which he occupied for his business to the north west of the site. In the face of this opposition the council turned their attention to the southern part of the combined site resulting in the 1978 permission which involved the site including the slightly larger than the red land. Further adjustments to the northern boundary ultimately produced the site which became the red land.

The reaction of Mr Corbett to an approach from a potential developer in respect of the whole of the red and yellow lands can only be a matter of conjecture. His concern about the effect of the proximity of the housing development to his business appeared to have diminished by 1990 when he obtained permission for development of the yellow land for residential purposes, by which time the value of the land had increased considerably. Whether for security reasons or because he wanted to develop the land himself, I can see nothing in the evidence which suggests that an approach to Mr Corbett by a developer seeking to buy both the red and the yellow lands would have met with any more success than the approaches by the council. It is however a matter of fact that Mr Corbett was prepared to sell the red land and the balance of probabilities is that a developer, aware that the red land was available, faced with the possibility that Mr Corbett would not sell the yellow land (and had his own plans for the development of the yellow land), knowing that the council had been consistently opposed to access from Chester Road and did not want to see a through link between Chester Road and Mercer Close (which would in effect provide a "mini bypass" round the town centre) and that if permission was refused for a Chester Road access an appeal to the Secretary of State would be necessary which could take time and would have an uncertain result, seeing a ready made access from Mercer Close and knowing that he could complete the access by buying the blue land, albeit at a ransom value, would have concluded that the red land offered the most certain opportunity of achieving a viable residential development with no undue delay.

For the above reasons I conclude that even in the no-scheme world the most likely development would have taken place on the southern part of Mr Corbett's land. The area would not necessarily be identical with that of the red land but I see no reason to assume that Mr Corbett would have been prepared to agree a different boundary between the red and the yellow lands to that which he actually agreed with the council and particularly since any adjustments of the area would involve valuation consequences which were not considered by either party during the hearing I propose to accept that the claimants are correct in contending that, as far as the valuation of the subject land is concerned, a development on the red land only need be contemplated.

It is not in dispute that the price a developer would pay for the red land would be reduced by the price he had to pay for any necessary access. Having regard to the condition requiring access from Mercer Close the claimants regard the red land in ransom strip parlance as a "single access" case and have not considered the possibility that access could have been obtained other than from Mercer Close. The acquiring authority, while contending that in the no-scheme world the red and yellow lands would have been the natural development unit and that the access would have been from Chester Road thus relegating the Mercer Close access to at best a secondary entrance, nevertheless argues that even if the development of the red land had been a possibility in the no-scheme world it would still not have been a single access site.

Considerable time was taken during the hearing in discussing the council's three alternative means of access to the red land or from Chester Road and referred to by Mr Hall as route 1 (through the yellow land), route 2 (through the fire station land) and route 3 (through British Telecom land). I can dispose very shortly of route 2. When Mr Hall wrote somewhat belatedly in September 1992 to the owners of the fire station site, Cheshire County Council, canvassing support for his contention that alternative routes were available and asking what the county council's views would have been in 1990 to an approach asking for access through the council's land the county property manager replied on September 15 1992:

> I do not think the County Fire Officer would have granted permission for the sale or letting of an access strip through this authority's land to your scheme in case this jeopardised the future development of the new fire station

and

> If an access way had been granted across this authority's land . . . the County Fire Officer is likely to have required considerable remuneration for such an arrangement since the land would obviously have been valuable to your authority

and concluded

> if your authority had requested permission to create an access road from Chester Road to your sheltered housing scheme, it is very unlikely that this

authority would have agreed to sell or lease an appropriate access strip across the new fire station site at Chester Road.

In cross-examination Mr Hall accepted that the county council's views would probably have been the same at all times since the site was acquired in 1985. In addition Mr Hall said in evidence that since February 1990 there had been ongoing discussions between the county and the city councils regarding a joint use of the fire station site for both fire station and public car park purposes. Mr Roberts in responding to a letter from Mr Hall in September 1992 regarding access and referring to the proposed car park said:

> I would not consider this access would be suitable to serve residential development: I would not support the concept of any sort of residential development being accessed through a car park.

On the evidence I do not think that any developer would have considered access through this site as even a remote possibility and I can see no point in examining it further.

Route 1 was clearly physically possible although it would have required Mr Corbett's co-operation, additional land would have been needed for site lines across the adjoining British Telecom land and there was a risk of almost certain opposition from the planning authority both in respect of the access itself and of the "mini bypass" effect. It is wholly a matter of conjecture as to what Mr Corbett's attitude would have been as the possibility was never raised with him. Any access to the red land would have reduced the area of the yellow land and consequently reduced the number of plots if the remainder of the yellow land was subsequently developed for residential purposes. In cross-examination Mr Roberts tried to distance himself from the view he had expressed in September 1992 by saying there was no public car park proposal in 1990, apparently unaware of Mr Hall's evidence to the contrary and conceded that he would still have been of the same opinion at the earlier date. I cannot see that this access would have been pursued seriously by a developer.

In respect of route 3, Mr Hall also wrote to British Telecom in September 1992 on similar lines to his letter to Cheshire County Council in respect of the fire station site. British Telecom responded by saying:

> During early investigations, following your initial request [a reference to the negotiations for temporary access in 1988] we had considered the possibility of selling part of Malpas Telephone Exchange site to you subject to BT being able to meet its own operational requirements, and although now hypothetical, BT would have been prepared to enter into negotiations with the City Council to assist the provision of an access off Chester Road.

Whether the same answer would have been forthcoming in 1990 if the application had been from a private developer is conjectural. Assuming however that British Telecom would have been prepared to make sufficient land available, not only would potential objections from the planning authority to a Chester Road access have to be overcome and land for site lines acquired from the county council and from Mr Corbett, but in view of the fire station/car park proposals there was a problem of spacing of junctions which could have resulted in strong objections from the highways authority. Again from the evidence I cannot see that a potential developer would consider this route to be worth more than a passing consideration.

I am not however persuaded that Mr Guise's "single access" approach is wholly correct. From what I have heard I am in no doubt that prospective developers in the no-scheme world would have concluded that the red land presented the most readily attainable and viable site for access to the residential development on the red land and that the Mercer Close access would have been the preferred access. Providing the cost of acquiring the subject land was effectively paid by the owner of the red land, by an appropriate reduction in the price of the latter, there would have been little incentive for a developer to investigate alternative means of access all of which in the present case appear to have presented problems which were additional to any question of ransom value. Nevertheless a prudent developer would have investigated the other means of access and the possibility, however remote, that even with the attendant problems, one of the other access routes would produce a lower overall cost cannot be totally disregarded.

Previous decisions by this tribunal on valuation issues tend to be regarded as evidence when similar issues arise in subsequent cases and the decision in *Stokes* v *Cambridge* is regularly referred to as a benchmark in ransom strip cases. However in the second Court of Appeal decision in *Wards Construction (Medway) Ltd* v *Barclays Bank plc* Nourse LJ said at p399:

> Again it must be emphasised that the *Stokes* v *Cambridge* principle, if such indeed it is, is one of valuation not of law. It was for the member to decide whether or not it was a useful starting point for his valuation of the order land.

Since in the present case Mr Guise is of the view that a developer of the red land would be prepared to pay one-third of the development value to the owner of the subject land on the basis that the latter was the only possible access and Mr Hall, taking a multi-access view (in so far as the red land only was concerned), accepted that a developer would pay one quarter of the development value I can derive little assistance from previous cases other than to the extent which is already reflected in the parties' valuations that they establish as a generality that in ransom value terms a single access is likely to be more valuable then an access which is one of several alternatives.

In the present case the evidence suggests that the Mercer Close access was far and away the most likely access and such few doubts which I have would at worst result in some small reduction in the estimate by Mr Guise that one-third of the development value was the correct figure. On balance I would put the figure at 30%.

Mr Hall initially thought that a developer would have paid £11,000 for route 1 and £19,450 for route 2. During the hearing he conceded that he had made no provision for the likely cost of the land which would actually be required for access and his final valuations were £105,571 for route 1, £82,804 for route 2 and £98,004 for route 3 (which had not featured in his earlier valuations). Mr Hall regarded the Mercer Close access as route 4 and again he had originally equated the value of the subject land with the cost of physically providing the access at a figure of £6,500. The actual cost element was subsequently agreed at £3,500, but taking a *Stokes* v *Cambridge* view, he subsequently conceded that the value was £77,379 having had to double an earlier valuation of £38,690. Even making allowance for the fact that Mr Hall's alternative route valuations were his "fallback" position to refute Mr Guise's single access approach and that his (Mr Hall's) preferred valuation was arrived at on a totally different basis, I find that his erratic approach to the valuation of the alternative routes casts some doubt as to the weight that I can give to his views on the valuation issues generally. I particularly have in mind that Mr Hall's valuation in respect of route 4 where, having in the closing stages of the hearing reduced his valuation by 50%, he was still left with a valuation in which, having settled on 25% as the proportionate share of development value to reflect his multi-access approach, he had further reduced his 25% by one-third "to reflect the risk of a Chester Road access being chosen". Since I have concluded that Mercer Close was the most likely access rather than the least likely access, I think his valuation for route 4 is completely unrealistic. The consequences of removing the one-third deduction and making an adjustment for his admitted error in making a final deduction of 50% in respect of a share of development value to the owner of The Leech for which he had already made an allowance when arriving at his calculation of the total development value would be to show that of the four access routes to the red land the Mercer Close access produced the highest value for the subject land rather than the lowest value as he had initially contended.

Mr Guise's valuation approach has been consistent from the outset of his involvement in this case and in the circumstances I propose to adopt his valuation format. The development value of the red land and the existing use value are both agreed and are not in issue. The circumstances surrounding the acquisition of The Leech by the council and the effect of the acquisition are somewhat confused but it was accepted, again in the context of the red land only, that the development prospects for The Leech were enhanced by the acquisition of the subject land and I see no reason to depart from Mr Guise's assessment of the additional value, ie £49,667. Mr Guise added a nominal sum of £100 to his *Stokes* v *Cambridge* share to reflect the value of the land taken; it seems to me however that if one is using the *Stokes* v *Cambridge* approach to arrive at a value of a ransom strip then as long as the resulting value is greater than the existing use value the latter value becomes irrelevant. Although the amount involved is trivial I would exclude it as a matter of principle. Although the title issue was initially a matter of concern to the acquiring authority it was eventually agreed that the claimants' title insurant approach was acceptable and that the cost of such insurance was a single premium of £2,680.

Incorporating my conclusions the valuation of the subject land becomes as follows:

Development site (red land)

	£
1.42 acres at £345,000 per acre	489,900
Less: existing use value @ £2,500 per acre	3,500
	486,350
Additional development value in respect of the Leech	49,667
	536,017
Share due to the owner of the access (blue land) 30%	160,805
Deduct: Title insurance premium	2,680
	158,125

The resulting figure, £158,125, is the amount of my award.

This decision determines the substantive issues raised between the parties, and the tribunal's award is final. The parties are invited to make such submissions as they are advised as to the costs of the hearing, and a letter accompanies this decision as to the procedure for submissions in writing. The tribunal will, in due course, incorporate an order as to costs in an *addendum* to this decision. Rights of appeal under section 3(4) of the Lands Tribunal Act 1949 and Rules of the Supreme Court Ord 61 will not accrue until the decision has been thus completed, ie from the date of the *addendum*.

Costs addendum

I invited submissions as to costs from the parties. I received a submission from the claimant. Having opened two sealed offers made by the acquiring authority dated August 18 1993 and November 8 1994 I find that both offers are less than the amount of my award. I order that the costs for the claimant shall be paid by the acquiring authority, such costs, if not agreed to be taxed by the registrar of the Lands Tribunal on the High Court scale.

Appendix 2
Final valuation by Mr MA Guise
A. Development site (Land edged red)

	£
1.42 acres @ £345,000 per acre	489,000
Less: existing use value as agricultural land (£2,500 per acre)	3,550
Development value	486,350
B. Additional development value in respect of The Leech	49,667
	536,017
Share due to owner of access (one-third)	178,672
C. Value of land taken (spot)	100
	178,772
D. *Deduct*: Title insurance premium	2,680
	176,092
Say:	£176,000

Appendix 3
Final alternative valuations by Mr PJ Hall
A. Development Site

	£
1.42 acres @ £345,000 per acre	489,900
Less: existing use value as agricultural land (£2,500 per acre)	3,550
	486,350
B. Additional value of The Leech Rather than the complex hypothesis of Mr Guise, the actuality was that the council discounted the market value of £230,000 and settled at £215,000 therefore	15,000
	501,350
Share due to owner of access (whichever access) (as *per Bachelor* case re compensation for a "multi access" site): Ransom value as one of several alternative accesses: 25%	£125,338

Costs of alternative means of access to red land only

Route 1 (Corbett)

		£
Share of development value (ie 25%)		125,388
Less: BT splay @ 5% max	£6,267	
Construction costs	£13,500	19,767
Maximum amount available to pay Corbett:		105,571

Route 2 (Fire station)

Share of development value (ie 25%)		125,388
Less: BT splay @ 5% max	£6,267	
House splay @ 5% max	£6,267	
Construction costs	£30,000	42,534
Maximum amount available to county		82,804

Route 3 (BT)

Share of development value (ie 25%)		125,338
Less: County splay @ 5% max	£6,267	
Corbett splay @ 5% max	£6,267	
Construction costs	£14,800	27,334
Maximum amount available to BT:		98,004

Route 4 (Mercer Close)

Share of development value (ie 25%)		125,338
Deduct one-third to reflect risk of a Chester Road access being chosen		41,779
		83,559
Less: Title insurance	£2,680	
Construction costs	£3,500	6,180
		77,379

Wickham Growers Ltd *v* Southern Water plc

September 20 1996

P H Clarke FRICS

Estates Gazette April 26 1997

[1997] 17 EG 167

***Compensation** — Apple trees grubbed out for water main — Claimants grubbed out fruit trees on adjacent land post-valuation date with benefit of recently introduced grant — Whether events occurring after date of entry relevant in assessing compensation under Water Industry Act 1991*

On May 17 1993 the compensating authority, following the service of a notice under section 159 of the Water Industry Act 1991, entered the claimants' fruit farm and grubbed up 212 apple trees before laying a water main. On March 16 1995 the parties agreed compensation, including a sum for the consequences, and the loss of, the apple trees. In March 1995 the claimants grubbed out 50 acres of apple trees on adjacent land with the benefit of a government grant. The compensating authority refused to pay the agreed compensation on the ground that the claimants had suffered no loss, for the purposes of disturbance compensation, from the destruction of the trees because they would have been grubbed out with the other trees in 1995. The tribunal formulated a preliminary point of law as to whether, for the purposes of assessing compensation under Schedule 12 to the 1991 Act, regard may be had to events occurring after the date of entry, the occurrence of which could not have been predicted as at that date.

***Decision*: In assessing disturbance compensation for compulsory purchase, facts occurring after the date of valuation may be taken into account, even though the assessment must be made as at the valuation date. This enables the actual losses to be considered. However there is unpredictability in the subsequent events and this raised questions of causation, remoteness or mitigation of loss which could not be determined on a preliminary point of law in the absence of evidence.**

The following cases are referred to in this report.

Birmingham Corporation v *West Midland Baptist (Trust) Association (Inc)* [1970] AC 874; [1969] 3 WLR 389; [1969] 3 All ER 172; (1969) 67 LGR 571, HL

Bwllfa & Merthyr Dare Steam Collieries (1891) Ltd v *Pontypridd Waterworks Co* [1903] AC 426

Dhenin v *Department of Transport* (1990) 60 P&CR 349

London County Council v *Tobin* [1959] 1 WLR 354; [1959] 1 All ER 649; (1959) 57 LGR 113; 10 P&CR 79, CA

Munton v *Greater London Council* [1976] 1 WLR 649; [1976] 2 All ER 815; *sub nom Munton* v *Newham London Borough Council* (1976) 32 P&CR 269; 74 LGR 416; [1976] 2 EGLR 5; [1976] EGD 469; 239 EG 43, CA

John Hobson (instructed by Brachers, of Maidstone) appeared for the claimants; Miss AJ Blay, solicitor, represented the compensating authority, Southern Water plc.

Giving his decision, **MR PH CLARKE** said: This is a decision on a preliminary point of law in a reference to assess the compensation payable for the laying of a water main in agricultural land.

John Hobson of counsel appeared for the claimants; Miss AJ Blay, solicitor, appeared for the compensating authority.

Facts

The parties have prepared a statement of agreed facts and I find the following facts:

(1) Wickham Growers Ltd ("the claimants") farm 150 acres of land at Ash and Harpers Farm, Goudhurst, Kent ("the subject property"), mainly to top fruit (apples, blackcurrants) and hops.

(2) On December 8 1992, Southern Water Services Ltd, now Southern Water plc ("the authority"), served notice on the claimants that in pursuance of their powers under section 159 of the Water Industry Act 1991 ("the 1991 Act"), they intend to lay a resource main in the subject property.

(3) The authority entered on the subject property on May 17 1993 and laid the resource main during 1993 and 1994.

(4) In order to lay the main and provide the associated working area, the authority's contractors grubbed some 212 apple trees on the subject property. The loss of these trees form one item in the claimants' claim for compensation.

(5) The parties each instructed agents to negotiate the compensation payable to the claimants for the laying of the resource main in the subject property. In December 1994 the claimants' agent submitted a claim in the sum of £60,748.18, later amended to £66,501.46. In January 1995 the parties' agents agreed in principle the sum of £75 per tree compensation for the apple trees grubbed to lay the resource main. On March 16 1995 compensation was agreed for all items of claim (except hops) in the total sum of £40,000, comprising £3,500 for the pipeline easement and £36,500 for disturbance, plus professional fees.

(6) In March 1995 the claimants, with the benefit of a government grant, grubbed 50 acres of apple trees on part of the subject property adjacent to the route of the resource main.

(7) On March 24 1995 the authority's agent stated that the grubbing of these trees altered the nature and extent of the compensation payable to the claimants, which should now be on the basis that the grubbing grant should be taken into account because the apple trees grubbed by the contractors to lay the resource main would have been grubbed in any event.

(8) On December 18 1995 the claimants referred their claim for compensation to this tribunal for determination.

(9) On April 2 1996 the registrar, on the application of the claimants, ordered that there be a preliminary hearing on a point of law "to determine on what basis and by reference to what valuation dates are the applicants entitled to be compensated by the respondents in respect of the exercise by the respondents of statutory powers to lay a water main through the applicants' land".

Issues

Mr Hobson opened his case for the claimants by identifying two issues for my determination. First, whether the loss of the future productive capacity of the apple trees grubbed to lay the resource main is an element to be taken into account when considering the depreciation in the value of the land under para 2(1) of Schedule 12 to the 1991 Act, or whether it is loss or damage under para 2(2) of Schedule 12, sustained as a result of the laying of the main? Second, even if the claim for the loss of the productive capacity of the trees is other loss or damage, whether it is appropriate to take a later date for quantification than the date of entry for the laying of the main?

It became clear as Mr Hobson developed his submissions, however, that the answer to the first question would be fundamental to my decision. It appeared to be common ground that the category of the compensation would affect the assessment of that compensation and therefore my decision on the preliminary issue of law. Mr Hobson argued that this compensation represented depreciation in the value of the land. I invited the parties to agree this fundamental issue but Miss Blay, for the authority, declined to accept that the disputed compensation constituted depreciation in land value.

It appeared to me therefore that the preliminary point of law rested on an issue of fact, namely whether the compensation for the loss of the trees represented depreciation in land value or other loss or damage analogous to disturbance on compulsory acquisition. In the absence of agreement as to the nature of this compensation, I would need to make a finding of fact, but was unable to do so in the absence of evidence. In the light of this position, I invited the parties to reformulate the issue of law and they were able to do so in the following terms:

> Whether for the purposes of the assessment of compensation under paragraph 2(2) of Schedule 12 to the Water Industry Act 1991, regard may be had to events occurring after the date of entry, the occurrence of which could not have been predicted as at that date?

Claimants' case

Mr Hobson submitted that, when assessing compensation, it is wrong in law to have regard to events after the valuation date, particularly events which could not have been foreseen at that time. The assessment of compensation is not an open ended exercise: it must be tied to a particular date.

The realities of the situation, said Mr Hobson, are that possession was taken of part of the claimants' land on May 17 1993 and trees which formed a productive part of that land were uprooted to allow the resource main to be laid. At that time the claimants suffered an immediate and direct loss. Compensation was agreed but, before payment, the authority reneged on this agreement on the grounds that the claimants grubbed trees on adjoining land with the benefit of a government grant. This grant was introduced in July 1994. The authority argue that the claimants would also have grubbed the trees previously removed for the water main (with financial assistance) and would therefore not have suffered loss in respect of those trees. Mr Hobson said that the later grubbing was carried out in the knowledge that the compensation payable included recompense for the trees previously removed.

Mr Hobson said that in Schedule 12 to the 1991 Act provisions of the Land Compensation Act 1961 are expressly applied to the assessment of compensation for pipe-laying and therefore compensation must be assessed as at a specific date. This is the date of entry by the authority: see *Birmingham Corporation* v *West Midland Baptist (Trust) Association (Inc)* [1970] AC 874, at p894G. All heads of compensation must be assessed as at that date. Even though some of the compensatable loss arises after the valuation date, it is referred back to that date. Mr Hobson referred to *West Midland Baptist (Trust) Association (Inc)* at pp896G, 897A, 899B and 911G, *Munton* v *Greater London Council* [1976] 2 All ER 815 at p818h and *Halsbury's Laws of England* vol 8(1) para 295. Mr Hobson said that fruit trees were growing on the claimants' land at the date of entry and a purchaser would have paid a price reflecting the future productive capacity of those trees. This was the basis of compensation agreed. If, with the benefit of hindsight, and having regard to events which occurred after the valuation date and which could not have been foreseen at that date, the parties' valuers would agree a different figure, this is irrelevant. The scheme which led to the future grubbing of trees on the claimants' land was not in existence at the date of valuation and should not be taken into account. It is an unforeseen event. The position is similar to the situation where, after compensation has been agreed, an unforeseen hurricane blows down the trees or the market for apples collapses so that the trees become less valuable than originally thought. These are risks that a purchaser at the date of valuation has to take.

In support of these arguments, Mr Hobson referred to *Dhenin* v *Department of Transport* (1990) 60 P&CR 349 and distinguished *Bwllfa & Merthyr Dare Steam Collieries (1891) Ltd* v *Pontypridd Waterworks Co* [1903] AC 426 and *London County Council* v *Tobin* (1959) 10 P&CR 79. These latter decisions, he said, preceded the *West Midlands* case; this takes precedence and prescribes a fixed date for the assessment of compensation.

Authority's case

Miss Blay said that she could agree with the claimants' contention that "the clock stops" at the date of entry when assessing compensation for depreciation in land value under para 2 (1) of Schedule 12, but not in respect of loss or damage under para 2(2). This latter compensation represents disturbance and is intended to reflect the actual loss suffered after the date of dispossession, or in this case, the date of entry. Miss Blay said that para 2(2) has nothing to do with the fixing of a valuation date, it is concerned with compensation for "loss or damage" in consequence of pipe-laying works. This is not amenable to calculation at a particular date. If compensation is assessed and paid before the loss is actually suffered but in anticipation of that loss, then elements of prediction are involved. But where, as in this case, the extent of the loss is actually known before compensation is paid, then it should be based on the known facts: see *Bwllfa* and *Tobin*.

Miss Blay referred to *Halsbury's Laws of England* vol 8(1) para 295, which she said supports the view that loss or damage is a question of fact and is not attached to a particular date. She also referred to *Munton* at p189. The principle that a claimant must mitigate his loss is to be taken into account when assessing disturbance compensation. If steps are taken in mitigation, which reduce the loss suffered, then this is to be set off against compensation. In this case, the claimants would have grubbed the apple trees on the line of the resource main with the benefit of the grant, in the absence of the main, and would not have suffered the loss which they now claim. It is not the purpose of compensation to give claimants a windfall gain.

Miss Blay distinguished the decision of this tribunal in *Dhenin* on two grounds. First, that the compensation in that case was governed by the statutory code in Part 1 of the Land Compensation Act 1973, which differs from Schedule 12 to the 1991 Act. Second, *Dhenin* relates to the depreciation in land value, where the *Bwllfa* principle does not apply and where value is fixed at a particular day, and not to disturbance (or loss or damage under para 2(2) of Schedule 12) where events after the date of entry may be taken into account.

Decision

I look first at the relevant provisions in the 1991 Act. Compensation in respect of pipe-laying and other works is to be assessed in accordance with Schedule 12 to the Act. Para 2 of that Schedule deals with pipe-laying works in private land and provides for three categories of compensation:

(1) for the depreciation in value of any interest in land (para 2(1));
(2) for any other loss or damage (para 2(2));
(3) for damage to, or injurious affection of, other land (para 2(3)).

Any betterment to contiguous or adjacent land held by the claimants is to be deducted from the compensation.

In this reference, and as the question of law was reformulated, we are only concerned with compensation under para 2(2).

This provides as follows:

> Where the person entitled to an interest in any relevant land sustains loss or damage which —
> (a) is attributable to the exercise by any relevant undertaker of any power to carry out pipe-laying works on private land;
> (b) does not consist in depreciation of the value of that interest; and
> (c) is loss or damage for which he would have been entitled to compensation by way of compensation for disturbance, if his interest in that land had been compulsorily acquired under section 155 of this Act, he shall be entitled to compensation from the undertaker in respect of that loss or damage, in addition to compensation under subparagraph (1) above.

The term "relevant land" means the land where the power to carry out the pipe-laying works is exercised or land held with that land (para 2 (5)). The "power to carry out pipe-laying works on private land" includes the power under section 159 of the Act (para 2 (6)), which is the relevant provision in this reference. Para 3(1) of Schedule 12 provides that disputed compensation under para 2 shall be determined by the Lands Tribunal and that the provisions of sections 2 (procedure) and 4 (costs) of the Land Compensation Act 1961 shall apply, subject to any necessary modifications.

The effect of the above provisions is that the owner of an interest in land affected by pipe-laying works is entitled, in addition to compensation for any depreciation in the value of his land, to compensation for "loss or damage", which is to be assessed as if it were compensation for disturbance on compulsory acquisition. It is therefore to the law relating to the assessment of disturbance compensation that we must look to find the answer to the question in this reference.

The agreed date of entry on to the claimants' land for the pipe-laying works is May 17 1993 and the dispute between the parties is as to whether, when assessing the disturbance element in the compensation under para 2(2) of Schedule 12, regard may be had to events occurring after that date?

In my view, it is settled law that, in the assessment of disturbance compensation on compulsory purchase, and leaving to one side matters such as causation and remoteness of loss, facts occurring after the date of valuation may be taken into account, even though the assessment must be made as at the valuation date for the land taken. The position is concisely stated in *Halsbury's Laws of England*, vol 8(1) para 295:

> The normal date for the assessment of compensation for disturbance is the same as the date for the assessment of the value of the land itself, namely the date of entry or the date of the valuation itself, whichever is the earlier [*Birmingham Corporation* v *West Midland Baptist (Trust) Association (Inc)*]; but it is possible, where appropriate, to deal with the assessment of compensation for disturbance after the assessment of the value of the land itself, in order to enable the actual losses to be considered [*Munton* v *GLC*].

In *Munton* the position was explained by Lord Denning MR. He said that it was established under the Lands Clauses Act 1845 that only one sum of compensation was awarded, for the value of the land. He then said (p819a):

> But although only one sum is awarded, it is very proper, in assessing it, to divide it into two parts: (1) the land itself and (2) disturbance. Starting with the

Acquisition of Land Act 1919 and repeated in the Land Compensation Act 1961, Parliament itself has made a division between the two. In s 5, r (6), it says:

> "The provisions of rule (2) [that is about the value of the land] shall not affect the assessment of compensation for disturbance or any other matter not directly based on the value of land."
>
> Since those Acts, the practice always has been for the compensation for disturbance to be assessed separately from the value of the land. That is as it should be. The value of the land can be assessed whilst the owner is still in occupation. The compensation for disturbance cannot properly be assessed until he goes out. It is only then that he can tell how much it has cost him to move, such as to get extra premises or to move his furniture.

The principle that, when making an assessment of disturbance compensation, evidence is admissible of facts occurring after the date of valuation is supported by the decisions in *Bwllfa* and *Tobin*. In the former case, the owners of coal mines were prevented from working coal under or near the respondents' works and claimed compensation under the Waterworks Clauses Act 1847. This was to be assessed as at the date of the respondents' counternotice preventing the mining of the coal. In an arbitration to determine the amount of compensation, the question arose as to the admissibility of evidence to show that the price of coal rose after the date of the counternotice. The matter proceeded through the courts and the House of Lords held that this evidence was admissible. Lord Macnaghton referred to the statutory provisions governing the assessment of compensation and then said (p431):

> My Lords, I cannot find in this group of sections any provision or even any suggestion to the effect that a line is to be drawn at the date of the counter-notice, and that all that occurs after that date is to be as if it had never happened.
>
> It appears to me that the case put forward on behalf of the respondents is based on a false analogy. The counter-notice by the undertakers following a notice of the mine owners under section 22 does not operate to make a contract or to transfer property. It is not even a step towards a contract or a step towards expropriation. The undertakers acquire no property in the minerals. The property remains where it was. The mine owner is prohibited from working, and the undertakers are bound to make full compensation. That is all. If the question goes to arbitration, the arbitrator's duty is to determine the amount of compensation payable. In order to enable him to come to a just and true conclusion, it is his duty, I think, to avail himself of all information at hand at the time of making his award which may be laid before him. Why should he listen to conjecture on a matter which has become an accomplished fact? Why should he guess when he can calculate? With the light before him, why should he shut his eyes and grope in the dark?

The *Bwllfa* principle was later applied to compensation for disturbance on compulsory purchase. In *Tobin* the Lands Tribunal assessed compensation for loss of goodwill following the claimants' move to alternative premises, by deducting from the value of the goodwill at the old premises, the value of the goodwill at the new premises. When making this assessment, the tribunal took into account the profits achieved in the alternative premises after the date of valuation. The Court of Appeal held that this was correct. Morris LJ said (p89):

> Had the Lands Tribunal been dealing with the claim in December, 1955 at the time of the reference to them, their task would have been indeed difficult. As they were dealing with the claim in February and June 1957, it was possible for the Lands Tribunal to have regard to the fact that the business had been continued and was continuing and to the known facts of performance and achievement at the new address; it was entirely proper for the Lands Tribunal to take these known facts into calculation (see *Bwllfa* and *Merthyr Dare Steam Collieries (1891) Ltd* v *Pontypridd Waterworks Co*).

I was referred to the decision of this tribunal (CR Mallett) in *Dhenin*, where *Bwllfa* was distinguished and post-valuation date facts were largely excluded from consideration. This decision concerned the compensation payable under the Land Compensation Act 1973 for depreciation in the value of a house caused by the use of a new motorway. The claimants sought to rely on actual traffic counts after the date of valuation, rather than projected figures estimated before that date. The tribunal held that the *Bwllfa* principle had no application to this case because the 1973 Act provided for the assessment of compensation as at a specific date, on the facts at that date, on prices then current at that date and on any intensification of the depreciation which could be anticipated at that date. Evidence after the valuation date could, however, be used to determine the accuracy of the assumptions which had to be made at that date. *Dhenin* can be distinguished from this current reference for the same reasons given by the member for distinguishing it from *Bwllfa*. *Dhenin* was concerned with the *value of land* where the governing statute is specific as to dates. It was not concerned with disturbance where a different principle applies.

In my view, I am bound by the decisions of the superior courts in *Bwllfa* and *Tobin* to answer "yes" to the question put to me. The only hesitation I have in giving this answer centres on the unpredictability of the subsequent events. This raises questions of causation and remoteness which were only briefly touched on by the parties at the hearing. The question is: did the acceptance of the government grant and the grubbing of the trees constitute a separate chain of causation, providing a benefit which should not be taken into account when assessing compensation? In the absence of full submissions on this difficult point, I think it is an area into which I should not go. Some unpredictability must have been present in *Bwllfa* (as to the future price of coal) and *Tobin* (as to the future profitability of the business) at the relevant dates but post-valuation date facts were admitted in both cases.

My answer to the question put to me is therefore "yes". This disposes of the preliminary issue of law, but does not dispose of the reference, although I hope that the parties may now be able to reach agreement on the disputed compensation. I would add that, in my view, it would have been expedient if the question of disputed compensation had been wholly dealt with at a substantive hearing, rather than initially as a preliminary point of law. At the start of the hearing it became clear that the question of law rested on a fundamental and disputed question of fact, namely the nature of the disputed compensation, whether land value or other loss or damage (disturbance). The issue had to be reformulated to enable me to deal with it in the absence of evidence on the fundamental question of fact. Alternative figures of compensation, dependent on my decision on the issue of law, had not been agreed and it may well be that questions of fact will arise in the agreement of a proper figure of compensation for the trees, particularly if questions of causation, remoteness or mitigation of loss are considered further.

I heard submissions on costs. The claimants will pay the authority's costs of the determination of this preliminary issue of law, such costs, if not agreed, to be taxed by the registrar of the Lands Tribunal on the High Court standard basis.

Stanford Marsh Ltd *v* Secretary of State for the Environment

January 17 1997

MR PH CLARKE FRICS

Estates Gazette April 26 1997

[1997] 17 EG 170

Compensation — Sealed offer — Costs — Liability for costs where sealed offer accepted on day of hearing

The claimants referred to the tribunal a claim for compensation arising out of the acquisition of their property by the Bristol Development Corporation. On December 5 1996 the acquiring authority sent a sealed offer by fax to the claimants' solicitors offering to pay £100,000 in full and final settlement, plus surveyor's fees,

A reasonable legal fees and interest. Immediately before the start of the hearing on December 16 the claimants accepted the sealed offer and agreement was reached on all issues except costs.

***Decision*: Section 4 of the Land Compensation Act 1961 did not apply to the reference because the issue of compensation did not proceed to a hearing, and no award was made. The sealed offer ought to have been accepted by the claimants by December 11, five working days after its receipt. Accordingly, the claimants were entitled to their costs up to and including December 11 and must bear their own costs and the costs of the acquiring authority after that date, including the cost of the hearing.**

B No cases were referred to in the decision.

Alun Alesbury (instructed by John Stallard & Co, of Worcester) appeared for the claimants; Robert Fookes (instructed by Bevon Ashford, of Plymouth) represented the acquiring authority.

Giving his decision, **Mr PH Clarke** said: This is a reference to determine the compensation payable for the compulsory acquisition of the leasehold interest in a light industrial/warehouse building known as Unit 4, Great Western Trading Estate, Temple Meads, Bristol ("the subject property"). Immediately before the start of the hearing, the claimants accepted an offer of compensation in the sum of £100,000
C plus surveyor's fees, legal costs and statutory interest and the only issue for my determination is the liability for the costs of this reference.

Alun Alesbury of counsel appeared for the claimants and Robert Fookes of counsel appeared for the authority.

Facts

The following facts are relevant:

(1) On July 4 1990 Bristol Development Corporation made the Bristol Development Corporation (Temple Meads/Kingsley Village) (Part I) Compulsory Purchase Order 1990, which included the subject
D property. The order was confirmed on January 28 1992. Notice to treat and notice of entry were served on November 30 1993 and the corporation took possession on April 18 1994.

(2) The claimants relocated their business to 5 and 6 City Business Park, Bristol.

(3) On January 10 1994 the claimants submitted a claim for compensation and this was amended from time to time up to December 9 1996.

(4) The claimants referred their claim to this tribunal on October 19 1995.

(5) On December 18 1995 the rights and liabilities of Bristol
E Development Corporation were vested in the Secretary of State for the Environment ("the authority") under the Bristol Development Corporation (Property Rights and Liabilities) Order 1995.

(6) On December 5 1996 the authority's solicitors sent a sealed offer by fax to the claimants' solicitors offering to pay compensation of £100,000 in full and final settlement of the claim plus surveyor's fees on Ryde's Scale, reasonable legal expenses (clarified at the hearing to include the costs of the reference) and statutory interest from the date of possession.

(7) Immediately before the start of hearing on December 16 1996, the claimants accepted this offer and agreement was reached on all
F issues except liability for the costs of the reference after the date of the sealed offer.

Decision

Mr Alesbury said that the authority should pay all the claimants' costs of the reference, up to and including the hearing. Mr Fookes said that the authority will pay the claimants' costs up to the date of the sealed offer and the claimants should pay the authority's costs after that date.

Mr Alesbury said that the claimants were unable to make a decision on the sealed offer until they saw a revised report from the authority's expert valuation witness, Mr Napier, and this was not received until lunchtime on Friday December 13 1996, the last working day before the start of the hearing on the following Monday. Discussions between G
the claimants and their advisers took place over the weekend and the offer was accepted on Monday morning December16, immediately before the start of the hearing. This was the first opportunity which the claimants had to accept the offer in the light of Mr Napier's revised report.

Mr Fookes said that this is a case where the normal rule as to costs under section 4 of the Land Compensation Act 1961 ("the 1961 Act") should apply. The claimants have accepted a sealed offer and will not receive compensation in excess of that offer. Thus, in accordance with section 4(1) of the 1961 Act, they should bear their own costs and pay the authority's costs from the date of the offer. There are no special H
circumstances justifying a deviation from this general rule.

In reply, Mr Alesbury said that the claimants' formulation of their response to the offer was dependent on the authority's case, as set out in Mr Napier's report, and this was not received until December 13, the last working day before the start of the hearing. This is a special reason why the normal rule in section 4 of the 1961 Act should not apply.

Both counsel referred to negotiations as to the amount of compensation, the claimants' fluctuating claim, the preparation of a statement of agreed facts and the exchange of reports. I have considered these matters but do not regard them as material and have J
not set them out in this decision.

The jurisdiction of the Lands Tribunal as to costs rests on section 3(5) of the Lands Tribunal Act 1949 and r 52 of the Lands Tribunal Rules 1996. Section 3(5) gives the tribunal power to order costs and, so far as it is material to this reference, this is more specifically set out in r 52(1) of the 1996 Rules as follows:

> Subject to the provisions of section 4 of the 1961 Act . . . ,the costs of and incidental to any proceedings shall be in the discretion of the Tribunal.

The material part of section 4 of the 1961 Act is as follows: K

> (1) Where either —
>
> (a) the acquiring authority have made an unconditional offer in writing of any sum as compensation to any claimant and the sum awarded by the Lands Tribunal to that claimant does not exceed the sum offered; . . .
>
> the Lands Tribunal shall, unless for special reasons it thinks proper not to do so, order the claimant to bear his own costs and to pay the costs of the acquiring authority so far as they were incurred after the offer was made . . .

Mr Fookes said that section 4 applies and the Secretary of State should have his costs after the date of the sealed offer. Mr Alesbury said that, in the circumstances, there is a special reason why this provision should not apply and the claimants should receive all their L
costs.

In my view, section 4 of the 1961 Act does not apply to this reference because the issue of compensation has not proceeded to a hearing. Although as agreed and for the administrative convenience of the parties, I record in this decision the amount of the agreed compensation, I have not been required to make an award of compensation after a contested hearing. In my view, therefore, section 4 does not apply to the circumstances in this reference.

The sealed offer is, however, important when considering liability for costs. The crucial question is when should the claimants have accepted that offer? They say that they could not evaluate it until they M
received Mr Napier's report on December 13. I do not agree. Most of their claim is compensation for disturbance where the facts are wholly within the knowledge of the claimants. As I understand the position, Mr Napier's revised report was prepared in response to a revised claim made on December 9 1996, but he had prepared an earlier report in March 1996. I do not think that the claimants have been prejudiced in their evaluation of the sealed offer by receipt of Mr Napier's revised report only one working day before the hearing. This must be seen in the context of the late submission of an amended claim and the period of this reference.

The claimants' solicitors received the sealed offer by fax on Thursday December 5 1996. In my view, the claimants should have

A accepted it on or before Wednesday December 11 1996, five working days after receipt including the day of receipt. This was sufficient time for the claimants' solicitors to discuss the offer with their clients and their two expert witnesses, having regard to the advanced stage of the proceedings in the reference. Under these circumstances, the claimants should pay the authority's costs after December 11 1996, the date when the sealed offer should have been accepted.

My determination is as follows:

(1) The compensation payable to the claimants for the compulsory acquisition of their leasehold interest in the subject property shall be the agreed sum of £100,000 plus surveyor's fees on the appropriate Ryde's Scale. The payment of legal costs on the transfer of the B leasehold interest and interest on the amount of compensation by the authority are referred to in the accepted offer and are governed by specific statutory provisions.

(2) The authority shall pay the claimants' costs of this reference up to (and including) December 11 1996, the date when the authority's sealed offer should have been accepted. The claimants shall bear their own costs and pay the costs of the authority incurred after that date, including the costs of the hearing on December 16 1996. All costs, if not agreed, are to be taxed by the registrar of the Lands Tribunal on the High Court standard basis.

For a further case on this subject see p245

LANDS TRIBUNAL
LANDLORD AND TENANT ACT 1987

A
Linkproud Ltd *v* Ballard

December 2 1996

JUDGE RICH QC, sitting as a member of the Lands Tribunal

Estates Gazette June 28 1997

[1997] 26 EG 156

Landlord and Tenant Act 1987 — Pre-emption rights — Block of flats — Acquisition of freehold — Landlord paid vendor purchase price and sum for arrears of rents and service charges — Whether
B ***sum payable by nominee purchaser should include additional sum***

On September 10 1993 the appellant landlords contracted to purchase the freehold of a building divided into flats. In accordance with clause 2(f) of the contract the appellants paid the vendor, in addition to the purchase price of £2,100, a sum of £6,434.50, which represented arrears of rents and service charges. The vendor not having complied with the requirement of Part I of the Landlord and Tenant Act 1987 to first offer the freehold to the tenants, the tenants served a purchase notice on the landlords, and became entitled to acquire the freehold. The leasehold valuation tribunal held that the sum payable by the nominee purchaser to the landlords for the acquisition of the
C freehold did not include the additional amount of £6,434.50. The landlords appealed. The tenants contended that some part of the additional sum was in respect of service charges which were not properly recoverable from them.

Decision: **The sum payable by the nominee purchaser to the landlords was £8,534.50. The subject-matter of the original disposal was not only the freehold reversion, but also those rights which were the specific subject-matter of clause 2(f) of the contract. Under section 12 of the 1987 Act, those together are the estate which the purchase notice may require the landlords to dispose of "on the terms which the original disposal was made". At the moment of transfer the landlords were required to pay**
D **£8,534.50; this was the consideration paid for the subject-matter of the original disposal. The degree of overpayment, if it be overpayment, is part of what the landlords are entitled to be repaid on the compulsory acquisition by virtue of the purchase notice served by the tenants.**

No cases are referred to in this report.

Graham Clark (instructed by Malthouse Chevalier) appeared for the landlords; James Pyke, solicitor, represented the tenants.

Giving his decision, **JUDGE RICH QC** said: The appellants
E contracted to purchase the freehold of 45 Wickham Road in accordance with the terms of an auction contract on September 10 1993. It is a property let in five flats. The price which the appellants bid was £2,100, but they contracted in accordance with the special conditions of sale relating to the auction in terms of clause 2(f) which provided as follows:

> The purchaser shall on completion pay to the vendor the amount of any arrears of rent, service charge or other contribution due from any lessee and shall pay to the vendor the amount of any expenditure which may be recoverable from any such lessee by way of service charge or any other contribution but which shall not on the date of completion have been recovered by or reimbursed to the vendor. The vendor shall if so required and at the cost of the purchaser, F
> assign to the purchaser the benefit of such arrears. The certificate of the managing agent of the property shall be conclusive as to the amount of such arrears or expenditure and the purchaser shall accept that any sums so certified are properly recoverable.

The appellants accordingly paid to the vendor not only the sum of £2,100, which was the price which they had bid at the auction, but a further sum of £6,434.50, which was the amount that was certified by the managing agent as being due in accordance with clause 2(f) of the special conditions of sale.

The flats contained in the building are let on terms such that the G
lessees had a right of first refusal on the sale of the reversion under Part 1 of the Landlord and Tenant Act 1987. The vendor did not offer them the opportunity to exercise that right before the auction. Accordingly, they were entitled to serve on the new landlords, the present appellants, notice under sections 11 and 12 of that Act. Having been provided with the relevant particulars of the transaction in accordance with section 11, they served a purchase notice on the appellants in accordance with section 12, requiring them to dispose of the freehold to them. In accordance with section 12(3)(b) they reserved the right for the terms of the disposal to be determined by the leasehold valuation tribunal.

H
The leasehold valuation tribunal determined that the provisions of clause 2(f) should not apply in respect of the purchase by the lessees of the present appellants' interest, so that they would be required to pay to the appellants only the bid price of £2,100. The leasehold valuation tribunal does not appear to have considered whether the appellants should, after their disposal of the freehold, continue to be entitled to enforce against the lessees the arrears to which they had succeeded on their purchase being arrears of rent and service charge properly due to their predecessors in title at the date of the transfer. I must in due course, turn to the provisions of section 141 of the Law of Property Act 1925 more fully. For the moment I merely comment that the effect
of that section is that, in the absence of special provision, upon J
transferring the freehold to the purchasing lessees, the appellants would no longer be entitled to claim against the lessees any sum due under the leases in respect either of the period before they purchased or which accrued during the period between their purchase and their further sale under compulsion to the purchaser lessees. Thus the effect of the decision of the leasehold valuation tribunal, unless some additional provision is made, would be that the present appellants would have bought the property for £2,100 plus paying £6,400 odd for the outstanding arrears of service charge and ground rents and would have become entitled in the meantime to additional sums in respect of K
service charge and ground rent, but would be compelled to transfer their interest in the freehold and to divest themselves of any rights in respect of service charge, etc, on being paid only £2,100. That grotesquely unfair result is justified, as I understand it, as being necessary in order to avoid what is suggested would also be an unfair result, if the purchase by the lessees was made subject to the provision of clause 2(f) of the auction contract.

The service charge and rent which was outstanding at the date of the transfer to the present appellants was at that date the subject-matter of dispute by the lessees. Just as the appellants' predecessor in title had failed to comply with its obligation under the 1987 Act to offer the freehold reversion first to the lessees before putting it up for auction, so also I am led to believe, and for the purposes of this decision, I assume, they had failed to comply with their obligations under the Landlord and Tenant Act 1985, which require consultation and proper tenders before expenditure if they are to be able to recover charges. Thus, I will assume for the purposes of this decision, that the lessees were not liable to pay to the appellants' predecessors as landlord, the whole or maybe any part of the £6,400 odd which the appellants were required under the auction contract to pay to the previous landlord as vendor under that contract. Mr James Pyke contends that it would be unfair if the lessees' opportunity to challenge their liability to make that payment to their landlord were be lost because they were compelled to pay the whole of the disputed sum to the appellants for the transfer to them of the freehold.

In fact it is the intention of the lessees that their purchase should be made in the name of a residents' company and the appellants are content to transfer the property to such company. This makes it possible to express the true position of the parties more accurately by referring separately to the lessees and to the residents' company. The reality is, that it is the residents' company that would have to pay the whole of the sum certified as due from the appellants to their vendor. As between the residents' company and each individual lessee, liability to make payment under the leases could still be challenged and would still be determined by reference to the terms of their leases and the provisions of the Landlord and Tenant Act of 1985.

Nevertheless, the lessees viewing themselves as a body, apprehend an injustice if they are required to pay as part of the consideration for their purchase of the freehold from the appellants not only what the appellants' bid but also the whole of the disputed service charges and rent claimed by the predecessor landlord. In respect of service charges and rent accruing during the period of the present appellants' ownership, they raise no objection to having to pay on completion that which would be due at that date to the appellant company, it being the result of section 141 of the Law of Property Act of 1925 that the purchasing residents' company would upon acquiring the freehold reversion at the date of transfer be entitled to recover what was due from the lessees.

In order to resolve the conflict between the parties, it is necessary to refer to the terms of the Act and only to the terms of the Act. It is not appropriate to speculate whether the vendors who failed to offer the freehold to the qualified lessees have by some device succeeded in recovering service charges not properly due to them. The leasehold valuation tribunal seems to have accepted that the price bid at auction should have reflected the obligations under clause 2(f) to pay for service charges which might not be recoverable. If it did not do so, it is for the lessees to decide whether they wish to exercise their right to acquire against the present appellants, as they have done.

I turn therefore to the words of section 12. It provides that in the circumstances which here exist that the qualifying tenants of the flats may serve on the new landlord a notice requiring him

> to dispose of the estate or interest that was the subject-matter of the original disposal on the terms on which it was made (including those relating to the consideration payable) to a person or persons nominated for the purposes of the section.

It is usual for the terms on which a freehold reversion is purchased to take account of the effect of section 141 of the Act of 1925. In a case where the outstanding sums due from the lessees are not in dispute, what is, as I understand it, a standard form of condition on the contract for sale of a reversion of this kind would meet the situation appropriately. That is such as is contained in the first sentence of clause 2(f) in the present conditions for sale:

> The purchaser shall on completion pay to the vendor the amount of any arrears of rent service charge or other contribution due from any lessee and shall pay to the vendor the amount of any expenditure which may be recoverable from any such lessee by way of service charge or other contribution but which shall not on the date of completion have been recovered by or reimbursed to the vendor.

If there were no dispute between the landlord and tenant, the purchaser would by that requirement be bound to pay the full price for all that to which he became entitled under section 141 of the Law of Property Act upon transfer of the reversion to him. Upon disposal under the Landlord and Tenant Act 1987 "the new landlord" would if identical terms were incorporated into his sale, as had been incorporated into his purchase, have had the opportunity to collect the money in and to expend money as necessary during the period of his ownership, and upon disposal to the lessee purchasers would receive the net amount due to him and the lessee purchasers would obtain under section 141 the right to collect that same sum from the lessees.

Because of their objection to the rest of clause 2(f) of the contract under which the appellants acquired, the lessees in this case invoked para (b) of section 12(3) of the Act 1987. This provides:

> A purchase notice —
>
> . . .
>
> (b) may instead of specifying the estate or interest to be disposed of or any particular terms on which the disposal is to be made by the new landlord (whether doing so expressly or by reference to the original disposal), provide for that estate or as the case may be for any such terms to be determined by a rent assessment committee [now the leasehold valuation tribunal] in accordance with section 13.

Section 13 merely provides the jurisdiction of the relevant tribunal to determine questions. It makes no further provision as to the basis on which questions are to be determined than is to be inferred from the terms of section 12.

I turn back therefore to the provisions of section 12(1). The purchase notice requires the appellants in this case to dispose of the estate or interest that was the subject-matter of the original disposal. Mr Pyke has drawn to my attention authority for the proposition that the original disposal is the actual transfer or conveyance of the property not the contract made at the auction for such transfer or conveyance. The exact terms of clause 2(f) of the contract are not incorporated directly into the transfer or conveyance. They none the less determined, because the completion statement on transfer was drawn in accordance with such terms, what were the actual terms on which the disposal was made. I must, however, before referring to those terms, identify more clearly what is the estate or interest which was the subject-matter of the original disposal. Section 141 of the Law of Property Act 1925 provides that:

> Rent reserved by a lease, and the benefit of every covenant or provision therein contained having reference to the subject-matter thereof, and on the lessee's part to be observed or performed . . . shall be annexed and incident to and shall go with the reversionary estate in the land . . .

Thus when the reversionary interest passed from their predecessor in title to the appellants, there went with it the right to rent reserved and the benefit of covenants, and it is made clear by the subsequent subsections of section 141 as construed by the courts, that consequently any rent and service charges which were outstanding at the date of transfer became the property of the transferee (the appellants in this case) and the transferor (their predecessor in title) was no longer entitled to enforce a claim for any outstanding charge or rent. Thus the subject-matter of the original disposal is not only the freehold reversion, but also those rights which were the specific subject-matter of clause 2(f) of the contract. These together are the estate which the purchase notice may require the appellants to dispose of to the lessee purchasers "on the terms on which the original disposal was made (including those relating to the consideration payable)". At the moment of transfer, what the appellants were required to pay for the transfer to the appellants of the reversionary interests which included the right to enforce the covenants up till that date, was not only the price that they had bid at the auction, namely £2,100, but also the amount of the arrears of expenditure as certified by the managing agent. It is accepted by the parties that these were in fact certified to be

£6,434.50 and that was the sum paid on completion. It follows that what was paid in consideration of that which was the subject-matter of the original disposal was not £2,100 but £8,534.50.

What the appellants will be bound to dispose of to the lessee purchasers under the purchase notice will be that same reversionary interest together with such rights as are annexed and incidental to it in accordance with section 141 at the date of transfer. The lapse of time will mean that those interests will to a degree have changed from that which was the subject-matter of the original disposal. The leases will be that much nearer to their expiry date and the amounts which may be due by way of rent or service charge may have been increased by the passage of time and expenditure by the appellant landlords or reduced by payments made by the lessees. Nevertheless it is only the interests as they exist at the time when the purchase notice comes to be complied with, that the appellant lessors can be in a position to dispose of. The Act must therefore be construed as requiring that interest as thus changed by passage of time, to be disposed of in return for appropriate payment fixed on the same basis as was the basis upon which the appellants acquired their interest.

The difficulty in this case arises from the effect of the third sentence of clause 2(f) of the contract, in that it requires the purchasers (the appellants in this case) to pay for rights assigned to them under section 141, more than those rights are worth if, as I have assumed, the lessees are not truly liable to make the payments certified. However that degree of overpayment, if it be overpayment, is part of the consideration which the appellants have in fact paid to their predecessor in title. It is therefore part of what they are entitled to be repaid on the compulsory acquisition, by virtue of the purchase notice served by the lessee purchasers, of the interest which they acquired for such consideration.

I have invited Mr Clark on behalf of the appellants to formulate a clause which will have the effect of ensuring that the appellants upon transferring their interest to the lessee purchasers, receive in respect of the interest and rights that they acquired, the full sum which they paid for them, namely the £8,534.50 to which I have already made reference.

He has settled a clause in which he has included the provision of the second sentence of clause 2(f) even although it is accepted as being wholly unnecessary because section 141 achieves the result required without a specific assignment. Mr Pyke on behalf on behalf of the lessees has taken no objection to the draft as having the result, which I have determined to be the appropriate term. Mr Clark's draft is as follows:

> The purchaser shall on completion pay to the vendor:
>
> (a) The sum of £6,434.50 being the amount that the vendor paid to the vendor on the original disposal, in respect of arrears of service charge and ground rent pursuant to clause 2(f) of the contract in that disposal and as to which the certificate of the then Managing Agents was conclusive, it being agreed that no credit is due in respect of that sum, and
>
> (b) The amount of any arrears of rent, service charge or other contribution due from any lessee since 6 December 1993 and shall pay to the vendor the amount of any expenditure since 6 December 1993 which may be recoverable from any such lessee by way of service charge or other contribution but which shall not at the date of completion have been recovered by or reimbursed to the vendor. The vendor shall if so required and at the costs of the purchaser assign to the purchaser the benefit of such arrears.

I therefore allow this appeal, and determine the terms of the disposal to the respondent accordingly.

A
Blackstone Investments Ltd *v* Middleton-Dell Management Co Ltd

LANDS TRIBUNAL

October 31 1996

MR M ST J HOPPER FRICS

Estates Gazette April 5 1997

[1997] 14 EG 135

B

Leasehold Reform, Housing and Urban Development Act 1993 — Acquisition of block of flats — Determination of price — Whether any value in right to receive insurance commission

The appellant was the freehold owner of a block of 12 flats let on leases for terms of 99 years from March 25 1975 at rising ground rents. Under the leases the landlord was obliged, *inter alia*, to insure the premises against certain risks, and was entitled to recover the premiums under the service charge provisions. Nine of the lessees gave notice to acquire the freehold under the Leasehold Reform, Housing and Urban Development Act 1993. The appellant appealed
C the decision of the West Midlands Leasehold Valuation Tribunal, that the price payable should be £11,750, contending that the price should include the value of the right to receive commission on the insurance premiums amounting to an additional £2,032.20.

***Decision*: The landlord under the leases was entitled to recover the full cost of the insurance premiums under the service charge provisions without having to deduct any insurance commission earned. The price which would be paid for the freehold in the hypothetical circumstances under the 1993 Act would include an element to reflect the expectation of the receipt of insurance commission. Where there is no evidence of sales of comparable**
D **properties, which would reflect any element of value attributable to the expectation of the receipt of insurance commission, the value of any insurance commission must be considered separately from the value of ground rental income. A YP of 8.5, equivalent to a yield of 11.76%, was appropriate which when applied to an annual commission of £101.61 gave an addition to the price of £863.68**

The following cases are referred to in this report.

Castlebeg Investments (Jersey) Ltd v *Lynch* (1988) 57 P&CR 493, LT

Malcom Bruce Pennycuick FRICS appeared for the appellants; Brian Poultney, solicitor (instructed by Colley & Tilley) represented the
E respondents.

Giving his decision, **MR M ST J HOPPER** said: This is an appeal by Blackstone Investments Ltd (the appellants) from a decision of the West Midlands Leasehold Valuation Tribunal (LVT) determining the price to be paid by Middleton-Dell Management Co Ltd (the respondents) for the freehold interest in a block of flats known as The Dell, 125-127 Middleton Hall Road, Kings Norton, Birmingham (the subject property), under the provisions of section 24 of the Leasehold Reform, Housing and Urban Development Act 1993 (the 1993 Act) in the amount of £11,750.

Mr Malcolm Pennycuick [FRICS] is a director of the appellant F
company; he appeared on their behalf and gave evidence; Mr Brian Poultney appeared for the respondents.

From documentary evidence produced, and oral evidence given, at the hearing I find the following facts:

1. The subject property comprises a single three-storey block of 12 flats constructed in about 1962 with brick walls and a flat roof.

2. The freehold interest in the subject property is owned by the appellants; the individual flats are let on leases for terms of 99 years from March 25 1975 at rent of £25 pa for the first 25 years, £50 pa for the next 25 years, £75 pa for the next 25 years and £100 pa thereafter.
Under the provisions of the leases, first, the lessor is liable "to insure G
and keep insured in the full replacement value the Building and Lessor's fixtures therein against loss or damage or fire and such other risks as the Lessors should deem desirable or expedient in some insurance office, or with underwriters of repute"; second, the lessees are liable "to pay to the Lessors without any deduction by way of further and additional rent a proportionate part of the expenses and outgoings incurred by the Lessors in the repair, maintenance, renewal and insurance of the Building and the provision of services therein and the other heads of expenditure as the same are set out in the Fourth Schedule . . . (hereinafter called the service charge)"; the fourth
schedule includes a reference to "the cost of insuring and keeping H
insured throughout the term hereby created the Building and all parts thereof the Landlord's fixtures and fittings therein and all the appurtenances, apparatus and other things thereto belonging . . .".

3. Nine of the lessees qualified under the 1993 Act gave initial notice to purchase the freehold interest in the subject property in April 1994, although only eight of the lessees served the notice. The respondents admitted that the participating lessees were entitled to exercise a right to collective enfranchisement, but the price was not agreed.

4. The disputed price was the subject of a reference to the West J
Midlands LVT which heard the parties on May 25 1995 and by a decision dated July 5 1995 determined the price at £11,750. They attached to their decision details of the valuation by which they arrived at the price which, summarised, was as follows:

	£
(a) Value of the rent receivable during the term	5,383.10
(b) Value of the reversion	92.34
(c) Shared marriage value	6,262.28
(d) Compensation for loss of management and insurance commission	Nil
	11,737.72 K
Say	11,750

5. The appellants accept the price determined by the LVT, except its exclusion of the value of the potential receipt of insurance commission which they quantify at £2,032.20.

6. Mr Pennycuick owns all the shares in the appellant company, which employs as managing agents for the subject property, a partnership known as Pennycuick & Brown. The appellants contract for the insurance of the subject property as required under the flat leases through brokers known as White Hart Insurance Agency (White

A Hart); Mr Pennycuick is a partner in White Hart and effectively controls it.

7. White Hart arrange insurance cover for three other blocks of flats owned by the appellants and a number of other blocks of flats owned by other owners. On behalf of all their clients they insure the buildings concerned with Royal Insurance Co at premiums which are discounted by 70%, because of the large amount of insurance involved. The benefit of this discount is passed on, *inter alia*, to the lessees of the flats in the subject property; plant and machinery in the relevant buildings is separately insured with Vulcan Insurance Co; no discount is obtained upon the premiums relating to plant and machinery. Both B insurance companies pay commission on the total of the premiums of all the blocks insured through White Hart in the amount of 15% of the premiums payable, after the 70% discount in the case of the building insurance. The appellants allow White Hart to retain the commission in respect of the subject property.

Appellants' case

Mr Pennycuick said that, in their decision, the LVT suggested that, although the appellants had the advantage of the commission arrangement, it would seem to be a personal circumstance which must be disregarded for valuation purposes. This matter had not been C discussed at the hearing and Mr Pennycuick had not therefore been able to make any representations about it.

This had been the first case to be heard by West Midlands LVT under the 1993 Act and quite rightly the matter had been approached with extreme caution. However, by the date of the LVT decision there had been cases decided by other LVTs which should have formed some sort of precedent and Mr Pennycuick drew attention to five such decisions. First, in the case of a large house at Bournemouth converted into eight flats, (LRI/94/D), the Southern LVT had included in the decided price a sum of £645 in respect of the value of insurance commission. Second, in the case of a house converted into three flats at Brighton (BTN/LR/5) the South Eastern LVT had included in the price D decided an amount in respect of the value of insurance commission. Third, the London LVT had made a similar decision in relation to a building containing five flats at Surbiton (LON/ENF/5). Fourth, Mr Pennycuick referred to the decision of this tribunal in *Castlebeg Investments (Jersey) Ltd* v *Lynch* (1988) 57 P&CR 493, where the price payable for the enfranchisement of a house under the Leasehold Reform Act 1967 (as amended) (the 1967 Act) was determined at £680, which included a specific amount of £109.20 in respect of the expectation of the right to receive insurance commission. Fifth, Mr Pennycuick referred to a decision of the South Eastern LVT (BTN/LR/3) in respect of a house converted into flats at Hove, in E which the decided price was based on 15 years' purchase of the ground rents payable, to include, *inter alia*, any profit from insurance commission.

Mr Pennycuick also referred to another LVT decision made since the LVT decision in the present case in respect of a block of 50 flats built about 15 years ago in Bournemouth. In its decision the LVT had included an amount of £1,200 (two years' purchase of £596) as compensation for the loss of insurance commission in the decided price.

Mr Pennycuick submitted that the first three cases to which he had referred followed the trend shown by the *Castlebeg* case and two other F decisions of the West Midlands LVT in cases under the 1967 Act.

In the present case it was inconceivable that the purchaser of the freehold interest would not make an arrangement under which he would receive commission from an insurance company; such commission formed part of the regular income to be derived from the property.

Mr Pennycuick drew attention to the definition of the price payable in para 3(1) of the Sixth Schedule to the 1993 Act, which refers to the amount the freeholders' interest might be expected to realise if sold on the open market by a willing seller (with neither the nominee purchaser nor any participating tenant buying or seeking to buy) on certain assumptions. He had been purchasing blocks of flats for many years and assured me that, if the leases allowed for insurance by the G landlord, then the value of the commission relating thereto was taken into account when forming an opinion as to the price to be paid; in other words the right to receive such commission had a commercial value.

It was clearly of great benefit to the lessees of the subject property that the commission arrangement had been entered into so that the benefit of the substantial discount could be obtained and passed on to the lessees.

During actions for non-payment of service charges and insurance costs Mr Pennycuick had been involved in lengthy and detailed examinations of service charge costs and fees and it had never been H suggested by either learned counsel or the judge that insurance commission should be disallowed.

Mr Pennycuick produced a copy of a lease of one of the flats at the subject property and an audited statement of the total annual cost in respect of the service charge for the year ended December 31 1994.

His valuation of the expectation of receiving insurance commission as put to the LVT was:

		£
1994-1995 Plant and machinery	£223.26 @ 15%	33.49
1994-1995 Buildings	£681.22 @ 10%	68.12
		101.61 J
Years' purchase in perpetuity @ 5%		20
		2,032.20

Mr Pennycuick made no deduction for the cost of collection as this was done as part of the collection of the service charge.

The 5% capitalisation rate reflected the great security of the income and the fact that the income was on a much reduced rate, ie it could have been as much as 30% of the full premium; the building element reflected the 70% discount. Mr Pennycuick used a commission rate of 10%, rather than the actual 15%, which is the lowest figure he knew of in the market place and must be available to any reputable freeholder and/or agency. K

In some of the cases to which he had referred the value of insurance commission had been reflected by adjusting the figure at which ground rental income had been capitalised. As a valuer Mr Pennycuick could not agree with this approach; while the end result might or might not be correct it was proper to look at the actual insurance figures involved in each case.

During 1995 he had purchased a block of flats at Henley-in-Arden and paid an enhanced figure over and above his estimate of the worth of the ground rent income receivable to reflect the value of insurance commission. The ground rents aggregated £1,202 pa, the purchase L price was £13,400, which represented 11.2 years' purchase, but that merely showed the result not how he had determined the value.

The idea of valuing insurance commission was hardly new, but, in the past, had mainly been employed in the enfranchisement of single houses. The value in relation to flats blocks was much more substantive and under the terms of the 1993 Act should, in Mr Pennycuick's opinion, be calculated by reference to the commission to be paid, as it would be in the open market.

Applying the multiplier used in the *Castlebeg* case and using the actual commission passing would produce:

M

		£
1994-1995 Plant and machinery	£222.36 @ 15%	33.49
1994-1995 Buildings	£681.22 @ 15%	102.18
		135.67
Years' purchase		6.5
		881.86

Mr Pennycuick considered that the collection of insurance premiums as additional rent under the terms of the flat leases made the income very secure. It must by its nature keep pace with inflation and to reflect this fact he had used a 5% yield rate in his original valuation; he had tempered this by using a very modest commission rate. He submitted that his approach in his original valuation properly reflected the price that might be paid in the open market, particularly having regard to the long-term nature of the investment.

In answer to me Mr Pennycuick said that a purchaser of the freehold interest in the subject property would not be able to get 15% insurance commission, but, on the other hand, would not be able to get a 70% discount and these two factors would offset one another.

There had been no challenge to the service charge in respect of the subject property.

Mr Pennycuick sought a determination that the price payable by the nominee purchaser should be £13,769.92, made up as to the LVT's valuation before rounding of £11,737.72, plus £2,032.20 in respect of compensation for loss of insurance commission.

Respondents' case

Mr Poultney referred to what the LVT had said in relation to insurance commission. It was not the case that a landlord could never retain commission; it was dependent upon the terms of the leases. Clause 3(3) of the leases of the subject property referred to the lessees' liability to pay a proportionate part of the expenses incurred by the lessors in insuring the building; the expense incurred was the net amount, ie after allowing for the receipt of commission. Para 3 of the fourth schedule to the leases, which detailed the expenses payable by them by way of a service charge, referred to the cost of insuring the building; again this meant the net cost, after allowing for the receipt of commission. Clause 4(c) required the lessors to insure the building with some insurance office or underwriters of repute, but not with the landlords' nominated insurers.

In the *Castlebeg* case the insurance clause in the lease provided that the lessee had to insure the house through the "Commercial Union Assurance Company or such other office nominated by the lessor through the lessor or his agent". In the case of the subject property there was no express provision for the landlord to require the tenants to insure with his chosen company; the LVT had differentiated between the circumstances in this case and the *Castlebeg* case.

If the landlord of the subject property had entered into an arrangement to put all decorative repair work through a particular contractor and secured a 25% discount on cost in return, if £100 worth of work had been done and the landlord paid £75, the amount recoverable from the tenants by way of the service charge would have been £75.

Mr Poultney submitted that only the net cost of insurance after allowing for the commission, was recoverable under the terms of the leases of the flats at the subject property.

Alternatively, if the insurance cost recoverable should include commission section 19(1) of the Landlord and Tenant Act 1985 (the 1985 Act) provided that relevant costs were only recoverable to the extent that they were reasonably incurred; in the present case, if they were regarded as incurred, they were not reasonably incurred and the lessees could challenge the landlord's nominated insurers under the Schedule to the 1985 Act. The Financial Services Act 1986 placed obligations upon the recipients of commissions to disclose them; statute law was tending against the recipients of commission.

Mr Poultney agreed that the insurance element of the service charges at the subject property had not been challenged, although there had been disputes about other elements.

The commission was received by White Hart, but this was immaterial because Mr Pennycuick was effectively the owner of the subject property and a partner in the agency. The relationship could have been re-jigged so that the appellants took the commission. In these circumstances, the appellants could only recover the net cost of insurance, ie net of the commission, through the medium of the service charge.

Mr Poultney referred to three decisions of the London LVT deciding prices payable for freehold interests under section 24 of the 1993 Act. The first (LON/ENF/1) related to a north-west London house converted into two maisonettes; the landlord's valuer was recorded as having said that the receipt of insurance commission might be less certain in the future by reason of a possible change in the law. In the present case the appellant valued the commission at 20 years' purchase; that was too much; the long term was uncertain. Even if the leases enabled the landlord to retain commission and the lessees could not challenge that as unreasonable, it did not follow that the total price determined by the LVT would be increased. In the case of the north-west London decision to which he had referred, the price was determined on the basis of applying a 13% yield rate to the ground rental income, reflecting any benefit of insurance commission, which was more favourable to the nominee purchaser than the 9.5% yield rate, excluding insurance commission, which was the basis of the LVT decision in the present case.

The second case was that referred to by the appellants in respect of the building comprising five flats at Surbiton; the LVT had adopted a yield rate of 12.5% to capitalise the ground rental income and reflect the value of insurance commission; again this was more favourable to the nominee purchaser than the yield rate adopted in the present case.

Last, Mr Poultney referred to the case of a west London house (LON/ENE/4) converted into four flats; a LVT decided a price based upon a 13% yield rate; they observed that auction sale evidence showed that such a yield rate reflected the value of insurance commission; again the yield rate was more favourable to the nominee purchaser than in the present case.

Mr Poultney submitted that the LVT had held against the appellants and the onus of proof was on them.

Appellants' response

Mr Pennycuick said that what was important was the view of a purchaser of the freehold interest. The receipt of insurance commission was not peculiar to the appellants; their actual arrangement was immaterial.

The LVT did not say that the lease provisions did not allow the landlord to take commission, but said that the arrangement enjoyed by the appellants may not be enjoyed by a purchaser. Mr Pennycuick did not know of any landlord of a block of flats who did not have a commission arrangement. Insurance commission would be in the mind of a purchaser of the freehold interest in the subject property in the open market. Insurance commission was not a hidden profit; commission arrangements were well known.

In the cases of the leases of the subject property, the service charge included the insurance element, was reserved as rent and thus a secure income; failure to pay rent could result in restraint or repossession; orders for such remedies had been granted in Birmingham County Court. Was the charge fair and reasonable? Where a large discount had been obtained a judge would say it was reasonable.

In the LVT decision relating to the north-west London house converted into flats there was mention of insurance commission income being uncertain. Mr Pennycuick, with his experience as an insurance broker, did not see how individual lessees would be able to arrange all the insurance direct, in cases involving plant and machinery and multi-occupied properties.

In the present case the insurance premium was recovered as part of the service charge; his capitalisation of it at 20 years' purchase reflected the security of the income and had regard to bank base rates.

LVTs had not yet agreed on how to approach the valuation of the expectation of the receipt of insurance commission; some valued it separately; others reflected it in the yield rate at which they capitalised the ground rental income. A purchaser would value the latter as such and then value the income from insurance commission and fix the price accordingly.

Decision

Section 32 and the Sixth Schedule to the 1993 Act provide that the price payable by the nominee purchaser for the appellants' freehold interest in the subject property shall be the aggregate of three elements; the element relevant to the present appeal is the amount which, at the valuation date, the freehold interest (with neither the nominee purchaser nor any participating tenant buying or seeking to buy) might be expected to realise if sold in the open market by a willing seller on certain assumptions, which include an assumption that the freehold interest is subject to any subsisting leases and that there are no statutory rights to acquire any interest in the subject property nor any new lease thereof; the other assumptions are not relevant to this appeal.

The issue raised in this appeal is whether the value of the freehold interest in the subject property, as defined by statute, would reflect any element of value attributable to the expectation of the receipt of insurance commission and, if so, the amount of that element of value.

What has to be determined is the likely view of potential purchasers and Mr Pennycuick is right in saying that the appellants' actual arrangements are immaterial, except in so far as they provide a practical example of how insurance commission may be earned.

Mr Pennycuick valued the expectation of the receipt of insurance commission to a purchaser by reference to the actual premiums paid, which, in so far as the buildings are concerned, reflected a 70% discount, unlikely to be available to a purchaser, and estimating the commission by applying 15% to the premium in respect of plant and machinery and 10%, rather than the 15%, which the appellants themselves obtained, in respect of buildings.

Mr Poultney says that, assuming a purchaser of the freehold obtained insurance commission by entering into agency arrangements with insurance companies, the amount chargeable to the lessees under the terms of the leases would be the amount of the premiums, less the amount of the commission. The fourth schedule to the leases, which specifies lessors' expenses which are to be included in the service charge, refers to "the cost of insuring". The leases contain no obligation upon the freeholder to enter into an agency arrangement and earn commission and, if he did not do so, the lessees would not appear to have any valid complaint against the full amount of the premiums being included in the service charge.

Further I find that the freeholder levies the service charge in his capacity as landlord under the leases, whereas, if he enters into an agency arrangement with an insurance company and receives commission from them as their agent for performance of duties under the agency arrangement, albeit that he is able to enter into such an arrangement because he owns the freehold, he should be regarded as acting in a capacity other than as landlord and that any commission earned should not be taken into account in calculating "the cost of insuring".

The "cost of insuring" included in the service charge in respect of the subject property has never been challenged, although it made no allowance for insurance commission earned, neither was there any evidence to suggest that the point had been taken in any other similar cases.

I therefore find that the "cost of insuring" under the terms of the lease should be interpreted as meaning the cost of the premiums, disregarding any insurance commission earned by the landlord through an agency arrangement with an insurance company.

In the event that I so found, Mr Poultney submitted that the cost of insurance within the service charge, if it did not give credit for any commission earned, could have been challenged under section 19(1) and the Schedule to the 1985 Act, which limited costs included in a service charge to those "reasonably incurred". However, if, as I have found, the commission is received by the freeholder as insurance company agent rather than as landlord, I do not think that this statutory provision would enable the lessees to challenge the "cost of insuring", because it did not give credit for any commission earned by the landlord as agent.

Mr Pennycuick is experienced in the valuation of freehold interests in buildings let in flats on long leases and their management. His evidence is that the purchasers take into account the potential to obtain insurance commission when considering what to pay for the freehold interest, as he himself does. His is the only evidence before me on this and I accept it.

I accept Mr Poultney's submission that the circumstances in the *Castlebeg* case differ from those in the present case, in as much as, in that case, the tenant had to insure with an insurance company of the landlord's choosing.

However the other LVT decisions cited by the parties all related to cases under section 24 of the 1993 Act. Apart from one case, I could not find any details of the lease terms relevant to the recovery of insurance costs. In the case relating to a west London house the terms would seem to have been similar to those in the present case and I would expect the relevant lease terms in the other cases to be broadly similar and not materially different from those in the present case.

I therefore find it significant that, although the way in which the value of the expectation of the receipt of insurance commission was reflected, the prices decided in each of the LVT decisions referred to expressly included an element of value attributable to such an expectation.

On the evidence I find that the price which would be paid for the freehold interest in the subject property, in the hypothetical circumstances to be envisaged under the provisions of the 1993 Act, would include an element to reflect the expectation of the receipt of insurance commission.

I turn to consider the quantification of the value of the expectation of the receipt of insurance commission.

Where there is evidence of sale prices of comparable properties, these will reflect any element of value attributable to the expectation of the receipt of insurance commission and I can understand the rationale of reflecting the value within an overall valuation, including the receipt of ground rent income, where the decision is based upon such evidence. However, where there is no such evidence, I accept Mr Pennycuick's view that the value of the expectation of insurance commission should be considered separately from the value of the ground rent income.

Mr Poultney compared the yield rates used to capitalise ground rent income, and reflect the value of the expectation of insurance commission, in the three LVT decisions referred to, with the 9.5% rate used by the LVT in the present case, excluding any such value; in each of the three cases the yield rate was higher than 9.5%. However, there is only one issue in this appeal, which is whether the price should reflect a value for the expectation of the receipt of insurance commission; clearly the price decided by the LVT did not do so and, apart from the lack of any evidence on the point, I find it is not open to me to disturb the value to be found, excluding the expectation of the receipt of insurance commission.

Mr Pennycuick's valuation commences with actual premiums paid, reflecting the 70% discount in the case of buildings. He takes the commission at the actual rate of 15% for plant and machinery and 10% for buildings, as against the actual rate of 15%, because a purchaser would not expect to achieve such a commission rate, but would be certain to assume a rate of not less than 10%. In so far as a purchaser would also be unlikely to achieve a discount as high as 70% Mr Pennycuick's valuation seems favourable to the nominee purchaser and I accept his estimated commission of £101.61.

Mr Pennycuick regards the commission as well secured income recoverable as rent, which would rise with any increase in building costs over a long period; he thought this justified the use of a low yield rate. Mr Poultney submitted that there was uncertainty as to the commission income and I find that, in several respects, such income is not as secure as ground rental income. In the *Castlebeg* case, which specifically related to the relative yield rates to be applied to rental income and insurance commission, the tribunal accepted evidence as to the uncertainties of commission income as including the following reasons:

(a) commission income was not supported by any covenant; it was supported by a mere contract under which the insurance company was obliged to pay commission which, no doubt, could be terminated upon notice being given;

(b) there was no certainty that the commission rate would always remain the same there was a possibility that it might be reduced.

As far as precedents are concerned the years' purchase applied where the expectation of the receipt of insurance commission was separately valued in the cases cited, range from two to 6.5 compared

A with Mr Pennycuick's 20. Given that the estimated commission did not reflect a substantially discounted premium, unlikely to be obtainable by a purchaser, I do not think the value of the expectation of the receipt of commission would be likely to exceed 6.5 years' purchase, reflecting a yield rate of 15.38%. However, as Mr Pennycuick's estimate of commission income reflected a substantially discounted premium, unlikely to be envisaged by a purchaser, I think a rather higher years' purchase is justified and adopt 8.5, equivalent to a yield rate of 11.76% and giving a value of £863.68 (£101.61 x 8.5) to which I add the LVT's valuation before rounding thus:

B

	£
LVT valuation	11,737.72
Value of expectation of the receipt of insurance commission	863.68
	12,601.14
Say	12,600

The appeal is therefore allowed to the extent that I determine the price payable by the respondents to the appellants for the freehold interest in the subject property at £12,600.

This decision determines the substantive issues raised between the parties, and the tribunal's award is final. The parties are invited to
C make such submissions as they are advised as to the costs of the hearing and a letter accompanies this decision as to the procedure for submissions in writing. The tribunal will, in due course, incorporate an order as to costs in an *addendum* to this decision. Rights of appeal under section 3(4) of the Lands Tribunal Act 1949 and the Rules of the Supreme Court Ord 61 will not accrue until the decision has been thus completed, ie from the date of the *addendum*.

Addendum as to costs

The parties made submissions as to costs.

The appellants claimed their costs which they said that they had endeavoured to keep to a minimum by not employing a solicitor to
D represent them at the hearing. They had made a written offer, without prejudice, to settle the matter by the payment of an additional £500, E
with each party bearing their own costs, in respect of the loss of insurance commission on February 15 1996.

The respondents submitted that each party should pay their own costs; in terms of the additional price payable, the decision was less than half the amount claimed. Further the respondents had been penalised by the delay occasioned to their enfranchisement due to the time taken by proceedings before both the LVT and this tribunal. This should be taken into consideration in making any award for costs. As to the appellants' *Calderbank* offer, this was made in a letter dated April 10 1996 and received by the respondents' solicitors on April 11 1996, which left only two clear working days prior to the date of the F
hearing in which to take instructions from a number of individuals comprising the participating tenants. There was insufficient time in which to explain comprehensively the consequences of the offer and to take full instructions prior to the hearing date. The respondents should not therefore be penalised because of the *Calderbank* offer.

Each party commented upon the other submissions; the appellants pointed out that, although their *Calderbank* letter was dated April 10 1996 it repeated a without prejudice offer made on February 15 1996.

Although the appellants succeeded on the issue of principle involved in the appeal, the additional price decided was less than half the amount claimed and, but for the evidence of the appellants' offer, I G
should have been minded to award the appellants only part of their costs. However, it appears to me that the appellants' offer, first made by a letter of February 15 1996, was a fair one and, on the documentary evidence, it seems clear that the respondents had adequate time to consider it, which their solicitors indicated, in a letter dated March 18 1996 refusing the offer, they had done.

In these circumstances I think that the appellants are entitled to their costs of the appeal. It is therefore ordered that the respondents should pay the appellants' costs of the hearing, such costs, in default of agreement, to be taxed by the registrar of the Lands Tribunal on the High Court scale. H

LANDS TRIBUNAL
RESTRICTIVE COVENANT

In University of Westminster's Application

December 3 1996

JUDGE MARDER QC, President

Estates Gazette May 31 1997

[1997] 22 EG 147

Restrictive covenants — Unopposed application under section 84 of Law of Property Act 1925 — Whether tribunal exercised discretion properly in modifying rather than discharging covenants

The applicant university applied to the tribunal under paras (*a*), (*aa*), (*b*) and (*c*) of section 84(1) of the Law of Property Act 1925 to discharge or modify restrictive covenants affecting one of its properties, and contained in two deeds of July 1966, to enable the property to be used for general educational purposes. In pursuing its application, the university complied with the Lands Tribunal Rules and gave notice of its application to those who might have had the benefit of the covenants. Only one objection was made, and that was withdrawn on July 19 1996 before a preliminary hearing. The President decided to modify the covenants and a draft order was sent to the applicants. Following a letter from the applicants' solicitors requesting that the covenants should be discharged rather than modified, the President held a hearing to receive submissions.

***Decision*: The application to discharge the covenants was refused and an order modifying them in the form of the original draft was issued. There is a discretion in the tribunal as to whether to grant relief under section 84. It is a matter of discretion to limit or cut down the form of relief sought. The discretion is not limited by reason of the fact that the application is unopposed. The discretion must be exercised judicially. Two matters were relevant in refusing to order a discharge of the covenants, rather than their modification. The form of notice to potential objectors made plain that the applicants were intending to continue the use of the land for educational purposes. Second, to discharge the covenants wholly would deprive those who may be entitled to enforce them of any opportunity to exercise control over future user of the property.**

The following cases are referred to in this report.

Driscoll v *Church Commissioners for England* [1957] 1 QB 330; [1956] 3 WLR 996; [1956] 3 All ER 802; (1956) 7 P&CR 371; [1956] EGD 259; 168 EG 521, CA

Farmiloe's Application, Re (1983) 48 P&CR 317, LT

Gee v *National Trust for Places of Historic Interest or Natural Beauty* [1966] 1 WLR 170; (1966) 17 P&CR 6, CA

Ghey & Galton's Application, Re [1957] 2 QB 650; [1957] 3 WLR 562; [1957] 3 All ER 164; (1957) 9 P&CR 1, CA

Goldsmith's Application, Re (1961) 13 P&CR 209, LT

Jones v *Rhys-Jones* (1974) 30 P&CR 451, CA

Ridley v *Taylor* [1965] 1 WLR 611; [1965] 2 All ER 51, CA

Taylor's Application, Re (1964) 16 P&CR 169, LT

Nicholas Taggart (instructed by Nabarro Nathanson) appeared for the applicants.

Giving his decision, the President, **JUDGE MARDER QC**, said: This is an application for the discharge or modification of restrictive covenants, which raises a novel point of some general significance as to the nature and extent of the tribunal's discretion under section 84 of the Law of Property Act 1925. The applicants are the University of Westminster, and the property in question is an extensive area of land and buildings fronting Marylebone Road and occupied by the university as freehold owners.

The subject property is affected by restrictions as to user imposed by two deeds entered into in July 1966. The first deed of release and covenant was made July 26 1966 between the Portman Family Settled Estates Trustees and the Greater London Council (the predecessors in title of the present applicants). This deed was expressed to be supplemental to an indenture of 1898 between Lord Portman and the Poor Law Guardians of the Parish of St Marylebone, whereby the guardians had covenanted with Lord Portman "and his successors in title for the time being to the unsold portion of the Portman Family Estates in Marylebone" to observe the stipulations in a schedule of restrictions on user of the land. The deed of release recited that the remainder of the Portman Family Estates was now vested in the trustees and the premises which had been conveyed to the guardians were now vested in the GLC (the council) who were desirous of altering the use of the premises. By the operative part of the deed, the trustees released the council from the covenants in the 1898 indenture, in consideration of the council covenanting with the trustees and their successors in title as trustees to construct buildings on the land in accordance with annexed plans, and specifications, and not to use the buildings except for the purposes there specified, namely "a School of Management Studies, a College of Architecture and Advanced Building Technology, a Hall of Residence for students, and office for a District Surveyor and for housing purposes".

The second deed of release and covenant was made on July 27 1966 between Lord Howard de Walden, trustees, and the Howard de Walden Estate Company, on the one hand, and the council, on the other. This deed was likewise expressed to be supplemental to earlier deeds made in 1898 and 1902 between the Howard de Walden Estate and the Poor Law Guardians imposing restrictions on land used or to be used by the guardians for the purposes, *inter alia*, of a workhouse. By this second deed of release, the council (in whom the land was by now vested) were released from the earlier restrictions in consideration of a covenant by the council to construct buildings in accordance with annexed plans, and not to use the buildings so constructed except for the specified purposes, these being identical with the specified purposes quoted above from the first deed of release.

It is as a result of these transactions that there came into being the college buildings and ancillary premises which now compose the University of Westminster.

The present originating application was made by solicitors on behalf of the university and is dated November 30 1995. The application contains the information required by r 19 of the Lands Tribunal Rules 1975 then in force, and form 5 prescribed by those rules.

The application was:

for an Order that the restrictions may be wholly discharged; alternatively that the restriction be modified by:

(1) The deletion of the words "a School of Management Studies, a College of Architecture and Advanced Building Technology, a Hall of Residence for Students, an Office for a District Surveyor and for Housing Purposes" and

(2) The substitution of the words "for general educational purposes and a Hall of Residence for Students and for Housing Purposes".

The applicants did not admit in the application that any given person was entitled to the benefit of the restrictions, but they set out in respect of each of the two deeds a list of persons "who may have the benefit" of the restrictions. The persons so named were all expressed to be freehold owners of property in the vicinity of the university buildings.

Under the heading "Particulars of Application", the applicants set out the grounds on which they proposed to rely, under para l (*a*) (*aa*) (*b*) and (*c*) in section 84(1). I see no need to set out these particulars in detail, but it is to be observed that the substance of the justification for the application was that the existing user restrictions were too narrow for the needs of a modern-day university. It was said that the "proposed modification of the restrictions will allow for alternative uses which have no greater perceptible impact upon the owners of the benefit of the restrictions and/or those in the area surrounding the property than the current permitted uses". It was further stated that the reasonable user of the land which the existing restrictions impeded was "use for general educational purposes other than and/or in addition to those specified in the restrictions", and that the existing restrictions were contrary to the public interest "because they restrict the use of the property . . . for a full range of educational purposes".

From the foregoing details it is evident that the express purpose of the application was to enable the university to widen their educational activities while retaining the use of the property for what they described as general educational purposes.

In accordance with the usual practice of the tribunal, directions were issued in December 1995 and in February 1996 which required service of the notice of the originating application in an approved form upon the original covenantees and on those persons listed by the applicants as those who may have the benefit of the restrictions. The notice as approved by the tribunal set out details of the application, reproducing in substance those matters in the originating application which I have summarised above. These notices were served in compliance with the tribunal's directions.

By way of response, only one objection was lodged, namely that of Howard de Walden Estates Ltd dated March 8 1996, in which the allegations in support of the originating application were disputed and substantial compensation was claimed in respect of damage to the value of the estate's properties in the event of discharge or modification of the restrictions. As they were entitled to do, the originating applicants challenged the entitlement of this objector to the benefit of the relevant restrictions, and an order was made on April 17 1996 pursuant to section 84(3A) that the question of entitlement be determined at a preliminary hearing. However, before the hearing date of the subsection (3A) proceedings, the objection was withdrawn by notice dated July 19 1996.

Since the only objection had been withdrawn the application was placed before me with the applicants' consent, to consider whether to make an order without hearing. This accords with the normal practice of the tribunal and complies with what is now r 17(2) of the Lands Tribunal Rules 1996.

When the case was thus referred to me, I decided to order modification of the relevant restrictions as set out in the originating application so as to permit the use of the restricted land (in addition to the permitted uses) "for general educational purposes and as a hall of residence for students and for purposes of providing housing accommodation", being satisfied on the information provided as to the grounds (*aa*) and (*b*) in section 84(1). A draft order was drawn reflecting this decision and sent to the applicants' solicitors for their approval or suggested amendments.

By letter dated October 9 1996 the applicants' solicitors indicated that their clients would much prefer that the restrictions be discharged rather than modified, and gave reasons for the proposition that discharge was the appropriate remedy in the circumstances. By a letter of October 11 1996 which I approved, the tribunal notified the applicants' solicitors that having considered the points made, I was not prepared to exercise the tribunal's discretionary powers so as to discharge the relevant restrictions rather than to modify them in the manner set out in the draft order. Subsequently, I agreed to an oral hearing of an application by the applicants that the restrictions be discharged rather than modified.

I held such a hearing on November 19 1996 when the originating applicants were represented by Mr Nicholas Taggart of counsel. Mr Taggart had prepared a written outline of his submissions, which I found particularly helpful. Before me, he presented a forceful, sustained and skilful argument in support of his client's contentions that I had misapplied the discretionary power of the tribunal, and that in the circumstances of the case the only appropriate course was to order discharge of the restrictions.

At the close of the hearing, I indicated that I would refuse the application for discharge of the restrictions, and issue the order for modification as previously drafted, and that I would set out my reasons in writing.

The substance of Mr Taggart's submission can be summarised (I hope and believe fairly) in the following propositions:

(1) The application was clearly made for discharge of the restrictions with proposed modifications only in the alternative;

(2) It would be quite clear to any recipient of the notice of application that the applicants sought discharge and only as an alternative variation of the restrictions;

(3) No objections had been made to the application so framed except the one objection which was later withdrawn;

(4) The tribunal had a discretion which must be exercised judicially, and could be exercised only on the evidence available to the tribunal;

(5) There was no evidence on the basis of which the tribunal could properly refuse the primary relief sought by the applicants;

(6) There was no evidence to support the reason given by the tribunal for refusing discharge, namely that adjoining owners who may be entitled to enforce the restrictions would lose control over future user of the subject property.

Mr Taggart cited the following cases in support of his submissions:

Re Goldsmith's Application (1961) 13 P&CR 209 (LT)
Re Taylor's Application (1964) 16 P&CR 169 (LT)

and on appeal

Gee v *National Trust for Places of Historic Interest or Natural Beauty* [1966] 1 WLR 170 (CA)
Jones v *Rhys-Jones* (1974) 30 P&CR 451 (CA)
Re Farmiloe's Application (1983) 48 P&CR 317 (LT)
Ridley v *Taylor* [1965] 1 WLR 611 (CA)

He also referred me to passages in *Scammell on Land Covenants* (lst ed 1996).

I have considered the statutory framework of section 84 and the authorities to which my attention has been drawn, and I remain unpersuaded that the discretion vested in the tribunal is so fettered as Mr Taggart has contended. I do not consider that the applicants are entitled to an order for discharge of the restrictions, either as of right or as the inevitable result of exercising the tribunal's discretion. Nor do I consider that discharge of the restrictions is the appropriate remedy to be granted in the circumstances of this case.

Section 84(1) of the Law of Property Act 1925 as amended provides so far as relevant "the Lands Tribunal shall . . . have power from time to time . . . by order wholly or partially to discharge or modify any such restriction on being satisfied . . ." of certain specified grounds set out in paras (*a*) (*aa*) (*b*) and (*c*).

A Section 84(1C) provides:

> It is hereby declared that the power conferred by this section to modify restrictions includes power to add such further provisions restricting the user of or the building on the land affected as appear to the Lands Tribunal to be reasonable in view of the relaxation of the existing provisions, and as may be accepted by the applicant; and the Lands Tribunal may accordingly refuse to modify a restriction without some such addition.

There is no doubt that these provisions operate so as to confer jurisdiction on the tribunal to discharge or modify a restriction upon being satisfied that one or more of the statutory grounds has been made
B out; but further that on being so satisfied there remains a discretion in the tribunal as to whether or not to grant relief: see *Driscoll* v *Church Commissioners for England* [1957] 1 QB 330 (CA) *per* Denning LJ at p342 and Morris LJ at p350; *Re Ghey & Galton's Application* [1957] 2 QB 650 (CA) *per* Evershed MR at p660; *Jones* v *Rhys-Jones* (1974) 30 P&CR 451 (CA) *per* Ormrod LJ at p461.

Furthermore, the power conferred is expressed to be "wholly or partially to discharge or modify any such restriction . . .". This wording, coupled with the discretion granted by subsection (1C) to impose further restrictions on user as appear reasonable in view of the relaxation of the existing provisions, and to refuse to modify a restriction unless such further conditions are accepted, indicates clearly
C that the tribunal's discretion extends to determining the form and extent of the relief to be granted. It follows that the tribunal cannot be bound in any case to grant the applicants the full extent of the relief applied for, but is entitled as a matter of discretion to limit or cut down the form of relief which the applicants have sought.

Moreover, there is nothing in the statutory provisions to indicate that the tribunal's discretion is limited or (as the applicants would appear to suggest) abrogated, by reason of the fact that the application is unopposed. Doubtless the information available to the tribunal in the case of an unopposed application is likely to be less than in a case fought out at a hearing, but the tribunal is still under a duty to exercise
D its discretion to grant or refuse a discharge or modification in whole or in part in the light of known relevant circumstances. Mr Taggart is of course correct that the discretion is to be exercised judicially, not capriciously nor arbitrarily, nor by reliance on irrelevant considerations. Thus several of the cases cited by him are concerned with the question as to what relevant circumstances may be taken into account. In *Ridley* v *Taylor* [1965] 1 WLR 611 the Court of Appeal reversed the tribunal's decision to grant a modification in a leasehold case, partly on the ground that the tribunal had taken account of immaterial facts. Russell LJ at p623 said:

> E I do not think that the personality of the applicant or his past behaviour is relevant to the exercise of the discretion. I refer again to the fact that tomorrow an assign may make the same application. I think that the decision (including the exercise of discretion) must be related to the property and its history as such.

In *Gee* v *National Trust for Places of Historic Interest or Natural Beauty* [1966] 1 WLR 170 the Court of Appeal allowed an appeal against a refusal to grant a modification, in a case where the tribunal F had exercised its discretion against the applicant on the grounds that to grant the modification would hinder the National Trust's fund-raising efforts and discourage landowners from covenanting land to the National Trust. Davies LJ at p176 said:

> There are . . . two answers to these points. First as a matter of fact there was no evidence at all to support either of those two conclusions . . . it is really all a matter of surmise on the part of the Lands Tribunal from the general impression of the evidence given by those witnesses . . . secondly, even if there had been such evidence, on an application of this kind I think that both of those matters would have been irrelevant matters which the Lands Tribunal ought not to have taken into consideration in deciding whether or not in the exercise G of its discretion to make this order.

I accept, as of course I must, that the discretion to grant or refuse a discharge or modification or to limit the terms of the grant, is to be exercised judicially in the light of such relevant facts as are apparent from the documents, which in this case includes the instruments which imposed the original restrictions and the deeds of release by which those restrictions were in effect later modified.

As was stated in the tribunal's letter of October 11 1996 I had regard primarily to two matters in deciding to grant a modification rather than a discharge. The first related to the form of application and the notice which was delivered to potential objectors. The recipient of H such a notice would naturally and inevitably wish to know what the applicant was proposing to do on the land. On a fair reading of the notice, it would be plain that the owners proposed to continue the use of the premises as a university, albeit with a wider definition of their educational purposes. Had there been any indication beyond the bare reference to "discharge" as an alternative, that the applicants sought to remove all restrictions on user, then there might well have been objections forthcoming from nearby property owners, whom the applicants acknowledge may have the benefit of the restrictions and may be entitled to enforce them. Second, to discharge the relevant user restrictions wholly would deprive those who may be entitled to enforce J them of any opportunity to exercise control over future user of the property.

It is not the function of the tribunal to enforce restrictive covenants, nor to determine who (if anyone) may be entitled to enforce them. Interpretation of the relevant deeds and their effect in transmitting the benefit of the restrictions is a task for another forum. But as I have pointed out, the applicants themselves provided an extensive list of nearby property owners who, in their words, may have the benefit of the restrictions. The terms of the deeds do not appear to render such a supposition impossible or absurd. It seems to me that in considering as a matter of discretion whether or not to accede to the application, I K ought to have regard to the applicants' proposals for the use of the property, and to the interest of those whom the restrictions may have been intended to protect.

For these reasons, I refuse the application to discharge the relevant restrictions, and an order modifying the restrictions will issue in the form of the draft already submitted to the applicants.

LEASEHOLD VALUATION TRIBUNALS

LEASEHOLD REFORM HOUSING AND URBAN DEVELOPMENT ACT 1993

A

Hordern *v* Viscount Chelsea and another

LEASEHOLD VALUATION TRIBUNAL FOR THE LONDON RENT ASSESSMENT PANEL

July 5 1996

LADY FOX QC, DR STEVENS FRICS FRVA and MRS SE TURNER

B

Estates Gazette February 15 1977

[1997] 07 EG 144

Leasehold Reform, Housing and Urban Development Act 1993 — New lease — Determination of premium — Lease expired by date of hearing — Whether deduction for Landlord and Tenant Act 1954, Part I rights

The applicant tenant held an underlease of a first-floor flat at a ground rent of £175 pa the contractual term of which expired on September 29 1995. Following a notice claiming a new lease, the
C tenant applied to the tribunal for the determination of the premium payable; her valuers spoke to a sum of £233,340. The landlord sought £350,000 and contended that a 10% deduction should be made from the value of the existing freehold interest to reflect the risk that the tenant would claim security of tenure under Part I of the Landlord and Tenant Act 1954.

***Decision*: The premium payable was £283,100. The value of a 90-year lease of the property would be £325,000 to which 1% should be added for equivalent freehold interest. The deferment and yield rates were 6%. A 25% deduction should be made to reflect the risk of the tenant claiming security of tenure under Part**
D I of the Landlord and Tenant Act 1954

The following cases are referred to in this report.

Lloyd-Jones v *Church Commissioners for England* [1982] 1 EGLR 209; (1981) 261 EG 471, LT

Spath Holme Ltd v *Chairman of the Greater Manchester and Lancashire Rent Assessment Committee* [1995] 2 EGLR 80; [1995] 49 EG 128

Vignaud v *Keepers and Governors of the Possessions Revenues and Goods of the Free Grammar School of John Lyon* [1996] 2 EGLR 179; [1996] 37 EG 144

Stanley Gallagher (instructed by Baileys Shaw & Gillett) appeared
E for the tenant; Anthony Radevsky (instructed by Lee & Pembertons) represented the landlord.

Giving the decision of the tribunal, **LADY FOX QC** said: This is an application under section 48 of the Leasehold Reform, Housing and Urban Development Act 1993 (the Act) by Mrs Joan Hordern, the tenant, for determination of the premium payable in respect of the grant of an extension of the lease of the first-floor flat at 45 Cadogan Place, London SW1 (the subject flat). The tenant held the subject flat pursuant to an underlease dated September 30 1968 for a term of 28 years less one day commencing on September 29 1967, and expiring on
September 28 1995, at a ground rent of £175 pa. The said underlease F
was granted under the terms of the headlease between Charles Gerald John Cadogan (commonly called Viscount Chelsea) and Cadogan Estates Ltd and NAG Wickham and was assigned finally in 1993 to Ms EJ Crean whose term of 33 years from September 29 1962 expired on September 28 1995. The intermediate landlord therefore no longer has an interest in the subject flat.

By notice dated May 2 1995 the tenant claimed the right to acquire a new lease of the subject flat setting out two additional clauses to be included in the new lease. By counternotice dated July 7 1995 the freeholder admitted the tenant's right, but asserted that the terms of the
new lease should be in the Cadogan Estate's standard form for statutory G
lease extension under the Act. By reason of the tenant's notice and pursuant to Schedule 12, para 5(1) to the Act, the term of the said underlease continues for the currency of the claim and for three months thereafter.

Valuation of Mr Roger Doncom ARICS on behalf of the tenant

	£	£	£	
A *Diminution in value of landlord's interest*				
a Effective reversion to 999-year lease				
90-year lease	300,000			
1% for value freehold	3,000			
	303,000			H
55% of vacant possession				
Reversion		166,550		
b Proposed interest				
Reversion to	303,000			
Deferred 90 years 9%	.0004	121		
Diminution			166,529	
B *Marriage value*				
Proposed interests aggregated				
(i) Freeholders	121			
(ii) Claimants	300,000			
		300,121		J
Existing interests				
(i) Freeholders	166,500			
(ii) Claimants		166,500		
Marriage value		133,621		
Share 50%			66,811	
Premium payable			233,340	

Valuation of Mr J Clark ARICS on behalf of the freeholder

Calculation of premium for new lease first-floor flat, 45 Cadogan Place, SW1

	£	£	£	£	
A *Diminution in value of estate's interest*					K
(a) Value of estate's existing interest					
Annual rent payable £175					
Years' purchase 0.25 years @ 5%		Nominal			
Reversion to value of freehold in possession:					
(i) WA Ellis valuation of lease with 90 years unexpired @ peppercorn rent in amount of	350,000				
(ii) Represents 95% of freehold equivalent @	368,421				
Say	370,000				

A (iii) Defer 0.25 years @ 6%	0.986			
	364,820			
(iv) Less for risk of tenant claiming security of tenure under the Landlord and Tenant Act 1954 Part I 10%	36,482			
		333,518		
			333,518	
(b) Value of estate's proposed interest				
(i) Reversion to value				
B of freehold in possession	370,000			
(ii) Defer 90 years @ 6%	0.005			
			1,850	
Diminution in value of estate's interest				331,668
B *Calculation of marriage value*				
(a) Sum of values of proposed interests				
(i) Estate's (from above)	1,850			
(ii) Claimant's (from above)	350,000			
		351,850		
(b) Sum of values of existing interests				
(i) Estate's (from above)	333,518			
C (ii) Claimant's (for 3 months unexpired)	0			
		333,518		
Gain on marriage			18,332	
(c) Attributed to estate @ 100%				18,332
Premium payable				350,000

Mr Stanley Gallagher of counsel, instructed by Baileys Shaw & Gillett, solicitors, appeared on behalf of the tenant. He stated that the situation was somewhat unusual as the tenant was holding over on the expiry of the underlease on September 28 1995. Prior to the service of the tenant's notice to enfranchise, the freeholder had served a notice D under the Landlord and Tenant Act 1954 section 4 proposing a statutory tenancy on the termination of the underlease at a rent of £8,750 pa, but by virtue of Schedule 12, para 2 to the Act pending the present determination, the procedure under the 1954 Act was suspended so as to give the tenant the opportunity to enfranchise. The parties were agreed that the basis of valuation was set out in Schedule 13, para 2; no compensation was claimed under para 2(*c*) by the freeholder and the issues in the valuation to be determined were the diminution in the freeholder's interest and the landlord's share of the marriage value. On the basis that the parties were in broad agreement on the terms of the lease, Mr Gallagher said that he was taking the E present hearing date as the date of the valuation.

Mr Brian D'Arcy Clark ARICS gave evidence on behalf of the tenant. He said he was a founding partner in 1980 in Chesterfield, estate agents specialising in sales and acquisitions in central London residential property with its office located at no 28 in the same terrace as the subject premises. 50% of flats sold by his firm were located on or around the Cadogan Estate. He valued the landlord's current interest in the subject flat as a 90-year lease at today's date and would increase that value minimally, by 1% to arrive at the freehold value. He had inspected the subject flat and described it as a *pied-à-terre* rather than a one or two-bedroomed flat. It had an awkward layout, enjoying an F attractive large reception room, a small main bedroom, cramped kitchen through which access was gained to the roof terrace and the second small bedroom. Under the schedule of dilapidations the fireplace was required to be replaced in the main bedroom further reducing the space, the bathroom was not *en-suite*, and there was little wall space for cupboards in the kitchen. The subject flat was old fashioned, in need of modernisation and redecoration. Rearrangement of the flat was difficult by reason of the roof terrace being located at the rear, and not via the reception room. Three planning applications were lodged in respect of the adjacent property 43/44 Cadogan Place and planning consent in respect of 46/47 Cadogan Place relating to substantial alterations meant it likely that for the next two years any occupant of the subject flat would be surrounded by building works. Under the terms of the new lease no caretaker in residence was G required and Mr D'Arcy Clark would take 5% off the market price on this account. He produced a schedule of some 20 transactions relating to flats in Cadogan Place between 1992 and 1995 giving prices achieved per sq ft. Relying on this schedule and bearing in mind that price was affected by the position in the building, the ability to change the flat to the standard of accommodation required by the tenant, and its condition at the time of sale, he had concluded that the value of the subject flat in its present condition on a new 90-year lease at a peppercorn rent, with the service charge to be agreed at a rate appropriate to the building, was £300,000.

In answer to cross-examination from Mr Anthony Radevsky on H behalf of the freeholder, Mr D'Arcy Clark agreed that Mr Doncom in a letter dated August 15 1995 had referred to his valuation of the subject flat "on a long lease in a fully modernised state" as "to the order of £350,000". He agreed that he had included in his schedule all the transactions relating to Cadogan Place even though they were not directly comparable; they helped to give the full picture and the extraordinary layout of the subject flat made it very difficult to find any direct comparable. He agreed that 18 Cadogan Place was located opposite the entrance to the underground car park, but considered this might add value to tenants with cars. In answer to questions from the tribunal he agreed that an overall price per sq ft was appropriate when J applied to the subject flat where the fine reception room should be valued at a higher rate than the rear rooms. There was no lift, and while he did not know the amount of the service charges he thought they would be about £1,500 pa. The roof terrace was not included in the demise, but the tenant had the exclusive use and no liability in respect of its repair.

Mr Doncom gave evidence on behalf of the tenant in support of his valuation. He had been engaged on predominantly central London residential property matters first with Chestertons and for the last 12 years as the head of the professional and valuation department at K Marsh & Parsons. In this period he had dealt with valuations and negotiations under the 1967 Leasehold Reform Act, the 1974 Housing Act and the present Act, as well as with registration and determination of fair and assured rents. He described the layout, condition and potential for modernisation of the subject flat. The provision of an internal galley kitchen next to the reception room would still leave unsolved the problem of access to the terrace by means of any resited bathroom. The schedule of dilapidations related to restoration of the fireplace, removal of security grilles to windows and redecoration and compliance would add little to the market value as any purchaser was likely to carry out a radical rationalisation of the flat. In making his L valuation he had adopted Mr D'Arcy Clark's figure of £300,000 as the market value of the 90-year lease, increased by 1% for the freehold value. In his experience the value of a flat subject to a sitting tenant is depressed to approximately 50% to 60% of the vacant possession value. This discount affords the purchaser room for holding costs and eventual profit in negotiating or waiting for vacant possession in due course. The percentage varied within modest parameters depending upon the age, health and circumstances of the tenant, rather than location. Although the tenant was now aged 86, despite the actuarial factors, her life expectancy could be significantly extended.

The tenant's daughter stayed in the subject flat during the week in M order to run her business in London and the prospect of a succession to the tenancy must be considered. As the tenant was financially in a position to meet a fair rent and pay off the level of dilapidations likely to be negotiated, any investor would have to anticipate a substantial wait for vacant possession. Mr Doncom accordingly adopted a discount of 55% to arrive at the value of the freeholder's current interest, and relied on a transaction at 57 St Quintin's Avenue where a payment to a statutory tenant of £47,000 represented 52.5% of the value of £90,000 of the flat with vacant possession. He referred to transactions effected by his firm at 39 Warwick Gardens, 21 Earls Terrace, 23 Upper Addison Gardens, and by David Smith and Boston Carrington at 2/14-15 Ennismore Gardens, SW7, which supported a landlord's share of the marriage value between 45% to 52% and justified a 50:50 split of the marriage value in the present case.

In answer to cross-examination from Mr Radevsky, Mr Doncom did not accept that a different analysis of the figures in his transactions would give the freeholder a share between 72 to 77% of the marriage value. Nor did he accept that the deduction for the risk of the tenant staying on under the Landlord and Tenant Act 1954 was a notional exercise and to be represented by a 10% deduction, as recently confirmed by the Lands Tribunal in *Vignaud* v *Keepers and Governors of the Possessions Revenues and Goods of the Free Grammar School of John Lyon* [1996] 37 EG 144. Mr Doncom said that since the lease had expired and the tenant stayed on on a statutory basis, the risk had become greater. In answer to questions from the tribunal Mr Doncom said that he estimated the cost of dilapidations as in the region of £35,000; even if the tenancy passed to the tenant's daughter as an assured tenancy at a market rent he considered there to be a strong possibility that the statutory tenancy would continue. He maintained that his analysis of the transactions concerning marriage value was not based on his personal interpretation of values worth as they had to satisfy the negotiators on both sides.

Mrs Virginia Stourton, the tenant's daughter, gave evidence that during the week she lived with her mother at the subject flat spending the weekends in a cottage in Gloucestershire. From her knowledge of her mother's resources she had no doubt that she could meet the cost of dilapidations, and from inquiries at Kensington Town Hall, she understood that housing benefit would be available to assist her mother to stay on in the subject flat if she were unable to meet the fair rent out of her own funds. Mrs Stourton said the family had no plans to move her mother who, though she did not now go out, was perfectly fit and able to continue at Cadogan Place. The family's intention was to enable her mother to have a peaceful old age. In cross-examination Mr Radevsky suggested that the tenant intended to effect a back-to-back sale so that any premium payable to the freeholder would be met by selling on to a third party. In reply Mrs Stourton said that somebody had expressed an interest in purchasing the subject flat and any money made out of such a sale would be used to enable her mother to go to an old people's home.

Mr Radevsky called Mr Michael Duncan, senior partner in WA Ellis, estate agents, to give evidence on behalf of the freeholder. Mr Duncan said he had had considerable involvement with leasehold reform cases ever since the 1967 Act, and his firm had standing instructions to advise the Cadogan, Ilchester and Portman Estates on value issues arising from leasehold reform. He had inspected the subject flat and valued the proposed 90-year lease with vacant possession as at mid-March 1996 at £350,000 (approximately £457.50 per sq ft). In support of his valuation he relied on four transactions, as follows:

1/56 Cadogan Place: Ground-floor two-roomed, refurbished flat sold in July 1994 for £265,000 on a 49-year lease.

6/66 Cadogan Place: Ground-floor three-roomed flat sold in September 1994 for £485,000 on a 49-year lease.

On the basis that a lease of 50 years was worth no more than 82% of the value of the same premises for 90 years, the prices for these properties adjusted to a 90-year lease would be £323,000 for 1/56 and £590,000 for 6/66 Cadogan Place.

11/35 Pont Street, SW1: Third-floor flat, three rooms, two bathrooms, sold in good order on a lease for 80 years in October 1994 for £450,000.

15 Lennox Gardens, SW1: First-floor, two rooms with contracts exchanged on a 105-year lease in March 1996 for £320,500.

His comparables in Cadogan Place were also relied on by Mr D'Arcy Clark for the tenant and in Mr Duncan's view were the most relevant in the latter's list of comparables. 18, 24 and 27 Cadogan Place had inferior locations in the more northerly terrace close to the hotel and car park; 81 and 84 Cadogan Place were large apartments located in the north facing terrace in a modern purpose-built block.

In answer to cross-examination by Mr Gallagher, Mr Duncan said that his valuation was based on a resident caretaker but he did not consider that it had a great impact on the value. The flat would sell on the strength of the front reception room but he had applied a conservative price per sq ft to take account of the weaker rear rooms. When asked why he had chosen the two most highly priced flats as comparables from Mr D'Arcy Clark's list of comparables, Mr Duncan said that they alone were located in the same terrace as the subject flat, which was a superior position. In answer to questions from the tribunal Mr Duncan agreed that his comparables were sold in refurbished condition and that he had valued the subject flat with the existing fittings but on the assumption it was in good decorative order. He did not consider the service charges as a relevant component of the price; but he would deduct 15% to 20% for an inferior location north of the intersection with Pont Street, and add 10% for a first-floor level. As to the level of fair rent likely to be achieved for the subject flat, a fair rent of £6,000 pa with effect from September 1995 had been registered for a two-roomed flat without roof terrace, in an inferior location in 10/61 Cadogan Gardens.

Mr Julian Clark, an associate of Gerald Eve, chartered surveyors, with six years' qualified professional experience gave evidence on behalf of the freeholder. He referred to the extensive knowledge and experience of his firm Gerald Eve in leasehold reform matters and in advising the Cadogan and other major central London estates. His valuation was based on the grant of a 90-year lease some three months after the determination by the leasehold valuation tribunal on broadly similar terms to the old lease and he had taken the valuation date as the first day of the tribunal's hearing, ie April 22 1996. He relied upon the evidence of Mr Duncan for the value of £350,000 for a 90-year lease. There was no market evidence relating to freeholds of flats but taking into account that a lessee remained subject to the obligations of the lease, with control of the building and right of entry to the subject flat remaining with the freeholder, he considered a 5% differential appropriate and accordingly valued the freehold interest at £370,000. He adopted 6% both for deferment of the freeholder's present interest for three months and for deferment of the reversion for 90 years; he adopted this rate as appropriate for a high value property in a prime central London residential location with long term security for investors. Such long term growth in central London residential property was demonstrated by reported high prices achieved for recent sales relating to the Chesham Place Estate, Chalcots Residential Estate at Swiss Cottage and the Smith Charity Estate to the Wellcome Trust. Some allowance had to be made to take account of the statutory assignable three-month unexpired term now enjoyed by the tenant with the associated rights to security of tenure under the 1954 Act on termination. Such rights suffered from a number of disadvantages; they were non-assignable, incapable of appreciation in value for purposes of capital gains or as security for a loan or improvements, and the rent which was reviewable every two years was payable out of income after tax; and by reason of *Spath Holme Ltd* v *Chairman of the Greater Manchester and Lancashire Rent Assessment Committee* [1995] 2 EGLR 80; [1995] 49 EG 128, likely to approximate to the level of market rents. Accordingly, he adjusted the value of the freehold in reversion by a 10% deduction on account of the tenant's rights and relied in support on the Lands Tribunal decisions of *Lloyd-Jones* v *Church Commissioners for England* [1982] 1 EGLR 209; (1981) 261 EG 471 and *Vignaud* v *John Lyon School*. Mr Clark stated there to be some 31 leases for flats and maisonettes in Cadogan Place for which the contractual term had passed in September 1955 and which enabled the lessees to qualify for rights of security of tenure under the Landlord and Tenant Act 1954. Of these, 13 leases, just under half, had submitted a claim under the 1993 Act either for a lease extension or for a collective enfranchisement. Of the remaining 18, the lessees were continuing in occupation while procedures under the Landlord and Tenant Act 1954 were progressing. No statutory tenancies had been granted by the Cadogan Estate and the ultimate intentions of these 18 remain unclear. Mr Clark gave it as his opinion that at least half and maybe more of the 31 lessees would discount taking up a statutory tenancy at a fair rent.

The 1993 Act contemplated situations when either by agreement or determination by the tribunal the landlord's share of the marriage value would exceed 50%. The present application was an exceptional case where the contractual term of the underlease had expired. In terms of a claim for lease extension under the Act this was the most extreme case with the tenant's existing interest reduced to the statutory three-month period. The freeholder in these circumstances would only be willing to enter into a transaction with the claimant at a premium figure for the new lease which had proper regard to the imminent opportunity of achieving vacant possession.

In cross-examination on behalf of the tenant Mr Gallagher challenged Mr Clark as to his lack of evidence to support his percentages for the differential for the freehold, deferment, the tenant's rights under the 1954 Act and the landlord's share of the marriage value; Mr Clark pointed out that 6% had been used by Daniel Smith in relation to the tenant's comparable at 2/14-15 Ennismore Gardens, and by Mr D'Arcy Clark in his rough calculations in support of his figure of £350,000 in the Marsh & Parsons's letter of August 15 1995; Mr Doncom had himself employed a 10% deduction for the Landlord and Tenant Act 1954 in his earlier calculations. The situation at 57 St Quintin's Avenue was to be distinguished as there was greater certainty as to the tenant staying on where the option for a statutory tenancy had already been exercised. Mr Clark did not consider the personal resources of the tenant or the possibility of housing benefit to meet the rent altered the proportion of 10% for rights under the 1954 Act. The Cadogan Estate took a long term view and it was not its policy to buy out sitting tenants. In answer to questions from the tribunal Mr Clark said he did not agree that the tenant's act in serving a notice to enfranchise the subject flat had released the marriage value. He could not say whether it was usual or the intention of the Act that on holding over a tenant should pay the full market value of a 90-year lease.

Mr Radevsky in his final submissions on behalf of the freeholder invited the tribunal to accept the values for the freehold and leasehold interests, and percentages for deferment, Landlord and Tenant Act 1954 and landlord's share of marriage value put forward by the freeholder's witnesses; they were better supported by reasons. In assessing the tenant's means as affecting her likelihood to continue in occupation, her capital, as well as her income, had to be taken into account; both were also relevant in assessing availability of housing benefit, which might be restricted to a proportion of the fair rent payable. As to marriage value, the 1993 Act established 50:50 as a floor; when the contractual lease had expired and the tenant had only a chance of a statutory tenancy the landlord's share must increase to 100%.

Mr Gallagher in his final submissions on behalf of the tenant urged the tribunal to have regard to the general range of comparables in Cadogan Place. The subject flat was unique and the freeholder's four comparables by reason of different level, layout, accommodation and location had to be distinguished. As to the appropriate percentage deduction for the tenant's rights under the Landlord and Tenant Act 1954, the *Lloyd-Jones* case had been refined by the decision in *Vignaud* to require the tribunal to have regard to the financial and other circumstances of the tenant. Here there were funds, either personal or from housing benefit, to ensure that the tenant would continue to occupy the subject flat, making a high probability that she would exercise her statutory rights. As regards the share of marriage value it was clearly in the landlord's interest, whatever the general policy of the Cadogan Estate, to get the tenant out and since this could not be done without her consent the split of marriage value should be 50:50.

Inspection

The subject first floor-flat was within a six-storey, including basement, terrace of former houses built about 1840. The exterior stucco was in fair decorative order although there were many unsupported wires hanging down the face of the building. Security grilles had been fixed to the windows and doors fronting on to the balcony. There was a portico entrance to the property.

The common parts had been decorated to a high standard and the main stairs had good quality carpeting. Although there was a lift within the building it did not serve the first-floor flat. Heating to the common parts was provided by a night storage heater situated on the ground floor.

The conversion appeared to be longstanding and provided a living room, two bedrooms one of which was on a mezzanine floor, together with a small kitchen and an internal bathroom/WC.

The living room was of good proportions, many of the original features, including the ceiling, cornices and wall mouldings having been retained. The flooring was of hardwood laid to herring-bone pattern.

The original windows and doors lead on to a small narrow balcony and overlook a large private garden. The rear bedroom which is of more modest proportions, the ceiling height having been lowered, overlooks the rear terrace. The small cramped kitchen enjoys good natural light and provides access to the rear balcony. It also provides narrow and awkward access to the bedroom (a former staff room) on the mezzanine floor. This room is small and irregular in shape but also has good natural light. The internal bathroom/WC has mechanical ventilation.

Decision and reason

The determination of the price payable for the extended 90-year lease is to be made in accordance with Schedule 13 to the Act. At the hearing the parties intimated that the terms of the lease would be agreed between them with a little more effort. By letter of May 3 1996 the solicitors for the freeholder confirmed that the parties had now agreed the terms. Accordingly, as provided in Schedule 13, para 1, the date of the valuation is May 3 1996.

The first contested element in the valuation is the vacant possession value of an extended 90-year lease of the subject flat, which the surveyors of both parties took to mean in a condition in accordance with the terms of the subsisting lease. From our inspection it is clear that the subject flat is a somewhat unusual unit to value, enjoying one excellent large living room of attractive proportions and aspect but with inferior supporting accommodation by reason of its size and awkward layout. We do not consider the rear terrace to add greatly to the value as it was not included in the demise, nor have we made any deductions in value for the loss of a resident caretaker. The tenant's surveyor proposed a figure of £300,000 and provided a schedule of all recent transactions relating to flats carried out in Cadogan Place. We found few of these to be of direct assistance; flats in properties of nos 18 to 27 were located in a northern section of the terrace, which was inferior by reason of the entrance to the underground car park. Most contained larger accommodation and floor areas with the most similar, a first-floor flat in 27 Cadogan Place having a noisy corner location on to Pont Street. Of the higher numbers in the terrace, nos 81 and 84 were in a north-facing section in buildings of different character. This left as comparables two raised ground-floor flats at 1/56 and 6/66 Cadogan Place which were also relied on by the freeholder's surveyor and a fourth-floor flat at D/63 Cadogan Place.

All these flats had been entirely refurbished and enjoyed central heating; location of D/63 on the fourth floor without a lift made it difficult to compare and 6/66 Cadogan Place was a much larger flat with four rooms and two bathrooms. We accept 1/56 Cadogan Place as providing general guidance as to level of prices, despite its different floor location, absence of a fine reception room, and better condition. The freeholder's surveyor also relied on comparable transactions at 11/35 Pont Street and 15 Lennox Gardens. The former with smaller rooms was located in a busy street in a building modernised to luxury standard of a different character. The latter, 15 Lennox Gardens, was of greater assistance since it too had a quiet location with reception room overlooking gardens and was on the first floor. In the light of these comparables we adopted £325,000 as the value of the 90-year leasehold interest in the property.

The freeholder proposed a 5% uplift for the value of the freehold, but we adopt the 1% put forward by the tenant's surveyor. As regards the rate for deferment we consider 6% as an appropriate rate having regard to the excellent location.

The final contested element was the proportion of the discount to be made to take account of the possibility that the tenant might exercise her rights to a statutory tenancy. In view of the now expired contractual term of the lease, that possibility is imminent.

The tenant's surveyor proposed a 45% discount and supported it by reference to a market transaction, at 57 St Quintin's Avenue; the freeholder's surveyor proposed a 10% discount and Mr Radevsky submitted that this was in line with the 10% discount determined by the Lands Tribunal in *Lloyd-Jones* v *Church Commissioners* (1981) 261 EG 471 and more recently in *Vignaud* v *John Lyon School* as the appropriate discount for the tenant's rights under the Landlord and Tenant Act 1954 Part I in respect of enfranchised houses under the Leasehold Reform Act 1967, as amended.

In the present case we are dealing with the extension of the lease of a single flat, not a house. Schedule 13, para 3 of the 1993 Act does not expressly direct, as does section 9(1)(A) of the 1967 Act, that the determination of the premium payable shall be made on the assumption that at the end of the tenancy the tenant has the right to remain in possession pursuant to the provisions of the Landlord and Tenant Act 1954, Part III. The absence of such an express assumption is not necessarily decisive as para 4 of Schedule 13 indicates that the making of assumptions other than those specified is not precluded, but it may well indicate that the treatment of flats is different from houses. Para 3(2) directs that the value of the landlord's interest is to be based on "open market" value, and such a value should clearly reflect — to the extent that the market would do so — the prospects of the tenant holding over on a statutory tenancy.

Hague explains the need for such an assumption in respect of houses in the higher rateable value under section 9(1)(A) of the 1967 Act by reason of the fact that a tenant of such a highly rated house rarely allowed his tenancy to expire so as to rely on his statutory rights: *Hague — Leasehold Enfranchisement* 1987, p189, para 9.51. The position as to the possibility of a tenant staying on in a flat seems to us to be different. Mr Clark in his evidence on behalf of the freeholder stated that at the present time there are 31 contractual tenants of flats on the Cadogan Estate who still remain in possession after their terms have expired. In these circumstances there would clearly appear to be an increased risk that a tenant will exercise his statutory rights and the market evidence relating to 57 St Quintin's Avenue suggests that such a risk may be quantified as 52.5%.

We are accordingly of the view that in respect of a flat where evidence is available a higher proportion than the 10% adopted in the Lands Tribunal's decisions may be appropriate.

A further difference also exists on the facts relating to the tenant's personal circumstances from the latest Lands Tribunal decision of *Vignaud*, though we have to express some reservation as to the extent to which valuation under Schedule 13 requires us to investigate the precise financial circumstances of an individual tenant. As we have said above, it is for the tribunal to see the matter (and its possible outcome) as the market would see it, and to make such discount as the market would do so in all the circumstances.

In *Vignaud* the tenant at 74 Maida Vale, London W9, the property in question, agreed that she was unable to pay the regulated rent or the sum of £30,000 required to comply with the dilapidations notice. In our case, however, we had evidence from the tenant's daughter, not the tenant herself, and of a somewhat ambivalent nature, that the tenant had financial resources to be able to meet the dilapidations and to afford to stay on (if necessary with the assistance of housing benefit) so long as she remained in good health. In these circumstances we consider the possibility of the tenant staying on to be greater than in the *Vignaud* case.

Accordingly, taking all these matters into account we adopt a proportion of 25% for the risk of the tenant exercising her statutory rights to stay on.

As regards the freeholder's share of the marriage value, in so far as any market evidence was produced, it came from the tenant's surveyor, Mr Doncom, and supported the statutory residual proportion of 50:50. Mr Clark on behalf of the freeholder urged us to accept the situation of a tenant holding over as the most extreme case where she had no bargaining chip in the negotiation. But we remain of the view that the tenant's initial notice to extend the lease is a necessary condition to enable the realisation of the marriage value under the Act and we, therefore divide the marriage value equally between the parties.

Having regard to all the above matters we make the following valuation:

Premium payable by the tenant .in accordance with section 48(7) and Schedules 12 and 13 to the Leasehold Reform, Housing and Urban Development Act 1993.

(a) Schedule 13 para 2(a) the diminution of the landlord's interest in the tenant's flat as determined in accordance with para 3.

Term of existing lease				
Ground rent	£175			
YP 0.25 years @ 6%		Nominal		
Value of lease for 90 years at a peppercorn	£325,000			
Add for equivalent freehold	£3,250			
	£328,250			
Deferred 0.25 years @ 6%	0.98			
	£321,685			
Less risk of tenant claiming security of tenure under Part I, Landlord and Tenant Act 1954 — 25%				
	80,421	78,813	£242,872	242,872
Deferred 0.25 years @ 6%	0.98			
Reversion to value of freehold in possession PV£1 in 90 years @ 6%	328,250			
	0.005		1,641	
				£241,231
(b) Marriage value				
Value of tenant's extended lease	325,000			
Value of landlord's reversionary interest	1,641	£326,641		
Less landlord's existing interest	242,872			
Tenant's existing interest	NIL	242,872		
Gain on marriage value		£83,769		
Landlord's 50% share				41,885
(c) Compensation for loss arising from grant of new lease (para 5)				NIL
				283,116
Premium payable by tenant			say	£283,100

Accordingly we determine the price payable by the tenant for the extended lease in respect of the first-floor flat at 45 Cadogan Place, London SW1, pursuant to section 48 and Schedule 13 to the Act is £283,100 (two hundred and eighty three thousand, one hundred pounds).

A

Moore and others *v* Escalus Properties Ltd

LEASEHOLD VALUATION TRIBUNAL

July 30 1996

GF BOWDEN FRICS, DR STEVENS FRICS and R SKINNER

Estates Gazette February 15 1997

[1997] 07 EG 149

B ***Leasehold Reform, Housing and Urban Development Act 1993, section 24 — Acquisition of freehold reversion to three flats — Rising ground rents — Whether open market value will reflect value of insurance commissions***

The nominee purchasers sought the determination by the tribunal of the price payable for the acquisition of the freehold in a building converted to three flats; they put forward a valuation of £915 and referred to the price of £750 paid in 1993 when the freehold was acquired by the present landlord. The landlord argued for £3,655 contending that its investment in a freehold ground rent was valued not only on the ground rent itself but also on the insurance commission, C benefits of management and subsidiary income.

***Decision*: The price was £1,885; this was based on a capitalisation rate of 11%. Although there may be special purchasers willing to pay a higher price for ground rents because of the possibility of earning insurance commission, this could not be accepted as a proper basis for establishing market value.**

No cases are referred to in this report.

The nominee purchasers, Mr AG Moore, Mr SJ Moore, Mr KA Stirling and Ms WM Pank, appeared in person; Graham Clark (instructed by Malthouse Chevalier) represented the respondent landlord.

D Giving the decision of the tribunal, **MR GF BOWDEN** said: This is an application made pursuant to section 24 of the Leasehold Reform, Housing and Urban Development Act 1993 (the Act) by the qualifying nominee purchaser for determination of the price to be paid for the freehold in 44 Lancaster Avenue, West Norwood, London SE27 (the subject premises). The participating tenants were as follows: Mr Adam Moore and Mr Simon Moore co-tenants, Ms Wendy Pank and Mr Keith Stirling. The subject premises are a four-storey end-of-terrace property converted into three self-contained units as follows.

Flat 1, semi-basement and garden, with car parking at rear; Mr A E and Mr S Moore.

Flat 2, hall floor, with car parking at front; Ms W Pank.

Flat 3, first and second-floor maisonette with car parking at front; Mr Keith Stirling.

By initial notice dated November 16 1994 the joint nominee purchasers gave notice of their claim to exercise the right of collective enfranchisement, proposing a purchase price of £750 for the freehold interest in the specified premises and £250 for the freehold interest in the additional property. The reversioner Escalus Properties Ltd, through its solicitors, Malthouse Chevalier, served a counternotice in which it admitted the participating tenants' right to exercise their right F to collective enfranchisement but contested the tenants' proposed purchase price, making a counter-proposal of £2,250 for the specified premises and £750 for the additional property. It was agreed at the hearing that the tenants qualified for the right of collective enfranchisement in respect of the subject premises and that the price of the freehold interest was the only matter for determination by the tribunal. The tenants submitted a valuation in the recommended form prepared with the assistance of Messrs J Trevor & Webster.

Valuation of the nominee purchaser

Lease term (remainder) — whole house

1996-2018	22 years	@ £75 pa	Ground rent
2019-2051	33 years	@ £150 pa	Ground rent
2052-2084	33 years	@ £225 pa	Ground rent

1. *Value of lease ground rent @13%*					G
				£623	
2. *Reversion*					
Current value of flats:	£	£	£		
Upper flat	60,000				
Middle flat	34,500				
Lower flat	<u>60,000</u>				
	154,500				
Increase: 5%	<u>7,725</u>				
	162,225				
Present value @ 13%				3	
3. *Marriage value*					H
Assumed					
New leases (999 years)					
Upper flat	60,400				
Middle flat	34,900				
Lower flat	<u>60,400</u>				
		155,700			
Less present					
Freehold interest	623				
Upper flat	60,000				
Middle flat	34,500				
Lower flat	<u>60,000</u>				
		<u>155,123</u>			J
Marriage value			577		
Landlord's share at 50%			<u>(288)</u>		
				<u>289</u>	
Total valuation				<u>915</u>	

The reversioner did not submit a valuation in the form provided for in the Act, but analysed transactions in a proof of evidence prepared by Mr Christopher Case, as expert witness, from which Mr Graham Clark, counsel for the reversioner, proposed a value of £3,655, and Mr Case, proposed a value of £3,565.

At the hearing Mr Keith Stirling, tenant of flat 3, spoke on behalf of himself and the other applicants. He put before the tribunal a proof of evidence from which he spoke. K He explained that the property had been a single dwelling-house that had been converted into three units in 1985, and the separate flats had been sold on 99-year leases from December 1984. The ground rent, which was the same for each flat, was £25 pa for the first 33 years, rising to £50 pa for the next 33 years and then £75 pa for the remainder of the term. Flat 1, the garden flat, was occupied by Mr Adam Moore and Mr Simon Moore who acquired the lease in 1987. This flat has sole use of the small garden with parking space at the rear of the flat, and has its own separate entrance on Ardlui Road at the side. The middle flat, flat 2, was occupied by Ms Wendy Pank, who acquired the lease in 1993. This flat is accessed by the main front door to a communal lobby and then its own front door L to the flat. It has a balcony at the back overlooking the garden. Flat 3, the top flat, which is on two floors, is occupied by Mr Keith Stirling who acquired the lease in 1985. It shares, with flat 2, the parking area in front of the property, where there is provision for parking some four cars.

The property had been converted by a Mr Joseph Nolan, who after the individual flats had been sold, took little interest in the property. It was the leaseholders themselves who took over the management of the property, arranging insurance and security measures. They attempted on a number of occasions to purchase the freehold by agreement, but for one reason or another, these negotiations never came to M completion. The leaseholders were advised in May 1993 by the landlord's agents, Cornerstone, that the reversion was to be sold at auction; Ms Pank, who at that time was in the process of buying the lease of flat 2, through her solicitors, informed Mr Nolan's solicitors that there had been no offer to sell the freehold to her. She was advised that the auction would go ahead unless she signed a contract to purchase, prior to the auction. She was advised to attend the auction and serve on the purchaser a section 11 and 12 notice under the Landlord and Tenant Act 1987. This was done, but since at that time, she was unable to make immediate payment, the leaseholders were not able to complete. The freehold interest was sold in July 1993 for £750.

A In October 1993 each leaseholder received a letter informing them that the new reversioner was Escalus Properties. The leaseholders did not attach any particular significance to this communication and took no action. Three months later they each received a demand for £782.05 to cover insurance of the property (cover which had in fact already been effected by the tenants), maintenance, audit and management fees. These demands from the landlord have increased in the intervening year, and while they have been paid in the past, the size of the maintenance fund has remained a matter of dispute.

The tenants on deciding to purchase the freehold made a study of the local market. They put before the tribunal 10 comparables, four
B freehold and six leasehold transactions, as the basis of calculations set out (above). They offered the sum of £1,000 for the freehold interest in the subject property.

The nominee purchaser put before the tribunal the following transactions for consideration.

With regard to *freehold values* the following sales:

38 Rosendale Road
Four self-contained flats of one to three bedrooms on a 99-year lease from 1985. Ground rent £45 pa/flat first 33 years rising to £90 and £135 after periods of 33 years.
Freehold price paid March 1995 £3,000.

C

27 Lancaster Avenue
Four self-contained flats of one to three bedrooms on a 125-year lease from 1985. Ground rent £400 pa in total for the whole house.
Freehold price paid September 1995 £3,000.

28 Thurlow Park Road
Three self-contained flats of one to two bedrooms on an 88-year lease from 1995. Ground rent £15 pa/flat.
Freehold price paid December 1995 £1,000.

44 Lancaster Avenue (the subject house)
D Three self-contained flats of one to two bedrooms on a 99-year lease from Christmas 1984. Ground rent £25 pa/flat first 33 years rising to £50 and £75 after periods of 33 years.
Freehold price paid July 1993 £750.

With regards to *leasehold values* they gave the following market indications:

27d Lancaster Avenue
One-bedroom flat with garden. Freehold.
Price paid September 1995 £31,500. (Freehold purchased simultaneously.)

E

28 Thurlow Park Road
One bedroom flat. 88-year lease. Ground rent £15 pa.
Price paid September 1995 £30,000.

33 Rosendale Road
One-bedroom garden flat. 116 year lease. Ground rent £25 pa.
Price paid March 1992 £44,995.

36 Turney Road
Three bedroom flat. 125-year lease. Ground rent £150 pa.
Price paid November 1995 £73,500.

F

100a Casewick Road
One-bedroom flat. 999-year lease (from 1991). Ground rent one shilling pa.
Price paid October 1995 £43,000.

44 Lancaster Avenue (in subject house)
Hall-floor flat. One bedroom. 90-year lease. Ground rent £25 pa (rising after 2017) etc.
Price paid May 1993 £34,500.

Mr Clark, counsel for the reversioner called Mr Christopher Case, a director of Empress Properties Ltd, who act as managing agents in respect of the subject property for the reversioner Escalus Properties Ltd. Mr Case submitted a written statement. He stated that in G managing his clients' properties he placed all the insurance business through Princess Insurance Agencies, which provided cover on the whole portfolio. The portfolio is insured with the Commercial Union Assurance Co and a 20% commission is received by the landlord. When valuing a freehold ground rent, account is taken of the commission recoverable as a result of the landlord being able to effect the insurance cover. The value to be attributed to the insurance commission in respect of the subject property was £826.18.

With regard to auction sales, Mr Case stated that it was his experience, that those freehold ground rents sold at auction were invariably problem properties, either subject to litigation, or arrears of H service charges which are disputed by lessees, or with defective leases. It was a risky investment to purchase freeholds at auction on the limited information that was usually provided. It is therefore dangerous to rely on auction results for the purposes of a valuation under the 1993 Act. Since the subject property is part of a well managed portfolio, if sold it would only be willingly sold as part of such a portfolio and would command a higher price than by sale at auction.

Appendix 5 of Mr Case's statement set out a schedule of some 21 freehold ground rent sales pursuant to the Act. The properties listed in this schedule ranged over a wide area of London, although none of them in the broad locality of the subject property. Details were given J of the unexpired terms, ground rents, insurance premiums and commissions and the purchase prices of these properties. On analysis, it was said that whether the freehold interest is based on a 7% or 14% return on ground rent, the total paid is, on average 264% of the freehold interest on the basis of 7 years' purchase and 6 years' purchase of insurance commission. On this basis Mr Case contended the freehold interest and the marriage value to the tenants of the subject property was in the region of £3,565.

Mr Clark, in his submission, emphasised that his client's investment in a freehold ground rent was valued on the basis of not only the ground rent itself, but also any insurance commission, benefits K of management and subsidiary income such as premiums for deeds of variation. The most likely potential purchasers, prepared to pay for such benefits as that of the interest in the subject property, would be the holders of other substantial ground rent portfolios. Mr Case's evidence to the tribunal was that his client would not sell at auction, since this is not where the best properties are sold. True valuation is to be ascertained by reference to private treaty transactions and analysis of those sales, (such as set out in appendices 2A and B of Mr Case's statement) shows that ground rent and insurance commission income should be considered separately. Analysis suggests that there should be 7 YP for the ground rent and 6 YP for the insurance commission and L this produces a starting value of £1,351 for the freehold without special features.

With regard to marriage value Mr Clark identified the elements giving rise to a marriage value as being control achieved by the lessees and the ability to grant new leases without premium or limit of term; such control is of value to the lessees. He contended that on the analysis of actual sales the true price is about £3,655.

Inspection

Lancaster Avenue was a wide tree-lined residential road of substantial properties, largely of the Victorian period, but some of a later date and with some newly built redevelopment. The subject M property was a three-storey, end-of-terrace, semi-detached house with semi-basement. There were some six steps up to an impressive front entrance with a porch supported by corinthian-style columns. The front elevation had a bay front to the hall floor and the facade was rendered and white painted. The side and rear elevations were brick. The overall external condition of the property was in reasonably good order. The site of the house and garden was triangular in shape, being wider on the Lancaster Avenue frontage and tapering away at the rear along the line of the adjacent Ardlui Road. At the front of the property was a garden area which could provide off-street parking space for some four cars. These spaces were used by flats 2 and 3. The rear yard had a side

A entrance, with garage gates from Ardlui Road, and provided off-street parking for flat 1. The garden flat, flat 1, a semi-basement, had its own separate entrance at the side in Ardlui Road. It comprised a livingroom, two bedrooms, kitchen and bathroom. The damp proofing arrangements of the original conversion had proved to be ineffective, and new works had been carried out by the tenant in 1990. Many of the windows had security bars and grilles. There was double glazing to the windows of the front room. The other room at the front, although of reasonable size, had a particularly small window offering only restricted light to the room. Glazed patio doors to the rear room gave access to the patio area and the strip garden beyond. Natural light to
B this room was impaired by the overhang of the balcony to the flat above. The bathroom was an internal room. The galley-style kitchen was small but well equipped. The flat had an independent gas fired central heating system.

Flat 2 on the hall, or upper-ground floor, comprised a livingroom, and one bedroom, kitchen and bathroom, with a balcony overlooking the garden at the back. Access was via the main front entrance, through a small shared lobby to its own front door. The front livingroom retained its original proportions and some of the original features from when the house was built. The remainder of the flat though planned with a convenient and compact layout, had a somewhat cramped
C confined feel with a narrow hallway, small lobby and galley kitchen. The bathroom was an internal room, ventilated by an extractor fan. The flat had an independent, gas fired central heating system.

Flat 3, the first and second-floor maisonette with its front door off the entrance lobby and well lit staircase, leading up to a landing with cupboarded boiler and tank. The flat comprised livingroom, two bedrooms, kitchen-diner and bathroom. The rooms were spacious, light and well proportioned. The master bedroom was on the upper floor, a good sized room with a deep dormer window with part of the ceiling following the sloping roof line. The flat was equipped throughout with good quality fixtures and fitments. The flat had an
D independent, gas fired central heating system.

Decision and reasons

The tribunal was urged by the nominee purchaser to adopt a capitalisation rate of 13% based on the opinion of Messrs J Trevor & Webster and the fact that this rate had been adopted in some recent tribunal decisions. However the tribunal felt that in this particular instance 13% was too high. Having regard to all the factors, including tribunal decisions in enfranchisement cases where the unexpired term of the leases were of similar length the appropriate capitalisation rate was 11%.

E In assessing the current open market value of the leases of the subject flats, the tribunal took note of the actual sale price achieved of £34,500, for the hall-floor flat (flat 2) in May 1993. They also took into account the details of comparable local sales transactions put before them, having made an external inspection of these properties. In striking a balance between the points of difference and those of similarity, they found them of assistance in coming to a view. They accepted that the tenants' valuation of the upper maisonette (flat 3) at £60,000 bore the right relationship in relative values to flat 2 and the other comparables. However they were unable to accept the tenants valuation of the semi-basement flat (flat 1) at £60,000. They felt this estimate was too high and reduced the figure to £45,000 to establish a
F proper balance of relative values between the three flats. With regard to marriage value the tribunal took careful note of Mr Clark's arguments regarding the "tension" which is inherent in the relationship between landlord and tenant. They recognised the benefits, some quantifiable, but others intangible, that tenants achieve by gaining control of the freehold interest; the opportunity it offers for granting themselves new leases; lease extensions varying any terms of the leases; supervision of services and management decisions. These must be given due weight in calculating the marriage value. The tribunal felt that the tenants, in their valuation had underestimated these benefits.

The landlord argued that his right to receive commission on insurance premiums was a significant factor that should be taken into account in valuing the freehold interest. The tribunal recognised that G
this may be a material consideration for a professional investor in freehold ground rents who had established such an arrangement with brokers and insurance companies. In such cases they may be seen as special purchasers willing, perhaps, to pay a higher purchase price. However the tribunal could not accept this as a proper basis for establishing an open market value.

The tribunal noted that the landlord had purchased the freehold interest at auction in July 1993 for £750. They also accepted the landlord's argument that auction prices could not be taken as a wholly reliable guide to the market, since these freehold ground rents sold at auction were often those of problem properties subject to disputes and H
arrears. However in the absence of a body of evidence of private treaty sales of better properties, the tribunal felt they did offer, if treated with some caution in analysis, some guidance of market trends.

The tribunal made the following valuation.

Purchase price payable by nominee purchaser in accordance with Schedule 6 to the Leasehold Reform, Housing and Urban Development Act 1993.

i. *Para 2(1)(a): The value of the freeholders interest in the premises as determined in accordance with para 3.*

Ground rent for 22yrs		£75		J
YP for 22yrs @ 11%		8.176	£613	
Ground rent for 33yrs		£150		
YP for 33yrs @ 11%	8.801			
PV £1 def'd 22yrs @ 11%	0.1007	0.886	£133	
Ground rent for 33yrs		£225		
YP for 33yrs @ 11%	8.801			
PV £1 def'd 55yrs @ 11%	0.0032	0.0281	£6	
Reversion to capital value				
Upper flat	£60,000			
Middle flat	£34,500			K
Lower flat	£45,000	£139,500		
PV £1 in 88 yrs @ 11%		0.0001027	£14	
			£766	

ii. *Para 2(1)(b): The freeholder's share of the marriage value as determined in accordance with para 4.*

New 999-year leases at peppercorn rents with vacant possession.

Upper flat	£61,000			
Middle flat	£35,500			
Lower flat	£46,000	£142,500		
Deduct				
Present freehold interest	£766			L
Present reversion	139,500	£140,266		
Marriage value	£2,234			
Freeholder's share at 50%			£1,117	
iii. *Para 2(1)(c): Compensation for other loss*			NIL	
			£1,883	
say			£1,885	

The tribunal, therefore determined that the sum to be paid for the freehold interest in 44 Lancaster Avenue, West Norwood, London SE27, is £1,885 (one thousand eight hundred and eighty five pounds).

M

Donath and another *v* Second Duke of Westminster's Trustees

LEASEHOLD VALUATION TRIBUNAL FOR THE LONDON RENT ASSESSMENT PANEL

April 14 1996

LADY FOX QC, PF PRIOR FRICS and PS ROBERTS RIBA

Estates Gazette March 29 1997

[1997] 13 EG 120

Leasehold Reform, Housing and Urban Development Act 1993 — Determination of price for acquisition of freehold of two maisonettes — Appropriate risk rates — Marriage values

By an initial notice dated December 15 1993 the tenants holding underleases of two maisonettes and two garages in 36 and 37 Eaton Mews South, London SW1, gave notice to acquire the freehold from the respondent landlords. That notice was admitted by the landlords in February 1994 and the parties agreed the terms of the proposed conveyance to the applicant nominee purchasers, save for the price for the freehold. The nominee purchasers contended for a price of £265,000; the landlords for £734,000. The unexpired terms of the underleases were 20.25 years for the maisonettes and 13.25 years for one of the garages. The ground rents for the maisonettes were respectively £150 and £100 pa.

***Decision*: The price payable for the freehold (the intermediate landlords having agreed to be represented by the freeholders) should be £583,300. The reversion to the maisonettes should be valued using a risk rate of 6%, to give £466,601. The freeholders were entitled to 50% of the marriage value.**

The following cases are referred to in this report.

Baptist v *Masters of the Bench and Trustees of the Honourable Society of Gray's Inn* [1993] 2 EGLR 136; [1993] 42 EG 287
Cadogan Estates Ltd v *Hows* [1989] 2 EGLR 216; [1989] 48 EG 167; [1991] RVR 132, LT
County & Metropolitan Developments Ltd v *Brewer (formerly Swanston) Trust* [1971] EGD 675; (1971) 219 EG 775; [1971] JPL 632, LT
Delaforce v *Evans* (1970) 22 P&CR 770; 215 EG 315, LT
Lake v *Bennett* (1971) 219 EG 945
Lloyd-Jones v *Church Commissioners for England* [1982] 1 EGLR 209; (1981) 261 EG 471, LT
Norfolk v *Trinity College, Cambridge* (1976) 32 P&CR 147; [1976] 1 EGLR 215; 238 EG 121; [1976] JPL 376, LT
Raja Vyricherla Narayana Gajapatiraju v *Revenue Divisional Officer, Vizagapatam* [1939] AC 302
Sked v *Towngran Developments Ltd* [1995] 1 EGLR 216; [1995] 05 EG 169, LVT

Anthony Radevsky (instructed by Bircham & Co) appeared for the nominee purchasers; Simon Berry QC (instructed by Boodle Hatfield) represented the freeholders.

Giving the decision of the tribunal, **LADY FOX QC** said: This is an application made pursuant to section 24 of the Leasehold Reform, Housing and Urban Development Act 1993 (the Act) by Mr George A Donath, Mrs Lidia A Donath and Mrs Angela J Morrison, the tenants, for the determination of the price to be paid for the freehold interest in 36 and 37 Eaton Mews South, London SW1 (the subject premises). The tenants are represented by the nominee purchaser, Mr GA Donath and Mrs AJ Morrison. The tenants hold the subject premises under leases from the trustees of the estate of the second Duke of Westminster, John James, Sir Richard B Wilbraham and JH Newsom, as follows:

36 Eaton Mews South and the west garage (now converted into living accommodation) held by Mrs AJ Morrison under a sublease dated May 22 1989 for a term commencing on May 22 1989 to June 24 2015 at a ground rent of £150 pa;

The east garage of 36 Eaton Mews South held by Mrs AJ Morrison under an underlease dated May 22 1989 for a term commencing on May 22 1989 to June 24 2008 at a rising annual rent as set out in the lease;

37 Eaton Mews South and the east garage held by Mr GA and Mrs LA Donath under an underlease dated July 28 1965 for a term of 51 years commencing on June 24 1964 at a ground rent of £100 pa;

The west garage of 37 Eaton Mews South was surrendered in 1991 to the reversioner, by Mr GA Donath pursuant to the terms of a lease.

By initial notice dated December 15 1993 the tenants gave notice of their intention to acquire the freehold of the subject premises and by counternotice dated February 15 1994 the reversioner admitted that the tenants were entitled to the right of collective enfranchisement of the subject premises. By the date of the first hearing on June 5 1995, the parties had agreed the terms of the proposed conveyance and February 8 1995 was agreed as the date of valuation. Accordingly, the unexpired terms of the leases of the maisonettes were 20.25 years, and of the lease of the east garage at 36 Eaton Mews South 13.25 years.

At the hearings held on June 5 and 6 and November 28 to 30 1995 the nominee purchaser was represented by Mr Anthony Radevsky, counsel, instructed by Bircham & Co, solicitors, and the reversioner by Mr Simon Berry QC, counsel, instructed by Boodle Hatfield, solicitors. At the first hearing on June 5-6 Mr Simon Berry stated that the intermediate leaseholders were represented by the reversioner and that the tribunal was not asked to apportion the price paid between them. The parties' representatives stated that the following facts were agreed:

(i) the subject premises qualified as a block of flats within the meaning of section 101(1) of the Act, the two properties being separate dwellings within the building and a material part of them lying above the two garages which were in separate demises;

(ii) the subject premises are the buildings known as 36 and 37 Eaton Mews South presently comprising two maisonettes and three garages;

(iii) the west garage of 37 Eaton Mews South is to be comprised in the freehold interest to be transferred;

(iv) the estimated rental value of the east garage, 36 Eaton Mews South is £1,900 pa.

Valuations of the price payable for the freehold interest totalling £281,000, revised to £265,000 at the hearing on November 28-30, and £734,400 respectively were put forward by Miss Jennifer Ellis FRICS of Langley Taylor, chartered surveyors, on behalf of the nominee purchaser and by Mr Ian MacPherson FRICS of Gerald Eve on behalf of the reversioner and are set out in full at schedules A and B. The parties' representatives agreed that the price to be determined pursuant to para 2(1)(a) and (b) and Schedule 6 of the Act comprised two elements:

(i) the value of the freeholder's interest in the subject premises, and

(ii) The freeholder's share of the marriage value. There was no compensation payable under para 2(1)(c).

The points at issue between the parties as revised in the course of the hearings were as follows:

	Nominee purchaser	*Reversioner*
Capitalisation rate	10.8%	6%
Deferment rate	10.8%	6%
Freehold capital value of each maisonette incorporating one garage.	£580,000	£670,000
Freehold value of each separate garage.	£20,000	£40,000
Present leasehold value of 36 Eaton Mews South with garage.	£415,500	£300,000
Present leasehold value of 37 Eaton Mews South with garage.	£415,750	£300,000
Present leasehold value of the east garage 36 Eaton Mews South	Nil	£3,000
Reversioner's share of marriage value	50%	75%

In addition the parties disagreed: (a) as to whether there was an increase in prices achieved in the central London residential property market between January 1993 and February 1995, with Mr George

A Pope ARICS, the reversioner's surveyor claiming a 20% increase compared to no increase claimed by Mr David Haines FRICS, the surveyor of the nominee purchasers; (b) as to whether long-term investment prospects in the next 20 years were good in Belgravia; and (c) as to the evidential value and weight of negotiated enfranchisement sales relied on by the reversioner and of auction sales of ground rents relied upon by the nominee purchasers.

A schedule of comparable residential properties in both sections of Eaton Mews South, 52 Eaton Mews West, 28 and 43 Eaton Mews North and 8 and 9 Eaton Row was agreed by the parties and is set out at schedule C.1. Further schedules agreed were produced of three
B garages at 10 Shafto Mews, London SW1, and Byron Court, Elystan Street, London SW3, and Rosemoor Street, London SW3, (see schedule C.2); and of eight residential properties on the market, see schedule D.

At the hearing on June 5 and 6 Mr Haines gave evidence in accordance with a written proof on behalf of the nominee purchasers as to property values. He stated that he was a partner in Langley Taylor, having previously worked for Knight Frank & Rutley in the country house and estate department and as a valuer in the district valuer's office. He had been closely involved with the Department of the Environment in the preparation of the 1993 Act and had lectured
C and written a book on the subject. He acted as surveyor to Eton College and other landlords, and also tenants on enfranchisement matters.

Mr Haines described the location, accommodation and condition of the two maisonettes and garages. He compared the subject premises by reference to location in the mews, size, number and layout of rooms, quality of conversion, features and fittings, length of lease and enfranchisement to the properties listed in the agreed schedule C.1. In view of the "wealth of comparables" he himself relied on comparables in Eaton Mews South only, not considering it necessary to look at properties in other locations. Of those in Eaton Mews South he relied
D in particular on 15, 17 and 33. Mr Haines submitted that the property market had remained in a depressed state since 1993 with no uplift and probably a downward drift and produced articles in *The Times*, *Sunday Times*, *Economist* and *Investor's Chronicle* in support. He accordingly adopted the figure of £580,000 as the freehold value of each subject mews house, a value which the reversioner's advisers themselves had been prepared to put forward in the negotiations in early 1994. Leasehold values of 20 years' length would, in his view, be 50% to 60% of the freehold value; there being little evidence in Belgravia of freehold sales he relied on sales of 75-year leases. As to enfranchisement, he thought, due to uncertainty as to the operation of
E the Act, the market had remained depressed and asking prices had fallen. He was cross-examined by the reversioner's counsel on the comparability and adjustments for differences for these properties and also gave evidence and was cross-examined on the other properties listed in the agreed schedule.

Improvements

Mr Haines referred to secondary glazing, fitted cupboards and bookshelves, tiling and fittings in bathrooms and kitchens, in particular "Gaggenau" cooking units, for which he felt a purchaser of a 20-year lease would be prepared to pay an additional sum in the region of £10,000-£20,000. He did not consider such fittings to be purely a
F matter of taste. In answer to a question from the tribunal he accepted that on the purchase of a 90-year lease or freehold the value of such improvements would be written off and he accordingly would make no deduction from the freehold value for tenants' improvements.

Mr George M Pope of John D Wood & Co gave evidence in accordance with a written proof on behalf of the reversioner. He stated his qualification and experience saying he had joined John D Wood in 1970 after eight years in the Coldstream Guards, becoming a partner in 1975 managing the Chelsea office specialising in residential property until 1988 when he moved to the head office as joint chairman and finance director. Since 1980 he had acted for the Grosvenor Estate in conjunction with Mr Ian MacPherson of Gerald Eve. In support of his value for the leasehold interest of £300,000 he relied on transactions at G
29 and 33 Eaton Mews South and 52 Eaton Mews West, and in support of his freehold value of £670,000, transactions at 17 and 25/26 Eaton Mews South. He drew attention to the comparables, similarities to and differences from the subject properties. In applying the comparables, all of which he had inspected internally, he drew attention first to the difference between the two sections of Eaton Mews South, rating the northern section as less attractive; second he noted the differences in date, stating that, from January 1993 to February 1995, he considered a 20% increase in price should be made to take account of the rise in values in the central London residential market, a rise to which attention had been drawn in the annual H
statement of the joint chairman of John D Wood in July 1994, and was confirmed by the research and survey conducted by Savills. Demand for residential properties in central London came from overseas and was not affected by the recession. He rejected the suggestion put to him by Mr Radevsky on behalf of the nominee purchasers that this was "talking up the market". As to relativity between leasehold and freehold values he considered a 20-year lease worth 50%, and a 75-year lease 90% of the freehold value and he considered these relativities to be supported by the many settlements for extended leases achieved by the Grosvenor Estate, as particularly in a schedule which he produced. He accepted that he had put forward in February 1994 a J
lower value for the freehold but this had been a negotiating figure, adjusted upwards in the light of the comparable transactions which had taken place in the mews in the latter part of 1994 to 1995.

Inspection

Subject dwellings

The tribunal inspected the subject properties externally, and internally in the presence of the tenants, during the afternoon of June 6 1995.

Eaton Mews South lies between Eaton Square and Chester Square, K
and is bisected by Eccleston Street which divides it into west and east sections. The subject properties are in the west section. This section is also delineated by having mainly three-storey dwellings on the north side (where the subject properties are located) and mainly two-storey on the south side.

The mews is a cul-de-sac, the approach from Eccleston Street being through an archway leading to a cobbled mews roadway that is wide enough for parking. The streetscape is varied, with some dwellings, like the subjects, being relatively recently built or substantially rebuilt, and others being conversions from the original mews buildings. The rears of the dwellings on both sides are dominated by the taller and L
more substantial houses that front Eaton Square (to the north) and Chester Square (to the south).

As well as being a prestigious location, Eaton Mews is a convenient one for access to public transport, high class shopping and the River Thames.

The subject dwellings were built in the mid 1960s on three storeys, and have the external appearance of being a pair of terraced houses with ground-floor garages and entrances. The accommodation layouts are similar, the principal difference being that no 36 has converted one of the garages into a ground-floor living-room.

In summary, the accommodation of each dwelling is arranged as M
follows:

Ground floor: entrance hall, rear room, bathroom/WC
Outside: single garage (no 36), two garages (no 37)
First floor: intercommunicating lounge/dining room, kitchen/breakfast room, study, cloakroom (no 36) utility room (no 37).
Second floor: master bedroom with en-suite bathroom, two other bedrooms, bathroom/WC.

In addition, on the ground floor no 36 has a front room, converted by the tenant from a former garage.

Neither dwelling has a garden, but the rear wall of each abuts the rear gardens of the houses behind, which front Eaton Square. The rear rooms of the subject dwellings have windows directly looking on to these other gardens.

Both dwellings have gas-fired central heating with radiators. The internal fittings of both have been modernised and improved by the tenants in comparison with what was likely to have been originally fitted when new.

Comparables

The comparable dwellings were mainly located in Eaton Mews South (both west and east sections), with one in Eaton Mews West (a short distance away leading off Elizabeth Street) and two in Eaton Mews North (on the other side of Kings Road leading off Lyall Street).

Eaton Mews South (west section)

No 46: This is opposite the subject dwellings. The tribunal was able to inspect it internally. It is a two-storey plain fronted original mews house comprising living-room, kitchen, utility room (with internal access to garage) on the ground floor, two bedrooms (one with dressing room leading off), bathroom/WC on the first floor. It was smaller than the subject dwellings and unimproved. There are no windows or outlook to the rear.

No 47: This adjoins no 46 and is similar in style, but was in better condition. It was inspected internally. On the ground floor it has an entrance lobby, split-level living-room (with internal access to garage), and kitchen; on the first floor it has three bedrooms (one with en-suite shower room), bathroom/WC and a dressing-room. Again there was no outlook to the rear.

No 33: On the same side as and close to the subject dwellings. It was inspected internally. It is a three-storey plain fronted original mews house comprising entrance hall, living-room, bedroom, bathroom/WC on the ground floor; landing, L-shaped living/dining room, kitchen on the first floor; landing, two bedrooms (one with en-suite bathroom/WC), bathroom/WC on the second floor. The house appeared to be the subject of recent extensive refurbishment.

No other dwellings were inspected internally but the following, all three-storey dwellings located on the same side as the subjects, were inspected externally:

Nos 25 and-26: This is a rebuilt and laterally converted property of similar size to the subject properties, with rear extension to no 25, having attractive elevations, but close to the Eccleston Street entrance and thereby in a noticeably noisier position. It was in very good external condition and the roof garden was noted (no 26).

No 29: This is a rebuilt property with garage, also in good external condition.

No 34: This is a rebuilt property with garage.

No 35: Adjoins no 36, a rebuilt property with a garage. It has an attractive front elevation and possibly a roof garden.

Eaton Mews South (east section)

By crossing Eccleston Street the east section (also a cul-de-sac) was entered. The tribunal was of the opinion that it was less attractive than the west section. Three comparables were inspected externally, all are three-storey dwellings:

No 15: A larger double-fronted rebuilt property with an attractive elevation.

No 17: Another rebuilt property having attractive elevation (garage in the middle), in good condition.

No 20: At the time of the tribunal's inspection this dwelling was being gutted.

Eaton Mews West

This was not a cul-de-sac having accesses from Elizabeth Street at one end and from Eaton Square (Eaton Place) at the other. It also had a substantial non-residential user as almost the whole of the south side is a commercial garage. Only one dwelling (immediately opposite the garage) was inspected externally:

No 52: A three-storey rebuilt dwelling with two garages. It has an attractive elevation.

Eaton Mews North

This is a long mews accessed from Lyall Street and Lowndes Place. Two dwellings were inspected externally:

No 28: A two/three-storey rebuilt dwelling, with a wide frontage and garage, and having an attractive elevation. In good condition externally.

No 43: A three-storey rebuilt dwelling, with slated mansard roof snowcemmed elevation and garage. Attractive elevation.

At the resumed hearing on November 28 to 30 1995, the parties called evidence to support the three elements of capitalisation rate, deferment rate and marriage value in their valuations. Miss Jennifer Ellis FRICS, partner in Langley Taylor, chartered surveyors, gave evidence on behalf of the nominated purchasers in support of her revised valuation set out in schedule A. She described that she had been trained as a pupil of Mr E L Norman FRICS, practised in Debenham, Tewson & Chinnocks (DTZ) until 1989, from 1989 to 1995 in Caws & Morris where she had been appointed a director in 1990 and, leaving in May 1995 to join Langley Taylor. She had been closely involved in the preparation of the 1993 Act, advising the Building Societies Association in 1984, serving on the RICS working party on the government's proposals and lecturing and writing on the 1993 Act. She was a member of the general council of the RICS, director of its trading company Business Services and external examiner to Nottingham Trent University in its BSc/BSc (Hons) estate surveying courses. She had advised both landlords and tenants, but since the passing of the Act had acted only for tenants on all the main London estates, including Smith's Charity, the Eyre, Portman, Cadogan, John Lyon, Harris and Ilchester estates.

Capitalisation rate

After setting out the agreed facts and her valuation method, Miss Ellis spoke in support of her rate of 10.8% for capitalisation. She said by reason of the ownership of the great estates in central London there was a dearth of evidence of central London ground rent sales and she had prepared a schedule of auction sales of ground rent sales, first as to England and Wales, and second confined to inner London postal districts, selected on the basis that they were published open market sales, effected in the 12 months prior to the valuation date, February 8 1995, of ground rents on leases of 75 years or over relating to small blocks of flats. The average yield of the sales in inner London was 10.8%, and she had accordingly adopted this percentage, revised from her earlier figure of 12.5%, as the capitalisation rate based on the available evidence. When compared to yields from other types of property and long date gilts, a rate of 10.8% appeared reasonable.

Discount rate

She had adopted the same rate for deferment as for capitalisation, namely 10.8% and had done so by reference to the prospects for growth in residential capital values in Belgravia over the 20 years to the end of the lease term. Relying on indices of house prices from the DoE, Halifax Building Society and DTZ, Miss Ellis had come to the conclusion that an investor would take a cautious view, expecting to see no more real growth in the value of houses in Belgravia than elsewhere. Her rate of 10.8% was supported by the current gross redemption yield of 9.1% to 9.6% on zero dividend preference shares issued by investment trusts.

Share of marriage value

In the Lands Tribunal decision of *Lloyd-Jones* v *Church Commissioners for England* (1981) 261 EG 471* the share of marriage value on enfranchisement of houses under the 1967 Act had been argued from first principles and reference to the *Indian*† case [1939] AC 302; the Lands Tribunal had held that: "neither (party) can unlock the marriage value without the other. In friendly negotiations they

---

*Editor's note: Also reported at [1982] 1 EGLR 209

†Editor's note: See *Raja Vyricherla Narayana Gajapatiraju* v *Revenue Divisional Officer, Vizagapatam*

A would agree to divide it equally". Miss Ellis said para 4(1) of Schedule 6 to the 1993 Act provided that the reversioner was to receive 50% of any marriage value unless a larger proportion would have been agreed on a sale on the open market by a willing seller. Open market in law could include a monopoly position: see *Baptist* v *Masters of the Bench and Trustees of the Honourable Society of Gray's Inn* [1993] 42 EG 287*; and "willing", according to the dictionary definition, meant "ready" or "intent to sell", not merely acquiescent. In her view, the transactions relied on by Mr MacPherson to support a landlord's share of marriage value in excess of 50% were not "open market" sales, in that they involved special circumstances and sales outside the 1993 B Act. Neither the Cadogan nor the Grosvenor nor the Ilchester estates displayed the willingness required in the statute, selling only when compelled to do so.

Mr Berry, on behalf of the reversioner, cross-examined Miss Ellis. In answer to criticism of her exclusion of other factors in arriving at her capitalisation rate, Miss Ellis said that she had been solely concerned to determine what an investor would pay for rental income stream. On this account she had deliberately confined her inquiry to ground rents for leases of 75 years or more, where income stream was the sole consideration for the investment. To have regard, as Mr MacPherson had done, to the shortness of the lease and capital growth C in the determination of the capitalisation rate, would, in her view, result in double counting. Miss Ellis agreed that she had been surprised in compiling her table of capitalisation rates from auction sales of ground rents to find an increase in interest rates for ground rents in London, but it supported her general thesis that location and type of property were irrelevant in determining the investment value of ground rents on their own. She considered the subject properties for investment purposes must be valued as flats, not houses, since this was how they were categorised under the 1993 Act.

In respect of the rate which she adopted for deferment, Miss Ellis said that recent statistics supported her view that the property market D would not always stay in boom and that the prime location of the subject premises in Belgravia did not insulate it from the likely decline in property values in the long term. She agreed that she had no first-hand evidence of any settlement achieved with the Grosvenor Estate. Miss Ellis was closely questioned on the open market transactions, particulars of which are set out in schedule C.1. She remained of the view that in a settlement the tenant was in a weaker bargaining position than the landlord and consequently subject to the *Delaforce* effect. She submitted that Mr MacPherson consistently applied 6% as his capitalisation rate, prior to entering any negotiation with a sitting tenant and imposed that rate from the start on the parties. In the revised E calculations made by Mr Marr Johnson and Mr Boston in relation to 25 and 26 Eaton Mews South and 13 Gerald Road respectively, they had employed 6% as a rate dictated by the Grosvenor Estate, and not because they thought it to be the appropriate rate.

In re-examination, Miss Ellis agreed that in a purchase of the freehold of flats two or more parties were involved on the part of the tenants with all the consequent differences in interest, management and repairing obligation. Investment in flats differed from investment in a single house.

In answer to questions from the tribunal Miss Ellis stated that the proportion of 1.4%, which the capitalised ground rents represented in F her valuation to the capital freehold value, had no relevance, in her view, when seeking to determine the capitalisation rate. She considered it inappropriate to compare such short term leases to the ground rent value of 75-year leases where there was no capital reversion value. She accepted that a tenant holding a short term lease in a prime area, such as Belgravia, might seek an extension, and be willing to pay a premium at some point during that term. This potential value should be taken into account but not at the stage of determining the capitalisation rate. Asked what evidence she would be prepared to accept to support a higher proportion of market value than 50%, Miss Ellis replied that the requirement of a willing seller precluded in most circumstances the G possibility of the tenant agreeing to a higher proportion. But possibly the signature of the tenant or his proposed adviser on the calculation sheets of the Grosvenor Estate or a personal statement by the tenant might support a higher proportion.

Mr Berry called Mr Ian MacPherson FRICS, partner in Gerald Eve, chartered surveyors, to give evidence in support of his valuation set out at schedule B. Mr MacPherson said he had 25 years' professional experience and had advised the Grosvenor Estate since 1975, the Howard De Walden Estate since 1981 and the Ilchester Estate since 1992 on leasehold enfranchisement valuations under the 1967 Act, as amended, and the 1993 Act. He had also advised the Fitzwilliam H (Wentworth) Estates, Letchworth Garden City Corporation and many tenants' associations and individual leaseholder/ claimants. He had inspected internally the two subject premises and garages. Before explaining the detail of his valuation, he wished at the outset to state the reliance which he placed on negotiated enfranchisement prices for houses on the Grosvenor and Cadogan Estates as evidence of the relevant land market for the determination of enfranchisement prices for houses under section 9(1)A of the Leasehold Reform Act 1967 as amended. This settlement evidence covered all the houses on the Grosvenor and Cadogan Estates for which enfranchisement prices had been obtained, except one, 60 Cadogan Lane, which had been J determined by the Lands Tribunal in *Cadogan Estates Ltd* v *Hows* [1989] 48 EG 167*. This settlement evidence had been presented to the Lands Tribunal in the *Lloyd-Jones* and *Cadogan Estates* cases and Mr MacPherson cited passages from these decisions to show that they had been accepted as providing "massive" support for the percentages shown in the calculations made by the landlord's surveyors. He said his practice, as stated in those two cases, was to follow the historic settlement, evidence which he had obtained from negotiating enfranchisement prices for other houses in similar locations on a consistent basis unless he became aware of more reliable evidence in a response to well founded points made on behalf of leaseholders. In K appendices Mr MacPherson produced 143 enfranchisement settlements on the Grosvenor and Cadogan Estates relating to section 9(1) valuations, 220 settlements on those estates relating to section 9(1) valuations, and one settlement relating to section 9(1)C under the 1993 Act.

Capitalisation rate

Mr MacPherson proposed 6% as the appropriate rate for capitalisation of the ground rents on the basis that: they were well secured; the landlord could look forward to a substantial reversionary value in 20 years' time; relatively low immediate yields were acceptable to investors in respect of long term benefits due on L substantial estates of good residential properties; and that the rate of 6% was supported by the land market evidence of comparable negotiated enfranchisement prices for other houses on the estate. In this connection he considered that the hypothetical assumptions to be applied by virtue of Schedule 5, para 4, approximated to those laid down for enfranchisement of houses in the 1967 Act, as amended. Accordingly, the pattern of yield rates for "flats", varying with the term unexpired, should conform with those which the settlements evidence revealed, namely a rate of 5% for a term of zero to 10 years unexpired, rising to 8% for 60 years or over unexpired, with a band of 6% for 20 to 30 years unexpired. Accordingly, in the same way as the M estate would have negotiated a settlement at a capitalisation rate of 6% for a house with an unexpired term of 20 to 30 years, he proposed a 6% rate for the subject "houses" with 20.25 years unexpired, and applied the same rate in respect of the east garage at no 36, (although it had an unexpired term of 13.25 years, with five-year rent reviews) because it was a garage not a dwelling.

Deferment rate

Mr MacPherson applied the same rate of 6% in deferring the reversion values on the basis that there was a long term trend for

---

*Editor's note: Also reported at [1993] 2 EGLR 136

*Editor's note: Also reported at [1989] 2 EGLR 216

A values in residential property, particularly in central London to rise, with a growth rate in excess of inflation. The keenness of leaseholders to enfranchise and the opposition of freeholders to the extension of enfranchisement indicated that both interests considered property in central London to constitute a sound investment. Reported sales of the Chesham Place Estate and Chalcots Residential Estate at Swiss Cottage provided proof of investor interest in central London. It was his view that the excellence of the location, quality of the property, and greater potential for realising the marriage value by the shorter deferment of the reversion, were all factors affecting the determination of the rate. Mr MacPherson again relied on the consistent application
B of 6% deferment rate in the negotiated valuations for enfranchisement of 364 houses on the Grosvenor and Cadogan Estates and he listed some nine decisions of leasehold valuation tribunals which had applied 6% to central London properties. He distinguished tribunal decisions at *7 Queensmead*, *58 Eton Avenue* and *23 The Little Boltons* as giving undue weight to previous tribunal practice, ignoring land market evidence arising from settlements which was available in the present case and in respect of the first two decisions, applying a higher deferment rate for flats. He maintained that the deferment rate should not increase solely by reason of the length of deferment: see *County & Metropolitan Developments Ltd* v *Brewer (formerly Swanston) Trust*
C (1971) 219 EG 775; *Lake* v *Bennet* (1971) 219 EG 945.

Unlike Miss Ellis, he considered evidence of auction sales of ground rents of blocks of flats with leases of 75 years or over to be of no relevance to the security and growth prospects on a 20-year lease; they were concerned solely with the rental income stream over 75 years, as was also true of the leasehold valuation tribunal's decision in *Sked* v *Towngran Developments Ltd* [1995] 1 EGLR 216; [1995] 05 EG 169. While he had arrived at the same percentage of 6% for both capitalisation and deferment on the basis of different reasoning, Miss Ellis appeared to have adopted the rate for deferment by reason of the auction sales evidence supporting her capitalisation rate; this could not
D be correct. Nor had she been consistent in her application of differing rates in respect of the east garage at no 35.

Marriage value

He referred to the terms of Schedule 6, para 4(2), and said after collective enfranchisement that each participating tenant would enjoy virtually similar freedom and opportunities to do what they liked with their property as if they owned the freehold of their own mews house. In the hypothetical situation of the existing freeholder voluntarily selling the freehold, outside the Act, to the tenant in occupation by means of a similar arrangement, there was no doubt that the tenant would be prepared to pay a share of the marriage value because it
E would enable him to increase the value of the property. Adopting the freehold and leasehold valuations assessed by Mr Pope, a marriage value of £330,522 was arrived at. By para 4(1)(a) as the proportion had not been agreed by the parties, it fell to the present tribunal to determine whether it should exceed the statutory minimum of 50% and this depended on whether in a sale "in the open market by willing sellers" it would exceed 50%. As to the meaning of "willing" seller he relied on the definition in the manual of the Royal Institution of Chartered Surveyors: RICS Practice Statement 4.1.7:

> a willing seller . . . is neither an over-eager nor a forced seller prepared to sell
> F at any price, nor one prepared to hold out for a price not considered reasonable in the current market. The willing seller is restricted to sell the asset at market terms for the best price attainable in the "open market" after proper marketing, whatever that price may be. The factual circumstances of the actual asset owner are not a part of this consideration because the "willing seller" is a hypothetical owner.

Mr MacPherson gave evidence and submitted detailed analyses of open market transactions to support a greater share than 50% being achieved by the freeholder. (Particulars of these transactions are set out at schedule D.) Just as the historic settlement evidence, relating to section 9(1) and 9(1)A enfranchisements provided in the analysis evidence of the practice relating to capitalisation, deferment and entirety rates so he submitted the working document prepared in these open market transactions also supported the adoption of a 6% rate, and G provided evidence that the landlord's share of the marriage value now exceeded 50%. He referred to a number of errors in the reanalysis of the transaction at 25 and 26 Eaton Mews South provided by Mr Marr Johnson who had acted for the tenant. In the past, in the absence of open market evidence, reliance had been placed on settlements on the Grosvenor and Cadogan Estates to support an equal division of marriage value between the parties, but, in the light of the open market transactions (set out in schedule D), Mr MacPherson invited the tribunal to award the reversioner a higher proportion than 50% of the marriage value and the proposed 75%.

In concluding his evidence-in-chief Mr MacPherson was asked to H consider Miss Ellis' valuation and approach. He said he did not think auction sales of property of very different type, location and capital value to the subject premises — differences which he illustrated by a plan, photographs, detailed comment and comparison of yields — could provide relevant evidence in the present inquiry; he had himself externally inspected or had arranged inspection of all properties listed in Miss Ellis' schedules, and considered them in no way comparable; even taking the properties nearest to Belgravia listed on the inner London schedule which included 21 Ongar Road, Fulham, 9 Pembridge Villas, Notting Hill, 23/24 River Street, Clerkenwell, it was difficult to imagine a prospective purchaser of the subject properties J who would be interested in them. He could recall in recent experience no negotiation in respect of an enfranchisement price where a surveyor acting for the leaseholder had contended for a rate as high as Miss Ellis' original figure of 12.5%. In fact he had recently agreed a price for a freehold for a house in Belgravia at 13 Gerald Road under section 9(1)(C) of the Act, as amended, with Mr Charles Boston FSVA IRRV MCIM, and produced correspondence relating thereto in which Mr Boston agreed to a capitalisation rate of 6% and that the landlord should receive more than 50% of the marriage value. Mr Boston's offer of 69% on May 22 had been considered too low and with regard to the price of £250,000 finally achieved the landlord's share analysed K out at 68%.

Long term investment prospects in residential property in Belgravia

Continuing, Mr MacPherson said that Miss Ellis had expanded her evidence at the November hearing to take an uncommonly pessimistic view about the growth potential of prime central London residential property. She had disregarded the substantial international, as well as domestic, demand for such property, overlooked that the dominant part of the investment was the reversion, and wrongly over emphasised the statutory categorisation of the subject properties as "flats", so as to compare them to blocks of flats and not individual houses. L

In cross-examination by Mr Radevsky on behalf of the nominee purchaser, Mr MacPherson denied that he had claimed the highest figure possible for enfranchisement in the reversioner's counter notice in April 1994; the price then sought, £826,500 was based on 1994 values, including a lower value of £585,000 for the freehold of each mews house. Although his present proposed figure of £734,400 was significantly lower, he considered it to be reasonable, particularly in the light of £653,000 achieved for the 75-year lease for 25 and 26 Eaton Mews South, and taking into account the greater worth of a freehold. When questioned on negotiated settlements, Mr MacPherson agreed that in every case a notice to enfranchise the property had been M served on the landlord prior to negotiations. He agreed that, in these negotiations, he applied a capitalisation rate according to a fixed formula derived from the unexpired term; he had not been persuaded, by reference to money market rates, to apply a different rate. Indeed, in the present case, when interest rates were at a historically low rate, had he taken account of them he would have applied a lower rather than a higher rate than 6%. When asked why he had not obtained the signature of the leaseholder's surveyor to his calculation of a negotiated price, Mr MacPherson said he did not ask for a signature because neither in the present case nor on earlier occasions would any surveyor have been prepared to sign. As regards the open market transactions, he accepted that the Grosvenor Estate had been looking

A for evidence to support the landlord obtaining a higher proportion than 50% of the marriage value, but he did not think that leaseholders' professional advisers would be prepared to assist the estate by signing the valuation sheets; the present case relating to the proportion of the marriage value which the landlord should receive was well known on the estate. Mr MacPherson said he relied on his correspondence with Mr Boston to show that he was not himself prepared to do a deal, where the landlord received no more than 50% of the marriage value, and his analysis of the transactions for 13 Gerald Road showed that a higher proportion had been agreed.

He was not claiming his analyses were agreed, but he did claim that B a large number of settlements were agreed, based on his figures.

Mr MacPherson said he had adopted a 6% rate for deferment by going direct to comparable transactions and not, as had Miss Ellis, by reference to general trends in the residential London market. After further inquiry he said that he understood mortgages were available to leaseholders and the Grosvenor Estate with terms of under 30 years, provided the mortgage was repayable with 10 years still unexpired of the term. Mr Radevsky questioned Mr MacPherson closely on how he justified his analyses of the transactions at 28 Eccleston Mews and 17 Eaton Mews South.

In answer to questions from members of the tribunal, Mr C MacPherson said that he did not think investment prospects were affected by the operation of the 1987 Act, which gave tenants in flats the first option collectively to purchase the freehold. As to long term growth rates he accepted that the appeal of property in Belgravia in part depended on the strength of the British economy and the attraction of property in the best residential district of a major capital city. When asked to comment on the requirement in Schedule 6, para 3, that a valuation of the freehold interest should disregard the value of tenant's improvements, and the omission in para 4 of any such requirement in respect of the valuation of the leasehold interest, Mr MacPherson stated that he considered that improvements should be disregarded in D valuation of both interests. Otherwise, by its effect on the marriage value, the value of the landlord's interest would be higher where the property was unimproved than where it was improved. In answer to a further question, he stated that the lateral conversion of 25 and 26 Eaton Mews South did not, in his view make it significantly different from the subject properties.

At the conclusion of Mr MacPherson's evidence, Mr Berry requested permission to recall Mr Pope to give further evidence to refute Miss Ellis' pessimistic forecast of investment prospects in Belgravia. In accordance with a further proof, Mr Pope maintained that Miss Ellis had taken insufficient account of the two-tier market which E had developed in residential properties. The Halifax index and other indices relied upon by her, related to houses in low value brackets and were not relevant to sales in Belgravia. He produced the November 1994 edition of Savills *Quarterly Residential Research Bulletin* which reported: "in Central London growth of 20% the year to September 1994 had been achieved"; and the September 1995 edition made reference to "a high rise in values of 33% from September 1992". He relied on a graph entitled "Return to Trend" as demonstrating the wisdom of investment in central London residential property in the long term. Cross-examined by Mr Radevsky on behalf of the nominee purchaser, Mr Pope said that the statement, in the 1995 Annual Report of the chairman of John D Wood, that "at present the market is F subdued", was explained by hopes in respect of the opening of the new office in Belgravia not being fulfilled. He agreed 34 Eaton Mews South had been on the market for two years and the price had now been reduced. He did not accept that Savills *Quarterly* was designed to talk up house prices nor that the placing of the line in the graph on p3 was arbitrary, and without any mathematical significance.

On the conclusion of the above evidence on behalf of the reversioner, Mr Berry requested leave to call evidence relating to the sale for £280m of some 2,300 residential properties by Smith's Charity to the Wellcome Trust and he produced a schedule of some 58 selected properties prepared by Mr Roland Cullum of Cluttons, the agents acting for the Smith's Charity, and newspaper reports of the transaction. He urged the tribunal to admit the evidence on the ground G that it represented a real market transaction. Mr Radevsky strongly objected to the transaction being admitted. He reminded the tribunal that in correspondence between the first and second hearing it had requested the parties to deliver all new evidence by November 20, that the nominee purchaser, but not the reversioner, had complied with this request, and that the advisers of the nominee purchaser had totally inadequate opportunity to analyse it, since the new evidence was produced late on Monday November 27 1995. After consideration of the matter overnight, the tribunal at the resumed hearing on November 30 requested counsel for both parties to address them *de bene esse* and on the assumption that the tribunal were of a mind to admit evidence H relating to the sale by Smith's Charity, as to the relevance and weight of that transaction. Mr Berry, on behalf of the reversioner, submitted that the schedules of residential properties included in the sale set out the ground rent, expiry date of the lease and values of the reversionary and freehold vacant possession interests of each property agreed by the parties; it was accordingly possible to analyse the yields which were the basis of an open market transaction between parties with professional representatives. It was a relevant similar transaction to support Mr MacPherson's proposed rate of 6%. As to weight, although it differed from the subject enfranchisement, being a large transaction, not a purchase of a single reversion, it indicated that other investors in J the market were willing to invest at this deferment rate, alternatively he submitted that account of the Smith's Charity transaction should be taken in assessing the potential of acquiring the freehold in the subject properties.

Mr Radevsky, on behalf of the nominee purchaser said the evidence admitted *de bene esse* of the Smith's Charity sale showed nothing. On the face of it, the only agreed figure was the value of the reversion; it was not stated that the freehold vacant possession values had been agreed. Some 117 properties out of a total 2,300 had been selected by Cluttons; this represented 5% of the whole but, on the basis the total price was £280m, represented 14.6% of the total value. No information K was given as to rent reviews or "hope" value arising from sitting tenants and, even if time was available, it would not be possible on the information disclosed to come to any conclusion about yields. 2,300 properties had been sold for one global price and the tenants of Smith's Charity believed that the global price had been apportioned so as to load those properties where the Landlord and Tenant Act 1987 was likely to apply. He again asked the tribunal not to admit the transaction, it being of no relevance or weight.

The chairman of the tribunal requested Mr Berry to explain whether reliance could be placed on a document which appeared to have been prepared to enable the Smith's Charity and the Wellcome Trust to L sidestep the statutory procedure laid down in Part I of the 1987 Act, whereby tenants were given the first option to buy the freehold. Mr Berry stated that he had been informed that the 1987 Act had been followed to the letter; one tenant however, was taking proceedings against the Smith's Charity for alleged violation of the 1987 Act. After taking instructions, Mr Radevsky informed the tribunal that on Monday November 20 there had been a court hearing in which the Smith's Charity was represented by leading counsel. A court order had been made on the following terms: (i) undertakings to be given by the landlord that it would serve, within 48 hours, section 5 notices under the 1987 Act upon the plaintiffs, (they being tenants with properties M included in the sale to the Wellcome Trust); (ii) no immediate completion of the sale to Wellcome Estate; (iii) expedition of its application for an estate management scheme to the London Leasehold Valuation Tribunal; (iv) leave being given to the tenants to reapply if Smith's Charity was dilatory; (v) leave to apply to the Court of Appeal. Far from being settled, litigation was continuing by leaseholders challenging the lawfulness of the disposal by Smith's Charity.

After an adjournment, the tribunal stated its decision not to admit evidence relating to the transaction by the Smith's Charity. We did so for the following reasons:

(i) That we were not satisfied that it was a completed transaction.

(ii) Nor that it was in conformity with the requirements of the 1987 Act.

(iii) That it appeared to be a portfolio transaction not comparable to the subject acquisition of a single freehold.

(iv) There appeared to be a mismatch as to the number of properties selected for analysis, 5% of the total, and their value proportionate to the total value, 14.6% on Mr Radevsky's analysis.

(v) There was no information relating to rent reviews in respect of the ground rents shown.

(vi) Finally, that late production of the information gave the nominee purchaser and its advisers no adequate opportunity to answer it.

Mr Berry, in his final submissions on behalf of the reversioner, reminded the tribunal that, in accordance with High Court rulings, it should be slow to arrive at a decision unsupported by evidence on the method put forward by the parties. He put forward Mr Pope as an experienced valuer with long first hand experience of the Belgravia market, who had inspected the subject properties internally, whereas Mr Haines' valuation was based on second hand knowledge. He instanced particulars in Mr Haines' evidence and cross-examination, which indicated uncertainty, hesitancy and superficiality in his approach. He urged us to prefer Mr Pope's comparables of 29 and 33 Eaton Mews South and 52 Eaton Mews West for the leasehold value and 17, 25/26, 28 and 47 Eaton Mews South for the freehold value, to those put forward by Mr Haines. He also urged the tribunal to accept Mr Pope's figure of £40,000 for the capital value of a garage, suggesting that Mr Haines' concentration on size of garage was an over refined argument. He also invited the tribunal to prefer Mr MacPherson's evidence to that of Miss Ellis, suggesting that she lacked experience of the Grosvenor Estate, was over technical and had valued on the basis of nationwide statistics rather than data applicable to the central London residential market. She wrongly dealt with the subject properties as flats and did not assess the 1993 Act in the context of the body of law which already existed and the potent evidence relating to settlements. He equally submitted there to be a body of comparable evidence of open market transactions to support a 75% share of the marriage value to the landlord. When asked why the figure of 75% was chosen, when the open market transactions showed differing proportions, Mr Berry said Mr MacPherson was the expert and his opinion of the balance of the bargaining power of the parties should be accepted.

Mr Radevsky, in his final submissions on behalf of the tenants, reminded the tribunal that Mr Haines' figure of £580,000, for the freehold value had originally been agreed with the landlord's surveyors, at a time when they were putting forward a claim for £826,500. The overall claim had now decreased by £100,000, yet there was no explanation why the freehold value had increased to £670,000 in less than 10 months, a period during which there was no *market* evidence as to any increase in prices. He urged the tribunal to disregard 25 and 26 Eaton Mews South as a comparable, since the figures were disputed, and the layout required the property to be considered as similar to one large house in Eaton Square. He submitted that Miss Ellis' approach to the deferment rate was based on market evidence, not the unreliable settlements put forward by the reversioner. He questioned why the Grosvenor Estate did not itself repurchase if, as Mr MacPherson maintained, long term growth in central London was to be expected. Mr Radevsky submitted that there was no evidence that the share of marriage value "would have been determined by an agreement" at more than 50% in a mutually willing collective enfranchisement registration with the Grosvenor Estate outside the 1993 Act. Despite two years since its enactment, the Grosvenor Estate had been unable, in any of the transactions, to put forward a signature on the part of the tenant to support such an agreement to a landlord's share of more than 50%. The nearest to evidence, which Mr MacPherson was able to produce, was a "without prejudice" letter by Mr Boston relating to the enfranchisement of 13 Gerald Road. The exchange of correspondence merely demonstrated that Mr MacPherson on behalf of the reversioner was as intransigent in negotiations as he had shown himself in giving evidence.

Decision and reasons

In this case we are asked to determine the price payable for the freehold of two mews houses and four garages on the Grosvenor Estate; although the houses appear, on inspection, and are in occupation enjoyed as two separate vertically divided units, they do not qualify, by reason of the arrangement of the demises, as enfranchiseable houses under section 9(1)A of the Leasehold Reform Act 1967, as amended in section 9(1)C of the 1993 Act; instead, by reason of the separate conveyances of the two houses as separate dwellings within the building and a material part of them lying above the two garages which were in separate demises, they qualify as a block of flats within the meaning of section 101(1) of the Act.

Miss Ellis' approach to valuation was, accordingly, based on the view that section 9(1)A and C had little or no relevance, and submitted that our valuation should be based on Schedule 6 to the 1993 Act relating to collective enfranchisement. In particular, she submitted that the tenant's right to a 90-year extension, which is not available in respect of an enfranchiseable house, requires a separate valuation, and an addition to the value of the vacant leasehold possession.

The reversioner's case, while accepting that Schedule 6 to the Act applies, is to value the freehold and leasehold values of the subject premises by reference to comparable sales of houses as "houses" in the vicinity, and Mr MacPherson seeks to apply to the subject premises the percentages for capitalisation and deferment (but not marriage value see below) which were consistently applied in the negotiated enfranchisements of all houses on the Cadogan and Grosvenor Estates under the 1967 Act.

It is the tribunal's view that, so far as the law is concerned, we must look to Schedule 6 to the 1993 Act. That Act has permitted the enfranchisement of a class of dwellings ("flats") that previously were not enfranchiseable. The reversioner deliberately made the dwellings non-enfranchiseable as houses under the 1967 Act by withholding one garage located under each "flat" from the principal lease. In contrast with the 1967 Act, the 1993 Act gives very specific rules in Schedule 6 for the basis of valuation, including a definition of the marriage value, and a direction as to the proportion to be awarded to the landlord. Accordingly, we are of the view that these rules govern our decision, and in the event of any conflict, must prevail over any valuation practice adopted in relation to houses enfranchised under the 1967 Act. We consider further the relevance of negotiated enfranchisement settlements below when discussing the parties' submissions relating to the appropriate rates for capitalisation and deferment.

However, we are of the view that determination in accordance with the 1993 Act is a more complicated process than Miss Ellis allows, in that in practice, the market would treat the dwellings as houses, because in reality that is what they are. Indeed both Mr Haines and Mr Pope, in arriving at freehold and leasehold values, have compared the subject premises to transactions relating to sales of houses and, from our inspection, we consider this to be appropriate. It is indeed probable that, after collective enfranchisement, the tenants in the present case will sever so that each becomes owner of the freehold and able to sell it separately in the market.

On this account we do not think Miss Ellis' claim for recognition of additional value in the tenant's right to a 90-year extension can be sustained. She calculates the leasehold value by first accepting Mr Haines' 999-year lease value of £580,000, and then deducting from it the present value of the freeholder's interest (using 10.8% as to term and reversion); she then applies an equal share of the marriage value based on these figures and adds the calculated cost of the lease extension comprising premium, professional costs and allowing for time and risk. She arrives at figures of £415,500 for no 36 and £415,750 for no 37. In our view, the addition of any enfranchiseable rights attributable to the leasehold are reflected in prices paid in the open market. The tribunal was accordingly unable to accept Miss Ellis'

A approach for valuing the leasehold interest and would rather rely on the direct evidence of market transactions as provided by both parties.

Freehold and leasehold capital values

We, therefore, turn our attention to the valuation of the freehold and leasehold capital interests in the subject property. Evidence in the main was given by the parties on this issue at the first hearing.

The freehold interest in the context of collective enfranchisement is strictly a misnomer, in that, for the purposes of calculation of the marriage value, an extended lease of unlimited term without payment of a premium is the interest to be acquired by each participating tenant
B which is to be compared to the current leasehold term which he holds. Both parties recognise this distinction in their valuations; Mr Pope on behalf of the reversioner values on the basis of new long leases of 999 years at a peppercorn rent and Mr Haines values on a similar basis, maintaining that there is no difference in value between a lease of 999 years and one of 110.25 years. We also accept this distinction and "freehold" should be read throughout this decision as referring to a 999-year lease at a peppercorn rent.

The tribunal gave careful consideration and inspected all the comparables listed in the agreed schedule, save 8 and 9 Eaton Row which was accepted by the parties as not relevant. Of the comparables
C outside Eaton Mews South, several of the transactions referred to were completed one year to 18 months prior to the valuation date, February 1995; Mr Haines on the nominee purchasers' behalf maintained that there had been no increase in prices in the market, and possibly a downward drift and he cited newspaper and journal articles to that effect. Mr Pope, on the reversioner's behalf, on the other hand argued in favour of a 20% increase in the central London residential market, which was to be distinguished from the national state of the market reported in the media, and he relied in support on a survey of Savills and the annual statement of himself as joint chairman of John D Wood in July 1994.

D While we treat with caution statements of estate agents as to the state of the market, we think the evidence supports some increase in prices over the period January 1993 to February 1995, though of more modest proportions than 20% and have taken this into account in applying the comparables.

We discarded 52 Eaton Mews West, as the location close to commercial garage premises changed, in our view, the whole character of a mews and so affected the value, of what otherwise appeared to be a pleasant mews house, as to make comparison unreliable. Eaton Mews North, however, was a residential mews similar to Eaton Mews South having the advantage of access by two separate roads; no 43, due to the particular circumstances of the transaction, was not at the
E hearing put forward by Mr Pope as a comparable, but he relied on no 28 and we consider it, when adjusted for differences, to provide useful support for Mr Pope's higher values for the freehold interest. The date of transaction for 28 Eaton Mews North closely approximates to the valuation date, the shorter duration of the term sold and any eccentricity of layout is balanced by the larger size, wider frontage and advantage of a roof garden. We accept it as a useful market check for prices achieved within the subject Mews. Here we preferred the west section, where the subject premises were located, to the east section, which was narrower, more overlooked, with tall buildings in office use on one side. No 20 on inspection was gutted and as its conversion
F appeared to be in connection with the development of 2A Eaton Square, we considered the price not comparable, being affected by external considerations unknown to us.

15 and 17 Eaton Mews South provided us with useful guidance; no 15 required considerable adjustment to take into account its larger size by a third ($250m^2$), four years' additional length of lease and its condition as a luxury specialist house and, even applying Mr Pope's order of rise in prices for the year's difference in dates, this transaction indicates, in our view, a lower figure for the leasehold value than that advanced by Mr Haines. We found further support for higher freehold values in the price of £625,000 achieved at no 17 for a 75-year lease at a rent of £1,250 pa. We regard this price as having elements of an open market transaction of a similar sized house in reasonable, but not G
luxurious condition.

Turning to the comparable properties in the west section of Eaton Mews South, where the subject premises are located, we reject nos 46 and 47 (a slightly better conversion than no 46) being two-storey mews houses opposite the subject premises which we were able to inspect internally. We agree with Mr Pope that these are inferior houses, with no rear windows, skylights, awkward layouts and unmodernised features. In our view, there are too many different factors in these properties to base a comparison and we set them aside. We find ourselves to hold the same view in respect of nos 25 and 26; here the prices achieved relate to a complex transaction involving the H
combination of two houses, with one having a rear extension with consequent ability to provide an extensive suite of reception rooms, possibly for commercial purposes, on the first floor; this transaction was rendered increasingly complex at the second hearing by the differing analyses of its component valuation elements put forward by the respective valuers. This leaves us with nos 29 and 33, both open market transactions, both relating to longer periods of leases, but also three-storey houses on the same side and close to the subject premises. No 33, in particular, was much relied on by both parties and we having inspected it internally, and taken into account its smaller size (one room and a garage less), its longer lease by eight years, consider it J
strongly to support a leasehold value midway between the figures advanced by the parties' valuers.

To sum up, adopting no 33 as the best comparable, being a sale of a property similar in size, character, and layout in the same location achieved only six months prior to the valuation date, supported by the prices achieved for 29 Eaton Mews South and 28 Eaton Mews North, we arrive at values of £350,000 of the leasehold interest and £670,000 for the freehold interest.

Improvements K

While at the first hearing the parties seemed to have differing points of view as to improvements: see para 7 above relating to Mr Haines' evidence, by the time of the second hearing they were both of opinion that the value of improvements should be disregarded throughout the valuation. Bearing in mind that Mr Haines puts improvements as no more than £10,000 to £20,000, in total value, a sum which can have little effect on the outcome, we accept this position as agreed by the parties.

Garages: freehold capital value

On December 12 1995 we inspected externally the three L
garages/parking spaces listed at 10 Shafto Mews, SW1; Byron Court, Elyston Street, SW3 and Rosemoor Street, SW3. At Byron Court, parking spaces, open and undercover, were available within a large block of flats and did not seem comparable to the acquisition of a single garage. In terms of location in a mews, 10 Shafto Mews was the most similar to the subject garages, but located, it would appear, as was the single garage in Rosemoor, in a district where garage spaces were scarcer. The garage at Shafto Mews was larger than the subject garages, although we questioned whether it would accommodate four cars at one time.

In assessing the freehold value, it would seem that the sale at 10 M
Shafto Mews, after reducing the special purchase figure to £80,000, and assuming only three cars could be accommodated, supports a figure in the region of £27,000. The exclusive nature of the subject garages and their freehold interest however requires some upward revision of this figure. Mr Haines considered the sale for £35,000 of a garage in Rosemoor Street to be the best comparable, but viewed the location of this garage, together with the length of lease, to ensure a greater demand than for a small awkward garage in Eaton Mews South. Taking all these factors into account, we concluded that £30,000 was an appropriate value for the freehold vacant premises in respect of garage west, 37 Eaton Mews South.

A *Capitalisation and deferred rates*

As to the capitalisation rate, Miss Ellis' method was to analyse market transactions of sales of long term ground rents (over 75 years) on a miscellany of blocks of flats and then to apply the average rate to both the income and the reversion of the subject dwellings. In cross-examination it was made plain that her comparables were not comparables in the physical sense of location and style. It was also made plain, and seems clear to us, that there is no comparison between an investment which has 20 years to run to its reversion to those which have 75 years or more. We do not consider that returns on zero preference shares and gilts support her argument for a rate of 10.8%. B Zero dividend shares which are intended to grow incrementally over a number of years to attain a fixed value, differ from property reversions, in that there is no inflation proofing, and there is the possibility that the shares will not attain the fixed value of the portfolio on which they are based, if it does not perform adequately. As to gilts, because interest rates have fallen in recent years, many of the securities granted are now above par value and will be repaid at a future fixed date at £100 which is *less* than they are now priced, this representing a capital loss in nominal terms, and an ever greater loss in real terms, because of the effect of inflation. Rates of return in these types of investment are motivated and take account of contingencies totally different from C those obtaining in the property market.

While the tribunal, therefore, was prepared to accept, particularly where, as here, the income is of marginal worth, Miss Ellis' contention that it was now established practice to apply the same rate to deferment as to capitalisation, they were unconvinced by her reliance on auction sales that as high a rate as 10.8% was appropriate.

Mr MacPherson, on behalf of the reversioner also accepted that the same rate should apply to deferment as to capitalisation and he put forward 6%. Although he supported his 6% rate by reference to principles of valuation, his main argument for the adoption of such a rate was its consistent use in negotiated enfranchisements and he relied D strongly, as evidence of the land market in Belgravia, on such negotiations. Before turning to principles of valuation it is, therefore, necessary to consider in some detail the nature of this evidence and the reliance which the tribunal may place upon it.

Negotiated settlements

Mr MacPherson relied strongly, as evidence of the land market in Belgravia, on the calculations prepared by the landlord's surveyors in support of the negotiated enfranchisement prices for houses on the Grosvenor and Cadogan Estates. These negotiations comprise all sales of freeholds made pursuant to the 1967 Act with its amendments, E except for one, 60 Cadogan Lane, which was determined by the Lands Tribunal; the actual figures given were 143 settlements under section 9(1), 220 under section 9(1)A and one under section 9(1)C. The consistent practice in these negotiations was described in *Lloyd-Jones* v *Church Commissioners for England* (1981) 261 EG 471 and appeared to have been followed in the case before us; namely after notification of the tenant's intention to enfranchise, the landlord's surveyor, (Gerald Eve, Mr MacPherson acting for that firm in the present case), had sought the assistance of another firm of chartered surveyors, (Mr Pope of John D Wood in the present case), to advise them on the freehold and leasehold vacant possession values of the F subject premises on the assumption that the lessee did not have any right to enfranchise under the 1967 Act. Then, the landlord's surveyor prepared a report to the estate setting out a full calculation; after consultation, resulting in some cases in amendment, these figures were presented to the lessee, and included: percentages for the rates of capitalisation and deferment; the landlord's proportion of marriage value; and where relevant, the proportion of the entirety attributable to the site; and the percentage to be deducted for the tenant's right to stay on after the lease expired. In the *Lloyd-Jones* case the member of the Lands Tribunal, Mr WH Rees FRICS, although noting that in the course of negotiations "the figures in the calculations may be altered somewhat", and that in the 57 settlements relating to the Cadogan Estate "the lessee did not always agree each figure" in the calculation, nevertheless concluded that "the actual figures shown in the G calculation", and not those merely derived from "the analysis" of the landlord's surveyors, provided "massive support" for the rates proposed. He further considered that, in circumstances where settlements were rapidly and widely known, and there was an association of tenants, that there was no *Delaforce* effect pressurising the tenants to settle at a higher figure than would be awarded by a tribunal.

In the one failed negotiation, 60 Cadogan Lane, which had been determined by the Lands Tribunal, the tribunal again accepted settlements as evidence of the land market: see *Cadogan Estates Ltd* v *Hows* [1989] 48 EG 167; in that case the tribunal relied on 134 H settlements on the Grosvenor Estate covering a period of 15 years in which 83% of the lessees had been professionally represented. Before us, Mr Radevsky on behalf of the nominee purchasers, cross-examined Mr MacPherson and advanced in his final submissions contentions similar to those put forward on behalf of the lessee in the *Hows* case which were rejected by the Lands Tribunal; namely that: the settlements' evidence only demonstrated consistency of method and intransigence on the part of the landlord's surveyors, with certain elements in the valuation exercise regarded as "non-negotiable"; the settlements were evidence of the final price and not of the individual elements in that price; they were tainted by the *Delaforce* (*Delaforce* v J *Evans* (1970) 215 EG 315) effect; and were self supporting and self perpetuating. It is also true to say that on a number of occasions, eg in the decision concerning Clabon Mews, London leasehold valuation tribunals have expressed themselves unwilling to rely on negotiated settlements evidence.

As a tribunal we have been much exercised as to the evidential value of these settlements. Mr Radevsky on behalf of the nominee purchasers makes a strong case against them. First, on legal grounds, that all the settlements relate to houses, where the tenant had given notice of the right to enfranchise under the 1967 Act, as amended, whereas we are concerned with a collective enfranchisement of two K residential dwellings under the 1993 Act which the estate had by its conveyancing arrangements chosen not to bring within the earlier Act. Second, that in the absence of statement or signature of the tenant or a professional adviser, no consent to the figures on the calculation sheets by all parties can be established and they remain a document solely made by and for the purposes of the landlord. Mr MacPherson very fairly admitted that there was little possibility of the tenant or his advisers being willing to sign the calculation sheets. Third, any consent may be challenged as not fully voluntary, given that the estate was the sole freeholder, occupying a dominant position, and only willing to sell when legally obliged to. L Factually Miss Ellis has shown the evidence to be flawed: only two settlements under section 9(1)A and one under section 9(1)C relate to transactions completed in the last five years; the interval between the date of the claim for enfranchisement and the date of the analysis is frequently lengthy, ranging from three to five years with no clear indication of the date when the price was agreed or paid, encouraging, by reason of the ever shortening term and in a rising market, the tenant to settle over the odds; the figures in the calculations are capable of different analysis according to the capital values adopted, and she exhibits a number of transactions which, reanalysed, can be made to support different rates of return. The very consistency is said to belie their market veracity; M the schedule shows rates increasing in bands of 10 years, with a rate of 6% given for terms of 20 to 30 years (arguably the present terms of 20.25 years approximate more closely to the lower band rate of 5% given for the 10 to 20 years band). It is said that market transactions would be much more variable.

Against these weighty arguments, we have the authority of two decisions of the Lands Tribunal; they relate to properties under the ownership of the same reversioner who appears before us and the Lands Tribunal approved as evidence the calculation sheets made by Mr MacPherson's former partner in the *Lloyd-Jones* case, and by Mr MacPherson himself in the *Hows* case. Whatever charges may be made against the Grosvenor Estate as the dominant partner in these transactions, the factual position cannot be ignored that over a period of 15 years a rate of 6% was consistently applied for capitalisation and deferment of settlements relating to 20 to 30-year leases, and sums paid for enfranchisement to the estate were calculable on that basis. As there were no other sales these transactions, even if conducted in an unusually consistent manner, effectively represented in fact the local land market. On balance, and mindful that we are an inferior tribunal to the Lands Tribunal, we consider *for the purposes of transactions arising on the same estate and in relation to the same type of enfranchisement* that we must accept in this case the negotiated settlements as evidence. This decision in no way determines the evidential value of settlements when applied to different estates, different landlords and different areas of London.

The present case relates to a collective enfranchisement, not a purchase of the freehold of a single house. None the less, as we have already stated with regard to capital values, the market is unlikely to distinguish these particular "flats" from mews houses. Accordingly, the rate of 6% shown in the calculation sheets relating to houses is evidence in support of the same rate for the subject premises.

However, should we be wrong to admit such evidence, we are convinced on valuation principles that such a rate is appropriate. We agree with Mr MacPherson that the excellence of location and quality of property support this lower rate.

As to deferment rate, the difference between the parties largely turns on their assessment of a 20-year investment in Belgravia. Miss Ellis, adopting a generally cautious approach, argued in favour of 10.8% by reference to the income stream derived from auction sales of ground rents in London and took no account of the added characteristic in reversionary residential investment of possible future occupation by the investor. Mr MacPherson, supported by Mr Pope, reminded us that Belgravia is unique, of London's position as a global city and the high demand for houses located in good quality central residential areas. In this connection we have to say we were not over impressed by Mr Pope's evidence relating to the chart in Savills' *Quarterly Bulletin* entitled "The Balloon Deflates"; this chart is statistically flawed and represents an attempt merely to extrapolate the post-war trend upwards by claiming it will continue.

In determining the deferment rate we apply the assumptions in Schedule 6, para 3(1), in particular that the tenant is not a purchaser and has no right to enfranchise. In a situation, as here, where a great estate clings tenaciously to its freeholds, we think an investor purchasing the reversion, faced with the subject leases, would be mindful that in 20 years' time he will obtain vacant possession and a large capital gain with the possibilities of selling the freehold, occupying the house himself or by a member of his family, or letting at a full market rent. Even before 20 years there is the possibility of the tenant offering a sizeable premium for extension or surrender of the lease. The investor has a range of options which are likely to produce a capital gain; in addition he has a "status" property derived from its location in Belgravia and rarity on the market with tax advantages over the receipt of income (though this aspect was not explored in evidence). Given these favourable aspects of the investment, we consider that even if the general property market including Belgravia fails to recover its former rising character, the reversion on the subject leases remains an excellent investment. We accordingly adopt 6% as the rate of deferment.

For the reversionary value of the garages the parties' surveyors applied the same rates of capitalisation and deferment as they used for the mews houses; we, however, think it to be a different type of investment, with increased risk and costs, and accordingly apply a rate of 8.5%.

Marriage value

Schedule 6 para 4(2) defines marriage value as: ". . . any increase in the aggregate value of the freehold . . . when regarded as being (in consequence of their being acquired by the nominee purchaser) as interests under the control of the participating tenants" as compared with the value of those interests held separately; and the increase in value is defined as that: "which is attributable to the potential ability of the participating tenants, once those interests have been so acquired, to have new leases granted to them without payment of any premium and without restriction as to length of term", and: "which, if those interests were being sold to the nominee purchaser on the open market by willing sellers, the nominee purchaser would have to agree to share with the sellers in order to reach agreement as to price". Two aspects of this definition are noteworthy. First, that by reason of the continued separation, after collective enfranchisement, of the freehold now held by the nominee purchasers, and the leasehold interests still held by the participating tenants, the marriage value is defined, not as under the 1967 Act by reference to direct merger of the freehold and leasehold interest in the enfranchising tenant, but by reference to the opportunity of the participating tenants to obtain without premium extended leases of unlimited term. Arguably, because of the extra transaction costs involved, the total sum attributable to marriage value should be less than a direct enfranchisement of a single house by one tenant. Second, unlike the 1967 Act which contains no definition of marriage value, Schedule 6 para 4, expressly refers to an "increase" attributable to having "to agree to share with the sellers in order to reach agreement as to price". The definition can thus be said to recognise an element of market pressure which brings about the sale and to rule out a situation where the marriage value accrues in entirety either to the seller or the nominee purchaser. Arguably, the Act does not envisage the award of 100% of marriage value to the landlord, a proportion which Mr MacPherson maintained was achieved in some of the transactions which he relied upon. It is agreed by the parties that there is a marriage value in the present case although the variation in their values for the leasehold interest before and after enfranchisement, rates of return and in method produce differing amounts. In our experience there is an inevitable tendency for the tenant's surveyor to arrive at a smaller sum for the marriage value than that achieved by the landlords' surveyor.

Miss Ellis relies on para 4(1) which provides a residuary definition of the freeholder's share of the marriage value as 50%, where the parties have not agreed otherwise and it cannot be proved in a sale in the open market a greater share would have been agreed by a willing seller. She applies, to the determination of the landlord's share of marriage value, the hypothesis accepted in *Lloyd-Jones* that the arbitrator in awarding compensation must ascertain, to the best of his ability, the price that will be paid by a willing purchaser to a willing vendor of the land with its potential, in the same way that he would ascertain it in a case where there are several possible purchasers. Similarly the principle of that case that "neither can unlock the marriage value without the other. In friendly negotiations they would agree to divide it equally" is to be adopted.

The reversioner proposes that 75% of the marriage value should be awarded to the freeholder and submits that evidence of relevant comparable open market transactions should lead the tribunal to determine under para 4(1)(a) that a proportion higher than 50% would have been agreed on a sale on the open market by a willing seller. Somewhat surprisingly, in view of the emphatic insistence on their relevance to the determination of the other disputed issues of the capitalisation and deferment rates, Mr MacPherson sets aside the 220 (221) settlements under section 9(1)A of the 1967 Act which show the landlords obtaining only 50% of the marriage value. He explained his present position as follows: "such settlement evidence has been used with great effect in the past, and rightly so in the absence of open market evidence available to prove the elements of valuation for enfranchisement. However, there is now evidence of direct relevance to the current case in five open market transactions". Miss Ellis, in reply, challenges these transactions as not truly "open market" and says that in each case Mr MacPherson's analysis of the transactions is disputed or there were special circumstances which meant the purchaser was not truly willing.

Counsel for the reversioner referred us to *Norfolk* v *Trinity College, Cambridge* [1976] 1 EGLR 215 and *Lloyd-Jones* v *Church Commissioners* which make it clear that far from being self evident, the existence of marriage value and attribution of any share of it to the landlord was a controversial issue in the first decade after the

A enactment of the 1967 Act. With the enactment of section 82 of the Housing Act 1969 the sitting tenant's bid was no longer excluded from valuation under section 9(1)A and such a bid gave rise to the likelihood that the merger of leasehold and freehold interests would produce an increase in value over the two interests when valued separately. In the *Lloyd-Jones* case the surveyor for the tenant argued that "the lessee would need to bid only a nominal amount more than any other purchaser to obtain the freehold" and consequently since the tenant's bid released the marriage value he argued in favour of only 25% of the marriage value being awarded to the landlord. The Lands Tribunal, accepting Mr Hopper's contention on behalf of the landlord
B that the parties were of equal bargaining strength, awarded 50% to the landlord. Marriage value was thus generated by the tenant's participation in the bidding, but the statutory requirement of a "willing" seller led the Lands Tribunal in both the *Norfolk* and *Lloyd-Jones* cases to conclude it must be shared equally. Broadly, then, a ceiling of 50% to each party was established and it would seem that the Grosvenor Estate's surveyors, including Mr MacPherson, acted on that assumption by writing it into all their settlement schedules for enfranchiseable houses under the 1967 Act. The 1993 Act appears to have changed the position, but, far from being a ceiling, the 50% share of marriage value given to the freeholder is now a
C "floor" which in the absence of agreement or evidence of open market sales is to be the landlord's share.

Looked at afresh, from first principles, therefore, and having regard solely to the position in the Grosvenor Estate, it would seem that given the declared unwillingness of the Grosvenor Estate to sell, the rarity of a sale by a freeholder other than the Grosvenor Estate, and the shortness of the term leaving the tenants with little option but to seek an extension, the balance of market power is in the freeholder's favour with a consequent possibility that he might expect to obtain more than 50% as his share of the marriage value. Is this reasoning supported by practice? It is true that Miss Ellis referred us to a recent South East Panel decision
D where, without giving reasons, the tribunal awarded on a lease extension a 60% share of the marriage value to the landlord. On the other hand, we have the full weight of the 220 negotiated settlements on the Grosvenor and Cadogan Estates where the marriage value has always been divided equally and no claim was advanced by the landlord to a higher proportion. Faced with this consistent practice under the 1967 Act, we feel that we could only determine a proportion greater than 50% if there is good market evidence to displace the statutory "floor".

Open market transactions supporting a greater share than 50% of the landlord's share of the marriage value

E Although Mr MacPherson refers in his first proof to five transactions, we have identified eight which in the course of the hearings have been relied upon (see schedule D). Five of these, 28 Eccleston Mews, SW1; 69 and 78 Addison Road; 35 Kinnerton Street and 13 Gerald Road, relate to sales of freehold, but only the last two were enfranchiseable properties under section 9(1)C. Three transactions, 126A Hamilton Terrace, 25 and 26 and 17 Eaton Mews South, relate to surrenders of existing leases and grant of new leases. From a careful consideration of the available evidence, and the parties' analyses of these transactions we would place them on a spectrum, ranging from, at one end, the transaction where special circumstances
F or unknown factors defeat any rational analysis of the outcome, to transactions displaying some elements of bargaining, to those approximating to open market transactions at arms' length. On such a spectrum, 28 Eccleston Mews and 126A Hamilton Terrace are the least helpful; the first is riddled with special circumstances, namely the purchaser of the freehold of this unenfranchiseable mews property being the tenant of the adjacent main house at 28 Eaton Place and the additional prior payment of £50,000 to link the two properties, so as to make the transaction unreliable; the evidence for the second is contained in a report made by Daniel Smith for the purposes of section 32(3) of the Charities Act 1993, not section 9(1)C of the 1967 and 1993 Acts, and is again coloured by special circumstances; in this case they are a company let, a surrender following a forfeiture with relief and the grant of a new lease on terms that put it outside section 9(1)C. G
The two transactions in Addison Road relate to unenfranchiseable properties let to companies where the sale of the freehold was ancillary and linked to the principal sale of the merged interests to a third party; the amount paid by the tenant for the freehold was affected by the overall price and profit which the tenant was deriving from the third party transaction and possibly eliminated any element of independent risk.

Similar, in character, though not so closely linked to a third party transaction, is the complicated surrender of four leases and grant of two new leases relating to 25 and 26 Eaton Mews South; effectively, this was a tidying up operation with a view to marketing the two H
houses as one unit, with the Grosvenor Estate retaining the non-enfranchiseability of the property by introducing ground rents over £1,000 pa in the new leases.

In the transaction relating to 35 Kinnerton Street there is a hard figure of £235,000 paid as a premium for a house enfranchiseable under section 9(1)C, but the share of the marriage value, based on estimated, not market established values, varies according to the analysis.

The surrender and grant of a new lease at 17 Eaton Mews South again has elements of a market transaction but again is flawed by the co-operation of both landlord and tenant to achieve a sale, while J
preserving the unenfranchiseability of the property. There also appeared to be some confusion over the precise circumstances relating to a mortgage leading up to the transaction. Finally, in 13 Gerald Road we have another sale linked to an outside third party transaction but in this case, far from co-operation, there was deliberate withholding of market information by the tenant's surveyor, Mr Boston, in order to obtain a favourable price.

With the assistance of the parties we have been able to analyse in some detail these eight transactions. They reveal the extraordinary variety of circumstances, of the structure of deals and of the motives of the parties; it is consequently difficult to ascertain at what point in the K
spectrum the transaction becomes an open market sale within the statutory requirement of Schedule 6, para 4(1)(a). That subparagraph requires the tribunal to have regard to a hypothetical agreement on the open market "at the date of valuation". Arguably this rules out Mr MacPherson's comparables of 35 Kinnerton Street, 126A Hamilton Terrace and 13 Gerald Road, which were all effected subsequent to February 1995. The paragraph also requires "such proportion" in excess of 50% to "have been determined by an agreement between the parties" and since it continues "on a sale", we construe the inclusion of the word "agreement" to refer specifically to the proportion given to the landlord, and not merely the overall price paid. Even accepting that L
the transactions relating to 35 Kinnerton Street, 25/26 and 17 Eaton Mews South and 13 Gerald Road, with all their discussed flaws, have elements of "sales in the open market by a willing seller", Mr MacPherson is unable to offer them as evidence for anything more than the prices achieved; the proportions in excess of 50% which he gives, 100%, 67.7%, 61.5% and 68% respectively, have not been agreed by the other party, and is reanalysed by Miss Ellis or other surveyors to give proportions of 48%, 60.5% and 50% respectively (no differing figure is given for 17 Eaton Mews South).

Taking into account all the above factors and arguments, we concluded that there was no satisfactory evidence before us to justify M
the adoption of a higher proportion for the landlord's share of the marriage value and accordingly apply the statutory 50% share for the landlord.

Taking all the above factors into account and applying our knowledge and experience our valuation is as follows:

36 & 37 EATON MEWS SOUTH, SW1

Purchase price payable by nominee purchasers in accordance with Schedule 6 of Leasehold Reform, Housing and Urban Development Act 1993

Value of freeholder's interest

No 36

Dwelling and west garage (converted into reception room)

	£	£	£
Term			
Ground rent	150		
YP 20.25 years @ 6%	11.54	1,731	
Reversion			
Freehold value	670,000		
PV £1 in 20.25 years @ 6%	0.3074	205,958	
East garage			
Term 1			
Rent receivable	750		
YP 3.25 years @ 8.5%	2.73	2,048	
Term 2			
Agreed rent	1,900		
YP 10 years @ 8.5%	6.56		
	12,464		
PV £1 in 3.25 years @ 8.5%	0.7676	9,567	
Reversion			
Freehold value	30,000		
PV £1 in 13.25 years @ 8.5%	0.3395	10,185	229,489
No 37			
Dwelling and east garage			
Term			
Ground rent	100		
YP 20.25 years @ 6%	11.54	1,154	
Reversion			
Freehold value	670,000		
PV £0 in 20.25 years @ 6%	0.3074	205,958	
West garage			
Freehold value		30,000	237,112
Value of freeholder's interest			466,601
Freeholder's share of marriage value			
Nos 36 & 37			
Value after marriage			
2 dwellings and garages			
@£670,000 each	1,340,000		
2 garages @ £30,000 each	60,000	1,400,000	
Value before marriage			
Freeholder's interest	466,601		
Leaseholders' interest			
2 dwellings and garages			
@ £350,000 each	700,000		
East garage no 36	nil	1,166,601	
Gain on marriage		233,399	
50% to freeholder		116,700	
Compensation for loss resulting from enfranchisement			nil
ENFRANCHISEMENT PRICE			£583,300

Accordingly we determine the price payable for the freehold interest in 36 and 37 Eaton Mews South and the four garages pursuant to section 24 of the Act as £583,300 (five hundred and eighty three thousand, three hundred pounds).

Valuation of Miss J Ellis FRICS

		£	£	£	£	£
A: Value of freeholder's interest						
1.	Dwelling & garage @ 36					
	Term					
	Rent receivable	150				
	YP 20.25 years @ 10.8%		8.10		1,215	
	Reversion					
	Capital value		580,000			
	PV £1 in 20.25 years @ 10.8%	0.125		72,693		73,908
2.	Dwelling & garage @ 37					
	Term					
	Rent receivable	100				
	YP 20.25 years @ 10.8%		8.1		810	
	Reversion					
	Capital value of house	580,000				
	PV £1 in 20.25 years @ 10.8%	0.125		72,693	73,503	
3.	Garage east, 36 Eaton Mews South					
	Term					
	Current rent receivable	750				
	YP 13.25 years @ 10.8%		6.88		5,160	
	Increase in 1998 (£1,900-£7,50)		1,150			
	YP 10.25 years @ 10.8%		5.94			
	PV £1 in 3.25 years @ 10.8%	0.717		4,256		4,894
	Reversion					
	Capital value of garage	20,000				
	PV £1 in 13.25 years @ 10.8%	0.26		5,140		15,194
4.	Garage west, 37 Eaton Mews South					
	Capital value					20,000
	Freeholders' interest					162,605
B:	**Marriage value**					
	After marriage					
	Freeholder's interest		nil			
	Lessees' interests:					
	Dwelling/garage @ 36		580,000			
	Dwelling/garage @ 37		580,000			
	Garage east @ 36		20,000			
	Garage west @ 37		20,000		1,200,000	
	Before marriage					
	Freeholder's interest*		162,605			
	Lessees' interests:					
	Dwelling/garage @ 36		415,500			
	Dwelling/garage @ 37		415,750			
	Garage east @ 36		nil		993,855	
	Gain on marriage of interests				206,145	
	50% to vendors					103,027
	Enfranchisement price					265,677
			Say			265,000

*This includes the vacant possession value of garage west @ 37

Valuation of 36/37 Eaton Mews South, London SW1 by Ian MacpPherson FRICS

	£	£	£	£
1. **Valuation of landlords' interest excluding marriage value**				
36 Eaton Mews South — house & west garage				
(converted to reception room)				
For remainder of existing lease term				
Rent receivable pa	150			
Years purchase for 20.25 years @ 6.0%	11.55			
		1,732		
For reversion to				
Estimated value of freehold in possession	670,000			
Deferred 20.25 years @ 6.0%	0.307			
		205,690		
			207,422	

		£	£	£	£
36 Eaton Mews South — east garage					
Until rent review @ 24/6/98					
Rent receivable pa		750			
Years purchase for 3.25 years @ 6.0%		2.88			
			2,160		
For remainder of existing lease term					
Estimated rent on review @ 24/6/98 pa		1,900			
Years purchase for 10 years @ 6.0%	7.36				
Deferred 3.25 years @ 6.0%	0.827				
		6.09			
			11,571		
For reversion to					
Estimated value of freehold in possession		40,000			
Deferred 13.25 years @ 6.0%		0.462			
			18,480		
				32,211	
37 Eaton Mews South — house & east garage					
For remainder of existing lease term					
Rent receivable		100			
Years purchase for 20.25 years @ 6.0%		11.55			
			1,155		
For reversion to					
Estimated value of freehold in possession		670,000			
Deferred 20.25 years @ 6.0%		0.307			
			205,690		
				206,845	
36 Eaton Mews South — west garage					
Estimated value of freehold in possession				40,000	
					486,478
2. *Calculation of marriage value*					
Estimate value of new 999-year leases at peppercorn rents for					
36 Eaton Mews South					
House and west garage				670,000	
East garage				40,000	
37 Eaton Mews South					
House and east garage				670,000	
West garage				40,000	
					1,420,000
Less					
Value of landlords' interest excluding marriage value				486,478	
Value of participating tenants interests excluding marriage value					
36 Eaton Mews South					
House and west garage			300,000		
East garage					
Estimated ERV pa	1,900				
Less rent currently payable pa	750				
Profit rent pa	1,150				
Years' purchase for 3.25 years @ 8% & 3%	2.64				
		3,036			
		Say	3,000		
			303,000		
37 Eaton Mews South					
House and east garage			300,000		
				603,000	
					1,089,478
Gain on marriage					330,522
3. Calculation of enfranchisement price					
Value of landlords' interest excluding marriage value					486,478
Marriage value				330,522	
Landlords' share @ 75%					247,892
Total enfranchisement price					734,370
				Say	734,400

Schedule C.1

Schedule of comparables — Houses

	Address	*Unexpired term*	*Accommodation*	*Garage*	*Agent*	*Sale price*	*Date of transaction*	*Area*	*Comment*
1	15 Eaton Mews South	25 years to 22.04.2019	4 beds, 3 baths, 2 receps, utility, cloakroom	Yes	Gascoigne-Pees	£390,000	May 1994	250m^2	Less second garage (17m^2) let under separate agreement. Not enfranchiseable. Excellent condition
2	17 Eaton Mews South	75 years to 20.09.2069	3 beds, 2 en suite baths, shower room, 2 receps, study/bed 4, kitchen, cloakroom,	Yes	Friend & Falcke	£625,000	May 1994	170m^2	Not enfranchiseable. Excellent condition
3	20 Eaton Mews South	27¾ years to 29.09.2021	3 beds, 2 baths, dressing room, drawing room, dining area, study, kitchen, utility room	On separate lease to 2009 at £2,200. Review in 1999	Aylesford	£365,000	January 1994	152m^2	Not enfranchiseable. Immaculate condition. Bought in connection with 2A Eaton Square
4	25 & 26 Eaton Mews South	19½ years to 25.03.2014	No 25 — 3 beds, 3 baths, 2 receps, study/bed 4, sauna	Garage each on separate lease		£650,000 for renewal to 24.06.2069	November 1994	25-152m^2	Lease extension after surrender of leases for 25 & 26 EMS due to expire on 25.03.2014. Lease for 25 ems gge due to expire on 24.12.96.
			No 26 — 3 beds, 2 baths, 1 recep, study/bed 4, kit/brkfast rm, cloakroom					26-198m^2	Lease at 26 EMS EMS gge has already expired - holding over.
5	29 Eaton Mews South	26½ years to 24.12.2019	3 beds, bath, shower, 3 receps, kit/brkfast rm, cloakroom	Yes	Friend & Falcke	£324,500	August 1993	155m^2	Enfranchiseable
6	33 Eaton Mews South	28½ years to 25.03.2023	3 beds, 3 baths, 1 large recep room, study/bed 4, kitchen	On separate lease to 1998 at £1,600 pa	Best Gapp	£345,000	September 1994	148m^2	Not enfranchiseable
7	47 Eaton Mews South	46½ years to 24.12.2040	3 beds, bath, shower rm, 1 double recep with dining area, study/bed 4, kitchen	Yes	Lurot Brand	£375,000	August 1994	105m^2	Not enfranchiseable. Only on 2 floors
8	28 Eaton Mews North	50½ years to 15.12.2045	3 beds, 3 en suite baths, 2 receps, kitchen, roof terrace	On separate lease at £2,000 pa	John D Wood & Co	£650,000	March 1995	197m^2	Including terrace. Not enfranchiseable. Garage Rent Review 1995
9	43 Eaton Mews North	36½ years to 25.12.2030	3 beds, dressing room, 2 baths, 2/3 receps, kitchen, cloakroom, basement "pub"	On separate lease to 1998 at £100 pa	Lurot Brand	£550,000	Feb 1995	140m^2	Enfranchiseable
10	52 Eaton Mews West	18½ years to 25.12.2012	2 beds, 2 baths, drawing room, kitchen	2, one on separate lease to 1997 at £2,200 pa	Hobart Slater	£205,000	April 1994	143m^2	Not enfranchiseable
11	8/9 Eaton Row	21 years	4/5 beds, 3 baths, 3 receps office/bed 5 plus staff flat. Patio & roof terrace	Yes	Knight Frank & Rutley	£385,000	December 1993	255m^2	Inclusive of roof terrace. Enfranchiseable. Requiring modernisation

Schedule C.2

Schedule of comparables — Garages/parking spaces

Address	*Tenure*	*Description*	*Agent*	
10 Shafto Mews, SW1	28 years	Double garage	Douglas & Gordon	£100,000 May 1995
Byron Court, Elystan Street, SW3	97 years	Secure spaces	Rategrove Ltd	Undercover £30-36,000) In open £23-25,990) 1990
Rosemoor Street, SW3	78 years	Single garage	John D Wood	£35,000 September 1994

A **Schedule D** E

Transactions supporting a higher share to the landlord of marriage value than 50%

Property	Type of transaction	Date	Price	Freeholder	Leaseholder	Unexpired term	G.R.	L/Ls share of M.V.
28 Eccleston Mews Mews House	Sale of freehold. Not enfranchiseable	Oct 1994	£150,000	Grosvenor estate	Freeholder of adjacent no 28 Eaton Place	Until 2050 56.25 yrs	£1,250 pa with rent reviews	100%
69 Addison Road House	Sale of freehold. Not enfranchiseable	March 1993	£480,000	Ilchester estates	Company	Until 2060 67 yrs	£1,000 pa with rent reviews	62.75%
B 78 Addison Road House	Sale of freehold. Not enfranchiseable	30 Dec 1994	£937,500	Ilchester estates	Company	Until 2040 $45^1/2$ yrs	£500 pa with rent reviews	98% (McP) 66% (Ellis) F
35 Kinnerton Street House	Sale of freehold under S9(1)C	Sept 1995	£235,000	Grosvenor estate		51.75 yrs	£100 pa rent review in 3 months	100% (McP) 48% (Ellis)
126A Hamilton Terrace Converted coach house and garage	Surrender & grant and grant of 99 yr lease at £1,200pa with rent reviews. Not enfranchiseable	July 1995	£180,000	John Lyon's Charity		34 yrs		89% (McP) 85% (Briant)
C 25/26 Eaton Mews South 2 mews houses and garages	Surrender & grant of two new leases for 75 yrs at £1,250 pa each with rent reviews. Not enfranchiseable	Nov 1994	£650,000	Grosvenor estate		2 leases expiring 2014	£60 pa	67.7% (McP) 60.5% (M.J.) revised G
17 Eaton Mews South Mews House	Surrender & new lease of 75 yrs at £1,250 pa with rent reviews. Not enfranchiseable	April 1994	£625,000	Grosvenor estate. Joint sale by L/L and Leaseholder		Expiring 2015 21.5 yrs	£35 pa	61.5%
D 1 Gerald Rd House	Sale of freehold under S9(1)C	May 1994	£250,000 (improvements £50,000)	Grosvenor estate		56.75 yrs	£600 pa	68% (McP) 50% (Boston) H

A Kay-Green *v* Twinsectra Ltd

LEASEHOLD VALUATION TRIBUNAL FOR LONDON

April 9 1997

LADY WILSON, DL EDGE FRICS and EP GUDGIN FCA ATII

Estates Gazette June 7 1997

B [1997] 23 EG 146

Landlord and Tenant Act 1987 — Purchase notice — Whether price to be increased by death of regulated tenant — Extent of property to be acquired — Determination of tenants' rights over gardens and amenity areas — Whether chargee or landlord entitled to the consideration

Tudor Court consists of four principal buildings surrounded by gardens and amenity land. One of the buildings is a modern block of flats; the other buildings contain flats or consist of terraced houses or bungalows. The dwellings are occupied by tenants under terms which
C gave them rights to enjoy the gardens and amenity land subject to notice of termination by the landlord. Following an offer to sell the properties by auction, most of the tenants joined in serving a purchase notice under section 12 of the Landlord and Tenant Act 1987. In earlier proceedings ([1996] 2 EGLR 43), the tenants abandoned their claim to the two buildings not containing any flats, and the Court of Appeal ordered that the purchase notice served by the tenants was valid in so far as it related to the buildings containing flats. Meanwhile one of the flats became vacant on the death of the regulated tenant. The leasehold valuation tribunal were asked to determine the following issues: (1) whether the price to be paid should be increased by reason of the death of the regulated tenant; (2) whether the applicant nominee purchaser
D was entitled to acquire the three houses forming part of one of the buildings; (3) what rights should the tenants have over the gardens and amenity land; and (4) should the transfer require the consideration to be paid directly to the bank holding a charge over the properties.

***Decision*: (1) The consideration must be increased under section 12(6) of the Landlord and Tenant Act 1987 to include the vacant possession value of the vacant flat. (2) The Act can apply to premises which consist of the whole of a building consisting partly of flats and partly of houses, accordingly the applicant was entitled to acquire the houses. (3) The tenants were entitled to the rights to use the gardens and amenity land, but not subject to termination.
E (4) The transfer should require the payment of the consideration to the respondent's bank on confirmation that the bank had a charge over the property.**

The following cases are referred to in this report.

Belvedere Court Management Ltd v *Frogmore Developments Ltd* [1996] 1 All ER 312; [1996] 1 EGLR 59; [1996] 05 EG 131
Denetower Ltd v *Toop* [1991] 1 WLR 945; [1991] 3 All ER 661; [1991] 1 EGLR 84; [1991] 20 EG 194, CA

Edward Denehan (instructed by JE Kennedy & Co) appeared for
F the applicant nominee purchaser; Simon Brilliant (instructed by Wallace & Partners) represented the respondent landlord.

Giving the decision of the tribunal, LADY WILSON said: Tudor Court consists of four buildings which have throughout the proceedings been referred to as buildings 1, 2, 3 and 4. Building 1 is a large, mainly Tudor house, forming three sides of a rectangle, which has been converted into seven flats (1, 1A, 1B, 2, 2A, 3 and 3A Tudor Court) and three terraced houses (4, 4A and 4B Tudor Court). Building 2, which faces building 1 across a courtyard garden, consists of two semi-detached bungalows (5 and 5A Tudor Court). Building 3 is adjacent, but not attached to, building 1, and contains two terraced houses and two bungalows (6, 6A, 6B and 7 Tudor Court). Building 4, which is adjacent to but separate from the south wing of building 1, is G
a modern purpose-built block of five flats (8, 9, 10, 11 and 12 Tudor Court). The gardens of Tudor Court merge with those of a large 19th-century house known as Tudor House, now converted into seven flats (1, 2, 3, 4, 5, 6 and 7 Tudor House). Adjacent to and accessible from the gardens of Tudor Court is what is described as the amenity land, on which are ponds and trees. Separated by a wall from the grounds of Tudor Court and Tudor House is a modern block of flats called Parr Court. Most of Tudor Court, though not buildings 3 and 4, is included in one Land Registry title. Most of the gardens, together with parts of Tudor Court, the whole of Tudor House, the whole of Parr Court and the amenity land are together included in another title. H

2A Tudor Court in building 1 was let under a regulated tenancy but is now vacant, the tenant having died in April 1996. 5 Tudor Court in building 2 has been vacant since before the date of the disposition which led to these proceedings. 9 Tudor Court in building 4 is let under a regulated tenancy. All the other properties in Tudor Court and Tudor House are let on 125-year leases at low rents on substantially the same terms.

In January 1992 Tudor Court, Tudor House and Parr Court were offered at auction in three lots or alternatively in one lot. The special conditions of sale stated that the vendor had not served any notice on the lessees and tenants of the properties under the landlord and Tenant J
Act 1987. They were bought as one lot on January 28 1992 by Twinsectra Ltd for £240,000; the sale was completed on February 25 1992 and Twinsectra was registered as proprietor of the freehold interest in the properties on April 13 1992. On May 1 1992 most of the tenants of Tudor Court and Tudor House gave notice to Twinsectra under section 11 of the Act requiring the company to give to their nominated representative, the present applicant, particulars of the disposal. Solicitors for Twinsectra replied to that notice, saying that their reply was without prejudice to their contention that the notice was invalid and that the premises were not premises to which Part I of the Act applied. On July 20 1992 a purchase notice under section 12 of the K
Act was served on Twinsectra by the applicant on behalf of most of the long lessees of properties, both flats and houses, in Tudor Court and Tudor House.

On November 25 1992 most of the tenants of Tudor Court and Tudor House issued proceedings in Staines County Court seeking declarations that Twinsectra was in default in failing to comply with the section 12 notice and that they were entitled to acquire the freehold of the four buildings which comprised Tudor Court and of the whole of Tudor House. On August 23 1994, after a three-day hearing (at which no oral evidence was taken), Judge Hull QC dismissed the applications, holding that the purchase notice did not comply with L
section 12(3) of the Act and that all the buildings and land comprised within the two titles together formed "premises" falling within Part I of the Act.

All the tenants named in the schedule to the originating application in the county court appealed to the Court of Appeal, and the appeal was heard on April 22 and 23 1996. Judgment was given on May 15 1996*. Before the Court of Appeal the applicants abandoned their claim to buildings 2 and 3 of Tudor Court (neither of which contained any flats at all). The Court of Appeal allowed the appeal. Aldous LJ, who gave the leading judgment, identified three issues: (i) whether a landlord, upon whom a valid purchase notice had been served, was M
under a statutory duty to comply with it; (ii) whether the purchase notice was served by the requisite majority of qualifying tenants; and (iii) whether the content of the purchase notice rendered it invalid. On all three issues the court found for the applicants. Dealing with issue (ii), Aldous LJ rejected the arguments of leading counsel for Twinsectra that "premises" in section 1 of the Act meant the complex as a whole, and that the premises had to be comprised within one registered title, and concluded that the question of whether a relevant disposal of premises had been made had to be considered on a building

---

*Editor's note: Also reported at [1996] 2 EGLR 43

by building basis, and that, in the case of Tudor House and of buildings 1 and 4 of Tudor Court, the relevant applicants constituted the requisite majority of tenants in the relevant building. Under issue (iii), Aldous LJ said that the purpose of a purchase notice was to give to the new landlord adequate notice that the qualifying tenants of the building wished to acquire the freehold upon the terms of the original disposal or upon terms to be decided by the LVT. He cited the then Vice-Chancellor, Sir Nicolas Browne-Wilkinson, in *Denetower Ltd* v *Toop* [1991] 3 All ER 661*, at p669:

> The details of the property to be acquired and the terms of the acquisition can in default of agreement be sorted out by the rent assessment committee on an application under section 13(1).

On July 4 1996, the applicant issued five applications to the tribunal. Application 1 was a composite application relating to buildings 1 and 4 Tudor Court, to Tudor House and to "the estate land" (identified on a plan attached to the proposed transfer as the whole of the garden and amenity land). Applications 2, 3, 4 and 5, issued, we were told, out of caution, were separate applications relating to the individual components included in application 1.

The order of the Court of Appeal, drawn by counsel and sealed on August 16 1996, provides, in so far as is relevant to the matters we have to decide, that upon the applicants' counsel acknowledging that the applicants have no claim to acquire the respondents' reversionary interest in the gardens and amenity land, it was declared that the purchase notice was valid in so far as it related to Tudor House, building 4 Tudor Court, and building 1 Tudor Court consisting of flats 1-3A inclusive (but without prejudice to the right of the applicants and of the respondent to argue before a leasehold valuation tribunal whether or not the three houses, 4, 4A and 4B Tudor Court, can be acquired pursuant to the Act).

The hearing before us consisted entirely of legal argument and neither party adduced any oral evidence. The figures contained in a valuation report given for the applicant by Mr NL Maloney FSVA were agreed by the respondent. This report was prepared on the assumptions that the common parts as defined in each of the tenant's leases would be transferred to the applicant or alternatively that the applicant would be granted perpetual rights over the common parts, and also that the three houses contained within building 1, Tudor Court were to be transferred to the applicant. (These assumptions were not agreed by the respondent.) Alternative valuations given in appendices A and B to the report, together with the valuation assumptions and comments contained in appendices C, D, H and J, are attached to this report in a bundle marked "1" [not reproduced here]. According to these valuations, the amount payable by the applicant is £43,215 (£22,710 for building 1, £19,439 for building 4 and £1,066 for Tudor House) on the assumptions that the three houses in Tudor Court are transferred, that the tenants have the rights over the gardens and amenity land reserved in their leases, and that the vacant possession value of flat 2A is not payable to Twinsectra. If the three houses are not to be transferred, the applicant pays £457 less, and if the vacant possession value of flat 2A is payable to Twinsectra, the applicant pays £56,950 more. Curtailment of the tenant's rights to use the gardens and amenity land in accordance, if such curtailment was in accordance, with the leases, would not in Mr Maloney's opinion affect the value of the reversion.

On February 27 we inspected the whole development. We did not inspect any of the properties internally.

The issues which we were asked to decide were as follows:

(i) was the price to be paid by the applicant increased by reason of the death in April 1996 of the regulated tenant of flat 2A Tudor Court?

(ii) can the applicant acquire the three houses which form part of building 1 Tudor Court?

(iii) what rights should the tenants have over the gardens and amenity land?

(iv) should the transfer require the consideration to be paid by the applicant directly to the bank holding a charge over the respondent's interest in the property?

Decision

(i) *Is the price increased by reason of the death of the tenant of flat 2A?*

Section 12(6) of the 1987 Act provides:

> Where the property referred to in subsection (4) has at any time since the original disposal increased in monetary value owing to any change in circumstances (other than a change in the value of money), the amount of the consideration payable to new landlord for the disposal by him of the property in pursuance of the purchase notice shall be the amount that might reasonably have been obtained on a corresponding disposal made in the open market at the time of the original disposal if the change in circumstances had already taken place.

Mr Edward Denehan argued for the applicant that section 12(6) applied only to changes prior to the date of the purchase notice. To apply it to a change of circumstances such as this, arising after the original vendor had knowingly disposed of the property in contravention of its statutory duty and the purchaser had employed delaying tactics to thwart the return of the building to the applicants would cause grave injustice and was contrary to the scheme of the Act. If there was an ambiguity we were entitled to look at later legislation to resolve it (as the Vice-Chancellor did in *Denetower Ltd* v *Toop* (at p666)). Part I of the Act had been amended, though not retrospectively and thus not so as to apply directly to this case, by section 92(1) of the Housing Act 1996, in order, as appears from section 92(2), to "restate the principal provisions of that Part so as to (a) simplify the procedures for the exercise of the rights conferred on tenants, and (b) apply those procedures in relation to contracts and certain special cases". Section 12(6) has been amended by para 12B(7), set out in Schedule 6 to the 1996 Act, to omit the words "at any time" which supported, he said, his view that some limitation must be put on section 12(6).

Mr Simon Brilliant said that "at any time" in the section governing this application was unambiguous and meant what it said. The death of the regulated tenant was a fortuitous event and it was irrelevant that Twinsectra had acquired the freehold without the tenants having the opportunity to purchase first. In *Belvedere Court Management Ltd* v *Frogmore Developments Ltd* [1996] 1 All ER 312* Sir Thomas Bingham, the then Master of the Rolls, considered, at p327, the effect of a new landlord letting a vacant flat (not having offered the tenants in the block the right of first refusal), which might affect the purchase consideration. He said that, if it was suggested that section 12(6) was ambiguous, we should refer to *Hansard*.

With Mr Denehan's consent, we referred to an extract from *Hansard* (Standing Committee A, April 23 1987, column 41 *et seq*) *de bene esse*, but, in our view, section 12(6) is clear and unambiguous. To restrict its operation as Mr Denehan invited us to do would require us to rewrite it, which we cannot do. We recognise that this causes some injustice to the tenants who, had they been offered and had exercised their statutory rights of first refusal, would have obtained flat 2A without any extra payment, but the statute has clearly sought to address the problem to some extent by requiring us to value the change in circumstances as if it had taken place in February 1992. We therefore conclude that the consideration to be paid by the applicant must be increased, in accordance with Mr Maloney's agreed figures, to include the vacant possession value of flat 2A.

(ii) *Can the applicant acquire the three houses within building 1?*

We inquired whether this issue had been raised at the Court of Appeal hearing. Counsel, both of whom had appeared at the hearing, agreed that the matter had been raised by leading counsel but had not been determined by the court, who clearly intended that this issue

---

*Editor's note: Also reported at [1991] 1 EGLR 84

*Editor's note: Also reported at [1996] 1 EGLR 59

should be decided by us, under the power, which appears to be wide, given to us under section 13(1)(a).

Mr Denehan said that section 1 of the Act provided for the nominated person to acquire premises as defined by section 1(2), namely that:

> (a) they consist of the whole or part of a building; and
> (b) they contain two or more flats held by qualifying tenants; and
> (c) the number of flats held by such tenants exceeds 50 per cent of the total number of flats contained in the premises.

He pointed out that, by section 1(3), Part I does not apply to premises falling within subsection (2) if:

> (a) any part or parts of the premises is or are occupied or intended to be occupied otherwise than for residential purposes; and
> (b) the internal floor area of that part or those parts (taken together) exceeds 50 per cent of the internal floor area of the premises (taken as a whole) . . .

Accordingly, he argued, the applicant was entitled to acquire the whole of the building. If, indeed, part of the building (provided that part occupied less than half the floorarea) had been occupied as commercial premises, the applicant would have been entitled to acquire the whole of the building. It was necessary to keep separate the definition of premises from the definition of qualifying tenants within section 3, who were required to be tenants of a flat as defined by section 60. There was nothing offensive to the policy of the Act for the applicant to acquire a building which contained houses as well as flats, just as he could acquire a building which contained commercial premises. The tenants of the houses had joined in these proceedings from the start. The policy of the Act was to give tenants control of their own building.

Mr Brilliant relied on the short title of the Act: ". . . to confer on tenants of flats rights with respect to the acquisition by them of their landlord's reversion; . . .". This, he said, must mean the reversion to those flats. There was no policy reason why tenants should become landlords of other people's houses. Section 1(2)(a) contemplated that premises to which Part I applied could consist of part of a building, and a building could therefore be severed into parts which amounted to premises within Part I and parts which did not. If the original disposal was only of the freehold of the three houses, that would not have been a relevant disposal. Even if Mr Denehan's argument was correct, the powers of the tribunal under section 13 were so wide that we could in our discretion leave the landlord with the houses.

In our view Mr Denehan is correct. Although the short title of the Act refers to rights conferred on tenants of flats, we have come to the conclusion from the wording of section 1(2) that Part I can apply to premises which consist of the whole of a building consisting partly of flats and partly of houses. We do not think that the scheme of the Act requires us to reach a different conclusion; indeed, as is implicit in the judgments in the present case in the Court of Appeal, the policy of the Act is to assist tenants whose rights have been ignored by landlords, and we see nothing inappropriate in the concept of the applicant, on the facts of this case, acquiring the freehold not only of the flats in building 1, but also of the houses which form an integral part of the same building.

(iii) *What rights should the tenants have over the gardens and amenity land?*

Under most of the leases put before us, the common parts include: "gardens gates access yards roads footpaths parking areas and means of refuse disposal (if any) and other areas included in the Title above referred to provided by the Lessors for the Common use of residents in the curtilage of the Estate and their visitors and not subject to any lease or tenancy, to which the lessors are entitled to the reversion". The "Title above referred to" in each of the leases put before us is MX 420465 and MX 304042 which includes, according to the judgment of Judge Hull, the whole of Tudor Court, Tudor House, Parr Court, the gardens and the amenity land. The lease of flat 6, Tudor House, defines the common parts by reference to a plan which similarly includes all the gardens (but not the garden in the courtyard of Tudor Court) and the amenity land. All the leases give to the tenants the right

> . . . at all times and for all purposes in connection with the permitted user of the demised premises to go pass and repass over and through and along the common parts . . . provided always . . . that the lessors shall have the right temporarily to close or divert any of the common parts . . . subject to leaving available reasonable and sufficient means of access to and from the demised premises and provided further that the lessors shall have the right at any time and from time to time on giving in each case at least three months written notice to that effect to the tenant to exclude from the easement right and liberty granted by this paragraph the use of all or any part or parts of the garden or land forming part of the common parts but not so as to make access to the demised premises impracticable.

It was not disputed that Twinsectra had by letters dated March 23 1993, sent by recorded delivery, purported to give to each of the tenants three months' notice terminating, in accordance with clause 2(1) of each lease, the use of all or any part or parts of the garden or land forming part of the common parts of Tudor Court or Tudor House as defined in their leases, save in so far as the same would make access to the demised premises impracticable. Copies of the letters were put before us.

Mr Denehan said that, at the time of the purchase notice, the right to use the gardens and amenity land was an appurtenant right enjoyed by the tenants and they were entitled to it still. They did not ask for the freehold of any of the land, but wanted perpetual rights over the common parts as they had been granted by their leases, such rights being appurtenances within the meaning of section 29(4) of the Act, (applied by analogy to applications under Part I in *Denetower Ltd* v *Toop* (at p668)). Rights, such as easements, can be appurtenances which can be acquired under Part I. In *Denetower* the Vice-Chancellor had said (at p668):

> As to the roadways and paths over which the tenants have either express or prescriptive rights of way, in my judgment they are appurtenances of the building. However it may well be that a rent assessment committee acting under section 13(1) may reach the conclusion that the appropriate provision should be that the tenants should not acquire the freehold in the land of the roadway and paths, but be granted perpetual rights of way over them.

The powers of the tribunal under section 13(1) were, Mr Denehan said, quite wide enough to permit us to grant perpetual rights in accordance with the leases, but not subject to the liability to have those rights determined.

Mr Brilliant said that the rights given by the leases were not appurtenances, because no part of the gardens or common land had ever been demised to the tenants. Part I of the Act strictly gave tenants the right to acquire only premises, which were buildings. In *Denetower* it had been decided that gardens could also be acquired *because they had been demised.* At the most, the tenants should be provided with suitable and appropriate rights over the paths leading directly to their respective flats. The tenants had chosen to enter into leases which gave them only terminable rights over the common land.

We have found this aspect of the case to be the most difficult question we have to decide, but, on balance, we have come to the conclusion that the wide powers given to us by section 13(1) — their width emphasised by the Vice-Chancellor in *Denetower* — allow us to include within the transfer the rights given to the tenants in their leases but not subject to termination by Twinsectra. We are satisfied from the reasoning in *Denetower* that easements are rights which can be compulsorily acquired as appurtenances under Part I, by analogy with Part III. If the original vendors had complied with their statutory duty (as it was found to be in this case in the Court of Appeal) to offer the relevant freeholds to the tenants, the tenants would have received with them the appurtenant right to use the gardens and amenity land in accordance with their leases. Had they done so, the owner of the freehold of the land would in our view have met with insuperable difficulties had it tried to restrict those rights, the right to determine them having passed to the applicant with the freehold.

A (iv) *Should the transfer require the consideration to be paid by the applicant to Twinsectra's bank?*

By Schedule 1, para 2(1) of the Act:

> Where, in accordance with section 12(4)(a) an instrument will operate to discharge any property from a charge to secure the payment of money, it shall be the duty of the nominated person to apply the consideration payable, in the first instance, in or towards the redemption of any such charge . . .

By section 12(4), unless the court directs otherwise, the instrument by virtue of which the property is disposed of by the new landlord to the nominated person operates to discharge the property from any
B charge to secure the payment of money.

Mr Denehan said that it was common practice to include within a conveyance or transfer the provision that the consideration should be paid to a bank, and the fact that Twinsectra were resisting the suggestion made the tenants suspicious. Mr Brilliant said that it was unnecessary to include the requirement in the transfer, because by Schedule 1 para 2 it was the duty of the nominated person to apply the C
consideration, in the first instance, towards the redemption of any charge to secure the payment of money.

In the exercise of our discretion we decided that the transfer should require the payment of the consideration to be made to Twinsectra's bankers, on confirmation from Twinsectra and from the bank that the bank had a charge over any part of the property to secure the payment of money.

Determination

We therefore conclude that the amount payable for the freehold of D
buildings 1 and 4 Tudor Court and for Tudor House is £100,165 (one hundred thousand, one hundred and sixty five pounds). We have invited the solicitors for the applicant and for the respondent to put before us for our approval a form of transfer which accords with our decision.

SUPPLEMENT

CASES NOT REPORTED IN "ESTATES GAZETTE"

LANDLORD AND TENANT

GENERAL

A **Reed Personnel Services plc *v* American Express Ltd**

CHANCERY DIVISION

January 24 1996

JACOB J

Landlord and tenant — Break clause — Whether underlessee reasonably performed and observed the covenants — Whether pre-condition to determination of underlease satisfied

B By an underlease dated March 17 1989 the plaintiff granted the defendant tenants an underlease of business premises at a rent of £120,000 pa, subject to a tenant's option to determine. The defendants covenanted to repair the interior and to redecorate during the last six months of the term. The break clause was conditional on the tenants reasonably performing and observing the covenants of the underlease. By two letters dated December 22 1992 and January 2 1993 the defendants served notice under the break clause determining the underlease on March 24 1994. By a letter dated December 20 1993 the plaintiff enclosed a notice setting out breaches of the repairing and other covenants and stating that a reasonable time for complying was eight
C weeks. The cost of carrying out the works was about £21,000 of which £8,000 and £7,000 approximately were attributable to decorations and carpet replacement respectively. By March 24 1994 the works had not been done. The plaintiff contended that the pre-condition in the break clause had not been satisfied and the underlease continued.

***Held*: Judgment was given to the plaintiff. The defendants had not satisfied the pre-condition in the break clause; the covenants in the underlease had not been reasonably performed and observed. The poor state of the common parts, for which the plaintiff was responsible, was not relevant. It was relevant to look at the total size of the obligation. Where there has been a total failure to**
D **comply with an important obligation in the underlease, it could not be said there has been reasonable compliance. The economic value of non-compliance is irrelevant to the question as to whether there has been reasonable performance of an obligation. Reasonable performance might be satisfied by one coat of paint where there should be two, or retaining a carpet in perfectly good condition where there is an obligation to re-carpet.**

The following cases are referred to in this report.

Bairstow Eves (Securities) Ltd v *Ripley* (1992) 65 P&CR 220; [1992] 2 EGLR 47; [1992] 32 EG 52

E *Bass Holdings Ltd* v *Morton Music Ltd* [1988] Ch 493; [1987] 3 WLR 543; [1987] 2 All ER 1001, [1987] 1 EGLR 214, CA

Bassett v *Whiteley* (1982) 45 P&CR 87

Gardner v *Blaxill* [1960] 1 WLR 752; [1960] 2 All ER 457

Trane (UK) Ltd v *Provident Mutual Life Assurance* [1995] 1 EGLR 33; [1995] 03 EG 122

This was an application by the plaintiff, Reed Personnel Services plc, to determine the validity of the exercise of a break clause in an underlease held by the defendants, American Express Ltd.

Kirk Reynolds QC (instructed by the solicitor to Reed Executive plc) appeared for the plaintiff; Jonathan Ferris (instructed by Cameron Markby Hewitt) represented the defendants.

Giving judgment, **JACOB J** said: On September 9 1988, at the F
height of the property boom, the plaintiff, Reed, took an underlease ("the headlease") from the Prudential. The headlease expires in the year 2003; Reed has no right to break it before then. It was and is of the basement, first, second and third floors of 104 New Bond Street.

On March 17 1989, still at the height of the boom, Reed granted a subunderlease ("the lease") of the second and third floors. The lease was granted to the defendants, American Express ("AMEX").

Clause 11.6 of the lease gave AMEX a once and for all opportunity to determine it on March 24 1994. They had to give at least six months' notice in writing and satisfy certain pre-conditions. Apart from that opportunity to break the terms of the lease, the lease expires G
in 2003, just before the headlease expires.

The question before me is whether AMEX have effectively availed themselves of the break opportunity provided by clause 11.6. The lease, as is usual, contained covenants by AMEX. Clause 5.4 of the lease provides in detail for repair, cleaning and decoration. The detailed provisions are as follows:

> 5.4.1 In every fifth year of the Term and also in the last six months thereof whether determined by effluxion of time or otherwise in a proper and workmanlike manner to prepare and paint with two coats of good quality paint all the internal parts of the Premises usually or requiring to be painted and in like manner to prepare and repolish or revarnish with best quality material all H
> the internal parts of the Premises usually or requiring to be polished or varnished and to prepare and paint as aforesaid and repaper the ceiling and walls as the same are now treated and to clean and treat in a suitable manner for their maintenance in good condition and appearance all internal parts of the premises capable of being so cleaned and treated and not usually or requiring to be painted polished varnished or repapered.
>
> 5.4.2 From time to time and at all times during the Term well and substantially to repair cleanse and keep in good and substantial repair and condition the Premises as often as required (and in any event at the expiration or sooner determination of the Term) to replace the carpeting with good quality carpeting of a similar nature quality and appearance damage by any of the Insured Risks excepted (save to the extent that payment of the insurance J
> monies or any part thereof shall be withheld by reason of any act or default of the Tenant) provided that any door or window leading from the Premises and requiring to be replaced under the provisions of this clause shall be replaced with a door or window of similar quality and appearance or with such other door or window as shall first be approved in writing by the Landlord (such approval not to be unreasonably withheld).
>
> 5.4.3 To clean the Premises and keep them in clean condition.
>
> 5.4.4 At the expiration or sooner determination of the Term.
>
> 5.4.4.1 quietly to yield up the Premises so painted polished papered treated washed repaired carpeted cleansed and kept as aforesaid together with all additions and improvements thereto.
>
> 5.4.4.2 if so required by the Landlord at the Tenant's expense to remove K
> anything constructed or placed on the Premises by the Tenant or any sub-tenant or any predecessor in title of the Tenant or any sub-tenant and make good to the reasonable satisfaction of the Landlord any damage caused to the Premises by such removal.

Some of these provisions require continuous compliance, but others only require compliance at the end of the term. An example of the former is the obligation to keep the premises clean (clause 5.4.3); an example of the latter is the obligation to re-carpet at the termination of the lease.

The break clause provides as follows:

A 11.6 *Tenant's option to determine.*

If the Tenant (here meaning American Express Europe Limited only and not including any successors in title of that company) wishes to determine this Lease on the 24th March 1994 and shall give to the Landlord not less than six months notice in writing and shall up to the time of such determination pay the rent reserved by and reasonably perform and observe the covenants contained in this Lease then upon the expiry of such notice the contractual term shall immediately cease and determine but without prejudice to the respective rights of either party in respect of any antecedent claim or breach of covenant provided however that if at the date of such determination the rent review effective from the 25th March 1993 shall not have been completed the Landlord and Tenant shall take all such steps and do all such things as shall be
B necessary to complete such rent review with all due expedition and provided further that

11.6.1 upon such determination of the contractual term as aforesaid there shall immediately be due and payable by the Tenant to the Landlord by way of compensation a sum equivalent to six months Rent payable hereunder such rent to be calculated on the rent payable effective upon the completion of the rent review on the 25th March 1993.

11.7 *Landlord and Tenant Act Exclusion*

Having been authorised to do so by an Order of the Westminster County Court made on the 20th of February 1989 under the provisions of the Landlord & Tenant Act 1954 Section 38(4) the parties agree that the provisions of the Landlord & Tenant Act 1954 Sections 24 to 28 (inclusive) shall be excluded in
C relation to the tenancy created by this Lease.

Before returning to these provisions, I should deal with the law as to preconditions in general. There is little dispute as to this, so I can state the law reasonably briefly. First, where a lease contains an option which may be exercised only upon fulfilment of a condition, it is essential that that condition (a so-called pre-condition) be fulfilled before any purported exercise of the option can be valid. *Bass Holdings Ltd* v *Morton Music Ltd* [1988] Ch 493* contains a detailed review of the history of this kind of clause.

Second, the proposition is the same whether the option is to break the lease or to renew it. Of course, in a rising market tenants want to
D renew and in a falling market to break.

Third, it is not an alternative to compliance with a pre-condition for a party who has not fully complied with such a pre-condition to say that the condition is trivial and that accordingly his non-compliance should be overlooked. A good recent example of this is *Bairstow Eves (Securities) Ltd* v *Ripley* [1992] 2 EGLR 47. There there had been failure to comply with a pre-condition that the premises should be painted in the last year; they had been painted just before the beginning of that last year and there was no practical difference between what had been done and what ought to have been done and there would be only nominal damages for the breach. None the less, there was no
E defence to the assertion of failure to comply with the pre-condition.

Fourth, there is an exception where a breach is wholly spent or remedied in the past. An example of that is *Bass* v *Morton*, where the effect of a breach of covenant not to apply for planning permission had become spent by virtue of the court having given relief from forfeiture in the past.

Fifth, nor is it relevant that a pre-condition clause also contains a proviso reserving rights even if the condition is fulfilled. That emerges from the decision of Judge Cooke in *Trane (UK) Ltd* v *Provident Mutual Life Assurance* [1995] 1 EGLR 33. The reason why such a proviso is irrelevant is threefold. First, that the landlord might not
F actually have knowledge of the breach at the relevant date; second, there might be waiver; and, third, such a proviso is normally mutual applying to breaches by landlord or tenant. It was accepted before me that this represented the law and, as Judge Cooke said, it probably represented the law at least up to Court of Appeal level.

Some pre-conditions are absolute, others import a requirement of reasonableness. An example of an absolute covenant is *Bairstow Eves*.

Two cases are concerned with pre-conditions which import the requirement only of reasonable compliance and these were the only two cases which the researches of counsel were able to produce. The first of these was a decision of Paull J, *Gardner* v *Blaxill* [1960] 1 G
WLR 752. The main part of his decision is not concerned with the pre-condition clause at all. But towards the end of his judgment Paull J considered the clause in detail.

I shall first remind myself of the actual wording of the clause he was considering. The vital parts of the wording were:

> . . . provided that the tenant has reasonably fulfilled the covenants hereinbefore mentioned he has the option of continuing for an extension of seven or fourteen years . . .

Paull J said at p760:

> The insertion of the word "reasonably" means that the parties have agreed that H
> the tenant may exercise his option provided he has behaved as a reasonable tenant might have behaved in relation to his covenants.

A little later he said:

> I have come to the conclusion that . . . by inserting the word "reasonably" the parties not only intended to mean, but must be deemed to have meant the tenant can exercise his option provided that he behaves during his tenancy in a way in which a reasonably minded tenant might well behave.

On the facts of that case, Paull J accepted the evidence given on behalf of the tenant by an expert to the effect that the tenant had taken J
reasonable steps to maintain the premises within the conception of the purpose for which they were there to be used. It is true there had been some breaches of covenant, but the judge found them to be minor and overall of a trivial nature.

The other case where the pre-condition was not absolute, is *Bassett* v *Whiteley* (1982) 45 P&CR 87. The tenant had deliberately withheld rent in order to force the landlord to comply with a very important repairing covenant; apparently the roof of a pub was leaking on to the cocktail bar. The Court of Appeal had no difficulty in holding that the tenant had behaved reasonably, even if he had not complied with the obligation to pay rent.

Both of those cases were cases of an option to renew a lease being K
exercised by a tenant.

Mr Jonathan Ferris suggested that where there is a reasonableness requirement in a pre-condition there may well be a difference between the case of an option to renew and the case of an option to break a lease. I do not think Mr Kirk Reynolds QC was inclined to dispute that; indeed, I think he was rather in favour of there being a difference. It must be right in principle. For example, an obligation to paint the premises on the inside is much more important if the tenant is leaving than if he is staying on.

The distinction between the two kinds of case was recognised by Bingham LJ (as he then was) in *Bass* at p538, where he said: L

> Of the leading cases, the first two concerned a tenant's right to determine a lease before the expiry of the contractual term. This is of significance, because the context coloured the judicial approach to the question of construction. Where a tenant wished to take advantage of a break clause, the landlord is not greatly concerned with the history of the tenant's performance before the break. The worse the tenant's performance, the readier the landlord might reasonably be to get rid of him. But whatever the tenant's defaults in the past, the landlord would be very much concerned that at the time of the break the rent should be fully paid (because he could no longer distrain) and the covenants fully observed (so that the property could be re-let or sold without delay or additional expenditure). M

Bingham LJ also commented on the decision in *Bassett* v *Whiteley*. He said at p540:

> The crux of the decision was the word "reasonably". It was plain that the tenants had acted altogether reasonably whichever period or point of time was looked at.

I turn to the facts here. I have already recited the relevant clauses of the lease. I propose to deal with the facts as asked by Mr Ferris in case the matter goes further, though I think many of the facts upon which he asked me to make findings are, at the end of the day, irrelevant. I will go through the numbered facts as provided by Mr Ferris.

---

*Editor's note: Also reported at [1987] 1 EGLR 214

First, in September 1988, 104 New Bond Street had been refurbished completely and the entrance and the lobby on the ground floor had also been done. The condition was as new.

Second, Reed went into occupation of the first and second floors. The first floor was occupied by an employment agency for temporary staff and the office was visited by prospective temps seeking jobs.

Third, in March 1989, AMEX, when they took their lease, with the permission of Reed, altered the as new premises, putting in partitions, an extractor fan, shelves and so on and took occupation of the third and fourth floors.

Fourth, in April 1991 AMEX went out of occupation of the third and fourth floors, which were left vacant. They have been vacant ever since. AMEX began a marketing campaign to sublet or assign their lease in those floors, which was unsuccessful.

Fifth, in July 1991, Reed left the first and second floors and began a marketing campaign to sublet those floors, which again was largely unsuccessful.

However, sixth, in July 1992, a company called ISO Group Ltd took a two-year lease of the second floor. That lease was subject to a schedule of conditions, the lease was extended for a further three months. ISO left without complying with the conditions. After they left, they were served with a schedule of dilapidations, they did some of the work and not other and paid Reed agreed compensation for the work they had not done. The plaintiff has done no work to the premises since and, as I have said, there have been no further letting transactions in relation to any part of the premises, which at the present moment are empty.

Mr Ferris asked me to make a finding that there had been no carpet replacement or cleaning or decorative work done to the first or second floors between 1988 and the end of 1994 and no decorative work or carpet replacement to the common parts between 1988 and the end of 1994. I decline to make any such finding, there being no actual evidence about it. Discovery documents rather suggest that at least some work was done, perhaps pursuant to a requirement by the head landlord, the Prudential.

Eleventh, as at March 1994, Reed were in breach of their own lease from the Prudential because of the state of the common parts. In particular there had been some water penetration to the third and fourth-floor WCs and rear staircase and there had been no general decoration of the common parts. It would be wrong to say that these amounted to perilous structural defects, but they were obviously not without some substance. The total cost of the works required to remedy the defects appears to be about £9,000.

I am asked to make a finding about the graffiti in the lift. There apparently was some. It sometimes is painted out and sometimes new graffiti reappears. I do not regard it as of any significance.

Next, I find that the common parts were dingy and unappealing. I likewise find that the demised premises were in the same state, namely dingy and unappealing. I do not find the state of the common parts is a major obstacle to any successful letting. I do not think the evidence bears that out. In particular, when AMEX were trying to let the demised premises they did not complain to Reed about the state of the common parts and none of the advice they received in writing from their agents referred to any difficulties with the common parts at all, except one slight reference to the state of the graffiti in the lift. It would take about two months to carry out the remedial work to the common parts.

I am asked to make a finding about the defendants' breach prior to termination, to which I will come in due course.

The total price of the remedial work required to cure the breaches of the repairing covenants and the like of which AMEX accept they are guilty is about £21,000. That seems to be the common view of the experts who gave evidence and, perhaps more significantly than that, that is the amount which AMEX themselves got by way of a quote from a builder at the relevant time. Some of that work required relates only to painting and carpet replacement. It is broken down, as Mr Ferris wants me to find (and I do) roughly speaking into £4,000 in reinstatement work, removing partitions, shelving and extractor fan and consequent redecoration, £8,000 on decoration, less than £1,000 on general repairs, £7,000 on carpet replacement, these all being approximate figures.

It is common ground that AMEX were not under any obligation to do structural or external repair work. That of course follows from the nature of the lease which they had taken: they were just the tenants of the two floors on the inside. None of the outstanding work which had not been done threatens the integrity of the building and none of the work is necessary to comply with any enactment or building bylaw. Moreover, it seems unlikely there was any worsening of condition of the defendants' premises between April 1991 and March 1994; and again it will take probably about two months to do the actual works and perhaps another month to organise it.

Mr Ferris asked me to make a finding that the defendants' decision not to carry out the work to the third and fourth floors prior to marketing them from 1991 was influenced by the condition of the common parts, including the WCs. I decline so to find. The defendants gave no evidence of their internal thought processes at all.

I am also asked to make a finding that by the end of 1993 when the plaintiff served a schedule, to which I will come, the poor condition of the common parts was a factor in AMEX's decision not to carry out the work to the demised premises. I utterly reject any such finding. As we will see, the defendants have been particularly coy about what actually happened at the end of 1993/beginning of 1994.

I am then asked to make some findings of fact about general market conditions and the way that landlords and tenants behave generally in the market in the West End. I believe these findings to be irrelevant, particularly because they are not concerned with the case where a tenant is trying to comply with a pre-condition so as to operate a break clause. All the evidence which I heard related to cases of the natural determination of a lease.

In relation to those cases, it is, however, often the fact that rather than comply exactly with repairing covenants tenants come to some accommodation with the landlord and often after the lease has come to an end. This may have an advantage for the landlord, who can use the money in some way which he chooses rather than the tenant or it may have the advantage that a landlord who has actually found a new tenant can then do up the premises as the new tenant wants it. It is also the case that where this happens often the landlord will have to suffer some penalty by way of giving his new tenant an increased rent-free period. It is also the case, self-evidently, that some would-be tenants would not be interested because the premises are not fit for occupation immediately.

I should say that I reject the somewhat partisan evidence given on behalf of AMEX to the effect that where premises have been newly decorated either by a landlord or by a tenant that any incoming new tenant in the West End will invariably redecorate, furl out the new carpets and put in his own. I just do not believe that to be the case. These are hard times and people do not have money to throw away, even in the West End.

In the end, the same gentleman who gave that evidence was a little reluctant to answer the question, "Is not the landlord better off if the premises have been re-decorated at the expense of the tenant than if he has to enter into negotiations after expiry of the lease with the outgoing tenant with all the complications by way of arguments that might arise under the Landlord and Tenant Act 1927 and arguments about whether works should or should not have been done?"

I now turn back to what actually happened at the end of 1993/beginning of 1994. First, it is to be observed that the defendants have given very clear notice of its intention to exercise the option and did so by two letters of December 22 1992 and January 2 1993; in each case steps were taken to ensure that Reed acknowledged receipt of the letters. Reed caused the premises to be inspected on October 1 and AMEX must have known that. They were inspected by an architect. He produced a document described as a "schedule of dilapidation" and that was sent by Reed's solicitors both to AMEX and AMEX's solicitors with a letter of December 20 1993. The letter read as follows, it having been sent by recorded delivery:

> A As solicitor for Reed Personnel Services plc, I enclose by way of service formal notice under Section 146 of the Law of Property Act 1925. Please acknowledge receipt.
>
> In so far as the breaches of covenant set out in the notice are capable of remedy my client considers that a reasonable time for complying with the enclosed schedule is eight weeks.

It may be noted that the letter refers to section 147 of the Law of Property Act. It is common ground that that was a misplaced reference. It is also not suggested that in any way that misled AMEX. The letter asked for acknowledgement; no such acknowledgement was received, so far as I can see. Indeed, I do not really know what happened next B within AMEX. All I know is that they went to a builder and got a quotation for the work to be done set out in the schedule of dilapidation sent with the letter. Who asked for this to be done and why it took so long, I do not know, but AMEX apparently received a quotation from these builders called Comtrax on February 25 1994. Still nothing happened and nothing did happen until April 7 1994, when Reed's solicitors indicated to AMEX's solicitors that the pre-condition had not been complied with and that accordingly, in Reed's view, the lease was still in being.

There was then a discussion between representatives of the parties, a Mr Strong and a Mr Lee. I think that at the time they agreed that the C job ought to have been done. Over the whole of this period, there was no communication until April 7 between AMEX and Reed, notwithstanding the letter of December 22.

Going back to that letter, it clearly makes the point that the job can be done and ought to be done in a period of about eight weeks.

In those circumstances, AMEX suggest that they are not in breach of the pre-condition of clause 11(6). They say that they have reasonably performed and observed the covenants contained in the lease. I reject this, but I must deal with the arguments in detail.

First, as I have already indicated, I do not think that it was reasonable for a tenant who had been presented with a list of defects which he accepted ought to be remedied by the date when he leaves D the premises not even to communicate with his landlord, not to say to the landlord, "Can we do some sort of deal on this and pay you some money for not doing it? What is it you, the landlord, actually want to happen to these premises?"

Of course, such a deal might have emerged, but it is not for the tenant to say, "I will leave the place with rubbish all over it, needing painting, needing re-carpeting", all of which is in breach of his obligations under the lease. I say that because that is one area where AMEX seem to have no excuse, and what is advanced on their behalf is a more general sort of argument. First, it is said that it is generally better for the landlord to get the money and do up the premises himself E in accordance with the normal practice when a lease comes to an end. The trouble with that situation is that the landlord will be involved himself, perhaps, in organising redecoration, which he may or may not want to be involved in; he will be potentially involved in arguments with the tenant about the effect of section 18 of the Landlord and Tenant Act 1927, which caps the amount of damage that a landlord can get for breach of a covenant. As I have already said, it is not for the tenant to say what is good for the landlord.

Next it is said there was either no or very little damage to the reversion. I rather think AMEX tried to change their arguments on this point because initially their expert gave evidence to the effect that F section 18(1) of the Landlord and Tenant Act 1927 applied to the whole of the premises and there was either no damage to the reversion or, at most, £8,000. He treated all the works as being works of repair within that Act. Of course, if that is right then Reed were much better off if the covenant had been complied with: they would have had £20,000 worth of works done for them and, what is more, the job would have been done by March 24 1994. So there was a bit of an attempt to resile from the argument and to say that for some of the works there could not be a claim for compensation because they were not caught by the 1927 Act. I think it is all beside the point. The purpose of a pre-condition is so that the landlord has not got to get involved in any such arguments — or one of the purposes at least.

Next, it is said, "Put the whole thing in context. Here is an annual G rent of £120,000. The works only cost £21,000 or so. In the whole circumstances, this is therefore not of great significance". I do not think that that is the right way to look at it at all. Here we have tenants of two floors and their job is to look after the inside of those two floors according to their lease and leave them in pristine condition. One must look at the breach in relation to the obligation broken. It may also be relevant to look at the total size of the obligation. If we were looking at something that only cost £100 one way or the other, then it might not matter; but where you have a total failure to comply with an obligation which is an important obligation in the lease, I cannot see how it can possibly be said there was reasonable compliance with the conditions H of the lease.

Next it is said that the obligations did not relate to structural matters and a breach of an obligation relating to a structural matter might be more serious than one in relation to a purely decorative state. I do not see that that is necessarily so at all. The obligation was to keep the decorative state or to leave the premises in a good decorative state; that was not done. It cannot be said to be a reasonable compliance with the obligation to say, "There might have been some other more serious obligation which might not have been complied with".

Next it is said, "Reed themselves were in default of their J obligations to their landlord, the Prudential, and also to AMEX". This is because there was an obligation in the headlease on Reed to keep the common parts in a state of good repair and, in the lease they had granted to AMEX, there was an obligation to comply with the terms of the headlease.

As I have already indicated, the state of the common parts does not seem to have affected the lettability in 1991 significantly enough for anybody to complain about it or write anything down about it. But, in my judgment, the state of the common parts has got absolutely nothing whatever to do with whether or not the tenants, AMEX, reasonably K complied with their obligations.

Next, it is said that the case is like Paull J's case, where a repairing obligation had been partially but not fully complied with by the tenant. But there are significant distinctions between that case and this. First, as I have already pointed out, that was a renewal case and the obligations are of less significance in the case of a renewal because it is the tenant who will have to put up, mainly, with the failure to comply with the obligations.

Next, on the evidence in that case, Paull J accepted the tenant's surveyor's evidence that the tenant had in fact complied reasonably with the conditions. Here I have no such evidence. Nor could I have L heard such evidence because there was no compliance. What the evidence of AMEX here really is attempting to show is that the economic value of the non-compliance was not all that great. Whether that is so or not is irrelevant to the question of whether there was reasonable performance of the obligation.

Of course, the word "reasonably" does qualify the word "perform" and there are plenty of examples one can think of here where AMEX might reasonably have performed the obligation without actually doing it to the letter. For example, there is a requirement that there should be two coats of paint: one coat might have been quite good enough. There M is a requirement of re-carpeting at the very end of the lease and it may be that if the carpet was in perfectly good condition that it would not have been necessary to comply with that obligation.

I am not concerned with that kind of case at all. I am concerned with a case of total non-compliance. So far as I can see — I do not really know — it was perhaps due to some sort of administrative blunder within AMEX's organisation. The fact that the job needed doing was plainly recognised in April 1994 by the manuscript writing, "Do it", on the building quotation.

I think the plaintiff's claim succeeds.

For further cases on this subject see p13

LANDLORD AND TENANT
BUSINESS TENANCIES

Shotley Point Marina (1986) Ltd *v* Spalding

COURT OF APPEAL

March 7 1989

SIR STEPHEN BROWN P, RALPH GIBSON and BUTLER-SLOSS LJJ

Landlord and Tenant Act 1954 — Interim continuation under section 64 — Application for extension of time for appeal refused — Whether proceedings on appeal for purpose of section 64 — Whether tenancy determined three months after county decision

On August 9 1988 the appellant tenant's application under Part II of the Landlord and Tenant Act 1954 for a new tenancy was dismissed on the ground that the respondent landlord satisfied section 30(1)(*f*) of the Act. Ultimately the tenant applied to the Court of Appeal for an extension of time, and for leave to appeal that decision. On January 24 1989 the Court of Appeal dismissed those applications. On February 17 1989 Judge Barnett made an order of possession against the tenant, deciding that there were no proceedings on appeal and that the tenancy came to an end three months after the county decision of August 9 1988 pursuant to section 64 of the Act. The tenant appealed contending that her applications for an extension of time and for leave for appeal should be construed as proceedings on appeal so that her tenancy was continued under section 64 until three months after the Court of Appeal's decision on January 24 1989.

Held: **The appeal was dismissed. There were procedural irregularities consisting of a failure to serve a notice of appeal in time and a failure to comply with the RSC Ord 59, r5, as to the manner of lodging an appeal. There never was an appeal. The proceedings in relation to the application for leave to extend time amounted to no more than an attempt to commence an appeal. If an extension of time for appealing is granted, then there are proceedings on appeal and section 64 will apply. But no extension was granted.**

The following case is referred to in this report.

Zenith Investments (Torquay) Ltd v *Kammins Ballrooms Co Ltd* (No 2) [1971] 1 WLR 1751; [1971] 3 All ER 128, CA

This was an appeal by the tenant, Patricia Joy Spalding, from a decision of Judge Barnett in Ipswich County Court, in possession proceedings by the landlord, Shotley Point Marina (1986) Ltd.

Edward Irving (instructed by Turner, Martin & Symes) appeared for the appellant; David Holgate (instructed by Edwin Coe) represented the respondent.

Giving judgment, SIR STEPHEN BROWN P said: This is an appeal from the judgment of Judge Barnett of February 17 1989 at Ipswich County Court. He then made an order granting the plaintiff, Shotley Point Marina (1986) Ltd, possession of the Peninsular Boatyard at Shotley within 10 days of his order.

The proceedings before the learned county court judge were the sequel to proceedings in relation to an application by the defendant, as tenant, for a new tenancy of the Peninsular Boatyard at Shotley pursuant to Part II of the Landlord and Tenant Act 1954. Those proceedings have a lengthy history and, indeed, they eventually came (in the context to which I shall refer in due course) to this court in January 1989. The history of the matter is briefly recited in the judgment of the learned judge. He sets out the relevant dates. It is relevant to state that after proceedings which extended over a substantial period of time the preliminary point which had been ordered to be tried in the original county court proceedings was decided on August 9 1988. The issue before the county court judge, Judge Sheerin, on that occasion was whether the landlords had made out the necessary grounds for opposing an application by the tenant for a new tenancy under the provisions of section 30(1)(*f*) of the 1954 Act. The tenant had opposed the landlord's case, relying in particular upon section 31A of the Landlord and Tenant Act 1954. The effective issue at the hearing before the learned county court judge on that occasion was whether the landlord had, in effect, defeated the defence raised under section 31A and had made out his case under section 30(1)(*f*). On August 9 1988 Judge Sheerin decided the matter in favour of the landlord and, effectively, that resulted in a decision against the application for a new tenancy by the tenant.

Thereafter the matter became embroiled in the procedural provisions relating to an appeal. The learned judge had in fact abridged the time for appeal to the Court of Appeal from four weeks to 15 days. The time for appeal in fact expired. The notice of appeal was not served upon the defendant's solicitor at the appropriate address in time, and although this defect in the circumstances was regarded as being of a highly technical nature the notice of appeal was nevertheless not in fact served in time. Thereafter a series of errors by the tenant's solicitor aggravated the situation. There was a failure on the part of the tenant's solicitor to comply with Ord 59 of the Rules of the Supreme Court which regulate the setting down of an appeal and the lodging of the necessary documents together with the payment of the appropriate fee to the Civil Appeals Office. Ultimately an application to extend time for appealing was made, but that application was not in fact made until some two months after the original time for the giving of notice of appeal and the lodging of the appropriate documents had expired. The learned registrar of Civil Appeals referred the application by the tenant to extend the time for appealing to this court, and it was that matter which was heard over a period of two days concluding on January 24 1989. I presided over the hearing of that appeal. The leading judgment was given by Glidewell LJ, with which the other members of the court agreed. The court dismissed the tenant's application for extension of time and leave to appeal.

It is right that I should say that in hearing the application by the tenant the court considered, in addition to the procedural questions, the merits of the proposed appeal and came to the conclusions and expressed the clear view, that there was no reasonable prospect of success in the appeal on the merits. There had been, so the court also held, prejudice to the landlords as a result of the delay that had ensued by reason of the tenant's failure to comply with the necessary procedural requirements. This was described as prejudice of substance not merely minimal.

The matter which came before the learned county court judge at Ipswich in February 1989 related to the application and effect of section 64 of the Landlord and Tenant Act 1954. That section provides for the interim continuation of tenancies pending the determination by the court of an application for a new tenancy. Section 64(1) provides:

> (1) In any case where —
>
> (*a*) a notice to terminate a tenancy has been given under Part I or Part II of this Act or a request for a new tenancy has been made under Part II thereof; and
>
> (*b*) an application to the court has been made under the said Part I or the said Part II, as the case may be; and
>
> (*c*) apart from this section the effect of the notice or request would be to terminate the tenancy before the expiration of the period of three months beginning with the date on which the application is finally disposed of,
>
> the effect of the notice or request shall be to terminate the tenancy at the expiration of the said period of three months and not at any other time.
>
> (2) The reference in paragraph (*c*) of subsection (1) of this section to the date on which an application is finally disposed of shall be construed as a reference to the earliest date by which the proceedings on the application (including any proceedings on or in consequence of an appeal) have been determined and any time for appealing or further appealing has expired, except that if the application is withdrawn or any appeal is abandoned the reference shall be construed as a reference to the date of the withdrawal or abandonment.

When this court, on January 24 1989, refused the application of the tenant for an extension of time in which to appeal from the judgment of Judge Sheerin, of August 9 1988, the position was that if there was no appeal and if the proceedings in relation to the application for an extension of time and for leave to appeal could not be construed as proceedings "on" appeal, then the three-month period provided for by section 64(1) of the Landlord and Tenant Act 1954 would have expired on December 7 1988. If, on the other hand, the proceedings which came before this court in January 1989 are properly to be construed as proceedings "on" an appeal within subsection (2) of section 64, then it is agreed that the three-month period would run as from January 24 expiring on April 24 1989. Thus the dispute before the learned county court judge on February 17 in effect involved consideration of the question whether the tenancy had finally come to an end on December 7 1988 or, alternatively, whether it would continue until April 24 1989. The issue before the learned county court judge involved the construction of section 64(2), and in particular of the words "including any proceedings on or in consequence of an appeal".

Mr Edward Irving in his well-formulated submissions to this court — which have been assisted by a carefully drawn skeleton argument — submits that the application for an extension of time should be considered as "proceedings on an appeal" within the meaning of the section. He has argued that merely because procedural failures had necessitated an application for an extension of time in order to bring an appeal before the court, this should not prevent the proceedings relating to the application for leave being regarded as proceedings *on* an appeal and thereby bringing into play the provisions of section 64(1) resulting in the continuation of the tenancy for a period of three months from the dismissal of the application for leave to appeal.

The learned county court judge held that in the circumstances, there were no proceedings on appeal. He expressed it in his judgment in the terms to be found at p10 of his judgment: "This was never successful" — that is the proceedings to launch the appeal — "which meant that the appeal was never more than inchoate". He accordingly rejected the arguments put before him by Mr Irving and made an order for possession upon the basis that the tenant's right to occupy the holding had already expired in December 1988 at the latest.

Mr Irving has submitted that it is important in considering section 64 to bear in mind the importance to the parties, both landlord and tenant, of their being able to be aware of the time when the tenant's right to occupy the premises will expire. He says that it would not be possible to know in advance when the continuation period would come to an end unless the interpretation for which he contends is given to the meaning of "proceedings on an appeal". He has submitted that anomalies would result from adopting a different approach. He cited the case of *Zenith Investments (Torquay) Ltd* v *Kammins Ballrooms Co Ltd (No 2)* [1971] 1 WLR 1751 in support of his contention that procedural irregularities should not be allowed to determine the position. In point of fact the issue in that case was not the issue which is before this court. As Sachs LJ said at p1756D of the report:

> The point for decision in this appeal is as to whether the tenant's application to the Torquay County Court of September 4, 1968, was "an application to the court made under . . . Part II" of the Act of 1954:

The matter which came before that court, eventually determined after an appeal to the House of Lords, concerned the original application for a new tenancy which was procedurally irregular. But nevertheless, as Russell LJ said at p1756 B of the report:

> The correct answer in this case is that the proceedings up to and including the House of Lords are correctly described as proceedings on an application under Part II . . . the application was, as it turned out, vulnerable to a defence because its making was by section 24 subsection (1) subjected to the provisions of section 29(3): but it was nevertheless such an application.

That was the issue which was then determined. There was no question before the Court of Appeal on that occasion which gave rise to consideration of subsection (2) of section 64. There clearly had been proceedings "on appeal" right up to the House of Lords in that case. The case is cited by Mr Irving in support of his proposition that irregularities, whether in the making of an original application under the Landlord and Tenant Act or irregularities in complying with the procedural rules regulating the initiation and conduct of appeals, should not be allowed to defeat the tenant's position.

He has also invited attention to the terms of section 58 of the Supreme Court Act 1981 and to the phraseology of subsection (1) of section 58:

> Any jurisdiction exercisable in any proceedings incidental to any cause or matter pending before the civil division of the Court of Appeal and not involving the determination of an appeal may, if and so far as rules of court so provide, be exercised (with or without a hearing) by a single judge of that court, whether in court or in chambers, or by the registrar of civil appeals.

His point on that is that that section provides for the hearing of proceedings incidental to any cause or matter pending before the civil division of the Court of Appeal, which would include applications for an extension of time. He seeks to argue that proceedings "*in*" the Court of Appeal amount to the same thing as proceedings "*on*" an appeal in the terms of subsection (2) of section 64.

He has valiantly sought to persuade the court that if it is of the view that the learned county court judge was correct there would be unforeseeable difficulties in the future because of the possibility that anomalous situations might be encountered or created. For my part, I find the issue to be a very short and succinct one. In this case an appeal was never in fact commenced. The Court of Appeal was never seized of any appeal. That situation came about because of procedural irregularities which required the would-be appellant to obtain leave to institute and pursue an appeal. Those procedural irregularities consisted, first, of a failure to serve a notice of appeal correctly in time, and, second, a failure to comply with the provisions of r5 of Ord 59 of the Rules of the Supreme Court as to the manner of lodging an appeal. That subsequently involved a delay of some two months before the application for an extension of time was in fact made. Because of those deficiencies the appeal never commenced and, in the result, at the hearing in January 1989 this court upheld the submissions of the would-be respondent and refused leave to appeal. Accordingly, in my judgment, there never was an appeal in this case. The proceedings in relation to the application for leave to extend time amounted to no more than an attempt to commence an appeal.

In those circumstances, I am quite unable to give to the words "proceedings *on* an appeal" the interpretation contended for by Mr Irving. The matter, in my judgment, was correctly decided by the learned county court judge, who gave a well-reasoned and careful judgment. For these reasons, I would dismiss this appeal.

Agreeing, **Ralph Gibson LJ** said: Section 64 subsection (2) refers to "the earliest date by which the proceedings on the application

A (including any proceedings on or in consequence of an appeal) have been determined and any time for appealing or further appealing has expired . . .". Mr Irving submitted that provided the tenant applies for extension of time for appealing within three months of the judgment of the county court they are "proceedings on an appeal" within the meaning of section 64(2) until that application for extension of time has been finally disposed of.

I do not accept that submission. It seems to me that, if an appeal is not commenced in due time, there is no appeal in being in which there can be any proceedings for the purposes of section 64(2) unless or until extension of time for appealing is granted. If an extension is granted,
B then, for the purposes of section 64, there is an appeal on which proceedings will or may take place.

Mr Irving contended that that conclusion must be wrong because, if it were right, it must follow that in a case where the period of three months mentioned in section 64(1) expired before the grant of extension of time, then any ensuing appeal would be purely academic. The tenancy must, he said, have terminated irretrievably at the end of the three-month period and could not be brought back into existence C
again by the grant of the extension of time for appealing.

I cannot accept that submission either. It seems to me that, upon proper construction of section 64(2), if this court grants an extension of time for appealing the appeal is then in being and in the event the tenancy would determine upon final disposition of that appeal. Difficulties may arise in a particular case, but they do not seem to me to be of such a nature as to require this court to apply any different construction of the provisions in question. The consequences of granting an extension of time in any particular case may be relevant to the exercise of the court's discretion in granting or withholding the extension. Accordingly, since there was no extension of time for leave D
to appeal granted by this court in the proceedings, there were no proceedings on an appeal. Therefore, for the reasons also given by Sir Stephen Brown P, I would dismiss this appeal.

Butler-Sloss LJ agreed and did not add anything.

Appeal dismissed.

For further cases on this subject see p73

LANDLORD AND TENANT
RENT ACTS

A
Northumberland & Durham Property Trust *v* London Rent Assessment Committee

QUEEN'S BENCH DIVISION

February 29 1996

MACPHERSON J

B

Landlord and tenant — Rent Acts — Assessment of fair rents — Whether rent assessment committee erred in considering evidence of market and fair rent assessments

In September 1995 the rent assessment committee determined the fair rent under the Rent Act of two flats in a property owned by the appellant landlords; the landlords had been contending for £3,900 and £4,160 respectively. Before the committee was evidence of a virtually identical flat in respect of which the committee had determined a market rent for an assured tenancy of £8,320, and evidence of seven fair rent assessments. The landlords appealed contending the committee had failed to look at the market rent and the discount from it
C for scarcity, in favour of the evidence of fair rent settlements.

***Held*: The appeal was dismissed. In the circumstances the committee was entitled to consider the seven fair rent assessments; it had not disregarded the market rent evidence. There is no need for a committee to show its mathematical workings. Principle required the committee to start with the market evidence and then go on to consider other relevant matters. It would be perverse to pay no attention to market rent, but also absurd to pay no attention to evidence of fair rents.**

The following case is referred to in this report.

D
Spath Holme Ltd v *Chairman of the Greater Manchester and Lancashire Rent Assessment Committee* [1995] 2 EGLR 80; [1995] 49 EG 128

This was an appeal by the appellants, Northumberland and Durham Property Trust Ltd, from a decision of the London Rent Assessment Committee, which had determined a fair rent of two flats on the application of the appellant.

Rajinder Sahonte (instructed by Morgan Bruce) appeared for the appellants; Michael Furness (instructed by the Treasury Solicitor) represented the committee.

E Giving judgment, **MACPHERSON J** said: This is an appeal by Northumberland & Durham Property Trust Ltd, the appellants, against a decision made by the London Rent Assessment Committee. That decision was dated September 8 1995. It determined what the committee concluded to be fair rents for the second and third-floor flats at 28 Coleherne Road, London SW10.

It is important, in my judgment, just to look at the figures, at the outset, remembering that the rent assessment committee is a body which hears appeals from decisions of rent officers determining fair rents for residential property. The initial rent, that is to say, the reigning rent at the time of the original application for the two properties, were, respectively, the second-floor flat, £2,100 and the F
third-floor flat, £2,280. The rent officer fixed £2,580 for the second floor and £2,760 for the third-floor flat. The landlords were dissatisfied with those conclusions. They were contending for £3,900 for the second-floor flat and £4,160 in respect of the third-floor flat. Taking the third-floor flat as an example, that is an uplift of £560 pa.

One vital ingredient of this case was the existence of what is described as a virtually identical flat on the third floor of 22 Coleherne Road. The rent assessment committee determined on May 15 1995 a market rent of £8,320 for an assured periodical tenancy of that flat. The contention was that a fair rent of the third-floor flat at no 28 would have been £7,072 pa, which allowed 15% for scarcity. That was the G
contention of the chartered surveyor, Mr Rupert Dickinson [ARICS], who put in written representations on behalf of the landlords before the rent assessment committee. It is therefore, to my mind, very surprising that the landlords were contending for a rent of £4,160 for the third-floor flat. It is wholly unexplained why they hit upon that figure because their surveyor argued that taking scarcity into account and discounting the full market rent for an assured periodic tenancy the figure should have been £7,072 so there is the best part of £2,500 between those two figures. That alone must have made consideration of the rent in respect of the virtually identical flat very difficult for the committee. As I say, the reasoning of the landlords who contended for £4,160 is obscure. That, however, is only one feature of the case. But it H
may be, in the end, to some extent, the key to the answer which the parties seek.

At the hearing before the committee a lady representing the Private Tenants Rights Project represented both tenants. It is important, in my judgment, to read the contention for which she argued.

I read para 3 of the committee's decision:

> A hearing took place which was attended by Miss Jo Owen. Dealing first with Mr Dickinson's representations, she conceded that the third-floor flat at 22 Coleherne Road was virtually identical to the third-floor flat we were considering and that determination in respect of it was relevant to our J
> determination. She argued however that the suggested discount of 15% for scarcity was far too low. She said it was impossible on the evidence available precisely to quantify scarcity in arithmetical terms. The most that could be said was that it was very heavy in this locality and for guidance you should look not only at the single committee decision in respect of a market rent at 22 Coleherne Road but also fair rents recently determined by committees in the terrace.

The determinations to which she was referring had been made on February 7 1995 in respect of seven flats, two of them at 22 Coleherne Road and three at no 26 and another one in flat 28. There were seven recently determined fair rents to which attention was drawn by the K
person appearing on behalf of the tenants.

The committee had before it, as I have indicated, the report from the landlords' expert. They inspected the premises. They listened to argument in respect of the condition of particularly one of the flats which was under discussion, namely the second-floor flat. A reading of their decision shows that they did take many, if not all, of the material matters concerning the individual properties into consideration. In para 6 they started to give their reasons for their decision. They said:

> When we considered the rents we had regard to all the submissions made to us.

A We took the view *in the circumstances of this case* that we should have regard both to the market rent evidence in respect of the third-floor flat at 22 Coleherne Road and to the fair rent so recently determined with seven other flats in the terrace, one of them in the same building as the flats before us. All the fair rents must have been determined in the light of the prevailing open market rents and scarcity. The two variables to which Morritt J referred to in his judgment.

In parenthesis that is a reference to the case which has been much ventilated before me of *Spath Holme Ltd* v *Chairman of the Greater Manchester and Lancashire Rent Assessment Committee* decided in the Court of Appeal*. The judgment being delivered on July 19 1995.

B The committee went on:

> To ignore the seven relevant fair rents of which we had knowledge in favour of the one undeniably irrelevant market rent would, in our view, have been as perverse as to have ignored the market rent. From our knowledge and experience of *similar cases in the locality* it is our opinion that the level of scarcity of residential accommodation, particularly of small unfurnished flats, is as high in the South Kensington and Earl's Court area as anywhere in London. While we do not feel able on the evidence before us to put a precise figure on scarcity, as Mr Dickinson invited us to do, we are quite satisfied that there is a high degree of scarcity from which a substantial discount from the market rent must be applied.

C The decision continues. It is not necessary for me at the moment to read the rest of it. But it is right that the committee referred expressly to section 70 of the Rent Act 1977. They considered the differential applied by the rent officer between the rents of the two flats under review to have been insufficient. They dealt with the condition of the second-floor flat. Finally, they said:

> Accordingly, we had regard to all the evidence before us, to our inspection, to our knowledge and experience and to the provisions of section 70 of the Rent Act 1977 and we determined the fair rents to be registered exclusively to the rates which are set out in the schedule to our decision.

D Those figures I have already given. It must be remembered that section 70 reads as follows:

> (1) In determining for the purposes of this part of this Act what is or would be a fair rent under a regulated tenancy of a dwelling house regard shall be had to all the circumstances other than personal circumstances and, in particular, to
> (a) the age, character, locality and state of repair of the dwelling house.

The considerations which the rent officer and a committee on appeal have to take into account are wide indeed.

Next it must be remembered that subsection (2) says:

> For the purposes of the determination it will be assumed that the number of
> E persons seeking to become tenants of similar dwelling houses in the locality on the terms of the regulated tenancy is not substantially greater than a number of such dwelling houses in the locality which are available for letting on such terms.

That is the case which must be looked at in equilibrium. That is the subsection which requires the scarcity element to be taken into account. Then there is the disregard in subsection (3) to which I do not need to refer to in detail.

The rent officers and the committee therefore have a wide remit and must take all the circumstances into account. What is argued in this case is very simple, in essence, in connection with the main ground of
F decision made by this committee. It is said by Mr Rajinder Sahonte that they have erred in law because they have failed to follow the rule which is that they should consider open market rent discounted as the primary measure of comparability. That is a short-circuit way of setting out what Glidewell LJ and his brothers said in the case of *Spath Holme Ltd*. It is perfectly true that is the basis upon which cases of this kind must be approached where comparables are under consideration. It is however, at the same time, very important to note, as Mr Michael Furness asks me to do, that in the *Spath Holme* case the committee ignored and cast aside the open market comparable and looked simply at the other alternative comparables which were put before them. It is G
not necessary to read any particular passage in order to make that good. But that, plainly, is the situation. I believe that both sides accept that is the basis upon which the decision was made.

In the present case both the surveyor for the applicants, Mr Rupert Dickinson, and the committee themselves, referred specifically to the *Spath Holme* decision. It must have been and was before the committee not necessarily in detail but certainly in essence. Indeed, the committee set out, as I have indicated already in para 2 of their decision, the words of Morritt LJ who said that:

> Market rent adjusted for scarcity is precisely what the fair rent is required to be. H

Mr Sahonte says that in this case the committee have gone astray because while they have indicated that they have looked at the comparable premises and the rent which was applicable to those premises, they also looked at the fair rents in respect of the seven other premises which they were not entitled to do. The heart of Mr Sahonte's complaint is that this committee allowed itself to be diverted from its main task which was to look at the market rent and the discount from the market rent and virtually to ignore the fair rents which had been fixed in respect of the other premises, one of which was in the very same block as one of the two flats under consideration. J

I do not believe that *Spath Holme* or any other case indicates that that is, as matter of law, the only approach which the committee can have to a case of this kind. Indeed, I look at p11 of the decision in *Spath Holme*, which I have used in transcript form, OBCOF 95/044 8 and 9. I find the judgment of Harrison J and his analysis of the method in which a determination should be made. His decision was upheld by the Court of Appeal. It is interesting to note that what he said was:

> In determining that issue and having regard to the decided cases which I have quoted earlier in this judgment I consider that the following principles are applicable:
> 1. A fair rent under section 70 is the market rent adjusted for the scarcity element under section 72 and disregarding the personal circumstances mentioned in section 71 and the matters specified in subsection 73. K
> 2. There are various methods of assessing the fair rent including the use of registered fair rent comparables and the use of assured tenancy comparables.

It seems to me that there is support in that careful analysis by Harrison J of the way in which a case of this kind should be approached that both can be considered. I fully understand the argument set out by Mr Sahonte and that put forward into *Spath Holme* that it is wrong for a committee simply to throw overboard the market rent comparable which is put before them in favour, clearly, of fair rent comparables. But that, in my judgment, is simply not what this L
committee did. This committee was faced with the oddity of the figures to which I have already referred. It was faced with apparently a very high market rent, discounted by 15%, even then still being £2,500 above the rent being argued for. It had before it and had considered, with the assistance of the advocate, seven relevant fair rents of which the committee indicated that it had knowledge. The committee said, as I have already quoted:

> To ignore the seven relevant fair rents of which we had knowledge in favour of the one undeniably relevant market rent would, in our view, have been perverse as to have ignored the market rent. M

They were indicating there not that they were throwing one out of the window and considering only the others but that they accepted that the market rent which was under consideration was undeniably relevant, but said that the fair rent must be relevant too. Once they had done that and once they had applied their own knowledge and experience of similar cases in the locality to the case, which they were fully entitled to do, they then went on to consider the question of scarcity when looking at the whole picture.

Mr Sahonte complains that their approach to scarcity can be criticised on the basis that they looked at the matter purely in terms of demand. I am unable to accept that argument. It seems to me that while

*Editor's note: Reported at [1995] 2 EGLR 80

A the language of the committee, by reference to small unfurnished flats in South Kensington and Earl's Court may be, to some extent, subject to criticism, their approach overall to the question of scarcity was perfectly fair. They knew what the principle was which had to be applied. The matter had been argued on a 15% basis in connection with the comparable by the surveyor on behalf of the landlords. They were experienced in applying the discount for scarcity to figures which were put before them because that is part of the experience of committees operating in this field. I see no error in law in their approach in connection with scarcity. How much they discounted in connection with scarcity is not identified. But, as the cases show, there
B is no need for a rent committee to show the mathematical working which they employ. Cases have been cited to me which, in my judgment, establish that beyond any peradventure. I do not need to name them because it seems to me that the basis of that argument on behalf of the respondents is unassailable. What the committee must do is to show that they have approached the case in the proper way. They must heed all the arguments that are put before them. They must follow the advice and instruction given to them in any case which is put before them. But, at the end of the day, provided they follow the principles set out and consider both the market rent discounted and the other comparables which they must unavoidably consider, they do not
C have to give fuller reasons than this committee gave for its own conclusions. At the end of the day what they are entitled to say is that they determined the fair rents as they conclude them to be in the final D
paragraph of their decision.

I am not impressed by the arguments therefore put before us by the appellants in this case. I do not accept that this committee blinded itself by ignoring part of the relevant consideration of comparables which it had to consider. It seems to me that both commonsense and, indeed, principle required them to start with the market rent discounted and then to go on to consider all other relevant matters. It would have been perverse to have refused to have paid any attention to market rent but it would also, in my judgment, have been absurd to pay no attention in relation to the fair rents in respect of seven properties, one of which as I have already indicated, was in the very house in which one of the E
flats under discussion was situated. I can find no basis for an attack in point of law upon this decision.

It is only if there is a legal error or a failure to approach the matter in the proper and legitimate legal terms under the Act that the decision can be attacked. In all the circumstances, while I am grateful to Mr Sahonte for his very clear argument, I do accept the points put forward on his behalf. I prefer the way in which the matter is put by Mr Furness both in his written and oral submissions. These appeals must be dismissed. F

Appeals dismissed.

A

Central Estates Ltd *v* Secretary of State for the Environment

COURT OF APPEAL

June 23 1995

GLIDEWELL and MORRITT LJJ and SIR JOHN MAY

B

Landlord and tenant — Rent review — Trigger notice — Interrelationship with break clause — Whether time of the essence for service of trigger notice — Whether landlord entitled to rent review

By a lease dated December 7 1971 office premises were let to the defendant tenant for a term of 42 years at a rent of £21,750 pa subject to review. The lease contained a break clause permitting either the landlord or the tenant to determine the lease by six-months notice at the expiration of the first 21 years of the term, namely September 28 1992. In respect of the relevant review period either landlord or tenant could require a review by serving a notice on the other party not less
C
than 12 calendar months before the commencement of the rent review period on September 29 1992. If the rent was not agreed then within six months of the service of the rent review trigger notice, the determination of the rent was to be referred to an independent surveyor. On April 14 1992 the landlord wrote a letter to the tenant requiring the rent to be reviewed. The tenant contended that time was of the essence, that the notice was too late and did not accept that the rent could be reviewed. The landlord's application for a declaration that time was not of the essence for having the rent review, and the letter was valid and effective, was granted by judge in the court below. The tenant appealed.

D

***Held*: The appeal was allowed. The trend of the cases involving an interrelationship between the rent review clause and a break clause was that time was of the essence for the rent review timetable. The terms of the lease did not give rise to meaningful distinctions from those cases. There was an interrelationship between the break clause and the relevant review period. Both the rent review, which could be upwards or downwards, and the break clause could be activated by either the landlord or the tenant. Following the triggering of a review for the relevant rent review period, either party could then operate the break clause. This mutuality required that time should be of the essence of the service of the rent review notice.**

E

The following cases are referred to in this report.

Al Saloom v *Shirley James Travel Service Ltd* (1981) 42 P&CR 181; [1981] 2 EGLR 96; [1981] EGD 404; 259 EG 420, CA

Coventry City Council v *J Hepworth & Son Ltd* (1982) 46 P&CR 170; [1983] 1 EGLR 119; 265 EG 608, CA

Legal & General Assurance (Pension Management) Ltd v *Cheshire County Council* [1984] 1 EGLR 102; (1983) 269 EG 40, CA

Metrolands Investments Ltd v *JH Dewhurst Ltd* [1986] 3 All ER 659; [1986] 1 EGLR 125; (1986) 52 P&CR 232; 277 EG 1343, CA

Richards (C) & Son Ltd v *Karenita Ltd* [1972] EGD 9; (1971) 221 EG 25

Samuel Properties (Developments) Ltd v *Hayek* [1972] 1 WLR 1296; [1972] 3 All ER 473; (1972) 24 P&CR 223; [1973] EGD 266; 225 EG 1749, CA F

Stephenson & Son v *Orca Properties Ltd* [1989] 2 EGLR 129; [1989] 44 EG 81

United Scientific Holdings Ltd v *Burnley Borough Council* [1978] AC 904; [1977] 2 WLR 806; [1977] 2 All ER 62; (1977) 33 P&CR 220; [1977] 2 EGLR 61; [1977] EGD 195; 243 EG 43 & 127, HL

This was an appeal by the tenant, the Secretary of State for the Environment, from the decision of Judge Baker QC, sitting as a judge of the Chancery Division, in an application by the landlord, Central Estates Ltd, seeking the determination of the validity of a rent review notice.

G

Michael Barnes QC and Mark Cunningham (instructed by the Treasury Solicitor) appeared for the appellant; David Neuberger QC and Erica Foggin (instructed by Harkavays) represented the respondent.

Giving the first judgment at the invitation of Glidewell LJ, **MORRITT LJ** said: This appeal from the judgment of Judge Baker QC, sitting as a judge of the Chancery Division, given on February 4 1994 concerns the question whether a notice to review the rent payable under a lease dated December 7 1971 given by Central Estates Ltd, as the landlord, to the Secretary of State, as the tenant, after the time expressly allowed for that purpose, is nevertheless valid as time was H
not of the essence of the provision of the lease. The judge decided that time was not of the essence for service of the notice so that it was valid. The Secretary of State appeals on the ground that the judge was wrong in law.

The lease was made on December 7 1971 between the predecessor in title of the landlord, Overly Properties Ltd, and the Secretary of State for the Environment and demised certain office premises in Gravesend to the Secretary of State for a term of 42 years from September 29 1971 at an annual rent of £21,750, subject to review in accordance with the provisions contained in the Schedule thereto.

The relevant provisions of the lease are as follows, clause 4(3): J

> If the Lessee shall desire to determine this Lease at the end of the first twenty-one years of the said term and of such his desire shall give not less than six calendar months previous notice in writing to the Lessors and shall pay all rents and substantially perform and observe all the covenants and conditions hereinbefore contained and on the part of the Lessee to be performed and observed up to such determination or if the Lessors shall desire to determine this Lease at the end of the first Twenty-one years of the said term and of such their desire shall give not less than six calendar months previous notice in writing to the Lessee then in either case immediately after the expiration of the said period of Twenty-one years this present Lease and everything herein contained shall cease and determine but without prejudice to any claim by K
> either party against the other in respect of any antecedent breach of any covenant or condition herein contained.

The schedule, so far as relevant, provides:

> The Lessors or the Lessee may require a review of the yearly rent to be paid by the Lessee to the Lessors during the period from the expiration of the fourteenth year of the said term to the expiration of the Twenty-first year of the said term (hereinafter after called "the second period") during the period from the commencement of the twenty-second year of the said term to the expiration of the twenty-eighth year of the said term (hereinafter called "the third period") during the period from the commencement of the twenty-ninth year of the said

A term to the expiration of the thirty-fifth year of the said term (hereinafter called "the fourth period") and during the period from the commencement of the thirty-sixth year of the said term to the expiration of the said term by effluxion of time (hereinafter called "the fifth period") respectively by serving notice in writing upon the other of them requiring such review not less than twelve calendar months before the commencement of the second period and/or the third period and/or the fourth period and/or the fifth period respectively PROVIDED THAT in case of failure to agree upon the yearly rents to be paid during each of such periods respectively within six calendar months after the service of any such notice then the said yearly rents shall be referred to the decision of an independent Chartered Surveyor.

B There then follow provisions directing that the rent should be assessed on the basis of an open-market rent as between a willing landlord and a willing tenant and it continues:

. . . and upon the agreement or determination of such yearly rents then the yearly rent payable during such respective periods shall be the sum of TWENTY-ONE THOUSAND SEVEN HUNDRED AND FIFTY POUNDS (£21,750) or the yearly rent so agreed or determined for the appropriate period whichever is the greater.

Upon the agreement or determination of the matters to be decided hereunder the parties shall endorse and produce to the other party true Memoranda of such agreement or determination on the original and counterpart of this Lease and such Memoranda shall at all times thereafter be conclusive evidence of the matters contained therein.

C The rent payable for the second period as so defined was increased to £65,000 as appears from the memorandum dated November 11 1985 indorsed on the lease. That period would expire on September 28 1992, which was also the date on which the term would determine if either party operated the provisions of clause 4(3). The interrelation of the break clause and the rent review procedure in respect of that date is shown by the following timetable:

September 29 1991 latest date for the service of a notice requiring a rent review.

March 29 1992 latest date for the service of a notice to determine the term and the date on which, if a notice activating the rent review D had been duly given, the period for agreeing an increased rent expires and thus the date on which a reference to the surveyor to determine the amount of rent for the third period occurs.

September 29 1992 the date on which the term determines or the third period commences.

On April 14 1992 the landlord wrote to the Secretary of State stating:

As you are aware, the rent payable under the lease falls to be reviewed with effect from 29 September 1992 for the ensuing seven year period. We shall be obliged if you will accept this letter as notice requiring such review to take E place.

The Secretary of State was not prepared to accept the letter on that basis on the ground that it was too late. Although a surveyor was appointed by the President of the Royal Institute of Chartered Surveyors on the application of the landlord he has taken no steps to ascertain the rent payable in respect of the third period in view of the contention of the Secretary of State.

The originating summons was issued on January 19 1993 and seeks declarations that time is not of the essence for service of a notice by the landlord requiring a review of the rent payable from the commencement of the 22nd year of the term and that the letter of April F 14 1992 was valid and effective for that purpose. Before the judge and this court counsel for the tenant accepted that the two go together in the sense that if time was not of the essence the notice was good even though given after the time by which a notice to determine the lease had to be served.

The judge referred to the principles considered and established in what he described as the classic decision of the House of Lords in *United Scientific Holdings* v *Burnley Borough Council* [1978] AC 904* and three subsequent reported cases, namely *Legal & General Assurance (Pension Management) Ltd* v *Cheshire County Council* (1983) 269 EG 40* *Metrolands Investment Ltd* v *JH Dewhurst Ltd* G [1986] 3 All ER 659† and *Stephenson & Son* v *Orca Properties Ltd* [1989] 2 EGLR 129. The reasons for his decision are succinctly set out at p15 of the transcript of his judgment in the following terms:

The real question, as I see it, is whether, as a matter of substance, and not of form, the break clause is correlated or interrelated with the review clause, or associated with it, as a relief for the tenant against the consequences of the review. It is not enough that the result of the review may have some impact on the tenant's decision to operate the break clause if the clause is there for some other purpose.

Here we have a long lease of 42 years with one mid-term break. It is significant that that break can be operated by either the landlords or the tenant. H That sets it off from the normal type of break clause associated with rent reviews. From that fact, one deduces that the purpose of that break clause is to allow either party to determine the lease for reasons which have no necessary connection with the rent review. One could easily visualise, in the context of a 42-year lease, that the tenant may no longer need the property or the landlord may wish to get it back because it would make it more profitable by some sort of refurbishment or redevelopment which the passage of time shows to be desirable.

It can really be conceded that at the time when the tenant come to exercise the break, the prospect of the amount of the rent for the next seven years may be a factor in influencing his decision. By no means will it be the sole or even most important factor in influencing his decision in this context. And if it is, here, exceptionally the tenant could trigger the review to find out the likely J new rent.

Although I have not overlooked the fact that the landlords are in default and served their notice after the time that the tenant could have broken the lease, nevertheless, as it seems to me, time is not of the essence and I shall make the appropriate declaration.

The Secretary of State contends that the judge was wrong because his conclusion is contrary to the general principle established by the House of Lords in the *United Scientific Holdings* case and the specific example of its application given by Lord Diplock in his reference to *Samuel Properties (Developments) Ltd* v *Hayek* [1972] 1 WLR 1296 and in a number of other authorities. The landlord contends that all K those cases are distinguishable and that the judge correctly concluded that time was not of the essence in relation to the service of a rent review notice under this lease.

It is common ground that time is of the essence for the service of a notice to determine the lease under clause 4(3). With regard to the notice triggering a rent review the basic principles are not in dispute. The starting point is the formulation of the principle in the speech of Lord Diplock in *United Scientific Holdings Ltd* v *Burnley Borough Council* [1978] AC 904 at p930G where he said:

So upon the question of principle which these two appeals were brought to L settle, I would hold that in the absence of any contra-indications in the express words of the lease or in the interrelation of the rent review cause itself and other clauses or in the surrounding circumstances the presumption is that the time-table specified in a rent review clause for completion of the various steps for determining the rent payable in respect of the period following the review date is not of the essence of the contract.

The Secretary of State contends that the presumption is displaced in this case because of the interrelation of the rent review clause in respect of the third period as defined by the lease and the break clause. Likewise it is not disputed that whether the interrelation is sufficient is a question of construction. As Slade LJ giving the judgment of the M court said in *Metrolands Investments Ltd* v *JH Dewhurst Ltd* [1986] 3 All ER 659 at p668G:

Throughout this debate, we think it must be borne in mind that the ultimate object of the court in construing a rent review clause, like any other contractual provision, must be to ascertain the parties' intention from the particular words which they have used to express those intentions, read, of course, in the light of any admissible evidence as to surrounding circumstances; albeit with the assistance of the guidelines as to construction afforded by earlier authorities. In

*Editor's note: Also reported at [1977] 2 EGLR 61

*Editor's note: Also reported at [1984] 1 EGLR 102

†Editor's note: Also reported at [1986] 1 EGLR 125

A many, perhaps most, cases, of which the present is one, the rent review (if any) can result only in an increase, and thus is only for the landlord's benefit.

Essentially, therefore, the question to which the court has to direct its mind is this: is the proper intention to impute to the parties, from the words which they have used, the intention that the landlord shall lose his right to a review if the stipulated timetable is not strictly adhered to in the relevant respects?

The Secretary of State submits that the decided cases show that time is of the essence where the lease lays down a timetable which gives to the tenant a period after the service of the rent review notice, whether or not the new rent is determined in that period, before he has to decide whether to serve a notice to determine the tenancy. That, it is
B submitted, constitutes a sufficient interrelationship and is present in this case.

For present purposes the cases to which we have been referred fall into three categories. In the first are those in which the rent review clause is triggered by a notice from the landlord only, is upward only and is followed by a period in which the tenant, to whom alone the right is given, may consider whether or not to serve a notice to determine the term. In this category fall *C Richards & Son Ltd* v *Karenita Ltd* (1971) 221 EG 25; *Samuel Properties (Developments) Ltd* v *Hayek* [1972] 1 WLR 1296; and *Coventry City Council* v *J Hepworth & Son Ltd* (1982) 46 P&CR 170*. In the second of those
C cases there was also a clause which extended the time for the giving of a notice to determine the lease until a period after the increased rent had been determined. Of the first two of these cases in *United Scientific Holdings* v *Burnley* Lord Diplock at p935H said:

> *Samuel Properties (Developments) Ltd* v *Hayek* [1972] 1 WLR 1296 may be regarded as the origin of the dichotomy between "option" on the one hand and "obligation" or "machinery" on the other: the word option having been used in the lease itself to describe the landlord's right to require the rent to be reviewed. It should be treated as overruled. There was a complication in that the rent review clause was associated with a break clause which gave to the tenant the right to surrender the residue of the term on any rent review day by
> D giving prior notice. The time-table in the rent review clause for the determination of the new rent was obviously correlated with the time by which the tenant had to give notice of his intention to surrender, so as to enable him to make his decision whether or not to exercise that right in the knowledge of what the new rent would be if he continued in possession after the review date. Had that been all, as it had been in the previous and rightly decided case of *C Richards & Son Ltd* v *Karenita Ltd* (1971) 221 EG 25, it would, I think have been sufficient by necessary implication to make time of the essence of the rent review clause because of its inter-relation with the time by which notice was to be given under the break clause — a time which, for reasons I have given earlier, I consider to be of the essence of the contract.
>
> In *Samuel Properties (Developments) Ltd* v *Hayek*, however, the break
> E clause itself contained a provision under which the period during which the tenant could exercise his right to surrender would be extended in the event of the reviewed rent not having been ascertained within the time stipulated in the rent review clause. So the implication that would otherwise have arisen from the association of the rent review clause with a break clause was negatived.

In the light of that observation in the third case, namely *Coventry City Council* v *Hepworth*, Lawton LJ said of *United Scientific Holdings Ltd* v *Burnley* at p176:

> It comes to this, that their Lordships (though the matter was not immediately in point and was technically *obiter dicta*) were all of the opinion that, where you
> F have a triggering off of a rent review provision started by the landlord followed by an option given to the tenant to break the lease if he so wishes, then time is to be presumed to be of the essence of the agreement unless there are contra-indications.

The second category of case is that in which the timetable laid down in the lease requires the review notice to be served not later than the last date on which a break clause might be served. In these cases the former might only be served by the landlord in respect of an upwards-only rent review and the latter might only be served by the tenant. The cases to which I refer are *Al Saloom* v *Shirley James* G
Travel Service Ltd (1981) 42 P&CR 181* and *Legal & General Assurance (Pension Management) Ltd* v *Cheshire County Council* (1983) 269 EG 40. In the former Waller LJ said at p185:

> As I have already said, both provisions about break and rent review in this case were in the same clause and closely allied to each other. The words "not less than six months' notice in writing prior to the expiration of the said third year of the term" were used for both break and rent review. Whether the word "interrelated", "correlated" or "associated" is used, the implication of the use of the same phrase in the same clause is overwhelming. The phrase must mean the same in each case, either that in both time was of the essence, or in neither. In my view the phrase could not change its meaning in the course of 10 lines H
> unless there were some qualifying phrases to make such a change clear. I have no doubt that in both cases here time was of the essence;

In the latter Dillon LJ said at p41:

> But I would add this. It is pointed out, and stressed, by Mr Colyer that where the dates of a break clause and a rent review clause are coincidental, there is always a possibility that the landlord's service of a rent review notice may come too late — even if in time — for the tenant to be able to take into account and act on it in serving a break notice. That is theoretically possible if the astute landlord wants to take the chance of serving his rent review notice at the last possible permitted moment, or if by some accident a notice properly served in accordance with the notice provisions of section 196 of the Law of J
> Property Act 1925 does not come into the tenant's hands in due time. But most landlords who wish to invoke rent review procedures are apt to serve in comfortable time before the expiration of the permitted period. Therefore the learned judge in the present case was fully entitled, in my view, to say, as he did in his judgment: "I think the tenant could reasonably expect that the notice would be given in sufficient time for him to decide whether he wanted to break or not."
>
> The reason for having coincident dates is obviously that the tenant should have in mind, when he decides whether he is going to break or not, whether or not the landlord is going to invoke rent review procedure, and the reasoning which has led the courts to hold that time is of the essence of a rent review clause where the tenant's break clause in the same lease is geared to a date later than the date prescribed for the service of the landlord's rent review K
> notice leads equally, I think, to the conclusion which this court reached in the *Al Saloom* case and I would reach in this case, that time was of the essence of the rent review clause in this lease.

The third category comprises *Stephenson & Son* v *Orca Properties Ltd* [1989] 2 EGLR 129 in which either party might trigger the rent review in which the rent might be increased or reduced but only the tenant might serve a notice to determine the lease. The date by which the two notices had to be served was the same so that this case was in that respect similar to those in the second category. The landlord served a review notice late and sought to distinguish the cases in the L
second category in the manner described by Scott J at p132J in the following terms:

> First, the rent review machinery in the lease with which I am concerned can be triggered off either by the landlord or by the tenant. In the other two leases only the landlord could do so. Second, the rent review under the lease in the present case may lead to a reduction as well as to an increase in the rent. In the other two leases a reduction in the rent was not a possible consequence of the rent review. Third, the juxtaposition of the rent review provision and the break option is different in the present case from the corresponding juxtaposition in the other two leases.
>
> In my judgment, however, none of these points is a basis of legitimate distinction of the two Court of Appeal authorities. M
>
> The third point is, I think, of no weight whatever. The first two points do not, in my view, incline against the conclusion reached by the Court of Appeal in the two cases I have mentioned that time should be of the essence of the rent review provision. The decisions in the two Court of Appeal cases require me, in my view, to hold that time is of the essence of the rent review provision in the present case. The construction and effect of leases, and in particular leases of commercial premises, ought not, in my view, to be subjected to variation caused by minor and apparently unimportant differences in language. Certainty is highly important. It would be a disservice to the law if I were, in reliance on such points of distinction as have been relied on in the present case, to give to

---

*Editor's note: Also reported at [1983] 1 EGLR 119

*Editor's note: Also reported at [1981] 2 EGLR 96

A the rent review clause in the present case a different effect to that given to the rent review clauses in the two Court of Appeal cases in question.

Metrolands Investments Ltd v *JH Dewhurst Ltd* [1986] 3 All ER 659 did not concern a rent review notice. In that case the right to determine the lease was given to the tenant during the period of three months commencing on the date by which the lease stipulated that the increased rent for the ensuing period should have been determined by an arbitrator. It was held that time was not of the essence of the determination of the rent for the interrelation between that determination and the break clause was not sufficient to displace the presumption. The essential ground for distinction was that the date of
B the determination of the rent by the arbitrator was not within the full control of the landlord. In giving the judgment of the court Slade LJ at p671C said:

> Though there is the clearest possible interrelation between the timetable embodied in the rent review clause and that embodied in the break clause, such interrelation in the context of this particular lease is not, in our judgment, a "contraindication" sufficient to rebut the initial presumption of construction (from which we think it right to start) that the timetable specified in the rent review clause for the obtaining of the arbitrator's decision is not of the essence of the contract. The principal reasons why it does not suffice for this purpose are two, which must be considered in conjunction with one another, and have
> C not been applicable in conjunction with any previous case which has been cited to us.
>
> First, the date of the relevant event, namely the actual obtaining of such decision, is one which is to a substantial degree outside the lessor's control. Second, any potential hardship to the tenant which might otherwise arise through tardy action by the landlord in initiating the rent review procedure can be eliminated or, at least, substantially mitigated, by the tenant initiating such action itself. The ultimate object of the court in construing clauses such as this, as we have already indicated, must be to ascertain the parties' intentions from the words which they have used. Weighing in the balance all the competing considerations that have been urged on us, we have come to the clear conclusion that the intention properly to be attributed to the parties to this
> D particular lease is the intention that time should not be of the essence as regards the obtaining of the arbitrator's decision.

In my view, the clear trend of all these cases supports the proposition for which the Secretary of State contends, namely that the presumption is displaced and time is of the essence if the timetable laid down by the lease allows a period after the service of the review notice in which to consider whether or not to serve a break notice for if time is not of the essence the period allowed for the service of the break notice as to which time is of the essence is eliminated or eroded.

The essential question is whether the terms of this lease give rise to any meaningful distinction. A number of points were relied on which I
E will seek to deal with in turn.

First it was emphasised that this lease provides for four rent reviews, but only one break. This certainly distinguishes this case from all the others to which I have referred. Counsel for the landlord relied on the observations of Waller LJ in *Al Saloom* v *Shirley James Travel Service Ltd*, which I have already quoted. If the question is one of construction how can the words change their meaning depending which review they are applied to. Counsel for the Secretary of State did not shrink from one consequence of this approach and accepted that if it were necessary to do so he would contend that time was of the essence on all reviews notwithstanding that there was only a link with
F the break clause on one of them.

I do not accept that this point of distinction justifies the conclusion that time was not of the essence for service of the notice to trigger a review of the rent payable during the third period. The principle described in the speech of Lord Diplock is whether the interrelation between the rent review clause and the other provisions of the lease displaces the presumption that time is not of the essence. The relevant interrelation only exists in respect of the third period. I see no reason why in relation to that period for which the context is different the same words should not bear a different meaning to that which they bear in respect of the other periods. The question is therefore whether in respect of the third period the interrelation is sufficient.

This raises the second point of distinction. In this case unlike the G
others to which I have referred both the rent review, which may be upwards or downwards, and the break clause may be activated by either the landlord or the tenant. It is submitted that as either party considering whether to determine the lease could require a rent review if knowledge of the future rent was required there was no reason why he should not initiate the review. In this connection reliance was placed on the second reason referred to by Slade LJ in *Metrolands Investments Ltd* v *JH Dewhurst* in the passage I have quoted. Thus, it was submitted, that party did not need the period for which the lease provided in order to ascertain whether the rent was to be reviewed or not. H

I do not accept this as a distinction which points to a different result either. A tenant will seek a rent review if he thinks that the rent should be reduced; likewise the lessor will seek a review if he thinks it should be increased. Neither will seek a review just for the sake of it. In each case the other party is entitled to give notice to determine the lease. Thus as the rent review may be both up and down and either party may trigger its operation and give notice to determine the term the effect of the timetable contained in the lease is to confer on both parties that which in the first category of case to which I have referred was conferred on one of them only, namely the tenant. In my judgment, this mutuality requires that time should be of the essence for the J
service of the rent review notice by whichever party it is served, in the same way and for the same reasons as in the cases in the first category, time is of the essence of the landlord's notice triggering the rent review because of its relation to the tenant's right to determine the term.

The remaining point is the same as that which failed to impress Scott J in *Stephenson & Son* v *Orca Properties Ltd*, namely juxtaposition. In my view, to conclude that this was a valid distinction would be to put form before substance. The point was not pressed and it is unnecessary to say any more about it.

Counsel for the landlord also sought to address arguments as to what the rent would be if time is of the essence so that there is no valid K
rent review for the third period. This point does not arise on this originating summons, but for the sake of the argument I will assume that the rent will revert to the initial rent of £21,750. In this event the landlord submits that the detriment to him and the incentive to the tenant to do nothing in the hope that the landlord will not serve notice in time is in each case that much the greater so that the intention to achieve that result by making time of the essence cannot be attributed to either party. I do not find this argument compelling for it is one of degree and is only valid if the market is rising; if the market is falling then the converse operates and produces a detriment to the tenant and an incentive to the landlord. L

For these reasons I accept the arguments for the Secretary of State. In my judgment, the judge reached the wrong conclusion. In my view, the reason he did so is that the principle he sought to apply, namely:

> . . . whether, as a matter of substance, and not of form, the break clause is correlated or interrelated with the review clause, or associated with it, as a relief for the tenant against the consequences of the review. It is not enough that the result of the review may have some impact on the tenant's decision to operate the break clause if the clause is there for some other purpose

is a much more stringent test than any of the authorities justifies. In particular, in all the cases in which time has been held to have been of M
the essence, the result of the review would do no more than have some impact on the tenant's decision whether to operate the break clause, and in most of them the tenant would know little more than that the landlord was seeking a review; but he would at least have the opportunity to obtain advice before deciding whether to serve a notice to determine the term. For the reasons I have tried to explain I would allow this appeal.

Agreeing, **SIR JOHN MAY** said: I gratefully adopt Morritt LJ's summary of the terms of the relevant lease in this appeal and of the material facts and relief sought. As he has said, the question for us, as with the learned judge below, is whether time was of the essence for

A the service of the notice by the landlord requiring a review of the rent payable from the beginning of the 22nd year of the term granted by that lease. Morritt LJ has also referred to the material decided cases to which our attention was drawn but to which I need not now refer further.

Were it possible to consider the question which arises in this appeal in isolation, then there was clearly a sufficient interrelation between the rent review procedures and the provisions of the lease entitling either party to give notice to lead to the determination of the term granted at the end of the first 21 years thereof, to require us to hold on the authorities that time was of the essence for the service of the B landlord's notice requiring a rent review at that time.

However if this would have been the correct answer in law to the limited question I have posed, then on equally well established principles time should also be held to have been of the essence in relation to the initiation of any other rent reviews under the terms of the lease. As a matter of pure construction, if time was of the essence with respect to one rent review notice under the lease, then *prima facie* at any rate, time should also be of the essence with respect to any other rent review notices under the same provisions of the same lease.

But it is equally clear, on the authority of the *United Scientific Holdings* decision in the House of Lords, that in the absence of any C contradictions in the express words of the lease, or in the interrelation of the rent review clause and other clauses, for instance the clause enabling either party to determine the lease after 21 years, or in the surrounding circumstances, then the presumption is that time is not of the essence in construing the timetable specified by the rent review clause.

In the present case, while there is without doubt a sufficient interrelation or association between the clause enabling the service of a notice to determine the lease and the rent review clause providing for a review on the same date, as I have said that interrelation or association does not exist in any real sense if one is considering the mechanics D leading to a rent review at any other date.

At the hearing of this appeal, counsel on each side clearly appreciated the logical difficulty to which the law and the arguments which I have outlined gives rise. Mr David Neuberger QC, for the respondent landlord, relied upon it as substantial support for his arguments. He asked rhetorically how time could be held to be of the essence of one particular rent review procedure but not of any other.

For the appellant tenant, Mr Michael Barnes QC, was understandably reluctant to commit himself about how this potential illogicality should be resolved. If pressed, however, I think that he was ready to accept any resolution of the illogicality provided that we held that time continued to be of the essence of the material procedure E taking effect at the end of the 21st year of the term.

Viewed as a question of construction there can only be three possible answers. One could conclude that the parties intended time to be of the essence of the contract in so far as the rent review procedures at the end of the 21st year of the term were concerned, but not so in so far as the other rent review procedures were concerned. For my part, I find it difficult to impute such a split intention to the parties.

One is therefore left with the two remaining possibilities. Either the parties' imputed intention was that time was of the essence on each occasion that the rent review procedures were operated, or it was that it F should never be of the essence in any such procedures.

In my experience, these cases involving questions of construction frequently give rise to artificiality, particularly when, as the law requires, one seeks to impute an intention to the parties when in truth they never applied their minds to the question at all. In reality it will be apparent that at least for the purposes of this and similar cases the courts have effectively developed certain rules of thumb to help them reach a decision about issues of construction. For my part, I think that this is a preferable approach, but for the time being I must continue the artificiality of seeking the parties' intention. Doing so I respectfully do not think that one is helped by a meticulous comparison of the facts, circumstances and contract terms with those of the many similar cases to which we have been referred.

Faced however with the stark choice between time being of the G essence on all or alternatively on none of the occasions when the rent review procedures were operated, I think that I am forced to the conclusion that had the parties been asked some 24 years ago what they really intended, they would have answered that time should be of the essence on all such occasions.

For these reasons I would, like Morritt LJ, allow this appeal with the results that he has suggested.

Also agreeing, **Glidewell LJ** said: At the conclusion of his speech in *United Scientific Holdings Ltd* v *Burnley Borough Council* [1978] AC 904 Lord Diplock said at p936G: H

> I would express the hope that your Lordships' decision in these appeals will reduce the number of occasions on which it will be necessary to have recourse to the courts in order to ascertain whether delay has deprived the landlord of his right to have the rent reviewed under particular rent review clauses.

This appeal and the decisions of this court and of courts below to which we have been referred during the course of argument show that that hope was over-optimistic. Nevertheless, there is no doubt that the principles to be applied in our decision are those to be derived from the speeches in that case. My lord, Morritt LJ, has already set out, and I do J not need to repeat *in extenso*, the quotations upon which I would rely from the speech of Lord Diplock at pp930G and 935 over to p936. I add, however, some passages from the speeches, first, of Lord Simon of Glaisdale at p946 where he said:

> However, where a rent review clause is associated with a true option (a "break" clause, for example), it is a strong indication that time is intended to be of the essence of the rent review clause — if not absolutely, at least to the extent that the tenant will reasonably expect to know what new rent he will have to pay before the time comes for him to elect whether to terminate or renew the tenancy (*cf Samuel Properties (Developments) Ltd* v *Hayek* [1972] 1 WLR 1296). That situation stands in significant contrast with those in the instant appeals. K

I refer also to Lord Fraser of Tullybelton who said at p962F:

> For these reasons I am of the opinion that the equitable rule against treating time as of the essence of a contract is applicable to rent review clauses unless there is some special reason for excluding its application to a particular clause. The rule would of course be excluded if the review clause expressly stated that time was to be of the essence. It would also be excluded if the context clearly indicated that that was the intention of the parties — as for instance where the tenant had a right to break the lease by notice given by a specified date which was later than the last date for serving the landlord's trigger notice. The tenant's notice to terminate the contract would be one where the time limit was mandatory, and the necessary implication is that the time limit for giving the L landlord's notice of review must also be mandatory. An example of such interlocked provisions is to be found in *C Richards & Son Ltd* v *Karenita Ltd* (1971) 221 EG 25.

Whether the application of those principles leads to the conclusion that the time specified in the lease by which a notice to review a rent must be served is or is not of the essence must of course depend on the terms of the lease in question. Nevertheless, in the light of the express approval by their lordships in *United Scientific Holdings* of the terms of the lease the subject of Goulding J's decision in *C Richards & Son Ltd* v *Karenita Ltd* (1971) 221 EG 25 as a paradigm of a rent review clause in which time is of the essence, the more nearly the terms of a M rent review clause in another lease approach to the lease in that case, the more likely it will be that the time for service of the notice will be held to be of the essence.

There are the following differences between the terms of the lease in the present case and those in *Richards* v *Karenita*.

First, the present lease provides for four rent reviews but only one break half-way through the term. Whether this leads to the conclusion that, if time is of the essence, it is of the essence only in relation to the review at the same time as the break clause, the review with which this appeal is concerned, (which, in my view, is a possible though unusual construction) or whether the conclusion must be as my lord, Sir John

May, prefers that it necessarily applies to all four reviews, for my part I do not find it necessary to decide. The fact is that the late service of the notice claiming the rent review has deprived the tenant in this case of any opportunity to consider the review before deciding whether to put an end to his lease. This is the very point referred to in the various speeches in *United Scientific* to which I have made reference.

The second point of distinction is that this lease provides for both review or a claim for a review of the rent and the opportunity to break the lease to be available both to the landlord and to the tenant. But like my lord, Morritt LJ, I regard that as a distinction which makes no difference. Either party in that case has the opportunity to serve a notice claiming a rent review if it is thought it would be of its benefit to do so. If one party serves such a notice late then the other is inevitably disadvantaged by having either less time or on the facts of this case no time at all to consider whether to exercise the right to terminate the lease at the half-way stage.

The point which appears to have been the main basis of Judge Baker's decision is that to which Morritt LJ has referred where the judge said at p15 of the transcript of his judgment:

> The real question, as I see it, is whether, as a matter of substance, and not of form, the break clause is correlated or interrelated with the review clause, or associated with it, as a relief for the tenant against the consequences of the review. It is not enough that the result of the review may have some impact on the tenant's decision to operate the break clause if the clause is there for some other purpose.

And then at the end of that page:

It can readily be conceded that at the time when the tenant comes to exercise the break, the prospect of the amount of the rent for the next seven years may be a factor in influencing his decision. By no means will it be the sole or even most important factor in influencing his decision in this context. And if it is, here, exceptionally the tenant could trigger the review to find out the likely new rent.

As Mr Barnes points out in his argument that situation must be common to all cases involving leases which include both a rent review clause and a break clause. For my part, therefore, I do not find it a proper basis for the conclusion to which the learned judge came.

In my judgment, unless I am bound by authority to the contrary, the pattern of the lease in this case is such that the clause governing the review of the rent for the third period from September 29 1992 is one in which the time specified for the service of the notice is of the essence.

Is there anything in the decisions of this Court subsequent to *United Scientific Holdings* which require me to reach a different conclusion? The only such decision in which this court concluded that time was not of the essence in somewhat similar circumstances was the decision in *Metrolands Investments Ltd* v *Dewhurst Ltd* [1986] 3 All ER 659. However the provisions of the rent review clause in that case are so far from those in the present case that Mr Neuberger disclaims any reliance upon the decision based on its facts. The principles of that decision are to be deduced from passages to which, most helpfully, emphasis has been given in the report itself. Slade LJ giving the judgment of the court said at p669C:

> . . . in a case where a lease contains a break clause as well as a rent review clause and the timetables of the two clauses are closely interlocked, the interrelation of the two clauses is *likely* to suffice as a contra-indication sufficient to rebut the ordinary presumption; though everything must depend on the wording of the particular lease.

Then at p670A:

> There is a basic difference in the provisions of the lease in the presence case from those under consideration in *C Richards & Son Ltd* v *Karenita Ltd* and in the *Coventry City Council* case. In those two cases, the leases had one most important feature in common, that is to say" (emphasis given in the judgment) *the event as to which time was held to be of the essence was one in the landlord's full control* (namely service of his trigger notice). In a case where an event included as part of the timetable in a rent review clause is one within the landlord's full control, it is by no means inconceivable that he should have been willing and should have intended to commit himself to a rigid adherence to that part of the timetable, even though the consequences of a failure to do so would be that of losing the entire benefit of the relevant rent review and even though the lease does not expressly state that time is to be of the essence.
>
> In contrast, the lease under consideration in the present case has this special and, we would think, unusual feature: the event as to which time is said to be of the essence is the (emphasis again) *actual obtaining of the arbitrator's decision.*

Here, the service of a rent review notice was in the landlord's full control (albeit that the tenant was also entitled to and thus had control over the question of whether he should serve the notice) but unlike the *Metrolands* case and distinguishing it from the reasoning to which I have just referred this case is one where the landlord was entirely able to activate the rent review.

I therefore find nothing in that decision which binds us to adopt a conclusion different from that which, seeking to apply the principles of *United Scientific Holdings*, I have already indicated I arrive.

For those reasons which do no more than add to those contained in the judgment of Morritt LJ, with which I agree, I would therefore allow this appeal.

Mr Neuberger has urged us to express a view about the proper interpretation of the lease in relation to the rent for the current period in the light of our judgment. As is apparent, I have not found it necessary to decide this question in order to reach my decision on the issue in the appeal, and I must therefore disappoint him by declining the request. The appeal is therefore allowed.

Appeal allowed.

For further cases on this subject see p99

LANDS TRIBUNAL
COMPENSATION

A
King and others *v* Dorset County Council

February 26 1997

JUDGE MARDER QC, President

Compensation — Land Compensation Act 1973 Part I — Alterations to road — Whether resurfacing of road an alteration — Whether evidence of acoustics experts helpful — Determination of compensation

B The claimants were the owners of two dwellings in Wimborne Minster. On July 14 1991 a distributor road was first opened to traffic which had been created by works of alteration to existing highways, including certain works outside the subject properties consisting of a new surface, provision of a regulating material and slight alteration to the profile of the road. The altered road was respectively 17.6m and 12.5m from each of the subject properties, the claims being £62,100 and £61,500 based on 90% of total depreciation in each case. Evidence on noise was given by the claimants and a noise expert. The compensating authority contended that that part of the altered road outside the properties had only been resurfaced and not altered within C the meaning of section 9(5) of the Land Compensation Act 1973.

Decision: **Compensation was awarded in the sums of £46,920 and £44,280, being 22% and 23% respectively of the unaffected values. The works done outside the subject properties were resurfacing; accordingly that length of road was not an altered carriageway and no account could be made of physical factors having their source of this length. However the altered highway was only a short distance away and discounts of 15% and 10% only were adequate and appropriate to reflect this. Little assistance was derived from the noise experts; it was the task of the tribunal to determine the depreciation (if any) in the value of D the claimants' interest and this was a matter for the market. Subjective impressions of the kind given in evidence of the increased traffic and noise would more strongly influence potential purchasers than scientific calculations of acoustics experts.**

The following cases are referred to in this report.

Broom v *Department of Transport* [1993] RVR 218, LT

Tucknott v *Department of Transport* LT Ref/206/93

Williamson v *Cumbria County Council* (1994) 68 P&CR 367;[1994] 2 EGLR 206; [1994] 47 EG 170; [1995] RVR 102, LT

E Gary Cowen (instructed by Jacobs & Reeve) appeared for the claimants; Anthony Ward (instructed by the solicitor to Dorset County Council) represented the compensating authority.

Giving his decision, **JUDGE MARDER QC** said: These are two claims arising under Part I of the Land Compensation Act 1973 which I heard together. They relate to Georgian town houses of some distinction situated on opposite sides of a street called West Borough, Wimborne Minster, and the claims arise in consequence of certain roadworks carried out by Dorset County Council in the vicinity of the subject properties.

At the hearing, Mr Gary Cowen of counsel represented the F claimants. He called the claimants Mr AJ Preselo ARICS and Mrs AV King; Mr Andrew Hurrell BSC MIOA Dipl in Acoustics; and Mr Robert Charles Gardner FRICS, a senior associate in Humberts, chartered surveyors of Salisbury. The compensating authority were represented by Mr Anthony Ward of counsel. He called Mr Gordon Frederick Cordier CEng MICE, chief engineer to Dorset County Council; Mr Simon Butler BSC MIOA Dipl in Acoustics; and Mr Richard Walker BSC ARICS, a valuer in the estates dept of the county council.

To a substantial extent, the background facts are not in dispute and may be summarised as follows:

(1) The two subject properties are 46 and 45 West Borough. No 46 G is owned and occupied by Mr and Mrs King and is situated on the eastern side of West Borough, just north of the junction of West Borough and Priors Walk. No 45 is owned and occupied by Mr and Mrs Preselo and is on the western side of West Borough, almost opposite no 46 but slightly further north from the junction of Priors Walk. Both houses are Grade II listed 18th-century buildings fronting directly to the street and both are in effect semi-detached. Dimensions and floor plans are agreed. No 46 is a three-storey building with three reception rooms and three bedrooms on the main two floors and a fourth bedroom and storage in the attic space. The total reduced covered area (hereinafter RCA) is 198m². There is an attractive rear H garden.

No 45 is a two-storey end of terrace house, the ground floor arranged with a reception room and kitchen at the front and a further reception room and kitchen at the rear, three bedrooms and a bathroom at the front of the first floor, with a further two rooms and bathroom at the rear. The RCA is 246m². No 45 has a large garden at the side and rear with a return frontage to School Lane, but about one-half of the garden land is held under a terminable licence from the National Trust.

(2) The subject properties are located in a predominantly residential area a short distance north of the town centre of Wimborne, being the area around the minster, The Square and the High Street. Plans were J submitted and are annexed to this decision as appendix 1 and 2 [not reproduced here], showing the direction of the traffic flows in the vicinity of the subject properties before and after implementation of the Wimborne Distributor Road scheme.

(3) Proposals for implementation of the Wimborne Distributor Road scheme were contained in the Colehill & Wimborne Local Plan dated February 1984. Under the heading Roads, Traffic and Public Transport, para 3.8 read:

> The congestion and environmental damage caused by the heavy flows of traffic through the centre of Wimborne has been partially relieved by the construction of the Wimborne By-pass, which has removed much of the east- K west through traffic. However, substantial flows of north-south traffic remain as does traffic to and from the town centre itself. These flows are likely to increase over the period of the plan, bringing a return of many of the former problems to town centre streets after a short term improvement following the opening of the By-pass.

In chapter 4, setting out "Aims and Objectives of the Plan", under the heading transport, the local plan said:

> To relieve existing traffic problems by providing roads of suitable standard for the main traffic flows.

> To improve the environment of Wimborne town centre by measures to control and manage the flow of traffic so that wherever possible the vehicular traffic is removed or reduced in the central shopping area.

And in chapter 5, under the heading "Main Traffic Routes", para 5.14 said:

> The measures include the provision of a route around the town centre for a large proportion of traffic movements, based upon the Lewens Lane/Rolands Hill/Hanham Road route and utilising this major past investment to the full. This route will be improved to take two-way traffic and will be extended by a link from Rodways Corner to the Poole Road. In addition a new link will be provided for Blandford Traffic from West Borough to the eastern end of Stone Lane.

Chapter 7 of the local plan set out the phasing of implementation of these proposals, but it is unnecessary to refer further to this aspect, since the parties were agreed at the hearing that significant changes in traffic flows came about only after completion of the whole scheme. An agreed plan which is annexed as appendix 3 [not reproduced here] indicates the routing of the completed distributor road and shows the pre-scheme and post-scheme traffic flows with 1991 forecasts. West Borough, from the Priors Walk junction to the junction with Stone Lane, became part of the distributor road as a result of this scheme, this being the section in which the subject properties are located.

(4) It is agreed that in implementing the distributor road scheme, public works of alteration to the carriageway of the highway were carried out at the junction of Priors Walk and West Borough. These works included realignment of the junction, construction of a new traffic island, the installation of traffic lights and pedestrian barriers, and ancillary works of kerbing and resurfacing. It is further agreed that the nearest point of these agreed works to 46 West Borough is 12.5m (42 ft), and to 45 West Borough is 17.6m (58 ft). There is however a dispute between the parties as to whether certain works carried out to the north of the junction and immediately adjoining the subject properties constituted alteration of the carriageway within the meaning of section 9(5) of the 1973 Act.

(5) It is agreed that the distributor road scheme opened to traffic on July 14 1991 and that the first claim day and the appropriate date for valuation is July 14 1992. The basis for valuation of the claims is agreed to be that provided by the 1973 Act, expressed as the depreciation in value of each subject property brought about by the physical factors, defined as noise, vibration, smell, fumes, smoke and artificial lighting, caused by the use of the public works, and the source of which is situated on the altered highway, subject however to the limitation in section 9, namely that the claim is limited to such depreciation as would not have been caused but for the alterations in the highway.

(6) Before scheme and after scheme traffic flows, so far as they rest on 16-hour traffic counts in West Borough north of the Priors Walk junction are agreed as follows:

June 1989	All movements	9,449
	HGVs (over 3.5t)	423
July 1992	All movements	10,907
	HGVs (over 3.5t)	529

(7) Plans, photographs and details of sale transactions etc in respect of other properties referred to in evidence were agreed.

Claimants' evidence

I heard first the evidence of the claimant Mr Preselo. He is a chartered surveyor employed by the district valuer at Bournemouth. He and his wife had purchased 45 West Borough from the National Trust in April 1988. Mrs Preselo's family had been tenants of the National Trust in that house for generations previously. Mr Preselo produced an extract from a local guide describing West Borough at the time of purchase as the finest residential street in the town, and illustrating the particularly fine doorcase of no 45 which was known as Gulliver's House, after a notorious Dorset smuggler of the 18th century. The house had been readily adapted to accommodate the needs of Mr and Mrs Preselo and of Mrs Preselo's retired parents.

Mr Preselo produced plans showing the principal pre-scheme traffic flows through the town. Particular constraints to movement of large vehicles were the layout of the junction of West Borough and Prior's Walk and the very narrow and tortuous Blind Lane to the north of his property. Hence traffic moving between the north west (Blandford, Yeovil and Warminster) and the south and south east (Poole and Bournemouth) travelled through the town centre of Wimborne but did not use that part of West Borough. The southern section of West Borough was part of a one-way system which carried through traffic into Prior's Walk. Mr Preselo described the northern section of West Borough in consequence as a "lightly trafficked street with an air of calm about it". Traffic was free flowing without undue noise or congestion and the proportion of HGVs was not significant.

Mr Preselo then outlined the consequences of the scheme. In 1984 the new link from West Borough to Stone Lane was completed, and West Borough began to be used as a secondary route to and from the Blandford direction. However despite an increased traffic volume, at the time of purchase in 1988 West Borough remained a relatively traffic free road and the degree of traffic nuisance was small.

Mr Preselo had carefully researched the progress of the scheme and produced detailed evidence as to the correspondence and meetings which had been held concerning the anticipated effects of the scheme. I find it unnecessary to reproduce this part of his evidence, save to note the contention of Mr Preselo that the council's forecasts of traffic growth and effect were misleading and inaccurate, and his comment that if accurate forecasts had been produced and a clear indication given of the increase in HGV traffic, he and his wife would have taken steps to sell the property before the public works commenced.

As to the position after the scheme became fully operational (when the improvement works at the West Borough/Priors Walk junction were completed) Mr Preselo said that he and his wife had been shocked by the extent of increased traffic and pollution. Increased levels of noise, vibration, fumes and lighting had ruined their enjoyment of the house and garden and living there had become an ordeal. Not only had traffic volumes increased, but the nature of the traffic had changed, with a significant increase in the number of HGVs using the route, of which a fair proportion were wholly unacceptable "juggernauts". The installation of traffic lights at the junction had led to interrupted traffic flows, with standing, stopping and starting vehicles outside the house, which compounded the nuisance.

Mr Preselo described the noise and vibration at times when traffic volumes was high as "very disturbing and inescapable". The noise varied from low frequency penetrating the house walls to high frequency noises more audible to the ear. During quieter periods relative calm would frequently be shattered by the arrival of a lorry, the noise of which was very distracting and unnerving. This was a frequent occurrence in the middle of the night or early hours of the morning.

Mr Preselo produced the results of his own survey on two consecutive week nights in December 1993. On December 15 he had been woken by a lorry at 4 am and had taken a count thereafter until 7 am. During that period of three hours 35 identified goods vehicles had passed including vans and HGVs. In addition two motorcycles and "probably 30-35 cars" had passed. On December 16 1993 he had taken a similar count between midnight and 7 am. During that period 41 identified vehicles passed, mostly articulated HGVs, vehicle transporters and other heavy lorries.

He said that his family had rarely enjoyed an undisturbed night's sleep since the scheme became fully operational. Before the scheme he could not recall ever being disturbed between 6.30 pm and 7.30 am. Mr Preselo also referred to the effects of vibration. Several panes of glass in the front sash windows had cracked and minor cracks were appearing internally at the front of the house. Sash windows and the front door rattled as heavy vehicles passed by, and sometimes bedroom furniture in the upper front rooms rattled during night time disturbance.

Mr Preselo also referred to the site visit in August 1992 by the city engineer, Mr Hutchinson. The latter had agreed on site that traffic conditions were environmentally unacceptable, and in a letter to Mr

Preselo dated August 3 1992 (produced) had emphasised that the council would do all in their power "to remove extraneous traffic from the town". In fact a temporary weight restriction banning all lorries over 7.5 tons had only been approved in February 1996 and had just come into effect at the time of this present hearing.

Mr Preselo also referred to installation of new "flood-lighting" as part of the scheme. 11 powerful new street lights were fitted to the walls of houses in the vicinity, some of which shone directly into the front bedrooms and disturbed the children. He referred also to air pollution by fumes and the smell of diesel and rubber. It was not practicable to have windows open, and he described the choice in hot weather as "between pollution and suffocation". Although secondary glazing had been installed this did not eliminate nuisance and added to discomfort in warm weather. Dirt and black particles on windows and curtains needed constant cleaning, and the external appearance of the house suffered, with deteriorating brickwork and paintwork and the inability of flowers in boxes and tubs to thrive.

Mr Preselo produced a large number of photographs, many of them illustrating the congestion and other problems he had spoken of caused by the increased number of HGVs and the manoeuvring of the larger vehicles through the West Borough/ Priors Walk junction.

Mr Preselo also gave evidence as to the form of resurfacing work carried out in the vicinity of his house. I do not reproduce his evidence, as this matter was dealt with later at the hearing by technical evidence.

In summary Mr Preselo said that the scheme had a major adverse impact on the quality of his home and family life, a character house and residential area had previously been a delight to live in, but had become unwelcoming and degraded. In his opinion the effect of the changes had been to depreciate the open market value of his property substantially. He acknowledged that the situation had been improved by the recent weight restriction order.

In the course of cross-examination Mr Preselo said he had known this property for over 30 years. He had lived a few miles away until 1987 and before that his wife's family had been tenants for generations. He was aware of "traffic proposals" at the time of his purchase, but had no inkling that the scheme would proceed in the way it did, though he knew of the opening of the new link to Stone Lane in 1984, and the subsequent small-scale residential development in the area of that new junction.

The agreed traffic count figures were put to Mr Preselo, but he was not challenged as to his counts of night-time traffic. He agreed that the HGV content of day-time traffic amounted to about one vehicle every two to three minutes, and was less than that in the small hours. He considered that 400 HGVs per 24-hour day, as suggested to him, was "probably a bit high".

Mr Cowen then called the claimant Mrs Ann Veronica King, who owns and occupies 46 West Borough. Mrs King said they chose to buy this house in 1976 because it was the most agreeable residential street in the town centre. It is a Grade II listed building, part of a pleasant group of 18th-century houses and cottages and at the time enjoyed a lovely vista to the towers of the minster. The carriageway of West Borough was spacious and traffic was light. The junctions were difficult and no HGV driver would choose to use West Borough as a route through the town. It was a backwater carrying only limited local traffic and it retained a pleasant village atmosphere.

The degradation of this desirable environment had been a consequence of the county council's deliberate policy in creating the distributor road. Local opinion had been critical, but they had been assured by the council's forecasts that the traffic increase would be small. However, as a result of the scheme the environment had deteriorated dramatically. The alterations made it possible for juggernauts to negotiate the route, which was now the preferred route between Blandford and the area further north and west, and Poole, Bournemouth and the Channel ferries. The road was now used by tankers, car transporters, double decker buses, animal transports, etc. She had been astonished to find passengers on the top deck of a bus staring into the bedroom. West Borough North had not previously been a bus route.

Mrs King said the most distressing feature of the scheme was the huge increase in noise which she and her family suffered. The noise was intermittent rather than continuous and reached a crescendo morning and evening. There were eight sets of traffic lights installed at the junction, and the noise of gear changing, airbrakes, accelerating engines and squelching tyres caused a racket immediately outside the house against which they had no protection. The main living rooms and bedrooms were at the front, and though secondary glazing had been installed upstairs, the construction of the ground-floor windows with original shutters made either secondary or full double glazing impracticable and impermissible in a listed building.

Mrs King also referred to increased pollution and dirt. Windows, doors and woodwork were covered with gritty dust within 24 hours of cleaning. Curtains became grimy and when obliged to keep windows open deposits of dust quickly covered furniture and internal surfaces. When windows were of necessity opened, the stench of diesel fumes was at times overpowering. She was concerned also about the effects of vibration on the structure of an old house. There were already instances of doors becoming ill-fitting.

Mrs King expressed the fear that the house would never be as desirable or as valuable as when they had bought it and conscientiously restored and renovated it. She accepted that the recent weight restriction order had brought some improvement since June 1996, but because of exemptions in favour of a distribution depot on the industrial estate in Stone Lane, which used very large vehicles, there was still an estimated 150-200 HGVs per day passing along West Borough. She produced a photograph taken in January 1996 of a stationary HGV outside her drawing room window, illustrating the problems.

Mrs King was not cross-examined.

Mr Andrew Hurrell is a research physicist with Precision Acoustics in Dorchester. At my request Mr Hurrell met the compensating authority's acoustics expert for discussion prior to giving evidence. In consequence they had reached agreement in certain respects *viz*:

(1) The withdrawal of Mr Hurrell's calculated pre-scheme noise levels;

(2) that measurements taken by Mr Hurrell were valid;

(3) that post-scheme traffic in West Borough is not steady flow, but the interpretation to be placed on that fact is not agreed; but

(4) that the addition of traffic controls will result in a change in the nature of noise levels.

Mr Hurrell expressed concern as resulting from these discussions as to the validity of his own calculations of pre-scheme noise level and had withdrawn them but nevertheless considered that it was less than the 72.5 dB(A) put forward by Mr Butler, which was equal to post-scheme levels at the busy Priors Walk junction.

In the light of the measure of agreement reached by the acoustics experts, it seems unnecessary to reproduce the whole of Mr Hurrell's comprehensive technical report, and what follows is a highly condensed summary.

There was a lack of information as to pre-scheme noise levels, and it was therefore only possible to compare calculated pre-scheme levels of noise based on limited traffic data, with measured post-scheme noise levels.

Mr Hurrell explained the methods of expressing the measurement of noise. The parameter in common use was the LA10 (18 hr) index which is the average A-weighted level exceeded for 10% of each hour in the 18 hours between 6 am and 12 midnight. He pointed out that the sound pressure level (SPL) would be below this level for 90% of the time. While the LA10 (18 hr) parameter was often found to correlate with subjective dissatisfaction, there were limitations to its usefulness. It was impossible for a single value to quantify the very transient nature of highway noise; in many instances it was unrepresentative of peak noise levels, and it had been designed specifically for use when the linear broad band SPL was less than 55 dB. In his opinion it was an inappropriate measurement parameter for this kind of highway noise and could be misleading and he had taken account of other evidence.

A There existed B and C weightings, the B weighting optimised for usage where the SPL was in the range 55-80 dB. The B weighting in his view was appropriately used in this instance, and the results of survey showed a larger increase in L10 (18 hr) levels post-scheme than is observed with the A-weighted results.

Mr Hurrell had also taken account of the Traffic Noise Index (TNI) which was another means of assessing the degree of community dissatisfaction. The use of this index suggested that the majority of local residents would be expected to be dissatisfied with current traffic noise levels.

When cross-examined Mr Hurrell accepted that the conventional B method used nationally for the assessment of traffic noise derived from the TRL publication "Calculation of Road Traffic Noise" (CRTN) and that the LA10 (18 hr) scale was widely used by government departments and local authorities including Dorset County Council. He did not deny the validity of the scale but regarded it as inappropriate as the sole guide in the present case. The literature made provision for exceptional cases and he considered this case to be exceptional. The index was of use for road design purposes, but failed to establish intermittent noise levels which were particularly disturbing, as in the intermittent stopping and starting of HGVs, or to reflect what happened at the peaks. He agreed that the CRTN required C measurements to be taken over a minimum three-hour period, whereas he had taken readings over a single two-hour period. However, his A-scale reading correlated well with those of the Council. He had taken B-scale readings because he considered the peak noise levels to be exceptional and not accurately reflected by the LA10 (18 hr). That scale was demonstrated to be inappropriate to this case. The figure of 74 dB(A) it produced compared with the 90dB which his readings showed could be reached at times. He regarded it as a duty to have regard to all available evidence.

Finally, Mr Cowen called Mr Robert Charles Gardner FRICS, a senior associate in Humberts, chartered surveyors practising from their D offices in Salisbury. He had 23 years experience since qualification in both private practice and the public sector. He had arrived in Salisbury in 1982 as deputy district valuer covering south Wiltshire and had joined Humberts in 1989. He had dealt with about 300 claims under Part I of the 1973 Act over the last 14 years, including several in respect of the Wimborne Distributor Road.

Mr Gardner described the policy documents leading to the scheme of public works comprising the Wimborne Distributor Road. The scheme had been carried out in two phases, Phase 1 consisting of the Stone Lane junction improvement and the link from West Borough to the Blandford Road. This was completed in 1984, but had no significant effect on traffic flows in the area of the subject properties, E because of the continued difficulty of negotiating the West Borough/Priors Walk junction, and the direction of flow had remained unchanged. Phase 2 consisted of the work to the south of Phase 1 including reshaping of the Priors Walk/West Borough junction, and was carried out in 1990/91. It was agreed that the scheme opened to traffic on July 14 1991.

The specific works carried out close to the subject properties included the widening of the junctions of West Borough with Priors Walk and with Stone Lane, the installation of traffic lights at each junction, and the provision of additional street lighting at the junctions and along the carriageway between the junctions. It was not in dispute F that these works constituted "public works" for the purposes of the Act. However, there was a dispute as to certain works carried out to the carriageway between the two junctions, the council contending that this work was not "improvement or alteration" but amounted only to "resurfacing" and was thus excluded from the definition of relevant public works. This work was carried out on the stretch of carriageway immediately outside the subject properties. The distance from no 45 to the nearest point of the agreed "improvements" was 17.6m (58 ft) and from no 46 was 12.5m (42 ft).

Mr Gardner referred to a specification of the work carried out and to a sectional diagram now produced by the county council's witness. He said that the work carried out was additional surfacing designed to change the level and alter the profile of the road in order to make it G capable of carrying more and heavier traffic as part of the overall scheme. In his view this was not mere resurfacing but an alteration of level within the terms of the statutory provision. However he had prepared his valuation evidence on alternative bases having regard to this area of dispute.

Mr Gardner gave detailed descriptions of the subject properties with the assistance of photographs and floor plans. The historic centre of the town was around the 11th-century minster and The Square. The subject properties were about 220m north of The Square in West Borough, which was regarded as one of the finest residential streets in the town. It was conveniently situated and contained a number of very H attractive period houses including the subject properties which were both Grade II listed.

Mr Gardner described the significant change in traffic flows and routing which followed completion of the distributor road scheme in July 1991. The pre-scheme position in West Borough north of the Priors Walk junction was such that it was located away from busy through routes and carried mainly free-flowing light vehicles. Apart from brief peaks, the road was relatively quiet with no congestion or noise arising from the junction. Completion of the scheme and the changed routing, together with the installation of traffic lights, had produced a radical change. This part of West Borough was now part of J the through route, widening of the junction now accommodated larger heavy vehicles, and the traffic lights caused continual interruptions in traffic flow. The agreed pre-scheme and post-scheme traffic flow figures represented an increase of 22.3%, and according to the council the system was designed to cater for an anticipated increase of up to 37% by 2006.

However, although there had been a large increase in vehicle numbers, the most significant change was in the character and content of the traffic. The proportion of heavier vehicles had increased markedly, which was of course the intention of the Distributor Road scheme. The traffic lights caused continual stops and starts. South K bound vehicles slowed down to low gear, braked and then queued outside the subject properties with engines idling, then pulled away in low gear. North bound traffic pulled away from the junction in low gear then accelerated through the gears as they passed the subject properties. Although there was no data for pre-scheme HGVs of 7.5 tons or more, he considered the deterrent effect of the junctions such that there were unlikely to be any apart from the occasional most determined driver in this part of West Borough.

Mr Gardner said that the agreed traffic counts were for a 16-hour day and took no account of the movements between midnight and 6 am spoken to by Mr Preselo, which the witness believed related to night ferry sailings to and from Poole docks. L

The consequence of this significant alteration in traffic movements was to increase greatly the nuisance arising particularly from heavy vehicles. The effects of the physical factors of noise, smell, fumes and vibration have all worsened, in addition to the effect of increased artificial lighting.

Mr Gardner referred to the estimates of noise levels which other witnesses dealt with, but stated that from his own experience of frequent visits to West Borough, the general noise had caused an intolerable level of intrusion. The front of no 45 was only 2.5m (8 ft) from the carriageway and no 46 was only 3.5m (11 ft 6 ins) from the carriageway. He pointed out that a resident would be in bed only M 10-12 ft away from an HGV producing up to 45 dB above the background level. Vibration was also very noticeable, as were the fumes and smell of diesel, petrol and rubber. The new street lighting caused the area to remain illuminated throughout the night. In view of these factors, it was inappropriate to leave doors or windows open at the front of the properties at any time.

Mr Gardner expressed the opinion that these physical factors had materially affected the character of West Borough. It was no longer the fine residential street it had been and had thus lost its premium value. Furthermore, he considered there had in consequence been a significant depreciating effect on the value of the subject properties.

A Mr Gardner set out the statutory basis of compensation, and said that his valuation approach was to measure depreciation in value of the subject properties as the difference between the "unaffected value" and the "affected value" by reference to prices current on the first claim day, July 15 1992. "Unaffected value" was the price a purchaser would pay with the public works in existence and use but with the physical factors no worse than before the scheme. "Affected value" was the price a purchaser would pay with the public works in use with the present and anticipated effects of the physical factors.

He relied in support of his valuation on a number of open market sales of properties in Wimborne, which he tabulated as follows:

B

		£
16 Avenue Road	May 1986	68,500
	21 April 1992	185,000
	August 1993	175,000
44 West Borough	August 1986	115,000
	18 August 1992	160,000
48 West Borough	August 1986	105,000
The Mill House Knobcrook Road	August 1993	172,000
106 West Borough	August 1994	120,000

Mr Gardner described 16 Avenue Road as a detached Victorian property situated in a good residential street and unaffected by any C physical factors of the sort affecting the subject properties. He set out details of the accommodation and produced a photograph. He estimated the reduced covered area (RCA) as 187m². There were three relevant sales transactions detailed above.

44 West Borough is a detached house adjoining no 46. It is a Grade II listed building of similar age and style; he described the accommodation. The estimated RCA is 311m². There were two relevant sales, of which he gave particulars. The sale in 1992 had included carpets, curtains and other chattels with a value estimated by the vendor at £12-15,000. He had adopted £12,000 to give an adjusted net sale price of £148,000.

D 48 West Borough was a terraced town house adjoining 46 West Borough and of similar age and style. Mr Gardner set out details of the accommodation and the estimated RCA was 195m². The sale in August 1986 had included a garage and hard standing in School Lane, stated by the purchaser to be valued at £8,000 giving a net sale price of £97,000.

The Mill House was a mid-19th-century detached family house in a secluded position at the northern edge of the town, with frontage to a narrow stretch of the river, immediately facing commercially run tea-rooms on the opposite bank. The RCA was 255m². Mr Gardner produced documentary evidence that the purchase price of £172,000 E paid in August 1993 had taken account of disrepair valued at £5-£6,000, and that an offer of £200,000 had been made by another prospective purchaser in July/August 1992.

106 West Borough is a detached Victorian house situated north of the subject properties and close to the Stone Lane junction. The RCA was 238m². It has a small front garden and the building is about 5m from the highway boundary. He produced sale particulars relating to the sale in August 1994 for £120,000. The council tax banding had been reduced due to the road scheme from band G (£160,000+) to band F (£120-160,000) as at April 1 1991. The house had been substantially extended to the rear by a two-storey extension which Mr Gardner characterised as "incongruous".

F Mr Gardner also referred to 35 West Borough. This is a three-storey terraced house situated at the Priors Walk junction. The building abuts the pavement and has an RCA of 150m². He had agreed compensation with the county council in the sum of £13,000, calculated on an unaffected value of £120,000 and an affected value of £100,000 as at July 15 1992. The total depreciation was therefore £20,000 or 16.67% of unaffected value, and apportionment in respect of physical factors was 65% of total depreciation, or 10.8% of unaffected value, ie £13,000. In his consideration of this settlement, Mr Gardner pointed out:

(a) This was a smaller, less well appointed property. In his opinion the value of smaller less valuable properties was affected by physical factors to a lesser extent than better appointed more valuable G properties.

(b) Pre-scheme traffic flows affecting 35 West Borough were much higher than the traffic flows affecting the subject properties. In consequence the impact of the scheme measured as the difference pre-scheme and post-scheme was not as great.

(c) The apportionment of 65% of total depreciation reflected the severe non-compensatable effects of the scheme, in particular the visual intrusion of new traffic lights and of traffic coming directly towards the property and to within 3m, which had not been the case pre-scheme.

(d) The agreed compensation was a compromise reluctantly H accepted by the claimant, and the best achievable in the light of her urgent needs.

(e) The agreement was made prior to the decision in *Tucknott* v *Department of Transport* LT Ref/206/93 at Bridport, on the basis of which the settlement might have been higher.

Mr Gardner's view was that the pre-scheme transactions he referred to clearly demonstrated pre-scheme relative values. He produced a schedule which is annexed as appendix 4 analysing the three transactions in 1986; in order to achieve a common date in August 1986, he had adjusted the price of 16 Avenue Road (May 1986) to allow for the steeply rising housing market at that time. J Upon the basis of a price per square metre for the three transactions in the schedule, it was demonstrated that the fine houses in West Borough attracted a significant location premium compared to a fine house in Avenue Road. This premium averaged 10%. His analysis of subsequent open market sales showed not only that this location premium had been lost as a result of the scheme, but also that West Borough properties had suffered further depreciation in value.

Mr Gardner produced a further schedule which is appended as appendix 5 which analysed post-scheme transactions in unaffected properties (16 Avenue Road and The Mill House) adjusted to the valuation date in July 1992, as a basis for an "unaffected" valuation of K West Borough properties post-scheme, including the subject properties. Adjustment of this evidence to the valuation date took account of a falling housing market of the order of 5% pa, and adjusted also for garages, or for any carpets or curtains included in a sale price. The falling housing market during the relevant period was shown by the sales of 16 Avenue Road in April 1992 and August 1993 on identical terms, and by the offer made of £200,000 for the Mill House in July 1992 compared with the subsequent sale in August 1993 for £172,000. In determining the unaffected valuation he added a 10% location premium to the West Borough properties as indicated by his earlier findings. L The schedule showed 106 West Borough with RCA 191m². This was subsequently shown to be wrong, the RCA being measured at 238m². Mr Gardner, nevertheless, regarded the "unaffected" valuation shown as still being accurate, having regard to the unusual feature of the incongruous modern addition at the rear.

By this means, as the appendix 5 showed, Mr Gardner arrived at an "unaffected value" of the subject properties as at July 15 1992 as follows:

45 West Borough	£211,500
46 West Borough	£192,500

M

In the final stage of his valuation, Mr Gardner produced a schedule annexed as appendix 6 in which an analysis of market transactions in affected properties led to his opinion of the "affected value" of the subject properties as at July 15 1992 as follows:

45 West Borough	£142,500
46 West Borough	£131,000

In consequence Mr Gardner identified the total depreciation post-scheme as:

45 West Borough	£69,000	32.6%
46 West Borough	£61,500	31.9%

A which was consistent with the evidence of market transactions in the case of 44 West Borough and 106 West Borough as appendix 6 indicated.

If he is held wrong in the assumption that the work adjoining the subject properties is "altered highway", then in Mr Gardner's opinion the short distance between the subject properties and the admitted public works required only a minor adjustment which he assessed as 15% for no 45 and 10% for no 46. This adjustment produced total depreciation:

45 West Borough	£69,000 x 0.85	=	£58,650
B 46 West Borough	£61,500 x 0.9	=	£55,350

As to the proportion of total depreciation attributable to the compensatable physical factors, he considered that non-compensatable factors, ie visual impact of the works and possible loss of privacy, were of little significance. He referred to the tribunal's decisions in two cases at Bridport (*Broom* v *Department of Transport* and *Tucknott* v *Department of Transport*) which illustrated the approach. He considered the present case more akin to the Tucknott decision than to the *Broom* decision. As in the Tucknott case, it was clear that most of the depreciation resulted from the compensatable physical factors, noise, fumes, etc in consequence of the scheme's operation. He C therefore adopted 90% of the total depreciation, and the resulting valuation was expressed in the alternative, depending on whether or not the work north of the junction constituted an altered highway:

		£	£
45 West Borough	Total depreciation	69,000 @ 90%	=62,100
	Adjusted x 0.85	58,650 @ 90%	=52,785
46 West Borough	Total depreciation	61,500 @ 90%	=55,350
	Adjusted x 0.9	55,350 @ 90%	=49,815

The adjusted figures represented approximately 25% of unaffected value in each case and unadjusted figures represented approximately D 29% of the unaffected value.

Mr Gardner said that the council's valuer Mr Walker had relied on evidence of settlements, which he regarded as unreliable for several reasons. Claimants may have motives for accepting a settlement figure even where the sum is not justifiable, such as an urgent need for the money to move elsewhere or for some other purpose. He cited the examples of the owners of 35 West Borough and 102 West Borough, both of which he had himself negotiated; in both instances the owners were desperate to receive the compensation quickly. With the possible exception of 106 West Borough, all settlements referred to were inferior properties, smaller and less provided, and situated at one or the E other of the junctions, where more non-compensatable factors such as visual impact had to be taken into account. Mr Walker had also cited 42 West Borough, which was a large guest house with only minor residential use and many complicating factors. In at least two cases Mr Walker had reached settlements with a valuer from Cardiff whose knowledge of the scheme and of Wimborne values was likely to be tenuous.

In the case of no 45 Mr Walker had made deductions from his estimate of unaffected value: (a) for the fact that part of the garden was held under licence; and (b) for "rearranging" the two kitchen layout; both these adjustments were unnecessary in Mr Gardner's opinion.

F In cross-examination, Mr Gardner explained that he had adopted an analysis based on RCA of large elderly houses to produce a valuation of space. He accepted that the method took no account of factors such as the size of garden or the convenience of room layout, but he considered this was the most appropriate and useful method to use to compare one with another when all were substantial family houses of character and similar style. He agreed that West Borough and Avenue Road were not of the same character, and accepted that Avenue Road could be called a "good suburban residential area". He was aware that planning permission had been granted in 1993 for conversion of a coach house at the rear of 16 Avenue Road to a holiday flat, but he did not consider this influenced the sale price. In that position there was no significant development potential, the prospective income being G unlikely to exceed the costs of conversion etc.

He agreed that 16 Avenue Road carried a blue plaque to Thomas Hardy who had lived there for about two years, but he did not think this had a significant effect either way on value. Some bidders might see it as a matter of kudos, others might consider it a potential burden.

He had allowed £20,000 for modernisation instead of the £10,000 estimated by a builder. This was a matter of judgment. The lower figure would have the effect of giving a greater location premium to West Borough than he thought could be supported. The figure of £20,000 was not "plucked out of the air", but was a reasonable and informed estimate of what the purchaser would probably be called H upon to spend. He had taken 16 Avenue Road because it had been sold three times during the relevant period, and in his view provided the best evidence of market value for an unaffected property which was comparable with the subject properties.

With regard to Mill House Mr Gardner accepted that the offer of £200,000 was not the best evidence of value, but it was more reliable than an asking price. He understood that the offer was refused because the vendor wanted more and thought he could get more. He had taken account in using the Mill House transaction of the cost of repairs, and of the adjoining commercial building, which he thought was imposing and to an extent overpowering. The river frontage had pros and cons; J there was a lack of privacy because of the proximity of the tea rooms on the other bank. On reflection Mr Gardner thought the existence of the tea garden at the other side of a narrow stream was more significant than the adjoining commercial building, but he had as a matter of judgment lumped the two factors together as the basis for a discount from the value.

Mr Gardner accepted that the settlement of 35 West Borough was the one case where he and Mr Walker had specifically agreed the components of the settlement, and had agreed to put them in the statement of agreed facts. It was the only case of a settlement which he considered relevant by reason of location. He had not suggested that K there was an improvement in conditions as a result of the scheme. The work at the junction had a severe impact on this property, but in the light of the pre-scheme situation the impact on value was not as great as it was at the subject properties. Furthermore, he had always believed that larger quality properties suffered more loss of value proportionally, as the market demonstrated.

Mr Gardner said that he did not accept the figures stated as "agreed no-scheme world values" in Mr Walker's evidence, with the exception of 35 West Borough where they had specifically agreed the build-up of the valuation. In other cases which he had settled, only the compensation figure was agreed, and he was surprised to see reference L in Mr Walker's proof to without prejudice discussions.

In answer to me, Mr Gardner said he believed that the more a purchaser was able to spend, the more discerning and exacting he would be in the choice of property and the less willing to accept lower standards of accommodation, amenity, etc. The properties he had used as comparables in West Borough and in Avenue Road were all family type, large period dwellings, and while tastes and requirements varied greatly, he considered that they were all within a broad bracket of the market.

Compensating authority's evidence

M Mr Ward called Mr Gordon Frederick Cordier, chief engineer to the council. He controlled the capital improvement works programme and budget and led the project managers in delivering this programme, which included the distributor road scheme. He described the scheme in detail, as summarised in the agreed facts.

Mr Cordier said that the contract documents for the works had made no provision for any works between the Priors Walk junction and the Stone Lane junction, that is to say the section of West Borough in which the subject properties are situated. Work had started in January 1991 on the eight-month contract and in the later stages, prior to commencement of final surfacing consideration was then given to whether this section of West Borough should be resurfaced. The area surveyor had bid separately for maintenance funds to resurface West Borough using a proprietary brand of surfacing material Ralumac. This was laid as a veneer and improved riding and skid resistance characteristics. Ralumac would not be used for reconstruction or upgrading.

Visual inspection had indicated no structural defect, but the surface showed signs of wear, and he had therefore decided to resurface this length of road with the surfacing material used in the main contract, and a price was negotiated with the contractor.

Mr Cordier described in detail the method of resurfacing adopted and the sequence of work. This is illustrated in a diagrammatic sectional drawing which is appended as appendix 7 [not reproduced here]. He said that this work was resurfacing, and rejected Mr Gardner's view that it amounted to more than resurfacing, such as was regularly carried out as a matter of maintenance throughout the county. If the road had required reconstruction as opposed to resurfacing, a different and more elaborate and costly process was involved, which he described. The road would then have been dug out to a considerable depth, probably disturbing gas, water and electrical installations, and a strengthening bituminous road base laid over a graded stone sub-base in turn over a capping layer. This process would not be undertaken unless there were signs of the road failing, which in this case there were not.

In cross-examination, Mr Cordier said that the distributor road was not regarded as taking significant volumes of freight traffic when compared to the trunk road network. In relation to carriageway design the volume of traffic was very small and the existing structure was more than adequate to take well in excess of the volume of freight using it now.

Asked about the material used in surfacing, he said it was a high stone content rolled asphalt, which represented good value for money. A balance had to be struck between cost and the ability to carry extra traffic over time without frequent resurfacing.

In answer to me, Mr Cordier said that the method illustrated in the sectional sketch (appendix 7) was a common technique. There was no point in removing all the old surface if it was generally in good order, but the result of adding the new material would be to raise the surface of the crown, as the sketch showed. The structure of the road was not altered at all.

Mr Simon Butler has 11 years experience in the field of transportation noise and is employed in a consultancy which is a division of the Transportation and Engineering Department of the council.

Mr Butler referred to the DTP memorandum "the Calculation of Road Traffic Noise" (CRTN) and to the use of the LA10 18-hour index as the most appropriate single number description of traffic noise levels, shown to correlate reasonably well with popular dissatisfaction. Measurements had been taken primarily to determine the eligibility of buildings under the Noise Insulation Regulations and where measurements were not taken levels were calculated by the use of the prediction method prescribed by CRTN. This prediction method was preferred since it is unaffected by variables such as weather conditions or unusual traffic flows. Measurements would only be taken by way of corroboration, and none were taken in respect of the subject properties. Pre-scheme and post-scheme levels for these properties were calculated by the prediction method, with results as follows:

Calculated readings (LA10 18 hour index)

46 West Borough	pre-scheme	73 dB(A)
	post-scheme	74 dB(A)
45 West Borough	pre-scheme	73 dB(A)
	post-scheme	74 dB(A)

It was noted however, that noise from the altered section of the highway does not make an effective contribution of 1 dB(A) to the overall noise level, and neither property was eligible for noise insulation work or grant under the Noise Insulation Regulations 1975.

Mr Butler offered a detailed critique of the approach of Mr Hurrell. He did not regard the situation in West Borough as exceptional, such as to justify rejection of the LA10 18-hour index, which was recommended by CRTN and was most commonly used. He accepted that the LA10 18-hour index, like any other single number index, had limitations, but no better method had yet been produced to assess people's reaction to traffic noise. The methods proposed by Mr Hurrell were, in his opinion, of little practical use in assessing the change in value of a property. He referred to Mr Hurrell's method as "quite extraordinary" and involving the addition of a "correction factor" which had no scientific basis. The "B" weighting advocated by Mr Hurrell was seldom used in practice because it offered no positive advantage over the customary "A" weighting.

In cross-examination Mr Butler maintained that the alternatives suggested by Mr Hurrell offered no advantages over the method of calculation recommended by CRTN, which was in universal use, and with which Mr Hurrell appeared to be unfamiliar. He acknowledged that LA10 18 hour was not perfect, and that other evidence could merit consideration; there was no simple way of defining the precise effect of traffic noise. He agreed that at night levels of HGV noise would be considerably higher than the ambient level, but the same was true prior to the scheme, though to a lesser extent. He accepted that 45dBA might be a reasonable measure of the sound of an HGV above ambient level. He did not however think that this was exceptional, other residential streets in Dorset experienced similar levels.

Mr Butler agreed that the introduction of traffic light controls changed the quality of noise and that the post-scheme traffic in West Borough was less steady-flow than before. He agreed further that the LA10 18-hour index did not cover the period between midnight and 6 am, and that neither peak nor trough levels could be derived from the calculation. He accepted that if the road adjoining the subject properties was held to be "altered highway", then the calculated increase in noise due to the scheme was ldB with a further 2dB resulting from the anticipated increase in traffic over a 15-year period.

Mr Richard Walker has 23 years' professional experience, both in private practice, and since 1985 as a valuer in the estates department of Dorset County Council. He has known Wimborne for 10 years and was involved in the acquisition of land for the Wimborne Distributor Road. Since 1992 he has dealt with all 54 of the consequent claims made under Part 1 of the 1973 Act.

Mr Walker's evidence contained a detailed description of the subject properties, of the components of the scheme, and of the basis for his valuation. He relied on the county surveyor's evidence as establishing the limits of the highway alterations, and on that basis stated that West Borough was resurfaced, but was not an altered highway as defined by statute. He set out the agreed traffic count figures, and referred to Mr Butler's evidence of calculated noise levels. Mr Walker noted that this evidence showed that noise from the altered section of highway (ie the Priors Walk junction) did not make an effective contribution of ldBA to the overall noise level. From this information and his inspection, he observed: (a) the physical factors affecting the subject properties emanate for the most part from vehicles directly outside and not on the altered highway; (b) other sources of physical factors are vehicles approaching from both directions on unaltered highway; (c) vehicles using the altered highway contribute only marginally; (d) traffic using the highway pre-scheme generated significant noise etc; and (e) additional traffic post-scheme generated only a marginal amount of extra noise, etc.

As the background to his valuation, Mr Walker referred to the Nationwide Building Society index as showing a drop in values in the South West region between the third quarter of 1991 and the third quarter of 1993 of 7.3%. Direct evidence of sales in Wimborne indicated a more local decline in values between 11 and 16% over the same period, and he referred to successive sales of 2 Ashdean Close, 1 Quarry Drive and 9 Riverdale.

Mr Walker said that his methodology was no different from any other valuation for compensation purposes. It was necessary in his view to know the area, to visit local agents, to consider available data, to have regard to the local market generally, and to identify comparables wherever possible. It was not the case that he had relied on settlement evidence. He had quoted settlements, but his valuation opinion was based on comparable market transactions, in particular the sales of 102 West Borough and 31 West Borough. He considered that the settlements cited gave an overall background picture against which to look as a guide mark, of particular assistance in securing consistency. He referred to the following transactions.

102 West Borough

Purchased in July 1988 for £147,000 and sold in August 1993 for £107,000. In his view only part of this loss of value was attributable to the scheme. Mr Walker had agreed compensation with Humberts at £9,100, representing 70% of total depreciation and 7.5% of no-scheme world value.

(Note: The phrase "no-scheme world value" is Mr Walker's; it appears to me to represent a pre-scheme value, not necessarily the same as Mr Gardner's "unaffected value").

Mr Walker said that the position of 102 West Borough in relation to the altered highway was similar to that of the subject properties, though he accepted that it was different in character and was not directly comparable with either no 45 or no 46.

35 West Borough

This property fronted directly on to the new traffic island and was directly opposite Priors Walk. A claim was agreed with Humberts at £13,000 representing 65% of total loss due to the scheme, and 10.8% of the no-scheme world value.

31 West Borough

This property is very similar to 35. A Part 1 claim was agreed with Humberts at £13,000, again representing 10.8% of no-scheme world value. These two properties were in Mr Walker's view arguably the most affected properties in the scheme.

42 West Borough

Although he had cited this in his report, Mr Walker placed little reliance on the settlement made, having regard to the complications of a commercial use and post-scheme parking difficulties, and a failure to agree with Mr Gardner the components of the settlement.

100 West Borough

Mr Walker had agreed compensation with a local agent at £6,000 representing 6.3% of no-scheme world value. Although this property was next door to 102 West Borough, it suffered to a lesser degree in his opinion because it was not directly facing the Stone Lane junction and this was reflected in the settlement figure.

104 West Borough

Compensation had been agreed with Humberts at £6,200 representing 6.2% of no-scheme world value. The property adjoined 102 West Borough but was affected to a lesser degree because most vehicles travelling down Stone Lane would turn right into West Borough and away from this property.

106 West Borough

Compensation had been agreed with William Ricketts of Cardiff at £4,000, representing 2.5% of a no-scheme world value which had been agreed as £160,000.

94 and 96 West Borough

A pair of modern houses in a terrace of three. In both cases compensation had been agreed (in one case by Mr Ricketts, and in the other by a firm from Yeovil) at £650 representing 1% of no-scheme values.

By reference to these settlements Mr Walker had concluded that a pattern emerged which related compensation to the distance between the altered highway and the affected property. Thus compensation had ranged from 10.8% in the case of the properties located at the Priors Walk junction to 1% in the case of 94 and 96 West Borough. This principle was adopted in practice in Dorset and to his knowledge elsewhere.

Turning to 46 West Borough, Mr Walker said there was discernible traffic noise and vibration, but the element emanating from the altered highway was difficult to isolate. The noise survey had shown only a marginal increase post-scheme and the altered highway made no effective contribution to any increase.

Based on his view of values derived from 35 and 102 West Borough, Mr Walker estimated the no-scheme world value of the subject property at £160,000, but he considered that the marginal effect of the scheme justified only a minimal figure for depreciation which he put at £5,000. Taking 70% of this as the compensatable factor, this produced a figure for compensation of £3,500 representing 2.18% of no-scheme world value. If contrary to his contention, it was held that West Borough was itself an altered highway then Mr Walker put forward an alternative valuation for compensation of £5,600 or 3.5% of no-scheme world value.

In the case of 45 West Borough, Mr Walker arrived at a no-scheme world value of £160,000, having made deductions in respect of the effect on value of part of the garden being subject to a terminable licence, and for the cost of rearranging accommodation to restore the house to single-family occupation. Adopting the same principles as in the case of no 46, he valued the compensation due as £3,500 representing 2.18% of his adjusted no-scheme value. If this section of West Borough was an altered highway, then his alternative valuation was £5,600 or 3.5% of no-scheme value.

In commenting on Mr Gardner's evidence, Mr Walker did not consider that a location premium for West Borough existed. The properties had generally higher values because of their intrinsic qualities and convenience, but such of those properties as were similar in style, character and size to properties in Avenue Road would have no location premium and would sell at similar prices. He did not endorse the use of RCA for the purpose of comparison. He considered it a blunt instrument with too many variables of location, arrangement, garden etc to be reliable.

He had adopted 70% generally in respect of compensatable factors because he considered the circumstances of this case very different from those at Bridport, where the Lands Tribunal had adopted 90%. As to the percentage of unaffected value claimed in this case, he regarded the figures as exceptionally high and unsustainable.

In cross-examination, Mr Walker agreed that open market transactions would be the best evidence where available, and would perhaps be more reliable than evidence of settlements. He insisted however, that he had not relied on settlements but had used them as background and tending to demonstrate a pattern.

He agreed that the evidence as to the trend in house prices was meagre. The Nationwide Index was not local but regional, and was inconsistent with the trend revealed by the few local transactions he had referred to. He had chosen to apply the Nationwide Index, but accepted that it could lead to arbitrary results, which he nevertheless was satisfied were correct.

In the case of 102 West Borough, Mr Walker had relied on successive sales in July 1988 and August 1993, but he was unable to say for certain whether the substantial rear extension to this property was added before or after the 1988 transaction. He agreed that if the extension had been added after the sale in 1988 then the two transactions were not "like for like".

As to 100 West Borough, Mr Walker agreed that the house was not really comparable with the subject properties; the character of West Borough changed broadly northwards from about no 80, but he had included these settlements at the northern end of West Borough to demonstrate the general principle of a decline in compensation related to distance from the altered highway. He agreed also that the settlement made with Ricketts in relation to 106 West Borough had been made prior to the sale of that property in August 1994 and that compensation would have been recalculated if that sale price had then been known.

Mr Walker did not agree with Mr Gardner's thesis that the purchasers of more expensive properties would not accept a lowering of quality and amenity to the same extent as the owner of cheaper property. He agreed that more expensive properties would tend to attract more compensation but not in percentage terms. He was aware that settlements had been made in Chichester up to a maximum 13 to 15% of unaffected values but he had no details. He had no personal knowledge of any other scheme where the relevant works were the width of a pavement from the claimant's house.

Submissions

Both counsel provided written summaries of their submissions which I found most helpful.

Mr Ward submitted that the road outside the subject properties was not an altered highway for the purposes of the 1973 Act. That part of the road was only resurfaced and therefore expressly excepted by section 9(5). He pointed out that the wording of section 9(5) was such that the Act clearly contemplated that a change in the level of the carriageway could take place which was caused by no more than resurfacing. He referred to the decision of the tribunal in *Williamson* v *Cumbria County Council* (1994) 68 P&CR 367*. The source of the noise etc must be located on the altered highway, and there must be a reduction in compensation the further that source is from the affected property.

Mr Ward rehearsed the figures for traffic flows and submitted that the amount and type of traffic resulting from the scheme had been exaggerated, as had the alleged effects. The only reliable evidence was in the agreed figures, and account must be taken of natural growth in analysing them. With regard to noise, the only acceptable evidence was that of Mr Butler, which clearly showed the exaggeration of the claimant's case. The increase in post-scheme noise of ldBA was insignificant and could have no real effect on the perception of dissatisfaction.

In relation to valuation, Mr Ward drew attention to the magnitude of the claims in terms of money and percentages. The sums claimed were between £50,000 and £62,000 and represented 25% and 29% of the unaffected values of the subject properties. This was 200% more than had so far been paid to any claimant in relation to this scheme. The highest figure mentioned in evidence had been the 15-20% referred to and which related to the A27 trunk road at Chichester. At Bridport, the Lands Tribunal had awarded 8.6% in one case and 10.8% in another, while settlements generally did not exceed 7%.

In order to justify the claimant's figures it had been necessary to exaggerate, and to seek to make out what Mr Ward called "a very very very exceptional and unique case". However, there were so many ifs and buts and assumptions and adjustments and judgements without any evidential basis that the whole of the claimant's valuation process was flawed and should not be relied on. Mr Walker's evidence was in his submission much sounder and based on proper valuation figures and assessments and comparables, and was therefore much more realistic.

In reply, Mr Cowen submitted that there was now only one issue of law. It was clear that it no longer mattered whether the Stone Lane improvement was or was not part of the scheme, since the agreed traffic figures showed that the radical change in the scale and nature of the flows followed from the completion of the second phase, ie the West Borough/Priors Walk junction reconstruction. Thus there only remained the issue as to the extent of the altered highway, and whether what was done immediately outside the subject properties was "resurfacing" within the meaning of section 9(5). The word "resurfacing" was not further defined in the Act, and he submitted that the question turned on the facts of an individual case. What had been done in this case might be "surfacing" in that a layer was added and the level altered, but "resurfacing" implied removal and replacement, doing something again. He considered the evidence in detail, and submitted that there had been an alteration of the profile and a restructuring of the carriageway and a strengthening so as to enable heavier loads to be carried. This was beyond what the Act referred to as "resurfacing". As to the traffic flows, the agreed figures showed an increase of 22% from pre-scheme to post-scheme, but if a base were taken in 1982, the overall increase 1982 to 1993 was over 50%. The HGV content (goods vehicles in excess of 3.5 tons) increased between 1989 and 1993 by 25%, but that increase became 100% if taken from the available figures 1987 to 1993 (264 to 529). Mr Cowen pointed out that all the figures were derived from 16-hour counts and took no account of Mr Preselo's evidence of night movements, which had not been challenged. It was not legitimate for the council to accuse the claimants of exaggeration when evidence on the impact of traffic by both Mr Preselo and Mrs King had not been cross-examined.

As to the technical evidence about noise, Mr Cowen described this as "interesting but perhaps of little value". A prospective purchaser would not make noise calculations or measurements, whether A or B weighted, but would have a perception of how disturbing it appeared to be. He submitted that in this respect also the evidence of Mr Preselo and Mrs King was the most valuable, and contrasted startlingly with that of Mr Butler, whose calculations had led him to the opinion of an "imperceptible" or "insignificant" increase in noise levels.

Mr Cowen presented a detailed critique of the LA10 18-hour index. It had severe limitations which were admitted. It did not measure peaks or troughs in sound; it presented only an average over 18 hours; it took no account of the difference between background level and peak noise; and it did not measure noise during the night hours when that difference would be at its greatest.

With regard to the valuation evidence, the approach of the two valuers had been radically different, and Mr Cowen submitted that the approach of Mr Gardner was more logical and conformed better to correct valuation principles. His evidence had been firmly based on market transactions, supported a location premium for West Borough properties, and demonstrated a clear pattern of severe depreciation caused by the scheme. Mr Gardner's adjustments if the adjoining highway was not "altered carriageway" were appropriate and were not challenged in cross-examination. The assessment of 90% in respect of physical factors was justified where there was no reasonable alternative cause for the depreciation in value, as in the *Tucknott* case at Bridport.

By contrast Mr Walker had relied almost exclusively on settlement evidence, and then relied on using percentages of unaffected value as a scale of comparables, which he applied to the subject properties without assessing any affected values at all.

Mr Walker criticised the large percentage figure which Mr Gardner's evidence produced. Mr Cowen submitted that this was completely wrong as an approach. If open market evidence supported the figure, then that was the appropriate compensation to award. Comparisons with the settled cases were unhelpful. The subject properties were the only properties in the scheme to be tested in the Lands Tribunal and in any event a substantial proportion of the depreciation resulted from wiping out the location premium.

Inspection

I carried out an accompanied inspection on November 11 1996, which was a fine sunny day. I inspected the two subject properties externally and internally, and was also able to enter 31 and 35 West Borough. I observed traffic conditions generally in the vicinity, and saw all the properties cited as comparables. Traffic flows appeared light and there were few goods vehicles, but I took account both of the time of day (around lunch time) and of the current weight restriction. Traffic noise was plainly audible in the front rooms of all the West Borough properties I entered, although partial secondary glazing in 45 West Borough had some effect in reducing the noise.

Decision

The first issue in dispute between the parties is the extent of the public works giving rise to a claim under the 1973 Act. Where the claim arises, not from the construction of a new highway, but from the alteration of an existing highway, section 9 of the Act has the effect of extending the right to claim compensation in respect of such depreciation as would not have been caused but for alterations to the carriageway, and "caused by the use of, and the source of which is

*Editor's note: Also reported at [1994] 2 EGLR 206

A situated on, the length of carriageway which has been altered". The relevant alterations are defined by section 9(5) in the following terms:

> For the purposes of this section the carriageway of a highway is altered if, and only if:
>
> (a) the location, width or level of the carriageway is altered (otherwise than by resurfacing); or
>
> (b) an additional carriageway is provided . . .

On the facts of this case it is agreed that the works carried out at the junction of West Borough and Priors Walk constituted alterations to the highway within that definition. The area of those agreed works B terminated at a point 12.5m short of 46 West Borough and 17.6m short of 45 West Borough. Further work was undoubtedly carried out to the length of road immediately outside the subject properties. The point at issue is whether that further work amounted to alteration of the carriageway "otherwise than by re-surfacing". The council contend that all that was done was resurfacing. The claimants say that what was done altered the level of the carriageway, and was more than resurfacing. The matter was ably argued by Mr Cowen, who pointed out that the word used was not "surfacing" but "re-surfacing" which implied no more than replacing what was there before. The evidence showed that this was not what happened. Instead as he contended, C there had been a change in the material used, a change in the profile of the road, and a change in the carrying capacity of the road to enable heavier vehicles to use it.

I appreciate the force of these submissions, but having considered the evidence, in particular the detailed evidence of Mr Cordier and his diagram which I have appended as appendix 7, I am satisfied that the work carried out outside the subject properties was resurfacing. The basic structure of the carriageway, its location and width all remained unchanged. Much of the existing surface was retained, but this was planed out to an extent at the carriageway edges. What was referred to by Mr Cordier as "regulating material" was then laid in a thin layer to D provide a smooth regular base for the final surface layer. The consequence was a marginal change in the profile of the carriageway, so that the new surface was 55 mm above the old at the crown, but remained unchanged in level at the kerbs. Mr Cordier told me that this was a routine process of resurfacing, and I accept that. The wording of section 9(5) plainly contemplates that there may be an alteration of the level of the carriageway which nevertheless comes within the scope of resurfacing.

It follows, in my judgment, that the length of roadway immediately outside the subject properties is not altered carriageway and that no account can be taken of the physical factors, noise, vibration, etc, E which have their source on this length of roadway. As I held in the case of *Williamson* v *Cumbria County Council* [1994] 2 EGLR 206; (1994) 68 P&CR 367 the grammatical construction of section 9(5) requires that the physical factors bringing about depreciation of value must be caused by the use of, and have their source on, the altered highway.

The consequence of this finding, in the circumstances of this case, is somewhat artificial. Because the distance between these subject properties and the altered highway is so short, it will inevitably mean that, from time to time, an occupier of the subject properties will be disturbed by noise from a large goods vehicle with one end of the F vehicle on the altered highway and the other end not on the altered highway. I suppose it is a reasonable assumption that an engine revving some 15m away from the front door will be less disturbing than one immediately outside; on the other hand, it may be that vehicles waiting at the traffic lights 15m away cause a large HGV to stand with its engine revving immediately outside the front door.

In these circumstances, I consider that the approach of Mr Gardner is appropriate and reasonable in valuation terms. In his view, if the road adjoining the subject properties were held not to be altered highway, then a small adjustment in his valuations is called for. He had adopted 15% for no 45 and 10% for no 46. I regard these discounts as adequate and appropriate.

I turn then to consider the technical evidence as to the effects on the G subject properties of the increased use of the adjoining roads as a result of the public works carried out. It is now common ground that phase 1 of the distributor road scheme did not significantly increase the traffic impact on West Borough, because there remained the bottleneck at the Priors Walk junction, which was only removed by phase 2. Thereafter there was undoubted increase in the scale and content of traffic flows, which requires consideration in terms of numbers and of physical effects, principally noise but also vibration, dust and fumes.

So far as concerns the number and type of traffic resulting from the scheme, there are agreed pre-and post-scheme traffic counts. In June 1989 the total movements in this part of West Borough over an 18- H hour day was 9,449 vehicles, of which 423 were goods vehicles over 3.5 tons. In July 1992, the corresponding figure for all vehicles was 10,907, of which 529 were goods vehicles over 3.5 tons. In percentage terms the growth in total movements during that period was over 15%, but the growth in HGV traffic was 25%. There was evidence also that as at the relevant date in 1992, traffic was anticipated to grow significantly over the foreseeable future, although in fact that growth has since been curtailed to an unknown extent by the weight restriction imposed in 1996. It is relevant to note that this restriction could not reasonably have been anticipated in July 1992.

However, this is by no means the only material available to make a J judgment as to the impact of traffic resulting from the distributor road scheme. The agreed traffic figures are based entirely on 18-hour counts and thus take no account of traffic between midnight and 6 am. I was impressed by the evidence of the claimants, in particular Mr Preselo, who had endeavoured to assess the number and type of vehicle movements during the small hours and whose evidence in this respect was unchallenged.

From all the evidence I heard on this issue, I conclude as matters of fact:

(a) that prior to completion of the West Borough/Priors Walk junction reconstruction, this part of West Borough was rightly K described as a quiet residential street with relatively light traffic including a small number of large goods vehicles;

(b) that in consequence of the scheme, large heavy goods vehicles, including what are popularly described as "juggernauts", were able to negotiate the junction, as indeed was the intention of the distributor road scheme to remove such traffic from The Square and town centre;

(c) that as a result there was a substantial increase in total traffic movements with a more substantial increase in the numbers of HGVs; and

(d) that the scheme facilitated night time movements of HGVs to or from the south coast ports, traffic which had a particularly disturbing L impact on local residents including the claimants.

The precise dimensions of this growth in traffic are not easy to determine, but, in my view, precise measurement is not necessary. It is sufficient to say that the growth of traffic over and beyond the anticipated natural growth, and of the proportion of heavy goods vehicles, in consequence of the scheme was substantial and for the most part would not have occurred but for the scheme. Both Mr Preselo and Mrs King told me that they had been surprised and shocked by the extent of the increased traffic resulting from the scheme, and I do not believe they were exaggerating. I accept that a particularly disturbing consequence of the scheme was the introduction of large lorries along West Borough and their transit past the subject M properties at all hours of day and night, in what previously had been a pleasant and quiet town street.

It is necessary for the purposes of these claims to isolate the "physical factors" defined by section 1(2) and which alone are compensatable. So far as material in the present case, those factors are noise, vibration, smell, fumes and artificial lighting. As in other claims made under Part I, I heard a great deal of evidence from acoustics experts upon the subject of noise measurement. For these claimants, as indeed is likely for most claimants in respect of a new road or a new road pattern, the additional noise generated by extra traffic is the most important of the physical factors underlying the claim. It should not be thought that the tribunal will disregard expert technical evidence from acoustics specialists as unimportant or irrelevant. Nevertheless, it should be borne in mind that the task of the tribunal is to determine the depreciation (if any) in value of the claimants' interest. This is a matter for the market, and as Mr Cowen observed the bidder in a residential market does not have an acoustics expert, not even a noise meter, at his elbow when making his bid.

In the event I derived little assistance from the evidence of the two experts in this case. Mr Hurrell had calculated pre-scheme noise levels which he subsequently acknowledged to be unreliable and withdrew. He advocated a method of noise measurement which he agreed was not in general use and was regarded as unconventional. He supported his opinion by reference to readings taken over an unacceptably short period, and by reference to obscure data of doubtful relevance.

Mr Butler on the other hand relied solely on the LA10 18-hour index, which is accepted to be in common use as the measurement shown to corelate best to popular dissatisfaction with traffic noise. The index is also used to determine entitlement under the noise insulation regulations. It is not the tribunal's function to denigrate, nor indeed to offer a critique, of this widely accepted method of measuring noise. Nevertheless, it is clear that the LA10 18-hour index has severe limitations in the context of such a case as the present, limitations which Mr Butler accepted and which, in my view, detract significantly from the force of his evidence. The LA10 18-hour index either measures or calculates the level of sound which is exceeded for 10% of any hour within a period of 18 hours. In this case the figures were calculated by an approved prediction method (ie related to predicted traffic flows and traffic speeds) and thus there were no actual pre-scheme or post-scheme readings. The method produces an "average" sound level, and therefore does not record peak or trough sound levels. Thus a particularly loud and disturbing noise which is of short duration or intermittent will not be reflected in the calculated result using this method. Perhaps more importantly the calculation relates to 18 hours only and therefore does not indicate what sound levels may occur between midnight and 6 am, the period when it may be assumed that ambient sound is at its lowest, and occasional traffic noise is proportionately greater and more disturbing.

For these reasons I find myself unable to accept Mr Butler's conclusion that the difference between pre-scheme and post-scheme calculated levels of traffic noise is so small as to be insignificant. The evidence of Mr Preselo and Mrs King was clear and I accept it. They and their families have been subjected to severe noise nuisance from the passage, and the stopping and starting, of large HGVs in particular as a result of the junction improvement and the introduction of traffic lights. Mr Preselo described the noise as "very disturbing and inescapable". Mrs King referred to it as "a racket against which we have no protection". In my judgment, it is subjective impressions of this kind which would more strongly influence the potential purchaser than would the scientific calculations of acoustics experts.

Furthermore, both Mr Preselo and Mrs King gave unchallenged evidence of other physical factors affecting their properties. I am satisfied that both properties have in consequence of the scheme been subjected to vibration, there is no evidence of significant structural damage beyond some slight movement of door frames and cracking of window panes, but both properties are of 18th-century construction and may therefore be vulnerable structurally, particularly in the eyes of a prospective purchaser. I am satisfied also that the advent of HGVs has resulted in significant nuisance by reason of dust and diesel fumes and smell, accentuated by the fact that both houses front directly on the street. I also take the view that the installation of new and relatively powerful artificial lighting has contributed, though to a small extent, to the environmental deterioration caused by the scheme.

I turn therefore to the contentious issue of valuation. For the claimants, Mr Gardner employed an elaborate method of analysis to establish what he called a "location premium" enhancing the pre-scheme value of the subject properties. In essence, he demonstrated this by comparing, on the basis of space measurement (RCA), substantial "character" houses in West Borough with a substantial "character" house in Avenue Road. He carried out a further analysis based on actual sales to determine the post-scheme or "affected" value. Mr Gardner then isolated from the total loss of value the proportion which he adjudged to be attributable to compensatable physical factors. He finally made an adjustment by way of discount on the assumptions (contrary to his own contention) that the road outside the subject properties was not "altered carriageway".

Mr Walker by contrast relied substantially on the agreed settlement of claims as demonstrating a pattern of compensation which he said was dependent primarily on the distance between the altered carriageway and an affected property. The pattern ranged from 1% to 10.8% of unaffected value. He then estimated the unaffected value of the subject properties, but considered that since the effects of the scheme were marginal, only "nominal" depreciation had occurred, which in each case he assessed at £5,000. Mr Walker then isolated 70% of this figure as representing compensatable physical factors, leaving £3,500 as due by way of compensation, representing in each case 2.18% of the unaffected or "no-scheme world" value.

Of the contrasting approaches to valuation, I have no hesitation in preferring that of Mr Gardner. It may be that in some circumstances, evidence based on negotiated settlements can be useful, but as Mr Gardner pointed out, the motives of claimants who agree to a settlement are many and varied, while the local knowledge and competence of those who represent them is also variable. It is clear for example, that the claims made and settled in respect of 94, 96 and 106 West Borough (two of these by a Cardiff firm, the third by a Yeovil firm) were very much undervalued if compared with adjoining properties where settlements had been reached between Mr Walker and Mr Gardner's firm. I bear in mind also Mr Gardner's evidence as to the pressure inducing settlements in some cases. Furthermore, the settlements which he had made were except for one instance without agreement as to the component figures.

As it seems to me, if there is evidence available of market transactions, such evidence should be used in order to assist in determining the pre-scheme and post-scheme value of the subject properties. Provided that evidence is reliable and properly used, it is more likely to produce the result required by the statute than is a reliance on a scale of settlements which is erratic and lacking in internal logic and consistency.

It is true, as Mr Ward pointed out in the course of his interesting submissions, that Mr Gardner's analysis involved numerous adjustments, for furniture included in a purchase, for the application of the location premium, etc. It is necessary therefore to consider it with caution. I have carefully considered the adjustments reflected in Mr Gardner's schedules together with his detailed explanations, and I reach the conclusion that Mr Gardner's use of the successive sales of 16 Avenue Road, of 44 West Borough and of 48 West Borough was a legitimate way of arriving at a reasonable "unaffected valuation" of the subject properties. His analysis of the sales of a fine period house in Avenue Road contrasted with the sales of fine period houses in West Borough showed that there was a location premium in favour of West Borough, which he applied by way of RCA comparisons to the subject properties. His view was reinforced by consideration of the 1993 sales of the Mill House, a property unaffected by the scheme. Accordingly, I accept that the unaffected value of 45 and 46 West Borough at the relevant date in 1992 would have been £211,500 and £192,500 respectively.

Mr Gardner likewise relied for his "affected valuation" on market transactions, the sale of 44 West Borough (an adjoining property) at a date very close to the valuation date being of particular significance. I see no reason to reject his view that the "affected value" of 45 and 46 West Borough at the relevant date was £142,500 and £131,000 respectively.

On this basis the total depreciation in value of the subject properties is shown to be £69,000 in the case of 45 West Borough and £61,500 in the case of 46 West Borough. The next stage is to take account of my finding that the altered carriageway does not include the stretch of road immediately adjoining the subject properties. As I indicated earlier, I consider that the discounts proposed by Mr Gardner of 15% for no 45 and 10% for no 46 adequately reflect this factor, and the total depreciation is adjusted accordingly to £58,650 for 45 and £55,350 for 46.

The final adjustment requires determination of the extent to which that depreciation in value is caused by compensatable physical factors. It is common ground that the proportion is substantial. Mr Gardner puts it at 90%; Mr Walker at 70%.

There are elements in the deterioration of West Borough which are not compensatable under the Act, principally the effects of natural growth of traffic which would have occurred if the distributor road scheme had not been implemented, and any effect on value of the visual intrusion caused by the scheme; this element of visual intrusion is not one of the "physical factors" to be taken into account, save in so far as it results from artificial lighting. Mathematical precision is not possible in calculating these effects, but doing the best I can, on the evidence available, I consider that Mr Gardner's 90% is rather too high and Mr Walker's 70% is rather too low, and I adopt 80% as the appropriate factor.

Accordingly, I determine that the value of 45 West Borough is depreciated within the meaning of the Act to the extent of £46,920 and that the value of 46 West Borough is similarly depreciated to the extent of £44,280. These figures represent in the case of 45 West Borough a depreciation of about 22% from the unaffected value, and in the case of 46 West Borough about 23%.

As Mr Ward pointed out with vigour in the course of his submissions, these percentages are far higher than had been paid on any other claim arising from this scheme. However, this is the first such case to be heard by the tribunal, and the tribunal must act on the basis of the evidence presented. The other claims have been settled, and I have drawn attention to some of the shortcomings of settlement evidence. I have had in this case to deal with two fine town houses abutting a street which has become part of a through route used by heavy goods vehicles. The character of the street has been transformed and the effect of the change on the occupiers has been drastic; in the particular circumstances it is not surprising that I have found the effect in terms of loss of value to be substantial.

The claimants Mr and Mrs Preselo are entitled to compensation in the sum of £46,920; the claimants Mr and Mrs King are entitled to compensation in the sum of £44,280, and these are my awards.

This decision determines the substantive issues raised between the parties, and the tribunal's award is final. The parties are invited to make such submissions as they are advised as to the costs of the hearing, and a letter accompanies this decision as to the procedure for submissions in writing. The tribunal will, in due course, incorporate an order as to costs in an *addendum* to this decision. Rights of appeal under section 3(4) of the Lands Tribunal Act 1949 and Rules of the Supreme Court Ord 61 will not accrue until the decision has been thus completed, ie from the date of the *addendum*.

Addendum as to costs

I have received and considered the submissions of the parties with regard to the costs of the proceedings. It is accepted on behalf of the county council that costs should be awarded in favour of the claimants, but it is submitted that some deduction or allowance be made for the fact that the council succeeded on the issue of the extent of the "altered highway".

I do not propose to make any such allowance. My decision as to the extent of the "altered highway" had a very limited effect on the valuation of the claims, and the time occupied by evidence and argument directed to the issue was a small proportion of the total hearing. The case was primarily one of valuation; on this basic point the parties were far apart, and my awards greatly exceeded the council's valuations, and the amount of an offer of settlement put forward by the council some weeks prior to the hearing.

The county council must pay the costs of the claimants, such costs if not agreed to be taxed by the Registrar on the standard High Court basis.

(continued overleaf)

A Appendix 4 G

ANALYSIS OF 1986 PRE-SCHEME MARKET TRANSACTIONS

Address	*Description*	*RCA*	*Date of sale*	*Sale price*	*Analysis*		*Remarks*
16 Avenue Road	Edwardian Detached House. *GF* Hall, S/R, D/R, Kit/Breakfast room, study, clkrm, conservatory. *FF* 5 bedrooms, bathroom. *Outside* Garage and outbuilding. Front and rear gardens.	$187m^2$	May 1986	£68,500	Basic value for $187m^2$ RCA Garage and Outbuilding Unmodernised Valuation adjustment for 5/86-8/86	£84,000 + £6,500 -£20,000 -£2,000 £68,500	
B 48 West Borough	Grade II listed Georgian town house *GF* Hall, S/R, D/R, study, kitchen. *FF* 3 bedrooms, 2 bathrooms. Rear walled garden.	$195m^2$	August 1986	£105,000	Basic value for $195m^2$ RCA Separate Garage LOCATION PREMIUM	£87,000 + £8,000 + £10,000 £105,000	11.5% of basic value H
C 44 West Borough	Grade II listed Georgian town house. *GF* Hall, S/R, D/R, Kit, store, utility. *FF* 5 bedrooms, bathroom, WC. SF 2 attic bedrooms, store. Rear walled garden.	$311m^2$	August 1986	£115,000	Basic value for $311m^2$ RCA LOCATION PREMIUM	£105,000 + £10,000 £115,000	9.5% of basic value J
16 Avenue Road	Edwardian detached house. *GF* Hall, S/R, D/R, kit/breakfast room, study, clkrm, conservatory. *FF* 5 bedrooms, bathroom. Front and rear gardens, garage and outbuilding.	$187m^2$	21/4/92	£185,000	Basic value at 7/92 Valuation adj to 4/92 Garage/outbuilding Carpets & curtains	£170,000 + 2,000 + 10,000 + 3,000 £185,000	
D The Mill House Knobcrook Road	Victorian detached house. *GF* Hall, 2 S/R, D/R, Kit, breakfast room. *FF* 4 bedrooms, bathroom, shower room, WC. Small garden with river frontage. Parking.	$255m^2$	8/93	£172,000	Basic value at 7/92 Valuation adj to 8/93 Disrepair Adjoining commercial properties	£195,000 -£10,500 -£5,000 -£7,500 £172,000	K

Appendix 5

UNAFFECTED VALUATIONS OF WEST BOROUGH PROPERTIES AS AT 15/7/92

Address	*Description*	*RCA*	*Valuation*		*Remarks*
E 44 West Borough	Grade II listed Georgian town house. *GF* Hall, S/R, D/R, kit, store, utility. *FF* 5 bedrooms, bathroom, WC. *SF* 2 attic bedrooms, store. Rear walled garden.	$311m^2$	Basic "unaffected" valuation as at 7/92 +10% LOCATION PREMIUM	£200,000 + £20,000 £220,000	L
45 West Borough	Grade II listed Georgian town house. *GF* Hall, 2 S/R, 2 kits, utility, WC, conservatory. *FF* 5 bedrooms, 2 bathrooms. Side and rear gardens. Vehicular access.	$246m^2$	Basic "unaffected" valuation as at 7/92 + vehicular access + 10% LOCATION PREMIUM	£190,000 + £2,500 + £19,000 £211,500	
F 46 West Borough	Grade II listed Georgian town house. *GF* Hall, 2 S/R, D/R, kit, WC. *FF* 3 bedrooms, 2 bathrooms. *SF* 1 bedroom, store. Rear walled garden.	$198m^2$	Basic "unaffected" valuation as at 7/92 + 10% LOCATION PREMIUM	£175,000 + £17,500 £192,500	M
106 West Borough	Victorian detached house. *GF*, 3 S/R, D/R, kit/breakfast room, clkrm, utility. *FF* 5 bedrooms, 2 bathrooms,shower room. Front and rear gardens. Parking.	$191m^2$	Basic "unaffected" valuation as at 7/92 + vehicular access + 10% LOCATION PREMIUM	£172,500 + £2,500 + £17,250 £192,250	

A Appendix 5, continued E

ANALYSIS OF EVIDENCE OF "AFFECTED" PROPERTIES AS AT 15/7/92

Address	*Description*	*RCA*	*Date of sale*	*Sale price*	*Analysis*	*Remarks*
44 West Borough	Grade II listed Georgian town house. *GF* Hall, S/R, D/R, kit, store, utility room. *FF* 5 bedrooms, bathroom, WC *SF* 2 attic bedrooms, store. Rear walled garden.	311m^2	18/8/92	£160,000	Basic "affected" value as at 15/7/92 £148,000 Carpets, curtains and equipment + £12,000 £160,000	Distance from highway and junction improvements 4m
106 West Borough	Victorian detached house. *GF* 3 S/R, D/R, kit/breakfast room, clkrm, utility. *FF* 5 bedrooms, 2 bathrooms, shower room. Front and rear gardens. Parking.	191m^2	8/94	£120,000	Basic "affected" value as at 15/7/92 £130,000 Valuation adj to 8/94 -£12,500 Vehicular access + £2,500 £120,000	Distance from highway 5m

Appendix 6

"AFFECTED" VALUATIONS OF WEST BOROUGH PROPERTIES AS AT 15/7/92

Address	*Description*	*RCA*	*Valuation*	*Remarks*
45 West Borough	Grade II listed Georgian town house. *GF* Hall, 2 S/R, 2 kits, utility room, WC, conservatory. *FF* 5 bedrooms, 2 bathrooms. *SF* 2 attic bedrooms, store. Side and rear gardens. Vehicular access.	246m^2	Basic "affected" valuation as at 15/7/92 £140,000 + vehicular access + £2,500 £142,500	
46 West Borough	Grade II listed Georgian town house. *GF* Hall, 2 S/R, D/R, kit, WC. *FF* 3 bedrooms, 2 bathrooms. *SF* 1 bedroom, store. Rear walled garden.	198m^2	Basic "affected" value as at 15/7/92 £131,000	

D H

For further cases on this subject see p167